Lecture Notes in Computational Science and Engineering

83

Editors:

Timothy J. Barth
Michael Griebel
David E. Keyes
Risto M. Nieminen
Dirk Roose
Tamar Schlick

For further volumes:
http://www.springer.com/series/3527

Ivan G. Graham • Thomas Y. Hou
Omar Lakkis • Robert Scheichl
Editors

Numerical Analysis of Multiscale Problems

 Springer

Editors

Ivan G. Graham
Robert Scheichl
Department of Mathematical Sciences
University of Bath
Bath BA2 7AY
United Kingdom
i.g.graham@bath.ac.uk
r.scheichl@bath.ac.uk

Omar Lakkis
Department of Mathematics
University of Sussex
Falmer
Brighton BN1 9QH
United Kingdom
o.lakkis@sussex.ac.uk

Thomas Y. Hou
Applied and Computational Mathematics
California Institute of Technology
MC 9-94,
Pasadena, CA 91125
USA
hou@cms.caltech.edu

ISSN 1439-7358
ISBN 978-3-642-22060-9   e-ISBN 978-3-642-22061-6
DOI 10.1007/978-3-642-22061-6
Springer Heidelberg Dordrecht London New York

Library of Congress Control Number: 2011937698

Mathematics Subject Classification (2010): 65N06, 65N08, 65N30, 65N12, 65N15, 65N55, 65N75, 65C30, 65F08, 65F10, 65F22, 65N20, 74Q05, 76S05, 76T10, 76F55, 76M35, 76M50

Cover illustration: The Editors would like to thank Patrick Jenny, Daniel Meyer and Manav Tyagi for providing the graphic which is used on the front cover of this book.

Cover design: deblik, Berlin

Printed on acid-free paper

Springer is part of Springer Science+Business Media (www.springer.com)

Preface

This book contains ten invited expository articles arising from the 91st LMS Durham Symposium on "Numerical Analysis of Multiscale Problems" which took place in the beautiful cathedral city of Durham in the UK from 5 to 15 July 2010. The Symposium was attended by 103 participants and was organised by Ivan Graham, Tom Hou and Rob Scheichl. The scientific programme highlighted novel research in theoretical numerical analysis and its applications to advances in areas such as oil reservoir modelling, high frequency scattering, data assimilation, waveguide modelling, uncertainty quantification, atomistic/continuum and polymer modelling. The selected articles in this book are written by some of the Symposium's speakers and their collaborators, including all those speakers who gave short courses of three lectures each. The topics of the articles have been chosen to give a good overall representation of the scope of the Symposium and to provide a resource for researchers who would like to learn more about contemporary progress in this area. The five themes of the Symposium, and all the speakers who supported them are as follows.

Theme 1: *Numerical analysis for multiscale PDEs.* This theme was anchored by short courses of Todd Arbogast and Mitch Luskin. Todd Arbogast surveyed multiscale approximation for elliptic PDEs (in particular mixed methods), motivated by problems in oil reservoir modelling, while Mitch Luskin described the numerical analysis of problems arising in atomistic-continuum modelling in solid mechanics. There was also a strong analysis content to the short courses of Andrew Stuart, Markus Melenk and Christoph Schwab, described under the other themes below. Yalchin Efendiev discussed approximation of high contrast diffusion problems motivated by reservoir modelling, and his talk also had strong connections to Themes 4 and 5. Endre Süli presented a talk on the fundamental analysis of high-dimensional PDE models arising in polymer modelling, related strongly to Theme 3. Assyr Abdulle's talk concerned adaptivity in the heterogeneous multiscale method, while the theme of atomistic/continuum modelling was continued in the talks of Ping Lin and Christoph Ortner. Richard Tsai discussed micro/macro-modelling in porous media flow. Arieh Iserles talked on asymptotic-numerical multiscale expansions, which was also related to Theme 2. Radial basis function approximation applied to multiscale modelling problems in the geosciences was presented by Holger Wendland, while Chris Budd presented moving mesh adaptive PDE methods applied to problems in numerical weather forecasting.

Theme 2: *Multiscale wave propagation problems.* This theme was anchored by the short course of Markus Melenk, and lectures by Liliana Borcea (inverse problems for waves in random media), Olof Runborg (time domain problems in heterogeneous media) and Zhiming Chen (on wave propagation problems in infinite domains). Bjorn Engquist and Martin Gander both gave lectures on different aspects of robust solvers for high frequency wave problems (see Theme 4). A session was also devoted to asymptotic/numerical hybrid methods for wave problems and included talks of Peter Monk, Simon Chandler-Wilde, Timo Betcke and Euan Spence.

Theme 3: *Stochastic problems.* This central topic was anchored by short courses of Andrew Stuart (on the Bayesian approach to inverse problems and application to data assimilation) and Christoph Schwab (PDEs with random input data and related high dimensional parametrized PDEs). A new approach for numerical analysis of PDEs with lognormal permeability fields was discussed by Marcus Sarkis. The short course of Patrick Jenny focussed on applications, but also had strong resonance with this Theme. Talks were also given by Mike Giles (multilevel Monte Carlo), Oliver Ernst (generalised polynomial chaos), Catherine Powell (solvers for high dimensional discretizations of stochastic PDEs) and Frances Kuo (quasi-Monte Carlo methods), while Raul Tempone gave a lecture on discretisation of kinetic Monte Carlo models in computational chemistry and Tom Hou considered the reduced basis approach in a stochastic setting. Viet Ha Hoang discussed related random hyperbolic problems.

Theme 4: *Efficient solvers and computational aspects.* This theme was anchored by the lecture of Pater Bastian on the DUNE programming environment, sponsored by the Centre for Numerical Algorithms and Intelligent Software (Edinburgh/Heriot-Watt/Strathclyde), and the lecture of Bjorn Engquist on preconditioning in FEM and matrix compression in BEM, both for the Helmholtz equation. Martin Gander also gave a survey of high-frequency Helmholtz FEM solvers, with emphasis on optimised interface conditions. Frédéric Nataf and Ludmil Zikatanov discussed solvers for heterogeneous media problems, providing a different perspective on problems introduced by Yalchin Efendiev (in Theme 1). Catherine Powell discussed multigrid solvers for high-dimensional systems arising from pdes with random data (linked to Theme 3).

Theme 5: *Application areas.* This theme was anchored by the short course of Patrick Jenny who presented probability density function methods with applications to problems such as turbulent combustion and CO_2 sequestration, and was also linked to Theme 3. Lou Durlovsky discussed the state of the art in uncertainty quantification in reservoir modelling in the real industrial context. Solutions to multiscale differential equations modelling cell biology and molecular dynamics were presented respectively by John King and Ben Leimkuhler. An embedded Industry Day included a number of external guests who joined with the participants of the full Symposium. Talks were given by Paul Childs (Schlumberger, seismic imaging), Anthony Baran (Met Office, scattering from ice crystals), Jill Ogilvy (BAE Systems, high frequency scattering), Roland Masson (Institut Français du Petrole) and Seong Lee (Chevron) (both on multiphase flow), Tim Payne (Met Office, data assimilation in weather forecasting), Grigory Vilensky (UCL, PDE modelling of ultrasound in cancer treatment), Peter Jimack (Leeds, lubrication problems) and Andrew Cliffe (Nottingham, uncertainty quantification in waste management). An additional talk by Xiao-Hui Wu (ExxonMobil) on uncertainty quantification in reservoir simulation was presented later in the Symposium.

The scientific organisation was made straightforward by the excellent local support at Durham, especially the help of our close colleagues James Blowey and Max Jensen, and Fiona Giblin and the rest of the team in the Maths office, as well as John Bolton and John Parker. Postgraduate students John Chapman, Aretha Teckentrup, Ray Millward and Tatiana Kim operated the video camera.

Finally we would like to thank the UK Engineering and Physical Sciences Research Council, the London Mathematical Society and The Centre for Numerical Algorithms and Intelligent Software for the financial support which made the meeting possible.

More information about the 91st LMS Durham Symposium is available at:

http://www.maths.dur.ac.uk/events/Meetings/LMS/2010/NAMP/

Bath   *Ivan Graham and Robert Scheichl*
California Institute of Technology   *Thomas Hou*
Sussex   *Omar Lakkis*

Contents

Multiscale Modelling and Inverse Problems

James Nolen, Grigorios A. Pavliotis, and Andrew M. Stuart

Abstract The need to blend observational data and mathematical models arises in many applications and leads naturally to inverse problems. Parameters appearing in the model, such as constitutive tensors, initial conditions, boundary conditions, and forcing can be estimated on the basis of observed data. The resulting inverse problems are usually ill-posed and some form of regularization is required. These notes discuss parameter estimation in situations where the unknown parameters vary across multiple scales. We illustrate the main ideas using a simple model for groundwater flow.

We will highlight various approaches to regularization for inverse problems, including Tikhonov and Bayesian methods. We illustrate three ideas that arise when considering inverse problems in the multiscale context. The first idea is that the choice of space or set in which to seek the solution to the inverse problem is intimately related to whether a homogenized or full multiscale solution is required. This is a choice of regularization. The second idea is that, if a homogenized solution to the inverse problem is what is desired, then this can be recovered from carefully designed observations of the full multiscale system. The third idea is that the theory of homogenization can be used to improve the estimation of homogenized coefficients from multiscale data.

J. Nolen
Department of Mathematics, Duke University Durham, NC 27708, USA
e-mail: nolen@math.duke.edu

G.A. Pavliotis
Department of Mathematics Imperial College London, London SW7 2AZ, UK
e-mail: g.pavliotis@imperial.ac.uk

A.M. Stuart (⊠)
Mathematics Institute, Warwick University, Coventry CV4 7AL, UK
e-mail: A.M.Stuart@warwick.ac.uk

I.G. Graham et al. (eds.), *Numerical Analysis of Multiscale Problems*, Lecture Notes in Computational Science and Engineering 83, DOI 10.1007/978-3-642-22061-6_1,
© Springer-Verlag Berlin Heidelberg 2012

1 Introduction

The objective of this overview is to demonstrate the important role of multiscale modelling in the solution of inverse problems for differential equations. The main inverse problem we discuss is that of determining unknown parameters by matching observed data to a differential equation model involving those parameters. The unknown parameters may be functions, in general, and they may have variation over multiple (length) scales. This multiscale structure makes the forward problem more challenging: numerically computing the solution to the differential equation requires very high resolution. The multiscale structure also complicates the inverse problem. Should we try to fit the data with a high-dimensional parameter, or should we seek a low-dimensional "homogenized" approximation of the parameter? If a low-dimensional parameter model is used, how should we account for the mismatch between the true parameters and the low-dimensional representation? After obtaining a solution to the inverse problem, one typically wants to make further predictions using whatever parameter is fit to the observed data, so it is important to consider whether a low-dimensional representation of the unknown parameter is sufficient to make additional predictions.

Throughout these notes the unknown parameters will be denoted by $u \in X$; typically u is a function assumed to lie in a Banach space X. We use $y \in Y$ to denote the data (for simplicity we usually take $Y = \mathbb{R}^N$) and z to denote the predicted quantity, assumed to be an element of a Banach space Z or, in some cases, a Z-valued random variable. The map $\mathscr{G} : X \to \mathbb{R}^N$ denotes the mapping from the unknown parameter to the data, and $\mathscr{F} : X \to Z$ (or $\mathscr{F} : X \times \Omega \to Z$ in the random case, where Ω denotes the sample space) denotes the mapping from the parameter to the prediction. We sometimes refer to $\mathscr{G}$ as the *observation operator* and $\mathscr{F}$ as the *prediction operator*. Both $\mathscr{G}$ and $\mathscr{F}$ are typically derived from a common solution operator $G : X \to P$ mapping $u \in X$ to the solution $G(u) \in P$ of a partial differential equation (PDE), where P is a Banach space. For example $\mathscr{G}$ may be derived by composing G with N linear functionals.

The ideal inverse problem is to determine $u \in X$ from knowledge of $y \in \mathbb{R}^N$ where it is assumed that $y = \mathscr{G}(u)$. In practice, however, the data y is generated from outside this clean mathematical model, so it is natural to think of the data y as being given by

$$y = \mathscr{G}(u) + \xi \tag{1}$$

for some $\xi \in \mathbb{R}^N$ quantifying model error[1] and observational noise. The value of ξ is not known, but it is common in applications to assume that some of its statistical properties are known and these can then be built into the methods used to estimate u. From (1) it follows that y is a random variable. Once the function u is determined by solving this inverse problem, it can be used to make a prediction $z = \mathscr{F}(u)$.

---

[1] Model error can be incorporated within the set of unknown parameters u and estimated using data; however this idea is not pursued here.

We illustrate three ideas that arise when attempting to solve the inverse problem defined by (1) in the multiscale context:

(a) The choice of the space or set in which to seek the solution to the inverse problem is intimately related to whether a low-dimensional "homogenized" solution or a high-dimensional "multiscale" solution is required for predictive capability. This is a choice of regularization.
(b) If a homogenized solution to the inverse problem is desired, then this can be recovered from carefully designed observations of the full multiscale system.
(c) The theory of homogenization can be used to improve the estimation of homogenized parameters from observations of multiscale data.

In Sect. 2 we consider in detail a worked example which exemplifies the use of multiscale methods to approximate the forward problems $\mathscr{G}$ and $\mathscr{F}$ for data and predictions; this example will be used to illustrate many of the general ideas developed in these notes, and the three ideas (a)–(c) in particular. Section 3 is devoted to a brief overview of regularization techniques for inverse problems, and to discussion of the idea (a). Section 4 is devoted to the idea (b). We study the problem of estimating a single scalar parameter in a homogenized model of groundwater flow, given data which is generated by a full multiscale model. This may be seen as a surrogate for understanding the use of real-world data (which is typically multiscale in character) to estimate parameters in homogenized mathematical models which do not represent small scales explicitly. Section 5 is devoted to the idea (c). We study the use of ideas from multiscale methodology to enhance parameter estimation techniques for homogenized models. The viewpoint taken is that the statistics of the error ξ appearing in (1) can be understood using the theory of homogenization for random media; when these statistical properties depend on the unknown parameter u the noise ξ is no longer additive and its dependence on u plays an important role in the parameter estimation process.

1.1 Notation

The following notation will be used throughout. We use $|\cdot|$ to denote the Euclidean norm on $\mathbb{R}^m$ (for possibly different choices of m). We let S^d (resp. $S^{d,+}$) denote the set of symmetric (resp. positive-definite) second order tensors on $\mathbb{R}^d$. If $\Gamma \in S^{d,+}$, we define the weighted norm $|\cdot|_\Gamma = |\Gamma^{-\frac{1}{2}} \cdot|$ on $\mathbb{R}^m$. Throughout the notes, X is a Banach space, containing the functions that we wish to estimate, and E a reflexive Banach space compactly embedded into X. When studying the inverse problem from a Bayesian perspective we will use Gaussian priors on X, defined via a covariance operator $\mathscr{C}$ on a Hilbert space $H \supseteq X$, with norm $\|\cdot\|_H$ and where the embedding is continuous. In this situation E will be the Hilbert space with norm $\|\mathscr{C}^{-\frac{1}{2}} \cdot\|_H$.

1.2 Running Example

We consider a model for groundwater flow in a medium with permeability tensor k, pressure p and Darcy velocity v (or the volume flux of water per unit area) related to the pressure via the Darcy law:

$$v = -\frac{k}{\mu}(\nabla p - \rho g \hat{e}_z) \tag{2}$$

where k is the permeability tensor field (discussed later) μ is the fluid viscosity, ρ is the fluid density, g is the acceleration due to gravity and $\hat{e}_z$ is the unit vector in the z-direction. We choose units in which $\mu = 1$. We also assume that we have a constant density fluid and redefine the pressure by adding $\rho g z$ (z is the vertical direction) to write (2) in the form $v = -k \nabla p$. Assuming that the Darcy velocity is divergence-free, except at certain known source/sink locations, we obtain the following elliptic equation for the pressure:

$$\nabla \cdot v = f, \quad x \text{ in } D,$$
$$p = 0, \quad x \text{ on } \partial D, \tag{3}$$
$$v = -k \nabla p$$

where $D \subset \mathbb{R}^d$ is[2] an open and bounded set with regular boundary, and f is assumed to be known. The permeability tensor field k, however, is assumed to be unknown and must be determined from data. In order to make the elliptic PDE (3) for the pressure p well-posed, we assume that the permeability tensor $k(x)$, for $x \in D$, is an element of $S^{d,+}$ and so we write it as the (tensor) exponential: $k(x) = \exp(u(x))$, $u \in S^d$. It is natural to view u as an element of $X := L^\infty(D; S^d)$ and to consider weak solutions of (3) with $f \in H^{-1}(D)$. Then we have a unique solution $p \in H_0^1(D)$ satisfying [18, ch. 6]

$$\|\nabla p\|_{L^2} \le c_1 \exp(\|u\|_X)\|f\|_{H^{-1}}, \tag{4}$$

for some $c_1 > 0$ depending only on d and D, and $\|u\|_X$ being the essential supremum of the spectral radius of the matrix $u(x)$, as x varies over D:

$$\|u\|_X = \text{ess} - \sup_{x \in D} \left(\max_{\substack{\xi \in \mathbb{R}^d \\ |\xi|=1}} |u(x)\xi| \right).$$

---

[2]For the application to groundwater flow, $d = 2$ or 3. However, the analysis presented in this paper is valid in arbitrary dimensions.

Thus we may define $G : X \to H_0^1(D)$ by $G(u) = p$. Now consider a set of real-valued continuous linear functionals $\ell_j : H^1(D) \to \mathbb{R}$ and define $\mathscr{G} : X \to \mathbb{R}^N$ by $\mathscr{G}(u)_j = \ell_j(G(u))$. The inverse problem is to determine $u \in X$ from $y \in \mathbb{R}^N$ where it is assumed that y is given by (1). Using (4) one may show that $G : X \to H_0^1(D)$ (resp. $\mathscr{G} : X \to \mathbb{R}^N$) is Lipschitz. Indeed if p_i denotes the solution to (3) with log permeability u_i then, we have

$$\|\nabla p_1 - \nabla p_2\|_{L^2} \leq (c_1)^2 \|u_1 - u_2\|_X \exp\left(2(\|u_1\|_X + \|u_2\|_X)\right) \|f\|_{H^{-1}}. \quad (5)$$

Study of the transport of contaminants in groundwater flow is a natural example of a useful prediction that can be made once the inverse problem is solved. To model this scenario we consider a particle $x(t) \in \mathbb{R}^d$ which is advected by the groundwater velocity field v/ϕ, where ϕ is the porosity (which is taken to be a constant) of the rock and v is the Darcy velocity field from (3), and subject to diffusion with coefficient 2η. Assuming that the contaminant is initially at x_{init} we obtain the stochastic differential equation (SDE):

$$dx = \frac{v(x)}{\phi} dt + \sqrt{2\eta}\, dW, \quad x(0) = x_{\text{init}}, \quad (6)$$

where $W(t)$ is a standard Brownian motion on $\mathbb{R}^d$. If we are interested in predicting the location of the contaminant at time T then our prediction will be the function $\mathscr{F}_\eta$ given by $\mathscr{F}_\eta(u) = x(T)$. Here for each fixed $\eta \in [0, \infty)$ the function $\mathscr{F}_\eta$ maps X into the family of $\mathbb{R}^d$-valued random variables.

2 The Forward Problem: Multiscale Properties

Some inverse problems arising in applications have the property that the forward model $\mathscr{G}$ mapping the unknown to the data will produce similar output on both highly oscillatory functions u and on appropriately chosen smoothly varying functions u. Furthermore, for some choices of prediction function $\mathscr{F}$ the predictions themselves will also be close for both highly oscillatory functions u and on appropriately chosen smoothly varying functions u. These properties can be seen from an application of multiscale analysis, and we illustrate them by considering the problem introduced in Sect. 1.2. There are many texts on the theory of multiscale analysis. For example, the basic homogenization theorems discussed here are developed in [7]. A recent overview of the subject, with many other references and using the same notational conventions that we adopt here, is [27].

We consider a multiscale version of the running example from Sect. 1.2 where the permeability tensor is $k = K^\epsilon(x) = K(x, x/\epsilon)$ where $K : D \times \mathbb{T}^d \to S^{d,+}$ is[3] periodic in the second argument, $\epsilon > 0$ a small parameter. For now we have assumed periodic dependence on the fast scale in K; however we will generalize this to random dependence in later developments.

With this permeability we obtain the family of problems

$$\nabla \cdot v^\epsilon = f, \quad x \text{ in } D, \tag{7a}$$

$$p^\epsilon = 0, \quad x \text{ on } \partial D, \tag{7b}$$

$$v^\epsilon = -K^\epsilon \nabla p^\epsilon. \tag{7c}$$

If we set $\eta = \epsilon \eta_0$, then the transport of contaminants is given by the SDE

$$dx^\epsilon = \frac{v^\epsilon(x^\epsilon)}{\phi} dt + \sqrt{2\eta_0 \epsilon} \, dW, \quad x^\epsilon(0) = x_{\text{init}}. \tag{8}$$

Standard techniques from the theory of homogenization for elliptic PDEs, e.g. [27, Ch. 12] can be used to show that for ϵ small,

$$p^\epsilon(x) \approx p^\epsilon_a(x) := p_0(x) + \epsilon p_1(x, \frac{x}{\epsilon}) \tag{9}$$

where p_0 and p_1 are defined as follows. First we introduce the *cell problem* for $\chi(x, y)$:

$$-\nabla_y \cdot \left(\nabla_y \chi K^T \right) = \nabla_y \cdot K^T, \quad y \in \mathbb{T}^d. \tag{10}$$

We can now define for each $x \in D$ the effective (homogenized) permeability tensor K_0

$$K_0(x) = \int_{\mathbb{T}^d} Q(x, y) dy, \tag{11}$$

$$Q(x, y) = K(x, y) + K(x, y)\nabla_y \chi(x, y)^T. \tag{12}$$

The effective diffusivity $K_0(x)$ is hence the average of $Q(x, y)$ over the fast scale y. This is *not* equal to the average of $K(x, y)$ over y, except in trivial cases. We denote by u_0 the logarithm of K_0 so that $K_0 = \exp(u_0)$.

The function p_0 is then defined as the solution of the (ϵ independent) elliptic PDE

$$\nabla \cdot v_0 = f, \quad x \in D, \tag{13a}$$

---

[3] $\mathbb{T}^d$ denotes the d-dimensional unit torus.

$$p_0 = g, \quad x \in \partial D, \tag{13b}$$

$$v_0 = -K_0 \nabla p_0. \tag{13c}$$

and the corrector p_1 is defined by

$$p_1(x, y) = \chi(x, y) \cdot \nabla p_0(x). \tag{14}$$

Note that (10) may be written as

$$-\nabla_y \cdot (Q^T) = 0, \quad y \in \mathbb{T}^d. \tag{15}$$

This shows that Q, which is averaged to give the effective permeability tensor, is divergence-free with respect to the fast variable y.

It is possible to prove that, in the limit as $\epsilon \to 0$, solutions to (7) converge to solutions to (13), the convergence being strong in $L^2(D)$ and weak in $H^1(D)$ [1, 11, 27]. However if we want to prove strong convergence in H^1 then we need to include information about the corrector term p_1. The following theorem and corollary summarize these ideas. For proofs see [1], or the discussion in the texts [11, 27].

Theorem 1. *Let p^ϵ and p_0 be the solutions of (7) and (13). Assume that $f \in C^\infty(D)$ and that $K \in C^\infty(D; C^\infty_{\mathrm{per}}(\mathbb{T}^d))$. Then*

$$\lim_{\epsilon \to 0} \| p^\epsilon - p_a^\epsilon \|_{H^1} = 0. \tag{16}$$

Corollary 1. *Under the same conditions as in Theorem 1 we have*

$$\| p^\epsilon - p_0 \|_{L^2} \to 0 \quad \text{and} \quad \| \nabla p^\epsilon - (I + \chi_y(\cdot, \cdot/\epsilon)^T) \nabla p_0 \|_{L^2} \to 0$$

as $\epsilon \to 0$.

In fact it is frequently the case that the convergence in Theorem 1 may be obtained in a stronger topology. Reflecting this we make the following assumption.

Assumption 2 *The function p^ϵ converges to p_0 in $L^\infty(D)$ and its gradient converges to the gradient of $p_0 + \epsilon p_1$ in $L^\infty(D)$ so that*

$$\lim_{\epsilon \to 0} \| p^\epsilon - p_a^\epsilon \|_{W^{1,\infty}} = 0.$$

In Appendix 2 we prove this assumption for the one dimensional version of (7). The proof in the multidimensional case will be presented elsewhere [25]. The proof of this assumption in the multidimensional case is based on the estimates proved in [2] (in particular, Lemma 16), see also [16, Lemma 2.1].

With these limiting properties of the elliptic problem (7) at hand it is natural to ask what is the limiting behaviour of x^ϵ governed by (8). To answer this question we define

$$\frac{dx_0}{dt} = \frac{v_0(x_0)}{\phi}, \quad x_0(0) = x_{\text{init}}. \tag{17}$$

Notice that this ordinary differential equation (ODE) has vector field v_0 which is defined entirely through knowledge of the homogenized permeability K_0: once K_0 is known, the elliptic PDE (13) can be solved for p_0 and then v_0 is recovered from (13c). If we can show that solutions of (8) and (17) are close then this will establish that the prediction of particle transport in the model (7), (8) can be made accurately by use of only homogenized information about the permeability.

In proving such a result there are a number of technical issues which arise caused by the presence of the boundary ∂D of the domain in which the PDE (7) is posed. In particular solutions of (8) may leave D requiring a definition of the velocity field outside D. These issues disappear if we consider the case where D is itself a box of length L and is equipped with periodic boundary conditions instead of Dirichlet conditions: we may then extend all fields to the whole of $\mathbb{R}^d$ by periodicity. In this case, the homogenization theory for (7) with (7b) replaced by periodic boundary conditions is identical to that given above, except that (13b) is also replaced by periodic boundary conditions. We write $D = (L\mathbb{T})^d$ and adopt this periodic setting for the next theorem, which is proved in Appendix 1

Theorem 3. *Let $x^\epsilon(t)$ and $x_0(t)$ be the solutions to (8) and (17), with velocity fields extended from $D = (L\mathbb{T})^d$ to $\mathbb{R}^d$ by periodicity, and assume that Assumption 2 holds. Assume also that $f \in C^\infty(D)$ and that $K \in C^\infty(D; C^\infty_{\text{per}}(\mathbb{T}^d))$. Then*

$$\lim_{\epsilon \to 0} \mathbb{E} \sup_{0 \le t \le T} \|x^\epsilon(t) - x_0(t)\| = 0.$$

In summary, this example exhibits the property that, if the length scale ϵ is small, the data generated from K^ϵ and K_0 may appear very similar due to homogenization effects. Therefore, when trying to infer parameters from data, it is difficult to distinguish between K^ϵ and K_0 without some form of regularization or prior assumptions about the form of the parameter. On the other hand, Theorem 3 shows that knowing only K_0 is sufficient to make accurate predictions of the trajectories of (8).

3   Regularization of Inverse Problems

In this section we describe various approaches to regularizing inverse problems, motivating them by reference to the multiscale example in the previous section. The approach to regularizing which is described in Sect. 3.2 is developed in detail in [6]. The Tikhonov regularization approach from Sect. 3.3 is developed in detail in [17,19]. Both of these regularization approaches are specific examples of the general set-up often called PDE constrained optimization, which we discuss in Sect. 3.4; this subject is overviewed in [20]. An overview of the Bayesian approach to inverse problems, a subject that we outline in Sect. 3.5, is given in [19, 21, 29].

3.1   Set-Up

Our objective here is to determine u, given y, where u and y are related by (1). In this section we consider the case where the date is finite and $Y = \mathbb{R}^N$. We assume that, whilst the actual value of ξ is not available, it is reasonable to view it as a single draw from a statistical distribution whose properties are known to us. To be concrete we assume that ξ is drawn from a zero-mean Gaussian random variable with covariance Γ: we write this as $\xi \sim N(0, \Gamma)$. We make the following continuity assumption concerning the observation operator $\mathcal{G}$.

Assumption 4 *For every $r > 0$ such that $u_i \in X$ satisfies $\|u_i\|_X < r, i = 1, 2$, there are constants $c_1, c_2 > 0$ such that*

$$|\mathcal{G}(u_1) - \mathcal{G}(u_2)| \le c_1 \exp(c_2 r)\|u_1 - u_2\|_X.$$

Note that this (local) Lipschitz condition also implies a bound on $|\mathcal{G}(u)|$ that is exponential in $\|u\|_X$.

In general the inverse problems such as that given by (1) with $\xi = 0$ are hard to solve: they may have no solutions, multiple solutions and solutions may exhibit sensitive dependence on initial data. For this reason it is natural to consider a least squares approach to finding functions u which best explain the data. In view of the assumed structure on ξ a natural least squares functional is

$$\Phi(u) = \frac{1}{2}|y - \mathcal{G}(u)|_\Gamma^2. \tag{18}$$

The weighting by Γ in the Euclidean norm induces a normalization on the model-data mismatch. This normalization is given by the assumed standard deviations of the noise in a coordinate system defined by the eigenbasis for Γ.

Example 1. Consider the running example of Sect. 1.2. Inequality (5) implies that Assumption 4 holds in this case, noting that $\mathcal{G}(u)_j = \ell_j(p)$ for some linear functional ℓ_j on $H^1(D)$, with the choice $X = L^\infty(D; S^d)$, provided $f \in H^{-1}$. We use this example to illustrate why inverse problems are, in general, hard to solve.

Assume that the linear functionals ℓ_j satisfy the property that $\ell_j(p^\epsilon - p_0) \to 0$ as $\epsilon \to 0$. This occurs if they are linear functionals on $L^2(D)$, by Theorem 1 or if

Assumption 2 holds, if they are linear functionals on $C(\overline{D})$. Writing this in terms of $\mathscr{G}$ we have $|\mathscr{G}(u^\epsilon) - \mathscr{G}(u_0)| \to 0$ as $\epsilon \to 0$. (Note that this occurs even though u^ϵ and u_0 are not themselves close.) Hence there is an uncountable family of functions (indexed by all ϵ sufficiently small) which all return approximately the same value of $\Phi(u^\epsilon)$ and thus simply minimizing Φ may be very difficult. Furthermore, there may be minimizing sequences which do not converge. For example fix a particular realization of the data given by $y = \mathscr{G}(u_0)$ where u_0 is the homogenized log permeability. Then $\Phi(u^\epsilon) \geq 0$ for all $\epsilon > 0$ and $\Phi(u^\epsilon) \to 0$ as $\epsilon \to 0$, since

$$|\Phi(u^\epsilon)| = \frac{1}{2}|y - \mathscr{G}(u^\epsilon)|_\Gamma^2 = \frac{1}{2}|\mathscr{G}(u_0) - \mathscr{G}(u^\epsilon)|_\Gamma^2 \tag{19}$$

On the other hand, u^ϵ does not converge in X as $\epsilon \to 0$.

In order to overcome the difficulties demonstrated in this example regularization is needed. In the remaining sections we discuss various regularizations, in general, illustrating ideas by returning to the running example.

3.2 Regularization by Minimization Over a Compact Set

Recall that E is a Banach space compactly embedded into X. We further assume that E is reflexive. Let $E_{ad} = \{u \in E : \|u\|_E \leq \alpha\}$. Then E_{ad} is a closed bounded set in E and, according to the Banach–Alaoglu theorem, any sequence in E_{ad} must contain a weakly convergent subsequence with limit in E_{ad} (see, for example, Theorem 1.17 in [20] where also the case where the consequences of E_{ad} being convex, in addition to being closed and bounded, are studied). Now consider the minimization problem

$$\overline{\Phi} = \inf_{u \in E_{ad}} \Phi(u). \tag{20}$$

Theorem 5. *Any minimizing sequence* $\{u^n\}_{n \in \mathbb{Z}^+}$ *for* (20) *contains a weakly convergent subsequence in* E *with limit* $\overline{u} \in E_{ad}$ *which attains the infimum:* $\Phi(\overline{u}) = \overline{\Phi}$.

Proof. This is a classical theorem from the field of optimization; see [20] for details and context. Since $\{u^n\}$ is contained in E_{ad} we deduce the existence of a subsequence (which for convenience we relabel as $\{u^n\}$) with weak limit $\overline{u} \in E_{ad}$. Thus $u^n \rightharpoonup \overline{u}$ in E. Hence, by compactness, $u^n \to u$ in X. By Assumption 4 we deduce that $\Phi : E \to \mathbb{R}$ is weakly continuous. By definition, for any $\delta > 0$ there exists $N = N(\delta)$ such that

$$\overline{\Phi} \leq \Phi(u_n) \leq \overline{\Phi} + \delta, \quad \forall n \geq N.$$

By weak continuity of $\Phi : E \to \mathbb{R}$ we deduce that

$$\overline{\Phi} \leq \Phi(\overline{u}) \leq \overline{\Phi} + \delta.$$

The result follows since δ is arbitrary.   □

Example 2. Consider the running example of Sect. 1.2. Let A denote a fixed symmetric positive-definite tensor A so that $\log(A)$ is defined. We define the subspace of tensor valued functions of the form $u' = uI + \log(A)$, for some constant $u \in \mathbb{R}$ noting that then $\exp(u') = \exp(u)A$. By Lipschitz continuity of $\mathscr{G}$ in $u' \in X$ we deduce (abusing notation) Lipschitz continuity of $\mathscr{G}$ viewed as a function of $u \in \mathbb{R}$. We define

$$E_{\text{ad}} = \{u \in \mathbb{R} : |u| \leq \alpha\}. \tag{21}$$

We may take the norm $\| \cdot \|_E = |u|$. Thus the problem (20) attains its infimum for some $\bar{u} \in E_{\text{ad}}$. The regularization of seeking to minimize Φ over E_{ad} corresponds to looking for solution over a one-parameter set of tensor fields, in which the free parameter is bounded by α. Note that such a solution set automatically rules out the oscillating minimizing sequences which were exhibited in Example 1.

3.3   Tikhonov Regularization

Instead of regularizing by seeking to minimize Φ over a bounded and convex subset of a compact set E in X, we may instead adopt the Tikhonov approach to regularization. We consider the minimization problem

$$\bar{I} = \inf_{u \in E} I(u), \tag{22}$$

where

$$I(u) = \frac{\lambda}{2} \|u\|_E^2 + \Phi(u). \tag{23}$$

Theorem 6. *Any minimizing sequence $\{u^n\}_{n \in \mathbb{Z}^+}$ for (22) contains a weakly convergent subsequence in E with limit $\bar{u}$ which attains the infimum: $I(\bar{u}) = \bar{I}$.*

Proof. This is a classical theorem from the calculus of variations; see [13] for details and context. Since $\{u^n\}$ is a minimizing sequence and $\Phi \geq 0$, we deduce that for any $\delta > 0$ there exists $N = N(\delta)$ such that

$$\frac{\lambda}{2} \|u_n\|_E^2 \leq \bar{I} + \delta, \quad \forall n \geq N.$$

From this it follows that $\{u^n\}_{n \in \mathbb{Z}^+}$ is bounded in E and hence contains a weak limit $\bar{u}$, along a subsequence which, for convenience, we relabel as $\{u^n\}$. The weak continuity of $\Phi : E \to \mathbb{R}$, together with weak lower semicontinuity of the function

$\| \cdot \|_E^2 \to \mathbb{R}$ implies the weak lower semicontinuity of $I : E \to \mathbb{R}$. Hence

$$I(\overline{u}) \le \liminf_{n \to \infty} I(u_n) \le \overline{I}.$$

Since $I(\overline{u}) \ge \overline{I}$, the result follows. $\qquad\qquad\qquad\qquad\qquad\qquad\square$

Example 3. Consider the running example of Sect. 1.2. Let $E = H^s(D; S^d)$ and note that E is compact in $X = L^\infty(D; S^d)$ for $s > d/2$. Thus the problem (22) attains its infimum for some $\overline{u} \in E$. As with the example from the previous section the regularization rules out highly oscillating minimizing sequences such as those seen in Example 1. The choice of the parameter λ will effect how much oscillation is allowed in any minimizing sequence.

3.4   PDE Constrained Optimization

The regularizations imposed in the two previous subsections involed the imposition of constraints on the input u to a PDE model and the resulting minimizations were expressed in terms of u alone. For at least two reasons it is sometimes of interest to formulate the minimization problem simultaneously over the input variable u, together with the solution of the PDE $p = G(u) \in P$: firstly computational algorithms which work to find (p, u) in $P \times X$ can be more effective than working entirely in terms of $u \in X$; and secondly regularization constraints may be imposed on the variable p as well as on u. If $J : P \times X \to \mathbb{R}$ then this leads to constrained minimization problems of the form

$$\min_{(p,u) \in P \times X} J(p, u) : \ p = G(u), \ c(p, u) \in \mathcal{K} \tag{24}$$

where $\mathcal{K}$ denotes the constraints imposed on both the input u and on the output p from the PDE model. Typically the observation operator $\mathcal{G} : X \to \mathbb{R}^N$ is found from G and then the information in Φ can be built into the definition of J.

Example 4. Consider the running example from Sect. 1.2 and assume that the observational noise $\xi \sim N(0, \gamma^2 I)$. Define

$$J(p, u) = \frac{1}{2\gamma^2} \sum_{j=1}^N |y - \ell_j(p)|^2 + \frac{\lambda_1}{2} \|u\|_{H^s}^2 + \frac{\lambda_2}{2} \|p\|_P^2$$

for some $s > d/2$. Choosing $\lambda_1 = \lambda$ and $\lambda_2 = 0$, together with $c(p, u) = (p, u)$ and $\mathcal{K} = P \times X$ we obtain from (24) the minimization from Example 3 in the case $\Gamma = \gamma^2 I$. Choosing $\lambda_1 = \lambda_2 = 0$, $c(p, u) = (p, u)$ and $\mathcal{K} = P \times E_{\mathrm{ad}}$ from Example 2 we recover that example. Choosing $\lambda_2 \ne 0$ or choosing the constraint set $\mathcal{K}$ to impose constraints on p leads to minimization in which the output p of the PDE model is constrained as well as the input u that we are trying to estimate.

3.5 Bayesian Regularization

The preceding regularization approaches have a nice mathematical structure and form a natural approach to the inverse problem when a unique solution is to be expected. But in many cases it may be interesting or important to find a large class of solutions, and to give relative weights to their importance. This allows, in particular, for predictions which quantify uncertainty. The Bayesian approach to regularization does this by adopting a probabilistic framework in which the solution to the inverse problem is a probability measure on X, rather than a single element of X.

We think of $(u, y) \in X \times \mathbb{R}^N$ as a random variable. Our goal is to find the distribution of u given y, often denoted by $u|y$. We define the joint distribution of (u, y) as follows. We assume that u and ξ appearing in (1) are independent mean zero Gaussian random variables, supported on X and $\mathbb{R}^N$ respectively, with covariance operator $\frac{1}{\lambda}\mathcal{C}$ and covariance matrix Γ respectively. By (1), the distribution of y given u, denoted $y|u$, is Gaussian $N(\mathcal{G}(u), \Gamma)$. The measure $\mu_0 = N(0, \frac{1}{\lambda}\mathcal{C})$ is known as the *prior* measure. It is most natural to define the measure μ_0 on a Hilbert space $H \supseteq X$. Under suitable conditions on $\mathcal{C}$, we have $\mu_0(X) = 1$. This means that under the measure μ_0, $u \in X$ almost surely so that $\mathcal{G}(u)$ is well-defined, almost surely. If $\mu_0(X) = 1$, it follows that the Hilbert space E with norm $\|\cdot\|_E = \|\mathcal{C}^{-1/2} \cdot \|_H$ is compactly embedded into X. The space E is known as the Cameron–Martin space. In the infinite dimensional setting, functions drawn from μ_0 are almost surely not in the Cameron–Martin space. See [10, 23] for detailed discussion of Gaussian measures on infinite dimensional spaces.

When solving the inverse problem, the aim is to find the posterior measure $\mu^y(du) = \mathbb{P}(du|y)$, and to obtain information about likely candidate solutions to the inverse problem from it. Informal application of Bayes' theorem gives

$$\mathbb{P}(u|y) \propto \mathbb{P}(y|u)\mu_0(u). \tag{25}$$

The probability density function for $\mathbb{P}(y|u)$ is, using the property of Gaussians, proportional to

$$\exp\left(-\frac{1}{2}|y - \mathcal{G}(u)|_\Gamma^2\right) = \exp\left(-\Phi(u)\right).$$

The infinite dimensional analogue of this result is to show that μ^y is absolutely continuous with respect to μ_0 with Radon–Nikodym derivative relating posterior to prior as follows:

$$\frac{d\mu^y}{d\mu_0}(u) = \frac{1}{Z} \exp\left(-\Phi(u)\right). \tag{26}$$

Here $\Phi(u)$ is given by (18) and $Z = \int_X \exp(-\Phi(u))\mu_0(du)$. The meaning of the formula (26) is that expectations under the posterior measure μ^y can be rewritten as weighted expectations with respect to the prior: for a function $\mathscr{F}$ on X we may write

$$\int_X \mathscr{F}(u)\mu^y(du) = \int_X \frac{1}{Z} \exp(-\Phi(u))\mathscr{F}(u)\mu_0(du).$$

Theorem 7. *([12]) Assume that $\mu_0(X) = 1$. Then μ^y is absolutely continuous with respect to μ_0 with Radon–Nikodym derivative given by (26). Furthermore the measure μ^y is locally Lipschitz in the data y with respect to the Hellinger metric: there is a constant $C = C(r)$, such that, for all y, y' with $\max \{|y|, |y'|\} \leq r$,*

$$d_{\text{HELL}}(\mu^y, \mu^{y'}) \leq C|y - y'|. \tag{27}$$

If μ, ν are probability measures that are absolutely continuous with respect to the probability measure ρ, then the Hellinger metric is defined as

$$d_{\text{HELL}}(\mu, \nu)^2 = \frac{1}{2} \int \left(\sqrt{\frac{d\mu(u)}{d\rho}} - \sqrt{\frac{d\nu(u)}{d\rho}} \right)^2 \rho(du).$$

For any function of u which is square integrable with respect to both μ and ν it may be shown that the difference in expectations of that function, under μ and under ν, is bounded above by the Hellinger distance. In particular, this theorem shows that the posterior mean and covariance operators corresponding to data sets y and y' are $\mathcal{O}(|y - y'|)$ apart.

The choice of prior μ_0, relates directly to the regularization of the inverse problem. To see this we note that since the operator $\mathscr{C}$ is necessarily positive and self-adjoint we may write down the complete orthonormal system

$$\frac{1}{\lambda}\mathscr{C}\phi_m = \sigma_m^2\phi_m, \quad m \in \mathbb{Z}^+, \qquad \lim_{m\to\infty} \sigma_m = 0. \tag{28}$$

Then $u \sim \mu_0$ can be written via the Karhunen–Loève expansion as

$$u(x) = \sum_{m\in\mathbb{Z}^+} \sigma_m \eta_m \phi_m(x) \tag{29}$$

where the η_m form an i.i.d. sequence of unit Gaussian random variables. We may regularize the inverse problem by modifying the decay rate of σ_m. For example, choosing $\sigma_m = 0$ for $m \notin \mathcal{M}$, where $\mathcal{M} \subset \mathbb{Z}^+$ is finite, restricts the solution of the inverse problem to a finite dimensional set, and is hence a regularization. More generally, the rate of decay of the σ_m (which are necessarily summable as $\mathscr{C}$ is trace class) will affect the almost sure regularity properties of functions drawn from μ_0 and, by absolute continuity of μ^y with respect to μ_0, of functions drawn from μ^y.

In the case that X is a subset of $H = L^2(D)$ with $D \subset \mathbb{R}^d$, the operator $\mathscr{C}$ may be identified with an integral operator:

$$\frac{1}{\lambda}(\mathscr{C}\phi)(x_1) = \int_D c(x_1, x_2)\phi(x_2)dx_2$$

for some kernel $c(x_1, x_2)$. The regularity of $c(x_1, x_2)$ determines the decay rate of σ_m [22]. For example, if D is a rectangle and $\mathscr{C} = (-\Delta)^{-\alpha}$ where $-\Delta$ has domain $H^2(D) \cap H_0^1(D)$ then the corresponding measure μ_0 has the property that samples are almost surely in the Sobolev space $H^s(D)$ and in the Hölder space $C^s(D)$ for all $s < \alpha - \frac{d}{2}$ (see [14] for more details). In particular, if $\alpha > d/2$, then $\mu(X) = 1$ when $X = L^\infty(D)$.

Priors which charge functions with a multiscale character can be built in this Gaussian context. One natural way to do this is to choose $\mathscr{M}$ as above so that it contains two distinct sets of functions varying on length scales of $\mathscr{O}(1)$ and $\mathscr{O}(\epsilon)$ respectively. A second natural way is to choose a covariance function $c = c^\epsilon$ which has two scales.

The formula (26) shows quite clearly how regularization works in the Bayesian context: the main contribution to the expectation will come from places where Φ is close to its minimum value and where μ_0 is concentrated; thus minimizing Φ is important, but this minimization is regularized through the properties of the measure μ_0. We now develop this intuitive concept further by linking the Bayesian approach to Tikhonov regularization and the functional I given by (23).

Given $z \in E$ and $\delta \ll 1$ define the small ball probability

$$J^\delta(z) = \mathbb{P}^{\mu^y}(\{u : \|u - z\| < \delta\})$$

Note that this ball is in X but centred at a point $z \in E$, with E (the Cameron–Martin space) compact in X. It is natural to ask where $J^\delta(z)$ is maximized as a function of z and placing z in E allows us to answer this question. Furthermore we then see a connection between the Bayesian approach and the Tikhonov approach to regularization. The next theorem shows that small balls centred at minimizers of (23) will have maximal relative probability under the Bayesian posterior measure, in the small ball limit $\delta \to 0$.

Theorem 8. *([15]) Assume that $\mu_0(X) = 1$. Then*

$$\lim_{\delta \to 0} \frac{J^\delta(z_1)}{J^\delta(z_2)} = \exp(I(z_2) - I(z_1)).$$

In the Bayesian context the solution of the Tikhonov regularized problem is known as the Maximum A Posteriori estimator (MAP estimator) [8, 19].

4 Large Data Limits

In the previous section we showed how regularization plays a significant role in the solution of inverse problems. Choosing the correct regularization is part of the overall modelling scenario in which the inverse problem is embedded, as we demonstrated in the running example of Sect. 1.2. In some situations it may be suitable to look for the solution of the inverse problem over a small finite set of parameters, whilst in others it may be desirable to look over a larger, even infinite dimensional set, in which oscillations are captured.

This section is devoted entirely to inverse problems where a single scalar parameter is sought and we study whether or not this parameter is correctly identified when a large amount of noisy data is available. The development is tied specifically to the running example, namely the PDE (3). For a fixed permeability coefficient generating the data, Fitzpatrick has also studied the consistency and asymptotic normality of maximum likelihood estimates in the large data limit [19]. Related work on parameter estimation in the context stochastic differential equations (SDEs) may be found in [3, 26, 28].

4.1 The Statistical Model

We consider the problem of estimating a single scalar parameter $u \in \mathbb{R}$ in the elliptic PDE

$$\nabla \cdot v = f, \quad x \in D,$$

$$p = 0, \quad x \in \partial D, \tag{30}$$

$$v = -\exp(u)A\nabla p$$

where $D \subset \mathbb{R}^d$ is bounded and open, and $f \in H^{-1}$ as well as the constant symmetric matrix A are assumed to be known. We let $G : \mathbb{R} \to H_0^1(D)$ be defined by $G(u) = p$. Then using the same linear functionals as in the running example from Sect. 1.2 we may construct the observation operator $\mathcal{G} : \mathbb{R} \to \mathbb{R}^N$ defined by $\mathcal{G}(u)_j = \ell_j(G(u))$. Our aim is to solve the inverse problem of determining u given y satisfying (1). For simplicity we assume that $\xi \sim N(0, \gamma^2 I)$ which implies that the observational noise on each linear functional is i.i.d. $N(0, \gamma^2)$. Since u is finite dimensional we will simply minimize Φ given by (18): no further regularization is needed because u is already finite dimensional.

Notice that the solution p of (30) is linear in $\exp(-u)$ and that we may write $G(u) = \exp(-u)p^\star$ where $p^\star$ solves

$$\nabla \cdot v = f, \quad x \in D,$$

$$p^\star = 0, \quad x \in \partial D. \tag{31}$$

$$v = -A\nabla p^\star$$

Note that $\mathscr{G}(u)_j = \exp(-u)\ell_j(p^\star)$ so that the least squares functional (18) has the form

$$\Phi(u) = \frac{1}{2\gamma^2} \sum_{j=1}^{N} |y_j - \mathscr{G}_j(u)|^2 = \frac{1}{2\gamma^2} \sum_{j=1}^{N} |y_j - \exp(-u)\ell_j(p^\star)|^2.$$

It is straightforward to see that Φ has a unique minimizer $\bar{u}$ satisfying

$$\exp(-\bar{u}) = \frac{\sum_{j=1}^{N} y_j \ell_j(p^\star)}{\sum_{j=1}^{N} \ell_j(p^\star)^2}. \tag{32}$$

It is now natural to ask whether, for large N, the estimate $\bar{u}$ is close to the desired value of the parameter. We study two situations: the first where the data is generated by the model which is used to fit the data; and the second where the data is generated by a multiscale model whose homogenized limit gives the model which is used to fit the data.

4.2   Data from the Homogenized Model

In this section we demonstrate that, in the large data limit, random observational error may be averaged out and the true value of the parameter recovered.

We define $p_0 = \exp(-u_0)p^\star$ so that p_0 solves (30) with $u = u_0$.

Assumption 9 *We assume that the data y is given by noisy observations generated by the statistical model:*

$$y_j = \ell_j(p_0) + \xi_j$$

where $\{\xi_j\}$ form an i.i.d. sequence of random variables distributed as $N(0, \gamma^2)$.

Theorem 10. *Let Assumptions 9 hold and assume that $\liminf_{N \to \infty} \frac{1}{N} \sum_{j=1}^{N} \ell_j$ $(p^\star)^2 \geq L > 0$ as $N \to \infty$. Then ξ-almost surely*

$$\lim_{N \to \infty} |\exp(-\bar{u}) - \exp(-u_0)| = 0.$$

Proof. Substituting the assumed expression for the data from Assumption 9 into the formula (32) gives

$$\exp(-\bar{u}) = \exp(-u_0) + I_1$$

where
$$I_1 = \frac{\frac{1}{N}\sum_{j=1}^{N}\xi_j\ell_j(p^\star)}{\frac{1}{N}\sum_{j=1}^{N}\ell_j(p^\star)^2}.$$

Therefore,

$$\mathbb{E}[I_1^2] = \frac{\gamma^2}{\sum_{j=1}^{N}\ell_j(p^*)^2} \leq \frac{2\gamma^2}{NL} \tag{33}$$

for N sufficiently large. Since I_1 is Gaussian we deduce that $\mathbb{E}I_1^{2p} = \mathcal{O}(N^{-p})$ as $N \to \infty$. Application of the Borel–Cantelli lemma shows that I_1 converges almost surely to zero as $N \to \infty$. $\qquad\square$

This shows that, in the large data limit, random observational error may be averaged out and the true value of the parameter recovered, in the idealized scenario where the data is taken from the statistical model used to identify the parameter. The condition that $L > 0$ prevents additional observation noise from overwhelming the information obtained from additional measurements as $N \to \infty$. It is a simple explicit example of what is known as *posterior consistency* [9] in the theory of statistics.

4.3 Data from the Multiscale Model

In practice, of course, real data does not come from the statistical model used to estimate parameters. In order to probe the effect that this can have on posterior consistency we study the situation where the data is taken from a multiscale model whose homogenized limit falls within the class used in the statistical model to estimate parameters. Again we define $p_0 = \exp(-u_0)p^\star$ and we now define p^ϵ to solve (7) with K^ϵ chosen so that the homogenized coefficient associated with this family is $K_0 = \exp(u_0)A$.

Assumption 11 *We assume that the data y is generated from noisy observations of a multiscale model:*
$$y_j = \ell_j(p^\epsilon) + \xi_j$$
with p^ϵ as above and the $\{\xi_j\}$ an i.i.d. sequence of random variables distributed as $N(0, \gamma^2)$.

Theorem 12. *Let Assumptions 11 hold and assume that the linear functionals ℓ_j are chosen so that*

$$\lim_{\epsilon\to 0}\limsup_{N\to\infty}\frac{1}{N}\sum_{j=1}^{N}|\ell_j(p^\epsilon - p_0)|^2 = 0 \tag{34}$$

and $\liminf_{N\to\infty}\frac{1}{N}\sum_{j=1}^{N}\ell_j(p^\star)^2 \geq L > 0$ as $N \to \infty$. Then $\xi-$ almost surely

$$\lim_{\epsilon \to 0} \lim_{N \to \infty} |\exp(-\bar{u}) - \exp(-u_0)| = 0.$$

Proof. Notice that the solution of the homogenized equation is $p_0 = \exp(-u_0)p^\star$. We write

$$y_j = \ell_j(p_0) + \ell_j(p^\epsilon - p_0) + \xi_j$$
$$= \exp(-u_0)\ell_j(p^\star) + \ell_j(p^\epsilon - p_0) + \xi_j.$$

Substituting this into the formula (32) gives

$$\exp(-\bar{u}) = \exp(-u_0) + I_1 + I_2^\epsilon$$

where I_1 is as defined in the proof of Theorem 10 and is independent of ϵ, and

$$I_2^\epsilon = \frac{\sum_{j=1}^N \ell_j(p^\epsilon - p_0)\ell_j(p^\star)}{\sum_{j=1}^N \ell_j(p^\star)^2}.$$

The Cauchy–Schwarz inequality gives

$$|I_2^\epsilon| \le \frac{\left(\sum_{j=1}^N |\ell_j(p^\epsilon - p_0)|^2\right)^{1/2}}{\left(\sum_{j=1}^N \ell_j(p^\star)^2\right)^{1/2}} \le \left(\frac{2}{NL} \sum_{j=1}^N |\ell_j(p^\epsilon - p_0)|^2\right)^{1/2}$$

for N sufficiently large. As in the proof of Theorem 10 we have, ξ-almost surely,

$$\lim_{N \to 0} |\exp(-\bar{u}) - \exp(-u_0) - I_2^\epsilon| = 0.$$

From this and (34) the desired result now follows.   □

The assumption (34) encodes the idea that, for small ϵ, the linear functionals used in the observation process return nearby values when applied to the solution p^ϵ of the multiscale model or to the solution p_0 of the homogenized equation. In particular, Corollary 1 implies that if $\{\ell_j(p)\}_{j=1}^\infty$ is a family of bounded linear functionals on $L^2(D)$, uniformly bounded in j, then (34) will hold. On the other hand, we may choose linear functionals that are bounded as functionals on $H^1(D)$ yet unbounded on $L^2(D)$. In this case Theorem 1 shows that (34) may not hold and the correct homogenized coefficient may not be recovered, even in the large data limit. An analogous phenomenon occurs in inference for SDEs where if the observations of a multiscale diffusion are too frequent (relative to the fast scale) then the correct homogenized coefficients are not recovered [26, 28].

5 Exploiting Multiscale Properties Within Inverse Estimation

In this section we describe how ideas from homogenization theory can be used to improve the estimation of parameters in homogenized models. We consider a regime where the unknown parameter has small-scale fluctuations that may be characterized as random. In this case, if we attempt to recover the homogenized parameter the error ξ appearing in (1) is affected by the model mismatch. This is because the simplified, low-dimensional parameter used to fit the data is different from the true unknown coefficient. So, even when there is no observational noise, the error ξ has a statistical structure. Nevertheless, homogenization theory predicts that this discrepancy between $G(u)$ and y associated with model mismatch will have a universal statistical structure which can be exploited in the inverse problem, as we now describe.

The specific ideas described here were developed by Nolen and Papanicolaou in [24] for one dimensional elliptic problems, including the groundwater flow problem that we study here. Bal and Ren [5] have employed similar ideas in the study of Sturm–Liouville problems with unknown potential. We begin by describing in Sect. 5.1 the homogenization and fluctuation theory for the case that the (scalar) permeability $k(x)$ is random. Then, in Sect. 5.2 we show how these ideas can be used to develop an improved estimator for the homogenized permeability coefficient. We conclude with numerical results in Sect. 5.3.

5.1 The Model

In this section we will present the approach of [24] in the simplest possible setting. We consider the two-point boundary value problem

$$-\frac{d}{dx}\left(\exp(u(x))\frac{dp}{dx}\right) = f(x), \quad x \in [-1, 1], \tag{35a}$$

$$p(-1) = p(1) = 0. \tag{35b}$$

This is, of course, (3) in the one-dimensional setting $d = 1$.

It is assumed that the coefficient $k(x) = \exp(u(x))$ is a single realization of a stationary, ergodic and mixing random field $k(x, \omega)$. Furthermore it is assumed that k^{-1} can be decomposed into a slowly varying non-random component, together with a random, rapidly oscillating component:

$$\frac{1}{k(x, \omega)} = \frac{1}{k_0(x)} + \sigma\mu\left(\frac{x}{\epsilon}, \omega\right), \tag{36}$$

where $\mu(x, \omega)$ is a stationary, mean zero random field with covariance

$$R(x) = \mathbb{E}(\mu(x+y)\mu(y)).$$

We assume that $R(0) = 1$ and $\int_{\mathbb{R}} R(x)\, dx = 1$. Thus, σ^2 and ϵ are the (given) variance and correlation length of the fluctuations. We are interested in the case where $\epsilon \ll 1$ so that the random fluctuations are rapid.

The solution $p = p_\epsilon(x, \omega)$ of (35) depends on $\epsilon > 0$ and on the realization of $k(x, \omega)$. However, in the limit as $\epsilon \to 0$, p_ϵ coverges to $p_0(x)$ which is the solution of the homogenized Dirichlet problem

$$-\frac{d}{dx}\left(k_0(x)\frac{d}{dx}p_0\right) = f(x), \quad x \in [-1, 1], \tag{37a}$$

$$p_0(-1) = p_0(1) = 0. \tag{37b}$$

Observe that the homogenized coefficient is the harmonic mean of k: $k_0(x) = \mathbb{E}[k^{-1}]^{-1}$. Moreover, in the limit as $\epsilon \to 0$, the solution p_ϵ has Gaussian fluctuations about its asymptotic limit [4]. Specifically, one can prove that

$$\frac{p_\epsilon(x, \omega) - p_0(x)}{\epsilon^{1/2}} \to \sigma \int_D Q(x, y; k_0)v_0(y; k_0)\, dW_y(\omega) \tag{38}$$

in distribution as $\epsilon \to 0$, where $W_y(\omega)$ is a Brownian random field, which is a Gaussian process. Here $v_0(x; k_0) = k_0(x)p_0(x)$, and the kernel $Q(x, y; k_0)$ is then related to the Green's function for the one dimensional system:

$$\begin{pmatrix} p_x \\ v_x \end{pmatrix} - \begin{pmatrix} 0 & 1/k_0(x) \\ 0 & 0 \end{pmatrix}\begin{pmatrix} p \\ v \end{pmatrix} = \begin{pmatrix} g_1 \\ g_2 \end{pmatrix}.$$

If the 2×2 Green's matrix for this system is $G(x, y; k_0) : D \times D \to \mathbb{R}^2 \otimes \mathbb{R}^2$, then $Q(x, y; k_0) = G_{1,1}(x, y; k_0)$. The important point here is that the integral

$$I(x, \omega) = \sigma \int_D Q(x, y; k_0)v_0(y; k_0)\, dW_y(\omega)$$

which appears on the right side of (38) is a centered Gaussian random variable with covariance

$$\mathbb{E}[I(x)I(z)] = \sigma^2 \int_D Q(x, y; k_0)v_0(y; k_0)^2 Q(y, z; k_0)\, dy.$$

This covariance depends on k_0. The asymptotic theory given by the limit theorem (38) gives us a good approximation of the statistics of $p_\epsilon(x, \omega)$ even when there is no observation noise, and shows that the fluctuations depend on k_0. In this simple case presented here, Q can be computed explicitly. In other cases, it can be computed numerically; see [24] for more details.

5.2 Enhanced Estimation

We now show how this asymptotic theory can be used to enhance estimation of the homogenized parameter $k_0(x)$. The inverse problem is to identify the parameter $k_0(x)$ in the model

$$-\frac{d}{dx}\left(k_0(x)\frac{d}{dx}p_0\right) = f(x), \quad x \in [-1, 1], \tag{39a}$$

$$p_0(-1) = p_0(1) = 0. \tag{39b}$$

We take the viewpoint that the data actually come from observations of $p_\epsilon(x, \omega)$, which is the solution of the multiscale model (35) with $k(x, \omega)$ given by (36), so there is a discrepancy between the model used to fit the data and the true model which generates the data. Now the outstanding modelling issue is the choice of statistical model for the error ξ in (1).

Suppose we make noisy observations of $p_\epsilon(x_j)$ at points $\{x_j\}_{j=1}^N$ distributed throughout the domain. Then the measurements are

$$y_j = p_\epsilon(x_j, \omega) + \xi_j, \quad j = 1, \ldots, N$$

where $\xi_j \sim N(0, \gamma^2)$ are mutually independent, representing observation noise. The limit (38) we have just described tells us that for ϵ small, these measurements are approximated well by

$$y_j \approx p_0(x_j) + \xi_j',$$

where $\{\xi_j'\}_{j=1}^N$ are Gaussian random variables with mean zero and covariance

$$C_{j,\ell}(k_0, \epsilon) = \mathbb{E}[\xi_j'\xi_\ell'] = \gamma^2\delta_{j,\ell} + \epsilon\sigma^2\int_D Q(x_j, y; k_0)v_0(y; k_0)^2 Q(x_\ell, z; k_0)\, dy \tag{40}$$

Therefore, we model the observations as

$$y_j \approx \mathscr{G}(k_0) + \xi_j', \quad j = 1, \ldots, N$$

where $\mathscr{G}(k_0) = p_0(x_j; k_0)$ with p_0 being the solution of (39). The modified statistical error ξ' has two components. The first term $\gamma^2\delta_{j,\ell}$ is due to observation error. The second term comes from the asymptotic theory and is associated with the random microstructure in the true parameter $k(x, \omega)$. Of course, if ϵ is very small, relative to γ^2, then the observation noise dominates (40). In this case, the observations of p_ϵ may be very close to observations of the homogenized solution p_0, and we might simply assume that $\xi' \sim N(0, \gamma^2 I)$, ignoring the error associated with the model mismatch. On the other hand, if γ^2 is small relative to ϵ then the statistical error ξ' is dominated by the model mismatch. In this case, homogenization theory gives us an asymptotic approximation of the true covariance structure of ξ', which is quite different from $N(0, \gamma^2 I)$. See [24] for a discussion of some properties of the covariance matrix $C(k_0, \epsilon)$.

Using the covariance (40), we make the approximation

$$\mathbb{P}(y|k_0) \approx \frac{1}{\sqrt{2\pi|C(k_0;\epsilon)|}} \exp\left(-\frac{1}{2}(y - \mathscr{G}(k_0))^T C(k_0;\epsilon)^{-1}(y - \mathscr{G}(k_0))\right),$$

where $|\cdot|$ denotes the determinant. The parameter $k_0(x)$ is a function, in general, and we may place a Gaussian prior μ_0 on $u_0(x) = \log k_0(x)$. Application of Bayes' theorem (25) (with k_0 replacing u) gives that

$$\mathbb{P}(k_0|y) \propto \frac{1}{\sqrt{2\pi|C(k_0;\epsilon)|}} \exp\left(-\frac{1}{2}(y - \mathscr{G}(k_0))^T C(k_0;\epsilon)^{-1}(y - \mathscr{G}(k_0))\right)\mu_0(\log k_0)$$

where the constant of proportionality is independent of k_0. The maximum a posteriori estimator (MAP) is then found as the function $k_0(x)$ which maximizes $\mathbb{P}(k_0|y)$ which is the same as minimizing $I(k_0) = -\ln(\mathbb{P}(k_0|y))$. The key contribution of homogenization theory is to correctly identify the noise structure which has covariance $C(k_0;\epsilon)$ depending on $k_0(x)$, the parameter to be estimated.

5.3   Numerical Results

In this section we discuss the results of a numerical computation that show some advantage to using homogenization theory as we have just described, compared to a simple least-squares approach. Given noisy observations of $p_\epsilon(x_j)$ we may compute the MAP estimator $\hat{k}_1$ using (41) with covariance $C(k_0;\epsilon)$ given by (40):

$$\hat{k}_1 = \text{argmax}_{k_0} \frac{1}{\sqrt{2\pi|C(k_0;\epsilon)|}}$$

$$\exp\left(-\frac{1}{2}(y - \mathscr{G}(k_0))^T C(k_0;\epsilon)^{-1}(y - \mathscr{G}(k_0))\right)\mu_0(\log k_0), \qquad (41)$$

On the other hand, we might ignore the effect of the random microstructure and simply use $C = \gamma^2 I$, accounting only for observation noise:

$$\hat{k}_2 = \text{argmax}_{k_0} \frac{1}{\sqrt{2\pi|\gamma^2 I|}} \exp\left(-\frac{1}{2}\gamma^{-2}|y - \mathscr{G}(k_0)|^2\right)\mu_0(\log k_0). \qquad (42)$$

Both estimates $\hat{k}_1$ and $\hat{k}_2$ are random variables, depending on the random data observed, but we should hope that $\hat{k}_1$ gives us a better approximation of k_0, since it makes use of the true covariance (40). Indeed for simple linear statistical models, it is easy to see that an efficient estimator, which realizes the theoretically optimal variance given by the Cramér–Rao lower bound, may be obtained by using the true covariance of the data. However, using the incorrect covariance may lead to an estimate with significantly higher variance than the theoretical optimum. See [24] for more discussion of this point. The present setting is highly nonlinear and the variance of the estimates $\hat{k}_1$ and $\hat{k}_2$ cannot be computed explicitly, since $C(k_0, \epsilon)$ depends on k_0 in a nonlinear way through solution of the PDE. Nevertheless the numerical results are consistent with the expectation that approximation of the true covariance (through homogenization theory) yields a MAP estimator that has smaller variance, relative to the estimate that makes no use of the homogenization theory (see Fig. 3).

In Fig. 1 we show one realization of the true coefficient $k(x, \omega)$ which was used to generate the data. The highly-oscillatory graph represents the true coefficient $k(x, \omega)$ with variation on many scales. The slowly-varying harmonic mean $k_0(x)$ also is displayed here as the thick curve; this function k_0 is what we attempt to estimate. The data was generated as follows. Using one realization of $k(x, \omega)$ and given forcing f, we solve the Dirichlet boundary value problem (35). The observation data involves point-wise evaluation of $p^\epsilon(x_j)$ at points $\{x_j\}_{j=1}^N$ spaced uniformly across the domain, plus independent observation noise $N(0, \gamma^2)$ at each point of observation. Using this data, we compute estimates $\hat{k}_1$ and $\hat{k}_2$ by minimizing (41) and (42), respectively. For the computation shown here, the function $k_0(x)$ is parameterized by the first three coefficients in a Fourier series expansion. So,

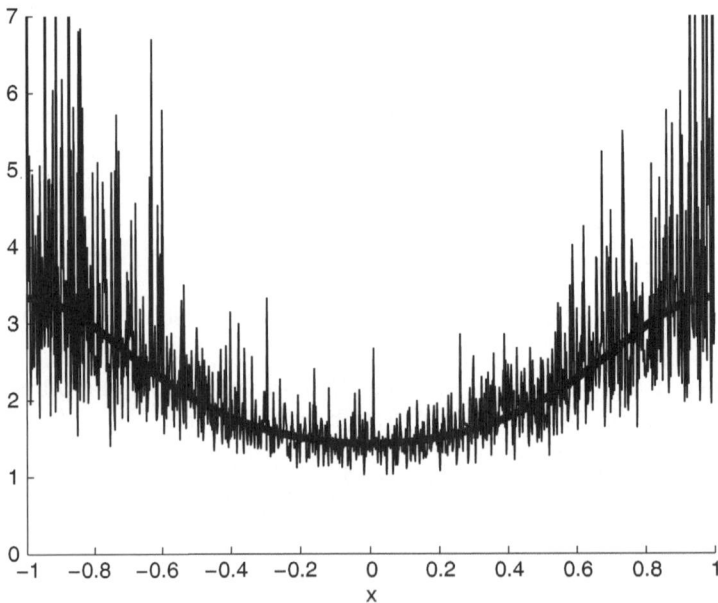

Fig. 1 The thin erratic curve is one realization of the true coefficient $k^\epsilon(x, \omega)$. The thick curve is the slowly-varying harmonic mean $k_0(x)$. This realization was used to generate the data

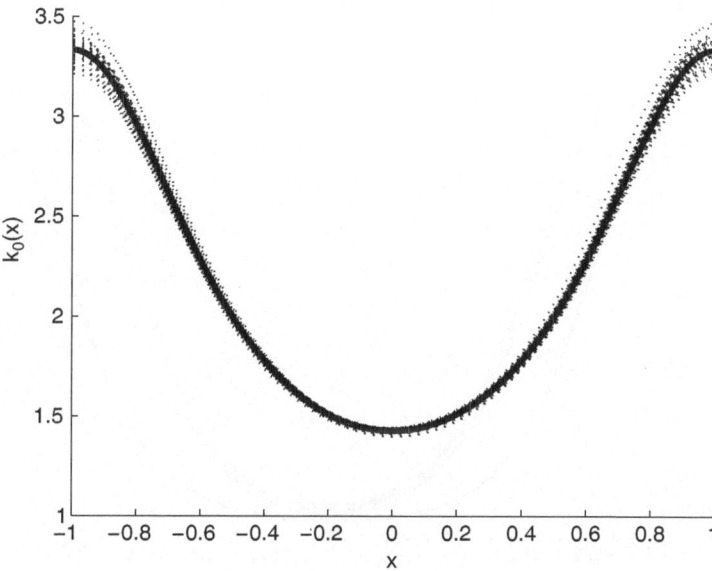

Fig. 2 The thick curve is the true k_0. The dashed series represent 100 independent realizations of the estimate $\hat{k}_1$

computing $\hat{k}_1$ and $\hat{k}_2$ involves an optimization in $\mathbb{R}^3$. To evaluate $\mathbb{P}(k_0|y)$ at each step in the minimization algorithm, we must solve the forward problem (39) with the current estimate of k_0, and in the case of $\hat{k}_1$ we must also compute $C(k_0, \epsilon)$. See [24] for more details about this computation.

Figure 2 compares the estimate $\hat{k}_1(x)$ with the true function $k_0(x)$. Since the estimate $\hat{k}_1(x)$ is a random function, we performed the experiment many times (generating new $k(x, \omega)$ to compute each estimate $\hat{k}_1$) and display the results of 100 experiments. The data for $\hat{k}_2$ is qualitatively similar. Nevertheless, the pointwise variance $\mathrm{Var}[\hat{k}_1(x)]$ is smaller than $\mathrm{Var}[\hat{k}_2(x)]$, as shown in Fig. 3. This is consistent with the linear estimation theory for which knowledge of the true data covariance yields an estimate with optimal variance.

Acknowledgements The authors thank A. Cliffe and Ch. Schwab for helpful discussions concerning the groundwater flow model.

Appendix 1

In this Appendix we prove Theorem 3 which, recall, applies in the case where (7b) and (13b) are replaced by periodic conditions on $D = (L\mathbb{T})^d$.

Theorem 13. *Let $x^\epsilon(t)$ and $x_0(t)$ be the solutions to (8) and (17), with velocity fields extended from $D = (L\mathbb{T})^d$ to $\mathbb{R}^d$ by periodicity, and assume that Assumption 2 holds. Assume also that $f \in C^\infty(D)$ and that $K \in C^\infty(D; C^\infty_{\mathrm{per}}(\mathbb{T}^d))$.*

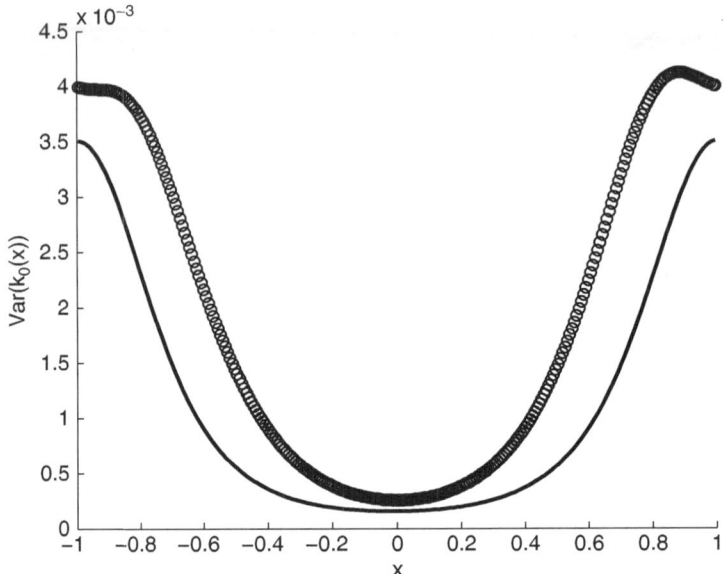

Fig. 3 The upper series (o) is the empirical variance $\mathrm{Var}[\hat{k}_2(x)]$. The lower series (−) is $\mathrm{Var}[\hat{k}_1(x)]$. Both quantities were computed using 500 samples

Then

$$\lim_{\epsilon \to 0} \mathbb{E} \sup_{0 \le t \le T} \|x^\epsilon(t) - x_0(t)\| = 0.$$

Proof. To simplify the notation we will set the porosity of the rock to be equal to 1, $\phi = 1$. Recall that $v^\epsilon(x) = K^\epsilon(x)\nabla p^\epsilon(x)$. Our first observation is that, for $p_a^\epsilon(x)$ given by (9),

$$K^\epsilon(x)\nabla p^\epsilon(x) = K^\epsilon(x)\nabla p_a^\epsilon(x) - \delta^\epsilon(x) \tag{43}$$

where

$$\delta^\epsilon(x) = -K^\epsilon(x)\nabla\Big(p^\epsilon(x) - p_a^\epsilon(x)\Big). \tag{44}$$

From Assumption 2 we deduce that

$$\lim_{\epsilon \to 0} \|\delta^\epsilon(x)\|_{L^\infty} = 0.$$

From the definition of $p_a^\epsilon(x)$ it follows that

$$K^\epsilon(x)\nabla p_a^\epsilon(x) = Q^\epsilon(x)\nabla p_0(x) - \epsilon\delta_1^\epsilon(x)$$

where

$$\delta_1^\epsilon(x) = -K^\epsilon(x)\nabla_x p_1(x, x/\epsilon), \quad Q^\epsilon(x) = Q(x, x/\epsilon). \tag{45}$$

From the definition of p_1 in (14) we see that

$$\|\delta_1^\epsilon(x)\|_{L^\infty} \le C.$$

Putting (43) and (45) together we see that

$$v^\epsilon(x) = -Q^\epsilon(x)\nabla p_0(x) + \delta^\epsilon(x) + \epsilon\delta_1^\epsilon(x)$$

and we see from (44) and (45) that the perturbations of $v^\epsilon(x)$ from $Q^\epsilon(x)\nabla p_0(x)$ are small; it is thus natural to expect a limit theorem for x^ϵ solving (8) which is Lagrangian transport in an appropriately averaged version of $Q^\epsilon(x)\nabla p_0(x)$. Furthermore, since $Q(x, y)$ is divergence free in the fast y coordinate, by (15), it is natural to expect that the appropriate average is the Lebesgue measure. We now show that this is indeed the case.

From (8) we deduce that

$$x^\epsilon(t) = x(0) + \int_0^t \left(-Q^\epsilon(x)\nabla p_0(x(s)) + \delta^\epsilon(x(s)) + \epsilon\delta_1^\epsilon(x(s))\right) ds$$

$$+ \sqrt{2\eta_0\epsilon}\, W(t). \tag{46}$$

Define now $V(x, y) = -Q(x, y)\nabla p_0(x)$ and consider the system of SDEs

$$\frac{dx}{dt} = \left(V(x, y) + \delta^\epsilon(x) + \epsilon\delta_1^\epsilon(x)\right) + \sqrt{2\eta_0\epsilon}\,\frac{dW}{dt}, \tag{47a}$$

$$\frac{dy}{dt} = \frac{1}{\epsilon}\left(V(x, y) + \delta^\epsilon(x)\right) + \delta_1^\epsilon(x) + \sqrt{\frac{2\eta_0}{\epsilon}}\,\frac{dW}{dt}. \tag{47b}$$

Since $y = x/\epsilon$ we see that $x(t)$, the solution of (47) is equal to $x^\epsilon(t)$ appearing in (46).

The process $\{x(t),\ y(t)\}$ is Markov with generator

$$\mathscr{L} = \frac{1}{\epsilon}\left(\left(V(x, y) + \delta^\epsilon(x)\right) \cdot \nabla_y + \eta_0\Delta_y\right)$$

$$+ \left(\left(V(x, y) + \delta^\epsilon(x)\right) \cdot \nabla_x + \delta_1^\epsilon(x) \cdot \nabla_y + \eta_0\nabla_x \cdot \nabla_y + \eta_0\nabla_y \cdot \nabla_x\right)$$

$$+ \epsilon\eta_0\Delta_x + \epsilon\delta_1^\epsilon(x) \cdot \nabla_x$$

$$=: \frac{1}{\epsilon}\left(\mathscr{L}_0 + \delta^\epsilon(x) \cdot \nabla_y\right) + \mathscr{L}_1 + \epsilon\mathscr{L}_2.$$

Consider now the Poisson equation

$$-\mathscr{L}_0\Phi = V(x, y) - v_0(x) \tag{48}$$

with (see (13)(c))

$$v_0(x) = \int_{\mathbb{T}^d} V(x, y)\, dy.$$

Equation (48) is posed on $\mathcal{T}^d$ with periodic boundary conditions. Notice that x enters merely as a parameter in this equation. The operator $\mathcal{L}_0$ is uniformly elliptic on $\mathbb{T}^d$ and the right hand side averages to 0, hence, by Fredholm's alternative this equation has a solution which is unique, up to constants. We fix this constant by requiring that $\int_{\mathbb{T}^d} \Phi(x, y)\, dy = 0$. We define $\Phi^\epsilon(x) := \Phi(x, x/\epsilon)$ and similarly for $\mathcal{L}_i \Phi^\epsilon(x)$. Applying Itô's formula to Φ and evaluating at $y = x/\epsilon$ we obtain

$$d\Phi^\epsilon(x) = \frac{1}{\epsilon}\left(\mathcal{L}_0\Phi^\epsilon + \delta^\epsilon(x) \cdot \nabla_y \Phi(x, x/\epsilon)\right) dt + \mathcal{L}_1\Phi^\epsilon\, dt + \epsilon\mathcal{L}_2\Phi^\epsilon\, dt$$

$$+ \sqrt{\frac{2\eta_0}{\epsilon}}\nabla_y\Phi^\epsilon\, dW + \sqrt{2\eta_0\epsilon}\nabla_x\Phi^\epsilon\, dW$$

$$= -\frac{1}{\epsilon}\left(V(x, x/\epsilon) - v_0(x) + \delta^\epsilon(x) \cdot \nabla_y\Phi(x, x/\epsilon)\right) dt + \mathcal{L}_1\Phi^\epsilon$$

$$+ \epsilon\mathcal{L}_2\Phi^\epsilon\, dt + \sqrt{\frac{2\eta_0}{\epsilon}}\nabla_y\Phi^\epsilon\, dW + \sqrt{2\eta_0\epsilon}\nabla_x\Phi^\epsilon\, dW.$$

Consequently,

$$\int_0^t V(x(s), y(s))\, ds - \int_0^t v_0(x(s))\, ds$$

$$= \int_0^t \left(\delta^\epsilon(x(s)) \cdot \nabla_y\Phi(x(s), x(s)/\epsilon) + \epsilon\mathcal{L}_1\Phi^\epsilon(x(s)) + \epsilon^2\mathcal{L}_2\Phi^\epsilon(x(s))\right) ds$$

$$- \epsilon\left(\Phi^\epsilon(x^\epsilon(t)) - \Phi^\epsilon(x^\epsilon(0))\right) + \sqrt{\epsilon}M^\epsilon(t),$$

where

$$M^\epsilon(t) := \int_0^t \left(\sqrt{2\eta_0}\nabla_y\Phi^\epsilon + \epsilon\sqrt{2\eta_0}\nabla_x\Phi^\epsilon\right) dW.$$

Since $\Phi(x, y)$ is periodic in both coordinates we have that

$$\|\nabla_y\Phi(x, x/\epsilon)\|_{L^\infty} \le C, \quad \|\Phi^\epsilon(x)\|_{L^\infty} \le C,$$

$$\|\mathcal{L}_1\Phi^\epsilon\|_{L^\infty} \le C, \quad \|\mathcal{L}_1\Phi^\epsilon\|_{L^\infty} \le C$$

and

$$\mathbb{E}\|M^\epsilon(t)\|^p \le C, \quad p \ge 1. \tag{49}$$

We combine the above calculations to obtain

$$x^\epsilon(t) = x(0) + \int_0^t v_0(x^\epsilon(s))\, ds + H^\epsilon(t) + \sqrt{\epsilon}\tilde{M}^\epsilon(t),$$

where

$$H^\epsilon(t) := -\epsilon\Big(\Phi^\epsilon(x^\epsilon(t)) - \Phi^\epsilon(x^\epsilon(0))\Big) + \int_0^t \big(\delta^\epsilon(x^\epsilon(s)) + \epsilon\delta_1^\epsilon(x^\epsilon(s))\big)\, ds$$

$$+ \int_0^t \Big(\delta^\epsilon(x(s)) \cdot \nabla_y \Phi(x(s), x(s)/\epsilon) + \epsilon\mathcal{L}_1\Phi^\epsilon(x(s))$$

$$+\epsilon^2\mathcal{L}_2\Phi^\epsilon(x(s))\Big)\, ds$$

and

$$\tilde{M}^\epsilon(t) = M^\epsilon(t) + \sqrt{2\eta_0}W(t).$$

Our estimates imply that

$$\lim_{\epsilon\to 0} \mathbb{E} \sup_{t\in[0,T]} |H^\epsilon(t)| = 0.$$

Furthermore, estimate (49), together with the Burkhölder–Davis–Gundy inequality imply that

$$\mathbb{E} \sup_{t\in[0,T]} |\tilde{M}^\epsilon(t)| \le C.$$

On the other hand,

$$x(t) = x(0) + \int_0^t v_0(x(s))\, ds.$$

Set $\theta(T) := \mathbb{E}\sup_{t\in[0,T]} |x^\epsilon(t) - x(t)|$. Because v_0 is periodic it is in fact globally Lipschitz so that we obtain

$$\theta(T) \le C \int_0^T \theta(t)\, dt + h^\epsilon(T),$$

where

$$\lim_{\epsilon\to 0} h^\epsilon(T) = 0.$$

We use Gronwall's inequality to deduce

$$\theta(T) \le h^\epsilon \big(1 + CTe^{CT}\big),$$

from which the claim follows.   □

Appendix 2

In this appendix we study the homogenization problem (7) in one dimension. In this case we can calculate the homogenized coefficient explicitly and to prove Assumption 2. More details can be found in [27, Ch. 12].

The Homogenized Equations

We take $d = 1$ in (7) and set $D = [0, L]$. Then the Dirichlet problem (7) reduces to a two–point boundary value problem:

$$-\frac{d}{dx}\left(\exp\left(u\left(x, \frac{x}{\epsilon}\right)\right)\frac{dp^\epsilon}{dx}\right) = f \quad \text{for } x \in (0, L), \tag{50a}$$

$$p^\epsilon(0) = p^\epsilon(L) = 0. \tag{50b}$$

We assume that $u(x, y)$ is smooth in both of its arguments and periodic in y with period 1. Furthermore, we assume that this function is bounded from above and below. Consequently, there exist constants $0 < \alpha \le \beta < \infty$ such that

$$\alpha \le \exp(u(x, y)) \le \beta, \quad \forall y \in [0, 1]. \tag{51}$$

We also assume that f is smooth.

The cell problem becomes a boundary value problem for an ordinary differential equation with periodic boundary conditions. Introducing the notation $k(x, y) := \exp(u(x, y))$, the cell problem can be written as

$$-\frac{\partial}{\partial y}\left(k(x, y)\frac{\partial \chi}{\partial y}\right) = \frac{\partial k(x, y)}{\partial y}, \quad \text{for } y \in (0, 1), \tag{52a}$$

$$\chi \text{ is 1–periodic}, \quad \int_0^1 \chi(x, y)\,dy = 0. \tag{52b}$$

Notice that the macrovariable x enters the cell problem (52) as a parameter. Since $d = 1$ we only have one effective coefficient which is given by the one dimensional version of (11),(12), namely

$$k_0(x) = \int_0^1 \left(k(x, y) + k(x, y)\frac{\partial \chi}{\partial y}(x, y)\right)dy$$

$$= \left\langle k(x, y)\left(1 + \frac{\partial \chi}{\partial y}(x, y)\right)\right\rangle \tag{53}$$

where we have introduced the notation $\langle \phi(x, y)\rangle := \int_0^1 \phi(x, y)\,dy$. The homogenized equation is then

$$-\frac{d}{dx}\left(k_0(x)\frac{dp_0}{dx}\right) = f, \quad x \in (0, L), \tag{54a}$$

$$p(0) = p(L) = 0. \tag{54b}$$

Explicit Solution of the Cell Problem

Equation (52a) can be solved exactly. After integrating the equation and applying the periodic boundary conditions, we obtain

$$\chi(x, y) = -y + c_1 \int_0^y \frac{1}{k(x, y)} \, dy + c_2,$$

with

$$c_1(x) = \frac{1}{\int_0^1 (1/k(x, y)) \, dy} = \langle k(x, y)^{-1} \rangle^{-1}.$$

Therefore, from (53) we obtain:

$$k_0(x) = \langle k(x, y)^{-1} \rangle^{-1}. \tag{55}$$

The constant c_2 is irrelevant. This is the formula which gives the homogenized coefficient in one dimension. It shows clearly that, even in this simple one–dimensional setting, the homogenized coefficient is not found by simply averaging the unhomogenized coefficients over a period of the microstructure. Rather, the homogenized coefficient is the *harmonic average* of the unhomogenized coefficient. It is quite easy to show that $k_0(x)$ is bounded from above by the average of $k(x, y)$. Notice that the homogenized coefficient can be written in the form

$$k_0(x) = e^{u_0(x)}, \quad \text{where} \quad u_0(x) = \log\left(\langle \exp(-u(x, y)) \rangle^{-1} \right). \tag{56}$$

Error Estimates in $W^{1,\infty}(D)$

The fact that we can obtain an explicit formula for the solution of the boundary value problem (50) as well as for the solution of the cell problem (52) enables us to prove Assumption 2.

Proposition 14 *Let $p^\epsilon(x)$ be the solution of the two-point boundary value problem (50) where the log permeability $u(x, y)$ is smooth in both of its arguments and satisfies (51). Let $k(x, y) = \exp(u(x, y))$ and define*

$$v^\epsilon(x) = k\left(x, \frac{x}{\epsilon}\right) \frac{dp^\epsilon}{dx}(x)$$

and

$$V(x, y) = k(x, y) \left(1 + \frac{\partial \chi}{\partial y}(x, y)\right) \frac{dp_0}{dx}(x),$$

where $p_0(x)$ is the solution of the homogenized equation (54). Then

$$\lim_{\epsilon \to 0} \|v^\epsilon(x) - V(x, x/\epsilon)\|_{L^\infty} = 0. \tag{57}$$

Notice that, by (14), the corrector $p_1(x, y) = \chi(x, y)\frac{dp_0}{dx}(x)$. Hence, using the bound (51) from below on a, together with the definition (9) of p_a^ϵ, this theorem delivers the following immediate corollary:

Corollary 2. *Under the assumptions of Proposition 14 we have*

$$\lim_{\epsilon \to 0} \|p^\epsilon - p_a^\epsilon\|_{W^{1,\infty}} = 0.$$

Proof of Proposition 14. We have that

$$\frac{d\chi}{dy}(x, y) = -1 + \frac{k_0(x)}{k(x, y)}.$$

Consequently

$$V(x, y) = k_0(x)\frac{dp_0}{dx}(x).$$

Define a function F by $F'(z) = f(z)$. We solve the homogenized equation to obtain

$$k_0(x)\frac{dp_0}{dx}(x) = -F(x) + c,$$

with

$$c = \frac{\int_0^L k_0^{-1}(z)F(z)\, dz}{\int_0^L k_0^{-1}(z)\, dz}.$$

Similarly, from (50) we obtain

$$k\left(x, \frac{x}{\epsilon}\right)\frac{dp^\epsilon}{dx} = -F(x) + c^\epsilon,$$

with

$$c^\epsilon = \frac{\int_0^L k^{-1}(z, z/\epsilon)F(z)\, dz}{\int_0^L k^{-1}(z, z/\epsilon)\, dz}.$$

From the above calculations we deduce that

$$\|v^\epsilon(x) - V(x, x/\epsilon)\|_{L^\infty} = |c - c^\epsilon|.$$

It suffices to show that $|c - c^\epsilon| = \mathcal{O}(\epsilon)$. This will follow from the fact that

$$\int_0^L k^{-1}(z, z/\epsilon)G(z) = \int_0^L k_0^{-1}(z)G(z)dz + \mathcal{O}(\epsilon)$$

for any smooth function G, as $\epsilon \to 0$. To see this, define integer N and $\delta \in [0, \epsilon)$ uniquely by the identity

$$L = N\epsilon + \delta. \tag{58}$$

Then note that, using the uniform bounds on $k(x, y)$ from below, together with uniform (in y) Lipschitz properties of $a(\cdot, y)$ and G, we have for $z_n = n\epsilon$,

$$\int_0^L k^{-1}(z, z/\epsilon)G(z)dz = \sum_{n=0}^{N-1} \int_{n\epsilon}^{(n+1)\epsilon} k^{-1}(z_n, z/\epsilon)G(z_n)dz + \mathcal{O}(\epsilon)$$

$$= \sum_{n=0}^{N-1} \int_{n\epsilon}^{(n+1)\epsilon} k_0^{-1}(z_n)G(z_n)dz + \mathcal{O}(\epsilon)$$

$$= \sum_{n=0}^{N-1} \int_{n\epsilon}^{(n+1)\epsilon} k_0^{-1}(z)G(z)dz + \mathcal{O}(\epsilon)$$

$$= \int_0^L k_0^{-1}(z)G(z)dz + \mathcal{O}(\epsilon).$$

This completes the proof. □

References

1. G. Allaire. Homogenization and two-scale convergence. *SIAM J. Math. Anal.*, 23(6):1482–1518, 1992.
2. M. Avellaneda and F.-H. Lin. Compactness methods in the theory of homogenization. *Comm. Pure Appl. Math.*, 40(6):803–847, 1987.
3. R. Azencott, A. Beri, and I. Timofeyev. Adaptive sub-sampling for parametric estimation of Gaussian diffusions. *J. Stat. Phys.*, 139(6):1066–1089, 2010.
4. G. Bal, J. Garnier, S. Motsch, and V. Perrier. Random integrals and correctors in homogenization. *Asymptotic Analysis*, 59:1–26, 2008.
5. G. Bal and K. Ren. Physics-based models for measurement correlations: application to an inverse Sturm-Liouville problem. *Inverse Problems*, 25:055006, 2009.
6. H.T. Banks and K. Kunisch. *Estimation Techniques for Distributed Parameter Systems.* Birkhäuser, 1989.
7. A. Bensoussan, J.-L. Lions, and G. Papanicolaou. *Asymptotic Analysis for Periodic Structures*, volume 5 of *Studies in Mathematics and its Applications*. North-Holland Publishing Co., Amsterdam, 1978.
8. J.O. Berger. *Statistical Decision Theory and Bayesian Analysis.* Springer, 1980.
9. P.J. Bickel and K.A. Doksum. *Mathematical Statistics.* Prentice-Hall, 2001.
10. V.I. Bogachev. *Gaussian Meausures.* American Mathematical Society, 1998.
11. D. Cioranescu and P. Donato. *An Introduction to Homogenization.* Oxford University Press, New York, 1999.
12. S.L. Cotter, M. Dashti, J.C. Robinson, and A.M. Stuart. Bayesian inverse problems for functions and applications to fluid mechanics. *Inverse Problems*, 25:115008, 2009.
13. B. Dacarogna. *Direct Methods in the Calculus of Variations.* Springer, New York, 1989.

14. G. DaPrato and J. Zabczyk. *Stochastic Equations in Infinite Dimensions.* Cambridge University Press, 1992.
15. M. Dashti and A.M. Stuart. Uncertainty quantification and weak approximation of an elliptic inverse problem. Preprint, 2011.
16. Y. R. Efendiev, Th. Y. Hou, and X.-H. Wu. Convergence of a nonconforming multiscale finite element method. *SIAM J. Numer. Anal.*, 37(3):888–910 (electronic), 2000.
17. H.K. Engl, M. Hanke, and A. Neubauer. *Regularization of Inverse Problems.* Kluwer, 1996.
18. L.C. Evans. *Partial Differential Equations.* AMS, Providence, Rhode Island, 1998.
19. B. Fitzpatrick. Bayesian analysis in inverse problems. *Inverse Problems*, 7:675–702, 1991.
20. M. Hinze, R. Pinnau, M. Ulbrich, and S. Ulbirch. *Optimization with PDE Constraints*, volume 23 of *Mathematics Modelling: Theory and Applications.* Springer, 2009.
21. J. Kaipio and E. Somersalo. *Statistical and computational inverse problems*, volume 160 of *Applied Mathematical Sciences.* Springer-Verlag, New York, 2005.
22. H. König. *Eigenvalue Distribution of Compact Operators.* Birkhäuser Verlag, Basel, 1986.
23. M.A. Lifshits. *Gaussian Random Functions*, volume 322 of *Mathematics and its Applications.* Kluwer, Dordrecht, 1995.
24. J. Nolen and G. Papanicolaou. Inverse problems. *Fine scale uneratinty in parameter estimation for elliptic equations*, 25:115021, 2009.
25. J. Nolen, G.A. Pavliotis, and A.M. Stuart. *In preparation*, 2010.
26. A. Papavasiliou, G. A. Pavliotis, and A. M. Stuart. Maximum likelihood drift estimation for multiscale diffusions. *Stochastic Process. Appl.*, 119(10):3173–3210, 2009.
27. G. A. Pavliotis and A. M. Stuart. *Multiscale Methods*, volume 53 of *Texts in Applied Mathematics.* Springer, 2008. Averaging and Homogenization.
28. G.A. Pavliotis and A.M Stuart. Parameter estimation for multiscale diffusions. *J. Stat. Phys.*, 127(4):741–781, 2007.
29. A.M. Stuart. Inverse problems: a Bayesian perspective. *Acta Numerica*, 19, 2010.

Transported Probability and Mass Density Function (PDF/MDF) Methods for Uncertainty Assessment and Multi-Scale Problems

Patrick Jenny and Daniel W. Meyer

Abstract For the simulation of fluid flows, probability and mass density function (PDF/MDF) methods have advantageous properties compared to moment-based approaches or purely deterministic methods and are applicable in different fields. For example, PDF and MDF methods are used for the quantification of uncertainty in turbulent or subsurface flows, and the simulation of multi-phase flows or rarefied fluids. In this chapter, differences of these methods compared to other solution techniques are discussed and illustrated by application examples. Moreover, the theory behind PDF and MDF methods is outlined. Finally, a PDF method for uncertainty quantification in subsurface flows and an MDF method for the simulation of rarefied fluid flows are discussed in more details.

1 Motivation for PDF or MDF Modeling

In many problems related to fluid dynamics, a statistical characterization is preferable compared to a deterministic description. For example in uncertainty assessment of contaminant transport in porous media, typically a statistical characterization of the soil parameters that govern the flow field is available. Completely characterized fields are unknown, though. Consequently, a methodology is applied that provides statistical predictions of dependent quantities in the form of space- and time-dependent probability density functions (PDFs) [19]. Similar conditions can be found in turbulent flows were uncertainty originates from initial or boundary conditions. PDF methods have a long tradition in turbulent combustion applications and were pioneered in this field by Pope and coworkers [12, 23, 24]. In problems that involve a wide range of length and timescales, a statistical description is

P. Jenny (✉) · D.W. Meyer
Institute of Fluid Dynamics, ETH Zurich, Switzerland
e-mail: jenny@ifd.mavt.ethz.ch; meyer@ifd.mavt.ethz.ch

I.G. Graham et al. (eds.), *Numerical Analysis of Multiscale Problems*, Lecture Notes in Computational Science and Engineering 83, DOI 10.1007/978-3-642-22061-6_2, © Springer-Verlag Berlin Heidelberg 2012

used to reflect fine-scale information that cannot be resolved due to computational limitations. Here, the mass fraction of fluid contained in a small-scale domain volume that is in a certain state is characterized by a mass density function (MDF; mass-weighted PDF times mean density) [23]. A similar concept are filtered density functions (FDFs) [5] that were introduced in the context of large eddy simulation (LES) to represent subgrid-scale effects. Direct numerical simulations (DNS) of turbulent flows, where all length and timescales are resolved are computationally very expensive and are performed only for idealized flows. For the solution of high-dimensional PDF or MDF transport equations, particle-based Monte Carlo techniques are typically applied [22]. Alternatively, if the joint PDF or MDF of all relevant quantities is well approximated by an idealized shape, a limited set of moment equations is sufficient to obtain predictions for the joint PDF or MDF. For example, a Gaussian PDF is fully characterized by means and covariances. Nevertheless, in many cases PDF, MDF, or FDF methods offer certain advantages over moment methods or purely deterministic approaches.

In the following, some of these advantages are discussed and illustrated with suitable examples from turbulent and subsurface flows and flows of rarefied gases. In Sect. 2, the Fokker–Planck equation is derived, which is at the heart of most PDF and MDF methods. More detailed application examples of PDF and MDF methods for subsurface dispersion and rarefied gas flows are discussed in Sects. 3 and 4, respectively.

Complex PDF Shape: In many problems and application cases, the PDF is arbitrary and a limited set of moments are insufficient to determine its shape.

Example Turbulent Mixing: An illustrative example is provided by considering mixing in turbulent flows. Here, the joint PDF of chemical species concentrations has usually a very complex shape that is resulting from the flow geometry, the mixing dynamics, and the governing reaction mechanism. In Fig. 1, simulation results of the mixing of three streams in a statistically two-dimensional turbulent channel flow are provided. The bi-variate composition PDFs are far from a joint Gaussian and virtually impossible to characterize with a small set of moments. In the joint PDF method, however, stochastic models are needed that account for unclosed terms in the PDF transport equation. For example, mixing models account for the molecular mixing term [21]. In Fig. 2, different mixing models are compared with a reference DNS.

Non-Linearities in the governing equations can lead to severe closure problems in the context of moment equations.

Example Turbulent Reactive Flows: The chemical reaction source term in a turbulent reactive flow, $Q_\alpha(\mathbf{Y}, T)$, is typically a highly non-linear function of the chemical composition vector $\mathbf{Y} = (Y_1, Y_2, \ldots, Y_n)^T$ and the temperature T. Accordingly, it is generally not suitable to estimate its ensemble mean, i.e., $\langle Q_\alpha(\mathbf{Y}, T) \rangle$, that appears in the mean conservation equation of Y_α by $Q_\alpha(\langle \mathbf{Y} \rangle, \langle T \rangle)$. Instead, a solution method is required that provides the full joint composition–

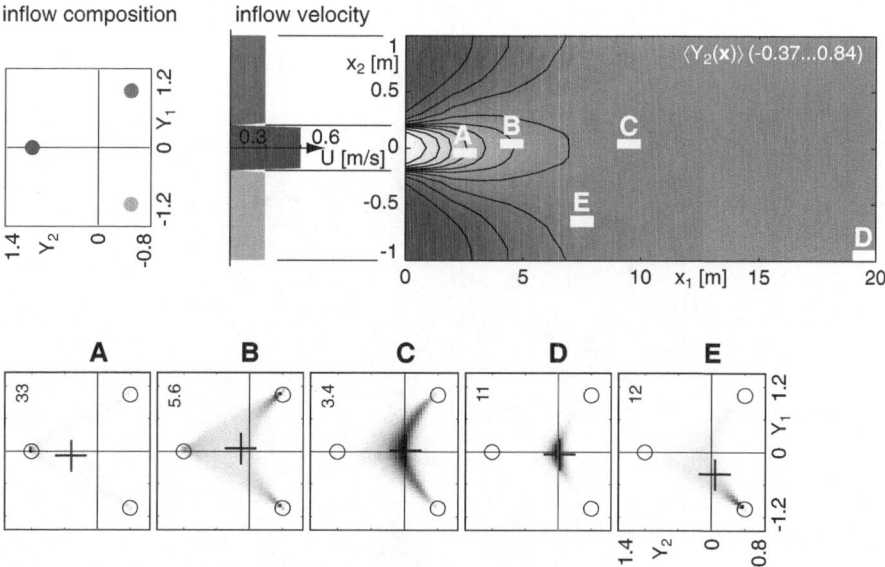

Fig. 1 Simulation results of the turbulent mixing of three streams (*red*, *green*, and *blue*) with different compositions in a channel are depicted. The composition of each stream at the left inflow boundary is specified by an according point in the bi-variate Y_1-Y_2 composition space. The grey-shaded contour plot shows the mean concentration of the second scalar $\langle Y_2(\mathbf{x}) \rangle$. Joint scalar PDFs at five different locations in the domain, i.e., A to E, are also provided. The crosses in the according density plots, represent the local mean composition $\langle \mathbf{Y}(\mathbf{x}) \rangle$. A joint velocity–composition–frequency PDF method was applied together with the PSP mixing model [17]

temperature PDF $f(\mathbf{Y}, T)$ since

$$\langle Q_\alpha(\mathbf{Y}, T) \rangle = \int_{\mathbb{R}^n} \int_0^\infty Q_\alpha(\mathbf{Y}, T) f(\mathbf{Y}, T) \, dT \, d\mathbf{Y}. \tag{1}$$

Moreover, the particle-based Lagrangian solution techniques that are applied to solve the PDF or MDF transport equations offer a natural approach to resolve non-linear modeling and closure problems.

Complex Temporal or Spatial Correlation Behavior may lead to difficulties in a moment-based framework.

Example Subsurface Flow: Simulation models for transport in the subsurface are relevant in applications that deal for example with the spreading of contaminants. The velocities of fluid particles that travel in the subsurface develop complex correlation behavior for increasing spatial variability of the hydraulic conductivity. This is illustrated in Fig. 3, where time series of fluid particle velocities are provided for porous media with different conductivity variance σ_Y^2. In the medium with high variability, i.e., $\sigma_Y^2 = 4$, fluid particles experience long correlated low-velocity

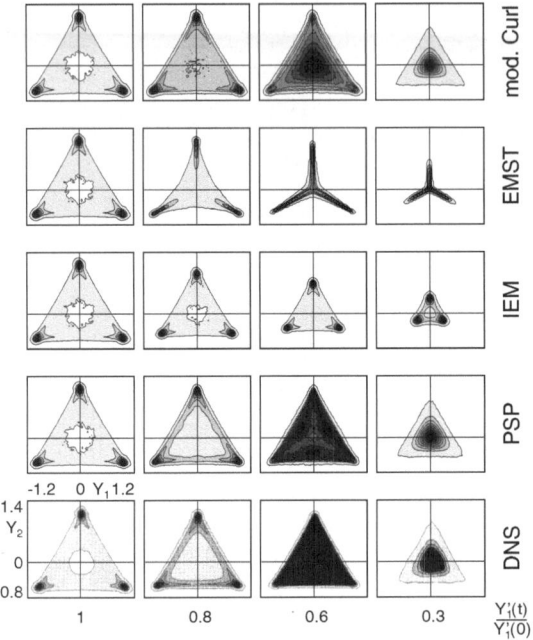

Fig. 2 Simulation of turbulent mixing in a periodic box. The flow statistics are constant in time and space and the composition statistics are constant in space. Depicted is the temporal evolution of the bi-variate composition PDF resulting from a joint composition–frequency PDF method [18] with different mixing models and from a DNS [15]. The plots in the first column represent the initial condition and the ones in subsequent columns mixing stages at later times for equally normalized r.m.s. values of the first concentration Y_1. The following mixing models are included: modified Curl [11], Euclidean minimum spanning tree (EMST) [27], interaction by exchange with the mean (IEM) [30], and parameterized scalar profile (PSP) [17]

periods interrupted by short sections where they travel at high velocities in highly conductive channels. Moment-based methods that are strictly valid for vanishing conductivity variability or small σ_Y^2 do not capture these effects accurately. Particle based methods typically used in PDF methods, on the other hand, can incorporate Lagrangian stochastic processes that reflect the outlined behavior accurately [19]. More details about a PDF method for uncertainty assessment of transport in heterogeneous porous media are provided in Sect. 3.

Multi-Scale Problems: MDF or FDF methods are advantageous for problems that involve a wide range of scales but where information about microscale variability is secondary and variations at a relatively large macroscale are ultimately sought after. If microscale processes depend on local properties only and macroscopic processes depend weakly on microscopic quantities, a scale separation is possible. This is an important precondition for the applicability of PDF or MDF methods. A weak dependence is given if macroscopic processes depend on the microscale information only in the form of statistical distributions that collect microscale data over

Fig. 3 Velocity time series, $u(t)$, recorded along fluid particle trajectories in porous media with multi-Gaussian log-conductivity distributions $Y(\mathbf{x})$. The heterogeneity level of the porous media is quantified by the spatial variance of the log-conductivity σ_Y^2. The particle travel time t was normalized by the mean velocity U and the log-conductivity correlation length l_Y

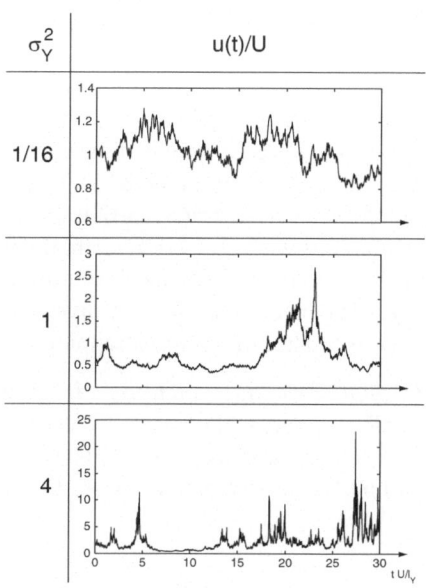

macroscopic volumes. A strong dependence is given, if a process at the macroscale depends on all details of the spatial distributions at the microscale. Variations in the macroscopic quantities are then resolved by a coarse computational grid. The subgrid-scale or microscale physics, however, are modeled with Lagrangian material particles whose dynamics are evolved with suitable stochastic processes (e.g., Brownian motion to model diffusion). The particle ensemble in each grid cell approximates a local MDF or FDF that reflects the variability or distribution of properties at the microscale. Macroscopic quantities are extracted from the MDFs or FDFs and influence processes at the macroscale. Where applicable, simulation strategies of this kind are superior to their conventional purely deterministic counterparts, which focus on the macroscale and offer little flexibility to model microscale effects. Deterministic methods that resolve the microscale, on the other hand, are applicable only for small problems because of their high computational cost. MDF or FDF methods honor microscale information in the form of statistical distributions, i.e., MDFs or FDFs, which leads to a simplified and computationally efficient coupling with macroscopic quantities.

Example CO_2 Sequestration in Saline Aquifers: In CO_2 sequestration applications, supercritical CO_2 is injected as an additional phase into a saline aquifer displacing brine. CO_2 and brine are immiscible phases. Saline aquifers have depths of the order of hundreds of meters and extend horizontally over several kilometers. Two trapping mechanisms act on the buoyant CO_2 as it migrates upward in the porous aquifer. CO_2 fractions get trapped in the porous rock due to capillary forces, which is called residual trapping. At the brine/CO_2 interface, CO_2 dissolves into the brine phase, thus the density of the CO_2-enriched brine increases, and fingers of dense brine form as is shown in Fig. 4. This mechanism is called dissolution trapping and it acts at lengthscales that are orders of magnitude smaller than the extension of the aquifer. A conventional Darcy-flow-based framework that relies entirely on macroscopic properties offers limited capability to incorporate enhanced dissolution trapping due to small-scale gravity fingering. In a particle-based MDF method, however, dissolution is easily modeled by mass exchange between CO_2 and brine particles. Moreover, heavy brine particles settle due to gravitational forces. Simulation results [29] where an MDF method with corresponding models is compared with a purely deterministic method are provided in Fig. 5. The deterministic method leads to lower concentrations of dissolved CO_2 at the bottom of the reservoir and therefore underestimates dissolution trapping.

Example Dynamics of Rarefied Gases: For rarefied gas flows or flows in very small structures, the mean free path length λ of gas molecules becomes large compared to a macroscopic reference lengthscale L_0. Consequently, the Knudsen number is no longer $\mathrm{Kn} = \lambda/L_0 \ll 1$ and the continuum assumption that is

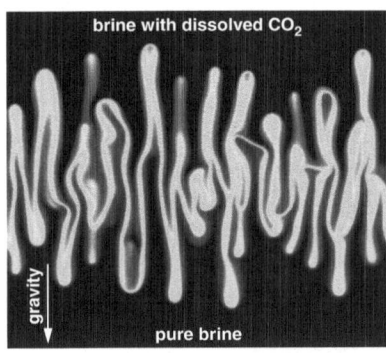

Fig. 4 Fully resolved simulation of miscible gravity-driven fingering of CO_2-enriched brine (*dark red*) in lighter pure brine (*blue*) [29]. At the beginning of the simulation, the enriched and pure brine fill the top and bottom half of the rectangular domain, respectively

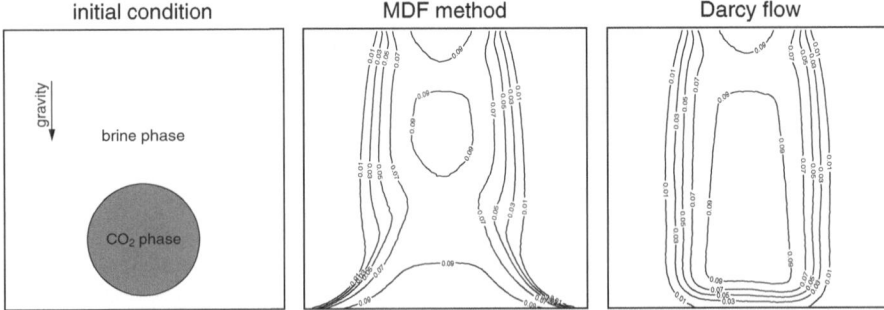

Fig. 5 Two-phase flow simulations [29] of the evolution of a circular CO_2 cloud released in porous media saturated with pure brine. Contour plots of the dissolved CO_2 concentration in the brine phase are provided. Note that the CO_2 phase is not depicted. Results from a conventional Darcy flow simulator including a dissolution model are compared with an MDF method that accounts for enhanced dissolution due to gravity fingering

at the heart of the Navier–Stokes equations becomes invalid. Collisions of gas molecules at the microscale start to matter. Nevertheless, macroscopic quantities like gas velocity or pressure distributions remain the focus of predictive methods. Fortunately, individual molecule interactions do not induce significant changes in these quantities and therefore a statistical treatment of molecule collisions seems sufficient. For example in the MDF method [13], microscale collisions of molecules or atoms are modeled by continuous stochastic processes for the particle velocities. Macroscopic properties like density, velocity, or temperature, on the other hand, are estimated on a relatively coarse grid based on an ensemble of stochastic particles that represent gas molecules. MDF-based simulation results of a channel flow with adiabatic walls and constant acceleration F in streamwise direction x_2 are provided in Figs. 6–8. Macroscopic velocity profiles for different Knudsen numbers are compared in Fig. 6. For Kn $\rightarrow$ 0, the velocity profile approaches the Poiseuille profile with no-slip condition at the walls and consistent with the Navier–Stokes equations. Joint PDFs of the wall-normal and streamwise velocity components of molecules near the wall and in the channel center are provided in Fig. 7 for different Knudsen numbers. For low Kn, the joint PDF is approximately Gaussian and consequently the PDF of the velocity magnitudes Maxwellian. An illustration of the Knudsen paradox is provided in Fig. 8: for constant acceleration F but different Kn, the overall mass flow has a minimum at Kn $\approx$ 1. This effect is known from experiments, e.g., [6], and is also predicted by the MDF method. More details about the MDF method for the simulation of rarefied gas flows will be provided in Sect. 4.

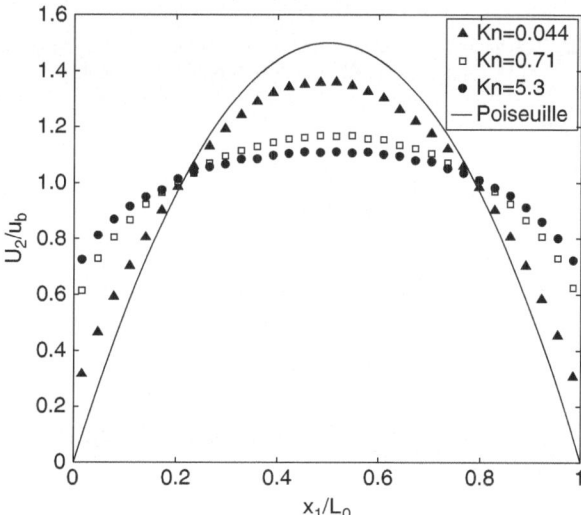

Fig. 6 Channel flow simulations with a Fokker–Planck equation-based MDF method for rarefied gases [13]. Depicted is the streamwise macroscopic velocity component U_1 normalized with the bulk velocity u_b resulting from the Navier–Stokes equations. The wall-normal coordinate x_1 is normalized with the channel width L_0. For vanishing Knudsen numbers Kn, the flow approaches a laminar Poiseuille flow as given by the Navier–Stokes equations

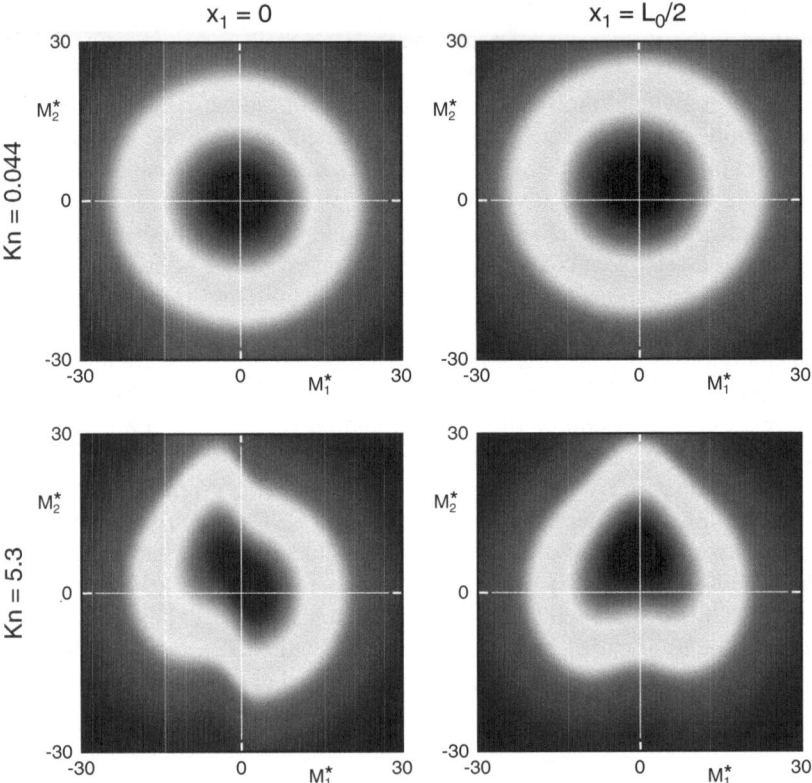

Fig. 7 Channel flow simulations with a Fokker–Planck equation-based MDF method for rarefied gases [13]. Joint velocity PDFs at different channel locations x_1 and for different Knudsen numbers Kn are depicted. The normalized velocity components M_i^* are defined as $M_i/(L_0 F \sqrt{\rho_0/p_0})$, where ρ_0 and p_0 are reference density and pressure, respectively

2 Basic Background

In this section, the basic mathematical concepts and theories are introduced that are used for the formulation and solution of PDF and MDF transport equations.

2.1 *Illustration with Brownian Motion*

Probably the first physical effect that was modeled with stochastic processes is Brownian motion: in 1827, the botanist Robert Brown [3] discovered under a microscope that pollen grains suspended in stagnant water perform a random motion. Several decades later, first Einstein [7] and later Langevin [16] provided

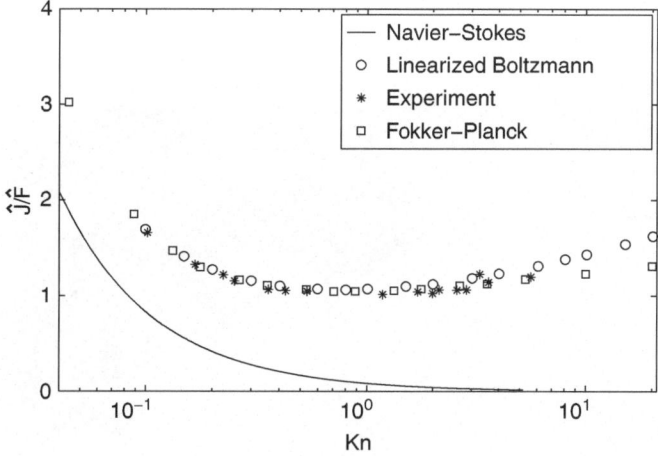

Fig. 8 Channel flow simulations with a Fokker–Planck equation-based MDF method for rarefied gases [13]. The ratio of the non-dimensional channel mass flow $\hat{J}$ and acceleration $\hat{F}$ is depicted as a function of the Knudsen number Kn. Experimental results were taken from [6]

a physical model for this so-called Brownian motion. Their work provides a probabilistic quantification of the distance traveled by a pollen grain over time and is based on a random process. Brownian motion is resulting from molecular collisions at the pollen grain's surface and Langevin proposed

$$dX_i = \sqrt{2\Gamma}\,dW_i \tag{2}$$

as a model for the pollen grain's motion $\mathbf{X}(t)$. Equation (2) is a so-called stochastic differential equation (SDE) and $W_i(t)$ is a Wiener process with $dW_i = W_i(t+dt) - W_i(t)$ being independent normally distributed random variables with $\langle dW_i \rangle = 0$ and $\langle dW_i dW_j \rangle = dt\,\delta_{ij}$. A statistically exact integration of the position $\mathbf{X}(t)$ is achieved with

$$\Delta X_i = \sqrt{2\Gamma\Delta t}\,\xi_i, \tag{3}$$

where ξ_i are independent standard normal random variables and Δt is the time step size. A sample trajectory of the random process given by (3) is depicted in Fig. 9.

Einstein showed that the evolution of the PDF of a pollen grain's position, $f_\mathbf{X}(\mathbf{x};t)$, is given by

$$\frac{\partial f_\mathbf{X}(\mathbf{x};t)}{\partial t} = \frac{\partial^2}{\partial x_i \partial x_i}[\Gamma f_\mathbf{X}(\mathbf{x};t)], \tag{4}$$

which is a Fokker–Planck equation and where x_i is the sample space coordinate that corresponds to the stochastic variable $X_i(t)$. The semicolon in the argument list of $f_\mathbf{X}(\mathbf{x};t)$ is used to point out that $f_\mathbf{X}$ is a density with respect to $\mathbf{x}$ but not t. Correspondence relations between SDEs, e.g., (3), and Fokker-Planck equations,

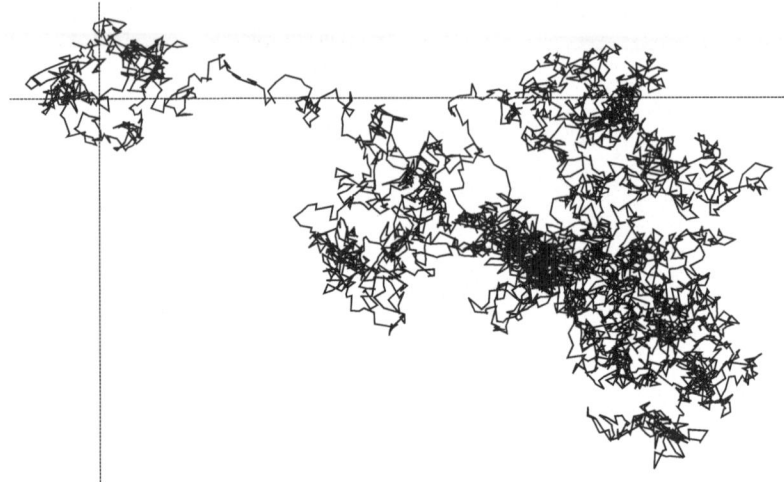

Fig. 9 Sample trajectory of the stochastic model process (3) for Brownian motion. 5,000 time steps were calculated and the origin is indicated by the dashed horizontal and vertical lines

e.g., (4), are provided in the next section. See [9] for a more detailed and complete account on stochastic diffusion processes.

2.2 General PDF Evolution Equations

2.2.1 Kramers–Moyal Expansion

Based on the sifting property of the Dirac δ-function, i.e.,

$$\int_{-\infty}^{\infty} \delta(x - a)g(x)\,dx = g(a), \tag{5}$$

and the definition of the statistical mean, i.e.,

$$\langle h[X(t)] \rangle = \int_{-\infty}^{\infty} h(x) f_X(x; t)\,dx, \tag{6}$$

one can write the PDF $f_X(x; t)$ of the time-dependent random variable $X(t)$ as

$$f_X(x; t) = \langle \delta[X(t) - x] \rangle. \tag{7}$$

The PDF of X at a slightly later time $t + \Delta t$ can be written based on a Taylor expansion as

$$f_X(x; t + \Delta t) = \langle \delta(X(t) - x + \Delta X) \rangle$$

$$= f_X(x; t) + \sum_{k=1}^{\infty} \frac{1}{k!} \left\langle \left(-\frac{\partial}{\partial x} \right)^k \delta(X(t) - x) \, \Delta X^k \right\rangle$$

$$= f_X(x; t) + \sum_{k=1}^{\infty} \left(-\frac{\partial}{\partial x} \right)^k \left[\frac{\langle \Delta X^k | x; t \rangle}{k!} f_X(x; t) \right], \quad (8)$$

where in the last step the derivative operator with respect to x and the ensemble mean were exchanged. By dividing the last equality in (8) by Δt and taking the limit $\Delta t \to 0$ we get

$$\frac{\partial f_X(x; t)}{\partial t} = \sum_{k=1}^{\infty} \left(-\frac{\partial}{\partial x} \right)^k \left[\underbrace{\lim_{\Delta t \to 0} \frac{\langle \Delta X^k | x; t \rangle}{k! \Delta t}}_{D^{(k)}} f_X(x; t) \right], \quad (9)$$

which is the Kramers–Moyal equation with coefficients $D^{(k)}$. It describes the temporal evolution of the PDF $f_X(x; t)$ based on an infinite number of conditional means that involve powers of the increments ΔX. These increments are provided by an according SDE for $X(t)$.

2.2.2 Theorem of Pawula

In a next step, it is demonstrated that for a certain class of stochastic processes $X(t)$ the first two terms of the infinite series of the Kramers–Moyal equation are sufficient to predict the PDF evolution. This derivation is based on the theorem of Pawula that is outlined together with a proof in [25]. Because of its high relevance in the present context, Pawula's theorem and its proof are provided here:

Theorem 1. *For $D^{(2r)} = 0$ with $r \geq 1$, we have $D^{(3)} = D^{(4)} = \ldots = 0$.*

Proof. The inequality of Schwarz relates statistical moments of two random variables α and β:

$$\langle \alpha \beta \rangle^2 \leq \langle \alpha^2 \rangle \langle \beta^2 \rangle. \quad (10)$$

For $\alpha = \Delta X^a$ and $\beta = \Delta X^b$ conditional on $X(t) = x$ and with $a, b \in \{1, 2, 3, \ldots\}$ we get

$$\langle \Delta X^{a+b} | x; t \rangle^2 \leq \langle \Delta X^{2a} | x; t \rangle \langle \Delta X^{2b} | x; t \rangle. \quad (11)$$

By multiplying the left hand with $(a + b)!/(a + b)!$ and the right hand side with $(2a)!/(2a)! \, (2b)!/(2b)!$ and by dividing through Δt and taking the limit $\Delta t \to 0$, expression (11) leads together with the definition of $D^{(k)}$ in (9) to

$$\left[(2a + k)! D^{(2a+k)} \right]^2 \leq (2a)!(2a + 2k)! D^{(2a)} D^{(2a+2k)}, \quad (12)$$

where b was substituted by $a + k$ with $k \in \{0, 1, 2, \ldots\}$. Equation (12) implies

$$D^{(2a)} = 0 \quad \Rightarrow \quad D^{(2a+1)} = D^{(2a+2)} = \ldots = 0 \tag{13}$$

and

$$D^{(2a+2k)} = 0 \quad \Rightarrow \quad D^{(2a+k)} = 0.$$

With $r = a + k \in \{1, 2, 3, \ldots\}$ we get from the previous expression

$$D^{(2r)} = 0 \quad \Rightarrow \quad D^{(2r-k)} = 0$$

with $k \in \{0, 1, \ldots, r-1\}$, which leads to

$$D^{(2r)} = 0 \quad \Rightarrow \quad D^{(2r-1)} = D^{(2r-2)} = \ldots = D^{(r+1)} = 0. \tag{14}$$

Equation (13) and repeated use of (14) eventually lead to

$$D^{(2r)} = 0 \quad \Rightarrow \quad D^{(3)} = D^{(4)} = \ldots = 0 \tag{15}$$

for $r \geq 1$. $\qquad\qquad\qquad\qquad\qquad\qquad\qquad\qquad\qquad\qquad\qquad\qquad$ □

In Sects. 7.2.1 and 7.2.2, Gardiner [9] shows that if the statistics of ΔX meet certain requirements, that essentially characterize a continuous stochastic process $X(t)$, the Kramers–Moyal expansion can indeed be truncated after the second term as implied by Theorem 1 of Pawula.

2.2.3 Fokker–Planck Equation

With the Kramers–Moyal expansion truncated after the first two terms we arrive at

$$\frac{\partial f_X(x; t)}{\partial t} = -\frac{\partial D^{(1)} f_X(x; t)}{\partial x} + \frac{\partial^2 D^{(2)} f_X(x; t)}{\partial x^2} \tag{16}$$

with

$$D^{(1)} = \lim_{\Delta t \to 0} \frac{\langle \Delta X | x; t \rangle}{\Delta t} \quad \text{and} \quad D^{(2)} = \lim_{\Delta t \to 0} \frac{\langle \Delta X^2 | x; t \rangle}{2\Delta t}.$$

So, for a given SDE for $X(t)$, the coefficients $D^{(1)}$ and $D^{(2)}$ are determined and the temporal evolution of the PDF $f_X(x; t)$ is given by (16), which is a Fokker-Planck equation. The corresponding Fokker–Planck equation for several random variables is crucial for transported PDF and MDF methods and is provided in the following box.

Multi-variate Fokker–Planck equation for the joint PDF $f_{\mathbf{X}}(\mathbf{x};t)$ of the random variable vector $\mathbf{X}(t) = (X_1, X_2, \ldots)^T$ with the sample space vector $\mathbf{x}(t) = (x_1, x_2, \ldots)^T$:

$$\frac{\partial f_{\mathbf{X}}(\mathbf{x};t)}{\partial t} = -\frac{\partial D_i^{(1)} f_{\mathbf{X}}(\mathbf{x};t)}{\partial x_i} + \frac{\partial^2 D_{ij}^{(2)} f_{\mathbf{X}}(\mathbf{x};t)}{\partial x_i \partial x_j} \text{ with} \qquad (17)$$

$$D_i^{(1)} = \lim_{\Delta t \to 0} \frac{\langle \Delta X_i | \mathbf{x};t \rangle}{\Delta t} \text{ and } D_{ij}^{(2)} = \lim_{\Delta t \to 0} \frac{\langle \Delta X_i \Delta X_j | \mathbf{x};t \rangle}{2\Delta t}$$

For increasing dimensionality of the probability space, conventional finite volume or finite difference methods become computationally too expensive for the solution of the Fokker–Planck equation (17). Alternatively, particle-based methods that evolve an ensemble of particles based on the SDEs, i.e., $\mathbf{X}(t)$, are used [22].

Example Brownian Motion

Having presented the relation between SDEs and the multi-variate Fokker–Planck equation, we return for illustrational purposes to the Brownian motion example introduced in Sect. 2.1. Langevin suggested to model Brownian motion with the SDE (3), i.e., $\Delta X_i = \sqrt{2\Gamma \Delta t}\, \xi_i$, which leads based on (17) to

$$D_i^{(1)} = \lim_{\Delta t \to 0} \frac{1}{\Delta t} \sqrt{2\Gamma \Delta t}\langle \xi_i \rangle = 0 \text{ and } D_{ij}^{(2)} = \lim_{\Delta t \to 0} \frac{1}{2\Delta t} 2\Gamma \Delta t \langle \xi_i \xi_j \rangle = \Gamma \delta_{ij}.$$

With these results and (17) we obtain the specific Fokker–Planck equation (4), i.e.,

$$\frac{\partial f_{\mathbf{X}}(\mathbf{x};t)}{\partial t} = \frac{\partial^2}{\partial x_i \partial x_i}[\Gamma f_{\mathbf{X}}(\mathbf{x};t)],$$

which is the diffusion equation found by Einstein. In a short excursion, it is verified if Theorem 1 of Pawula is applicable for one-dimensional Brownian motion. For $\Delta X = \sqrt{2\Gamma \Delta t}\, \xi$,

$$D^{(4)} = \lim_{\Delta t \to 0} \frac{1}{4!\Delta t}(2\Gamma \Delta t)^2 \langle \xi^4 \rangle = \lim_{\Delta t \to 0} \frac{1}{2}\Gamma^2 \Delta t = 0$$

and therefore based on Theorem 1, only the first two terms in the Kramers–Moyal expansion (9) are non-zero.

2.2.4   Ornstein–Uhlenbeck (OU) Process

A slight generalization of process (2) used to model Brownian motion is given by

$$dX = -\tau^{-1}(X - x_M)\,dt + \sqrt{2\Gamma}\,dW, \tag{18}$$

where τ is a relaxation timescale and x_M is a mean or equilibrium parameter. This process describes Brownian motion in a parabolic potential and is called Ornstein–Uhlenbeck (OU) process. The first term on the right hand side is a drift term that reduces fluctuations in $X(t)$ and leads to a relaxation toward x_M. The second term adds stochastic fluctuations and therefore counteracts the first term. In turbulent flows, OU processes are used to model velocity components of fluid particles. Also for this process $D^{(4)} = 0$ and therefore Theorem 1 is applicable. The corresponding Fokker–Planck equation is given from relation (17):

$$\frac{\partial f_X(x;t)}{\partial t} = \frac{\partial}{\partial x}\left[\frac{1}{\tau}(x - x_M)f_X(x;t)\right] + \frac{\partial^2}{\partial x^2}[\Gamma f_X(x;t)]. \tag{19}$$

For $t \to \infty$, $f_X(x;t)$ relaxes to a Gaussian equilibrium PDF with mean x_M and variance $\tau\Gamma$.

After having introduced the basic formulation of PDF methods, two different application examples are outlined in the next sections. The first example deals with a PDF method for uncertainty quantification in subsurface flow and transport, whereas the second example is about an MDF method for rarefied gas flows.

3   Illustration of PDF Method for Uncertainty Assessment in Subsurface Flow Transport

The spreading of a passive tracer in a saturated porous medium is described by the advection–dispersion equation (ADE)

$$\frac{\partial C}{\partial t} + u_i\frac{\partial C}{\partial x_i} = \frac{\partial}{\partial x_i}\left(D_{ij}\frac{\partial C}{\partial x_j}\right), \tag{20}$$

where $C(\mathbf{x},t)$ is the local tracer concentration and x_i and t are coordinates in space and time, respectively. $\mathbf{D}$ is the dispersion tensor that accounts for pore-scale dispersion (PSD), which includes mechanical mixing in the rock pores and molecular diffusion. The formulation (20) is based on a porous medium with constant porosity $n(\mathbf{x}) = n$, which is often a reasonable assumption. The flow velocity $\mathbf{u}(\mathbf{x})$, that quantifies advective transport in the fluid phase, is determined based on Darcy's law and the continuity equation, i.e.,

$$\mathbf{u} = -\frac{K}{n}\nabla h \text{ and } \nabla \cdot \mathbf{u} = 0,$$

respectively, where $h(\mathbf{x})$ is the hydraulic head and $K(\mathbf{x})$ is the hydraulic conductivity.

In groundwater aquifers that are shallow with large horizontal extensions, $K(\mathbf{x})$ fluctuates typically over several orders of magnitude. Furthermore, information about the hydraulic conductivity is sparse and therefore, it is considered to be a random variable and is usually characterized by geostatistical models [14]. In Monte Carlo (MC) simulations, realizations $K(\mathbf{x})$ are generated consistently with the selected geostatistical model. For each realization, the flow field $\mathbf{u}(\mathbf{x})$ is calculated and transport is simulated which is computationally expensive since variations in individual $K(\mathbf{x})$- and $C(\mathbf{x}, t)$-realizations have to be resolved [26]. For a sufficiently large ensemble, concentration statistics are obtained that eventually approximate the local concentration PDF $f_C(c; \mathbf{x}, t)$ with the concentration sample space variable c [4].

Alternative less expensive approaches are based on specific geostatistical models. The multi-Gaussian model is based on the assumption that the log-conductivity $Y(\mathbf{x}) = \ln[K(\mathbf{x})]$ at every point in space is described by a joint Gaussian distribution. Moreover, an exponentially decaying correlation structure with correlation length l_Y is typically used for $Y(\mathbf{x})$ and the variability of Y is characterized by a spatially constant mean and variance σ_Y^2. For a constant mean flow velocity $\langle \mathbf{u}(\mathbf{x}) \rangle = (U, 0)^T$ and σ_Y^2 close to zero, analytical or semi-analytical expressions can be derived for the first two concentration moments $\langle C(\mathbf{x}, t) \rangle$ and $\sigma_C^2(\mathbf{x}, t)$. Corresponding derivations are based on perturbation methods that neglect higher-order terms with respect to σ_C [8]. These methods imply Gaussian joint velocity PDFs $f_\mathbf{u}(\mathbf{v}; \mathbf{x})$ (where $\mathbf{v}$ is the sample space variable of $\mathbf{u}$), which is only a good approximation for $\sigma_Y^2 < 0.5$ [20]. For higher log-conductivity variances, the joint velocity PDF develops long tails that represent high velocity magnitudes (compare Fig. 3) and becomes skewed in longitudinal x_1-direction.

Recently, a transported PDF method was proposed [19] that accounts for non-Gaussian velocity effects and is applicable for large log-conductivity variances σ_Y^2. Fluid particles with properties position $\mathbf{X}$, velocity $\mathbf{u}$, and concentration C are considered. The SDEs that determine the evolution of these properties are given by

$$dX_i = u_i \, dt + \sqrt{2D_i} \, dW_{(i)},$$
$$du_i = a(u_i) \, dt - \delta_{i2}\omega_0^2 \Xi \, dt + b(u_i) \, dW_{(i)} \text{ with}$$
$$d\Xi = u_2 \, dt, \text{ and}$$
$$dC = M(C, \mathbf{u}) \, dt, \tag{21}$$

where $i \in \{1, 2\}$, and $a(u_i)$ and $b(u_i)$ are nonlinear drift and diffusion functions that are different for the longitudinal and transverse velocity components u_1 and u_2, respectively. Einstein's summation convention does not apply for bracketed repeated indices. Moreover, δ_{i2} denotes the Kronecker delta, ω_0 is a model parameter, and $M(C, \mathbf{u})$ is a dilution model that accounts for PSD effects. Similar models are used in PDF methods for turbulent flows to model molecular diffusion effects [18]. $\Xi(t)$ is an auxiliary random variable that leads to a partially negative autocorrelation function of the transverse velocity component in agreement with observations from MC simulations. The coefficients D_i denote the diagonal elements of the dispersion tensor whose off-diagonal elements are assumed to be zero. The functions $a(u_i)$ and $b(u_i)$ and the parameter ω_0 were calibrated with MC simulations [20] for σ_Y^2 ranging from $1/16$ up to 4. The resulting stochastic processes reproduce the complex correlation behavior and velocity PDF shapes observed in the MC simulations. Based on the SDEs (21) and relation (17), the transport equation

$$\frac{\partial f}{\partial t} = -v_i \frac{\partial f}{\partial x_i} + D_i \frac{\partial^2 f}{\partial x_i^2} - \frac{\partial}{\partial v_i}[a(v_i) f] \qquad (22)$$

$$+ \frac{1}{2}\frac{\partial^2}{\partial v_i^2}[b(v_i)^2 f] - \frac{\partial}{\partial c}[M(c, \mathbf{v}) f]$$

for the Eulerian joint velocity–concentration PDF $f(\mathbf{v}, c; \mathbf{x}, t)$ can be derived [19]. By integrating this PDF over velocity sample space we obtain the local concentration PDF $f_C(c; \mathbf{x}, t)$. From the PDF transport equation (22) and the ADE (20) moment evolution equations can be derived where the ones resulting from the ADE are exact reference equations. A comparison of the two equation sets leads to a consistency requirement for the dilution model, i.e.,

$$\langle M(C, \mathbf{v})|\mathbf{u}\rangle = 0 \quad \forall\, \mathbf{u}. \qquad (23)$$

This requirement implies that a suitable dilution model should not change the velocity-conditional concentration mean.

For a given initial condition, an ensemble of particles can be evolved by integrating the SDEs (21) over time. Statistical quantities like the concentration mean, $\langle C(\mathbf{x}, t)\rangle$, are estimated based on the particle ensemble on a relatively coarse grid as illustrated in Fig. 10. The estimation grid needs to resolve changes in the joint PDF $f(\mathbf{v}, c; \mathbf{x}, t)$. By assuming that the joint PDF is constant in each grid cell, one can use the particles in the grid cells to approximate the PDF locally. For example, $\langle C\rangle$ in a grid cell is estimated by averaging over the concentration values of particles in the corresponding cell. The grid resolution requirements for MC simulations, on the other hand, are considerably higher since gradients in individual $Y(\mathbf{x})$ and $C(\mathbf{x}, t)$ realizations have to be resolved.

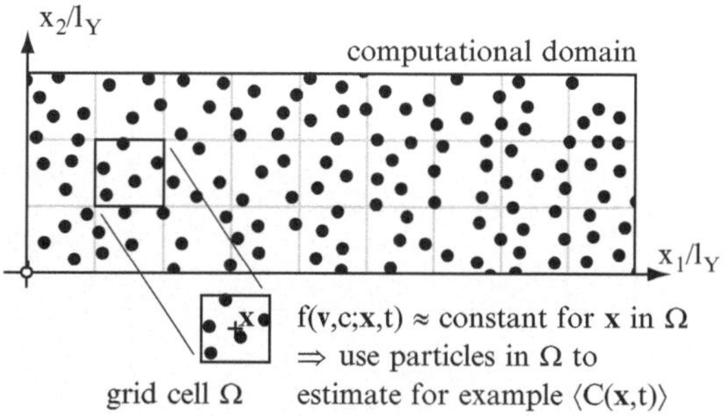

Fig. 10 Particle-based algorithm for the numerical solution of the Eulerian joint velocity–concentration PDF transport equation

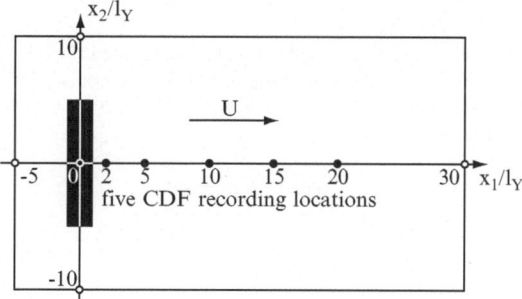

x_2/l_Y

computational domain

x_1/l_Y

grid cell Ω

$f(v,c;x,t) \approx$ constant for $\mathbf{x}$ in Ω
$\Rightarrow$ use particles in Ω to
estimate for example $\langle C(\mathbf{x},t) \rangle$

Fig. 11 Rectangular computational domain used in the joint PDF method with the initial tracer body (*black square*, $2l_Y \times 10l_Y$) and the locations where CDFs were sampled (*filled black circles*)

x_2/l_Y

10

U

-5   0   2 5   10   15   20   30 x_1/l_Y

five CDF recording locations

-10

Numerical simulations of the spreading of an initially rectangular tracer cloud depicted in Fig. 11 were performed with the outlined PDF method for a log-conductivity variance $\sigma_Y^2 = 2$ and a Péclet number Pe $= Ul_Y/D_1 = 1,000$. In most applications, Péclet numbers are of the order or larger than 100. An ensemble of ten million particles and a computational grid with 71×41 grid cells were used (grid spacing of $l_Y/2$). Comparisons with reference MC simulations reported by Caroni and Fiorotto [4] are provided in Figs. 12 and 13. The asymmetric shapes of the concentration mean and standard deviations depicted in Fig. 12 are a direct result of the non-Gaussian velocity PDF and the complex correlation behavior in the velocity statistics. Consequently, low-order approximations fail to predict this behavior. The cumulative density functions (CDF) provided by the joint PDF method and the MC simulations are compared at different locations $\mathbf{x} = (x_1, 0)^T$ (compare Fig. 11) and times $t = x_1/U$ in Fig. 13. Both CDFs and concentration moments are in good agreement and therefore, the joint PDF method provides a good alternative to low-order approximations and MC simulations.

4   Illustration of MDF Method for Rarefied Gas Flows

In rarefied gases or flows in very small structures, the continuum assumption and consequently the Navier–Stokes equations become invalid and the dynamics of molecules or atoms have to be considered. In this section, an MDF method

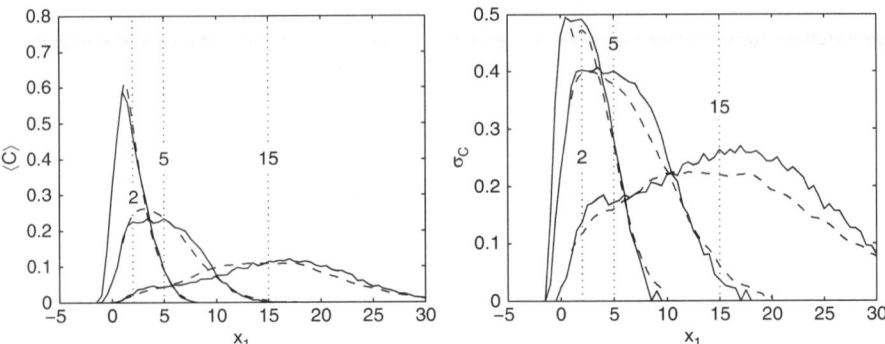

Fig. 12 Comparisons of concentration mean $\langle C[(x_1, 0)^T, t]\rangle$ and standard deviation σ_C $[(x_1, 0)^T, t]$ resulting from the joint velocity–concentration PDF method (*solid lines*) and the MC simulations (*dashed lines*) for $\sigma_Y^2 = 2$ and Pe $= 1,000$. Comparisons are provided for times $t\, U/l_Y = 2$, 5 and 15

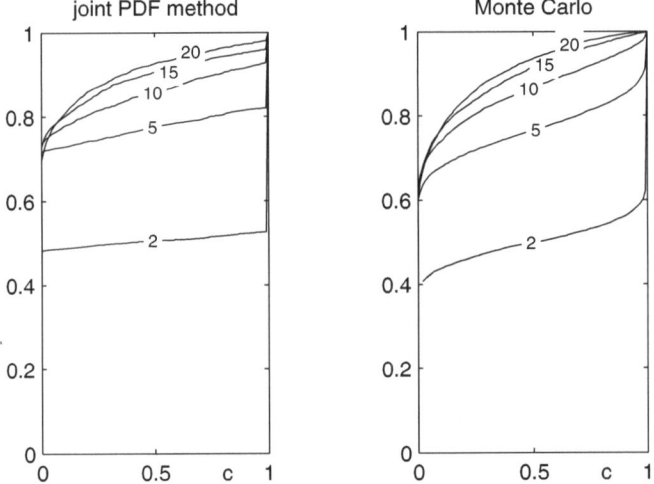

Fig. 13 Comparison of the concentration CDFs resulting from the joint velocity–concentration PDF method (*left graph*) and the MC simulations (*right graph*) for $\sigma_Y^2 = 2$ and Pe $= 1,000$. Comparisons are provided at locations $x_1/l_Y = 2$, 5, 10, 15, and 20 at times $t = x_1/U$. In both graphs, the abscissa represents the concentration sample space variable c and the ordinate the according cumulative probability

is outlined that takes molecular dynamics into account and that is based on the Fokker–Planck equation (17). An ensemble of computational particles are used that represent molecules or atoms. The particle motion that is subject to collisions and interactions with walls is modeled with suitable continuous stochastic processes. The outlined approach is, however, different compared to numerical solutions of the classical Boltzmann equation, where particle collisions are resolved. This approach is called direct numerical simulation Monte Carlo (DSMC) [2] and leads to stiff problems since changes in the macroscopic properties like velocity or internal energy take place at larger length- or timescales compared to the timespan between particle collisions. In the MDF method, on the other hand, particle collisions are modeled by employing continuous OU processes for the particle velocity components (compare SDE (18)). For these processes, a stable time integration scheme is devised that allows for time step sizes that are of the order of the macroscopic timescale, which leads compared to DSMC to a considerable reduction in computational cost.

4.1 The Boltzmann Equation is an MDF Transport Equation

A monatomic gas is considered. Each computational particle represents one atom or many identical atoms and has a velocity $\mathbf{M}(t)$ and a position $\mathbf{X}(t)$. The statistical distribution of atom velocities at location $\mathbf{x}$ and time t are given by the PDF $f(\mathbf{V}; \mathbf{x}, t)$, where $\mathbf{V}$ is the sample space vector of $\mathbf{M}$. It is assumed that in equilibrium $f(\mathbf{V}; \mathbf{x}, t)$ obeys a joint Gaussian distribution, which leads to a Maxwell distribution for the magnitude of the atom velocity $|\mathbf{M}|$. The gas mass in the same volume is determined by the gas density $\rho(\mathbf{x}, t)$. The mass density function (MDF) is then defined as

$$\mathscr{F}(\mathbf{V}, \mathbf{x}, t) = \rho(\mathbf{x}, t) f(\mathbf{V}; \mathbf{x}, t). \tag{24}$$

It is pointed out that this MDF definition is different from the one of Pope (equation (3.59) in [23]) proposed for the simulation of compressible low Mach number turbulent flows. There, the density ρ is also a random variable that depends on the scalar vector which is included in the joint PDF. Therefore, Pope's MDF contains mainly probabilistic information and integration of the MDF over the entire probability space leads to the mean density. The MDFs mentioned in this chapter, however, contain statistical information about unresolved microscales and therefore reflect deterministic quantities.

The Boltzmann equation is a conservation equation for $\mathscr{F}(\mathbf{V}, \mathbf{x}, t)$ in $\mathbf{V}$-$\mathbf{x}$-space and is given by

$$\frac{\partial \mathscr{F}}{\partial t} + V_i \frac{\partial \mathscr{F}}{\partial x_i} + \frac{\partial F_i \mathscr{F}}{\partial V_i} = S(\mathscr{F}), \tag{25}$$

where $\mathbf{F}$ is an external force, e.g., gravity, and $S(\mathscr{F})$ is the collision operator. The Boltzmann equation conserves mass, momentum and energy, the densities of which are the following macroscopic quantities

$$(\rho, \rho\mathbf{U}, \rho e_s + \tfrac{1}{2}\rho\mathbf{U}^2) = \int_{\mathbb{R}^3} \underbrace{(1, \mathbf{V}, \tfrac{1}{2}\mathbf{V}^2)}_{\boldsymbol{\Psi}_{\text{cons}}} \mathscr{F} \, d\mathbf{V}, \tag{26}$$

for a collision operator $S(\mathscr{F})$ that leaves these quantities unchanged, i.e.,

$$\int_{\mathbb{R}^3} \boldsymbol{\Psi}_{\text{cons}} S(\mathcal{F}) \, d\mathbf{V} = \begin{pmatrix} 0 \\ 0 \\ 0 \end{pmatrix} \tag{27}$$

for any $\mathcal{F}$, which is a minimal requirement for $S(\mathcal{F})$. In (26), $\mathbf{U} = \langle \mathbf{M} \rangle$ is the mean velocity and $e_s = \frac{1}{2}\langle (\mathbf{M} - \mathbf{U})^2 \rangle$ is the internal or sensible energy for monatomic gases. So, as implied by (26), multiplication of the Boltzmann equation (25) by $\boldsymbol{\Psi}_{\text{cons}}$ and integration over velocity sample space $\mathbf{V}$, leads for a collision operator that satisfies requirement (27) to the following conservation laws:

$$\frac{\partial \rho}{\partial t} + \frac{\partial \rho U_j}{\partial x_j} = 0, \tag{28}$$

$$\frac{\partial \rho U_i}{\partial t} + \frac{\partial \rho U_i U_j}{\partial x_j} + \frac{\partial p_{ij}}{\partial x_j} = \rho F_i, \text{ and} \tag{29}$$

$$\frac{\partial \rho e_s}{\partial t} + \frac{\partial \rho U_j e_s}{\partial x_j} + \frac{\partial q_j}{\partial x_j} + p_{jk}\frac{\partial U_j}{\partial x_k} = 0. \tag{30}$$

In the energy conservation equation (30), q_j is the heat flux and in the momentum conservation equation (29), $p_{ij} = \rho\langle (M_i - U_i)(M_j - U_j)\rangle$ is the molecular stress tensor. The latter can be written as a sum of an isotropic part and a deviatoric part, i.e., $p_{ij} = \delta_{ij} p + \pi_{ij}$, where $p = p_{ii}/3$ and δ_{ij} denotes the Kronecker delta. Comparing the definitions of p, p_{ij}, and e_s leads to the thermodynamic relation $p = \frac{2}{3}\rho e_s$ for the isotropic part of p_{ij}. This expression is consistent with the well-known expression for perfect gases, i.e., $p = (\gamma - 1)\rho e_s$, that are monatomic with $\gamma = 5/3$. The deviatoric part, π_{ij}, on the other hand, translates for very small Knudsen numbers or very short collision timescales into the viscous term of the Navier–Stokes equations.

4.2   Moment Equations

The set of moment equations (28), (29), and (30) provides the macroscopic quantities ρ, $\mathbf{U}$, e_s, and p. The deviatoric part of the molecular stress tensor π_{ij} and the heat flux q_i, however, remain unknown. Corresponding transport equations can readily be derived from the Boltzmann equation (25) by multiplication with $u_{\langle i} u_{j\rangle}$ (the angle brackets in the subscripts identify the deviatoric part of the tensor $u_i u_j$ with $\mathbf{u} = \mathbf{V} - \mathbf{U}$) and $\frac{1}{2}u_i u_k u_k$, respectively, and integration over velocity sample space. This leads to

$$\frac{\partial \pi_{ij}}{\partial t} + \frac{\partial \pi_{ij} U_k}{\partial x_k} + \frac{\partial m_{ijk}}{\partial x_k} + \frac{4}{5}\frac{\partial q_{\langle i}}{\partial x_{j\rangle}} + 2p\frac{\partial U_{\langle i}}{\partial x_{j\rangle}} + 2\pi_{k\langle i}\frac{U_{j\rangle}}{x_k} = P_{ij} \tag{31}$$

and

$$\frac{\partial q_i}{\partial t} + \frac{\partial q_i U_k}{\partial x_k} + \frac{1}{2}\frac{\partial R_{ik}}{\partial x_k} + p\frac{\partial(\pi_{ik}/\rho)}{\partial x_k}$$

$$+\frac{5}{2}\frac{k}{m}p_{ik}\frac{\partial T}{\partial x_k} - \frac{\pi_{ij}}{\rho}\frac{\partial \pi_{jk}}{\partial x_k} + \left(m_{ijk} + \frac{6}{5}q_{\langle i}\delta_{jk\rangle}\right) + q_k\delta_{ij})\frac{\partial U_j}{\partial x_k} = P_i, \quad (32)$$

where

$$P_{ij} = \int_{\mathbb{R}^3} u_i u_j \, S(\mathscr{F}) \, d\mathbf{V} \text{ and } P_i = \frac{1}{2}\int_{\mathbb{R}^3} u_i u_k u_k \, S(\mathscr{F}) \, d\mathbf{V} \quad (33)$$

encapsulate contributions from the collision operator. Regularized 13-moment-equations (R13) methods [28] solve the equation system consisting of five equations for ρ, **U**, and e_s, three equations for **q**, and five equations for π_{ij} ($6-1=5$ because the isotropic part of π_{ij}, p, is known from the thermodynamic relation). Closure models are, however, needed to determine unknown quantities that appear in (31) and (32).

4.3 Collision Operators

So far, little information was provided about the collision operator $S(\mathscr{F})$. Based on the assumption that velocities of colliding particles are uncorrelated and independent of position (Stosszahlansatz), Boltzmann proposed the operator $S^{(\text{Boltz})}$. Since individual collisions have to be considered in $S^{(\text{Boltz})}$, its applicability is limited due to the high computational cost involved. However, it proved to be accurate for a wide range of scenarios and therefore is considered as a benchmark. Less expensive alternative operators are tested for example by comparing the resulting P_{ij} and P_i with the reference expressions

$$P_{ij}^{(\text{Boltz})} = -\frac{1}{\tau_{\text{Boltz}}}\pi_{ij} \text{ and } P_i^{(\text{Boltz})} = -\frac{2}{3}\frac{1}{\tau_{\text{Boltz}}}q_i \quad (34)$$

that are implied by $S^{(\text{Boltz})}$. Here, τ_{Boltz} is a collision timescale that is inversely proportional to the fluid density and proportional to the atom mass. An alternative collision operator is for example the Bhatnagar–Gross–Krook (BGK) operator [1]:

$$S^{(\text{BGK})}(\mathscr{F}) = \frac{1}{\tau_{\text{BGK}}}(\mathscr{F}_M - \mathscr{F}), \quad (35)$$

where $\mathscr{F}_M$ is a joint Gaussian equilibrium MDF and τ_{BGK} is a relaxation timescale parameter to be determined. Therefore, the BGK operator enforces a relaxation of the MDF $\mathscr{F}$ to the equilibrium MDF $\mathscr{F}_M$ at a given rate τ_{BGK}. It leads to

$$P_{ij}^{(\text{BGK})} = -\frac{1}{\tau_{\text{BGK}}}\pi_{ij} \text{ and } P_i^{(\text{BGK})} = -\frac{1}{\tau_{\text{BGK}}}q_i. \quad (36)$$

In the remainder of this section, we will focus on a collision operator that is based on a Fokker–Planck (FP) equation [13], i.e.,

$$S^{(FP)}(\mathscr{F}) = \frac{\partial}{\partial V_i}\left[\frac{1}{\tau_{FP}}(V_i - U_i)\mathscr{F}\right] + \frac{\partial^2}{\partial V_i \partial V_i}\left(\frac{2e_s}{3\tau_{FP}}\mathscr{F}\right), \qquad (37)$$

where τ_{FP} is a relaxation timescale parameter. This operator is based on a continuous multi-variate OU process (compare (19)) and involves a drift term (first term on the right hand side) that reduces velocity fluctuations and a diffusion term (second term) that randomly adds fluctuations proportional to $\sqrt{e_s}$. Furthermore, it leads to a multi-variate Gaussian equilibrium MDF and accordingly to a Maxwellian distribution of the velocity magnitudes. With $S^{(FP)}$ we get from (33)

$$P_{ij}^{(FP)} = -\frac{2}{\tau_{FP}}\pi_{ij} \text{ and } P_i^{(FP)} = -\frac{3}{\tau_{FP}}q_i. \qquad (38)$$

Now, by comparing the P_{ij} from the less expensive BGK and FP operators with the Boltzmann reference expression (34), the relaxation timescale parameters are determined, i.e., $\tau_{BGK} = \tau_{Boltz}$ and $\tau_{FP} = 2\tau_{Boltz}$. With these choices the degrees of freedom in the two collision operator models are exhausted and the following expressions for P_i are resulting:

$$P_i^{(BGK)} = -\frac{1}{\tau_{Boltz}}q_i \text{ and } P_i^{(FP)} = -\frac{3}{2}\frac{1}{\tau_{Boltz}}q_i. \qquad (39)$$

These results are inconsistent with the reference expression (34) from the Boltzmann collision operator and imply also deviations in the resulting Prandtl numbers. Nevertheless, both operators do not need to resolve collisions of atoms and therefore are computationally less expensive. In fact, the time step of the FP operator is restricted only by timescales determined by changes in the macroscopic quantities as will be shown later.

4.4 Numerical Solution

Insertion of the FP-based collision operator (37) into the Boltzmann equation (25) leads to

$$\frac{\partial \mathscr{F}}{\partial t} + V_i\frac{\partial \mathscr{F}}{\partial x_i} = \frac{\partial}{\partial V_i}\left\{\left[\frac{1}{\tau_{FP}}(V_i - U_i) - F_i\right]\mathscr{F}\right\} + \frac{\partial^2}{\partial V_i \partial V_i}\left(\frac{2e_s}{3\tau_{FP}}\mathscr{F}\right). \qquad (40)$$

For the numerical solution of this MDF transport equation, a particle-based Monte Carlo method is used. The corresponding SDEs that describe the evolutions for the particle position $\mathbf{X}$ and velocity $\mathbf{M}$ are determined based on the general relation (17)

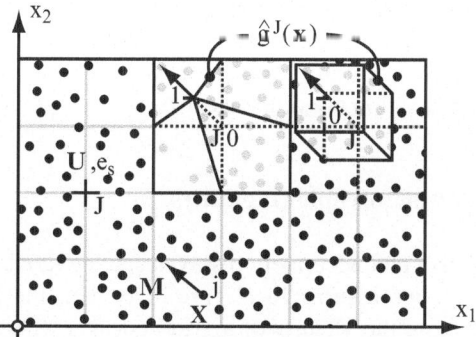

Fig. 14 An ensemble of particles with particle indices j evolving in the computational domain is depicted. A grid with nodes J and kernel functions $\hat{g}^J(\mathbf{x})$ are used to estimate macroscopic quantities like $\mathbf{U}(\mathbf{x}^J, t)$ and $e_s(\mathbf{x}^J, t)$ at the grid nodes $\mathbf{x}^J$. Both tent and top-hat kernel functions are depicted. Each particle j has a position $\mathbf{X}^j(t)$ and velocity $\mathbf{M}^j(t)$

and the MDF transport equation (40):

$$dX_i = M_i\, dt \quad \text{with} \tag{41}$$

$$dM_i = -\left[\frac{1}{\tau_{\mathrm{FP}}}(M_i - U_i) - F_i\right] dt + \sqrt{\frac{4e_s}{3\tau_{\mathrm{FP}}}}\, dW_i(t). \tag{42}$$

Based on these SDEs, an ensemble of computational particles that represent a large number of atoms can be evolved in a computational domain as illustrated in Fig. 14. Macroscopic quantities are estimated from the particles at grid nodes similar to the procedure outlined in Sect. 3 for the estimation of statistical moments. Accordingly, the computational grid has to resolve changes in the macroscopic quantities.

The solution algorithm for the MDF method with a second order integration scheme for the particle position involves the following steps:

1. $\mathbf{U}$ and e_s at time t are estimated at each grid node and interpolated to the particle positions.
2. The time step size Δt is determined.
3. A first half-time-step is performed to estimate the particle mid-points.
4. Mid-point boundary conditions are applied.
5. $\mathbf{U}$ and e_s at time $t + \Delta t/2$ are interpolated from the grid nodes to the particle mid-point positions.
6. The new particle velocities and positions are computed.
7. The boundary conditions are enforced.

In the remainder of this section, we focus on two aspects of this algorithm: first, it is shown how macroscopic quantities are calculated (point 1 in the algorithm) and second, an efficient time integration scheme is outlined that allows for time step sizes that exceed the relaxation or collision timescale τ_{FP} (points 3 and 6).

4.4.1 Estimation of Macroscopic Quantities

For a computational grid that accurately resolves spatial variations of macroscopic quantities like U or e_s we assume local ergodicity, i.e., the MDF $\mathscr{F}(V, x, t)$ is approximately constant in the vicinity of a grid node x^J, where J is the grid node index. Macroscopic quantities at each grid node are then estimated based on the particles surrounding the grid node. In the estimation process, particles that are located close to the grid node x^J have a higher weight versus particles that are further away. The particle weights are accounted for by kernel functions $\hat{g}^J(x)$. For example, to estimate a quantity at the grid node J, the weight of particle j located at X^j is determined by $\hat{g}^J(X^j)$. Tent and top-hat kernel functions are depicted in Fig. 14. For example, the macroscopic velocity U is then estimated by

$$U(x^J, t) = \frac{\mathscr{U}^J(t)}{\mathscr{W}^J(t)}, \text{ where} \tag{43}$$

$$\mathscr{U}^J(t) = \sum_{j=1}^{N_p} \left\{ \hat{g}^J[X^j(t)]M^j(t) \right\}, \mathscr{W}^J(t) = \sum_{j=1}^{N_p} \left\{ \hat{g}^J[X^j(t)] \right\},$$

and N_p is the number of particles. The internal energy $e_s(x, t)$ is estimated accordingly. Like in other Monte Carlo methods, a sufficient number of particles is required to keep the statistical error at an acceptable level.

For problems that are stationary, time averaging can be used to relax requirements for the particle ensemble size. The corresponding expressions for $\mathscr{U}^J(t)$ and $\mathscr{W}^J(t)$ are given by

$$\mathscr{U}^J(t) = \mu \mathscr{U}^J(t - \Delta t) + (1 - \mu) \sum_{j=1}^{N_p} \left\{ \hat{g}^J[X^j(t)]M^j(t) \right\} \text{ and}$$

$$\mathscr{W}^J(t) = \mu \mathscr{W}^J(t - \Delta t) + (1 - \mu) \sum_{j=1}^{N_p} \left\{ \hat{g}^J[X^j(t)] \right\},$$

where $\mu \in [0, 1[$ is a memory factor. For μ close to unity and zero the past and present are more dominant in the averaging procedure, respectively. It has been shown that this time-averaging technique allows to dramatically reduce the number of particles, which in general improves computational efficiency and reduces memory requirements for FP-based methods.

4.4.2 Accurate Time Integration Scheme

For the time integration of SDEs (41) and (42), an appropriate time integration scheme is required. Jenny et al. [13] have demonstrated that standard integration schemes may not conserve the internal energy and therefore caution is needed when selecting an integration scheme. In a brief derivation a scheme is presented that works for time step sizes Δt that exceed τ_{FP}. At the beginning of the time step (time $t = t_0$) the initial particle velocity and position are $M(t_0)$ and $X(t_0)$, respectively. After introducing the transformations

$$X_i'(t) = X_i(t) - X_i(t_0) - (U_i + \tau_{FP} F_i)(t - t_0) \text{ and} \tag{44}$$

$$M_i'(t) = M_i(t) - (U_i + \tau_{FP} F_i), \tag{45}$$

and the substitution

$$b = \sqrt{\frac{4e_s}{3\tau_{FP}}} \tag{46}$$

we arrive at the somewhat simplified SDEs

$$d\mathbf{y}(t) = -\mathbf{A}\mathbf{y}(t)\, dt + \mathbf{B}\, d\mathbf{W}(t) \tag{47}$$

with

$$\mathbf{y}(t) = \begin{pmatrix} M_i'(t) \\ X_i'(t) \end{pmatrix}, \mathbf{A} = \begin{bmatrix} \tau_{FP}^{-1} & 0 \\ -1 & 0 \end{bmatrix}, \mathbf{B} = \begin{bmatrix} b & 0 \\ 0 & 0 \end{bmatrix}, \tag{48}$$

and $\mathbf{W}(t)$ being a vectorial Wiener process where all components are statistically independent. Here, it was assumed that the macroscopic velocity $\mathbf{U}$, the internal energy e_s, and the force $\mathbf{F}$ are constant over one integration time step. The SDE system (47) is a bi-variate OU process (Sect. 2.2.4) for which Gardiner (Sect. 4.4.6 in [9]) has outlined an analytical solution of the form

$$\langle \mathbf{y}(t) \rangle = \exp[-\mathbf{A}(t - t_0)]\langle \mathbf{y}(t_0) \rangle \quad \text{and} \tag{49}$$

$$\langle \mathbf{y}(t), \mathbf{y}^T(s) \rangle = \langle [\mathbf{y}(t) - \langle \mathbf{y}(t) \rangle][\mathbf{y}(s) - \langle \mathbf{y}(s) \rangle]^T \rangle$$

$$= \exp[-\mathbf{A}(t - t_0)]\langle \mathbf{y}(t_0), \mathbf{y}^T(t_0) \rangle \exp[-\mathbf{A}(s - t_0)]$$

$$+ \int_{t_0}^{\min(t,s)} \exp[-\mathbf{A}(t - t')]\mathbf{B}\mathbf{B}^T \exp[-\mathbf{A}^T(s - t')]\, dt'.$$

For a known initial condition at the beginning of the time step ($t = t_0$), i.e., $\mathbf{y}(t_0) = (M_i'(t_0), X_i'(t_0))^T$, the exact solution of the system (47) for $t > t_0$ is a joint Gaussian PDF that is fully characterized by the mean $\langle \mathbf{y}(t) \rangle$ and the covariance matrix $\langle \mathbf{y}(t), \mathbf{y}^T(t) \rangle$. With the coefficients given in (48) and with the definition of the matrix exponential function, i.e.,

$$\exp(\mathbf{A}) = \sum_{k=0}^{\infty} \frac{1}{k!} \mathbf{A}^k,$$

the resulting moments for $X_i'(t)$ and $M_i'(t)$ are determined as

$$\langle M_i'(t)|\mathbf{y}(t_0)\rangle = \exp\left[-\frac{(t-t_0)}{\tau_{FP}}\right] M_i'(t_0),$$

$$\langle X_i'(t)|\mathbf{y}(t_0)\rangle = \tau_{FP} \left\{1 - \exp\left[-\frac{(t-t_0)}{\tau_{FP}}\right]\right\} M_i'(t_0),$$

$$\langle M_i'(t), M_i'(t)|\mathbf{y}(t_0)\rangle = \frac{b^2 \tau_{FP}}{2} \left\{1 - \exp\left[-2\frac{(t-t_0)}{\tau_{FP}}\right]\right\},$$

$$\langle X_i'(t), X_i'(t)|\mathbf{y}(t_0)\rangle = \frac{b^2 \tau_{FP}^2}{2} \left\{2t - 4\tau_{FP} \left\{1 - \exp\left[-\frac{(t-t_0)}{\tau_{FP}}\right]\right\}\right.$$
$$\left. + \tau_{FP} \left\{1 - \exp\left[-2\frac{(t-t_0)}{\tau_{FP}}\right]\right\}\right\}, \text{ and}$$

$$\langle M_i'(t), X_i'(t)|\mathbf{y}(t_0)\rangle = \frac{b^2 \tau_{FP}^2}{2} \left\{1 - 2\exp\left[-\frac{(t-t_0)}{\tau_{FP}}\right] + \exp\left[-2\frac{(t-t_0)}{\tau_{FP}}\right]\right\}.$$

These expressions are consistent with (74) and (75) in [13]. With $\Delta t = t - t_0$, (44) and (45), and the definitions $M_i^n = M_i(t_0)$, $X_i^n = X_i(t_0)$, and $\mathbf{y}^n = (M_i^n, X_i^n)$, this leads for $M_i^{n+1} = M_i(t)$ and $X_i^{n+1} = X_i(t)$ to

$$\langle M_i^{n+1}|\mathbf{y}^n\rangle = \exp\left(-\frac{\Delta t}{\tau_{FP}}\right)(M_i^n - U_i - \tau_{FP}F_i) + U_i + \tau_{FP}\dot{F}_i,$$

$$\langle X_i^{n+1}|\mathbf{y}^n\rangle = \tau_{FP}\left[1 - \exp\left(-\frac{\Delta t}{\tau_{FP}}\right)\right](M_i^n - U_i - \tau_{FP}F_i)$$
$$+ (U_i + \tau_{FP}F_i)\Delta t + X_i^n,$$

$$\langle M_i^{n+1}, M_i^{n+1}|\mathbf{y}^n\rangle = \frac{2e_s}{3}\left[1 - \exp\left(-2\frac{\Delta t}{\tau_{FP}}\right)\right], \tag{50}$$

$$\langle X_i^{n+1}, X_i^{n+1}|\mathbf{y}^n\rangle = \frac{2e_s \tau_{FP}}{3}\left\{2t - 4\tau_{FP}\left[1 - \exp\left(-\frac{\Delta t}{\tau_{FP}}\right)\right]\right.$$
$$\left. + \tau_{FP}\left[1 - \exp\left(-2\frac{\Delta t}{\tau_{FP}}\right)\right]\right\}, \text{ and}$$

$$\langle M_i^{n+1}, X_i^{n+1}|\mathbf{y}^n\rangle = \frac{2e_s \tau_{FP}}{3}\left[1 - 2\exp\left(-\frac{\Delta t}{\tau_{FP}}\right) + \exp\left(-2\frac{\Delta t}{\tau_{FP}}\right)\right].$$

Note that the components M_i and X_i are independent in the different coordinate directions and consequently, covariances with indices $j \neq i$ are equal to zero, for example, $\langle M_i^{n+1}, M_{j \neq i}^{n+1} | \mathbf{y}^n \rangle = 0$.

Integration of the SDEs (41) and (42) is performed as follows: for a given time step Δt and initial condition (M_i^n, X_i^n), two random numbers (M_i^{n+1}, X_i^{n+1}) are drawn from the bi-variate Gaussian PDF defined by (50).

Except for the fact that $\mathbf{F}$ and the macroscopic quantities $\mathbf{U}$ and e_s are assumed to be constant over Δt, the scheme defined by (50) is exact. This is an advantageous property since Δt does not need to resolve the collision or relaxation timescale τ_{FP}, but only variations in the macroscopic quantities and the external force.

4.5   Simulation Results

Simulations were performed for different flow geometries and Knudsen numbers [13]. Channel flow simulations with constant mean free path length λ but variable channel widths L_0 and corresponding Knudsen numbers, $\text{Kn} = \lambda / L_0$, were already discussed in Sect. 1 (Figs. 6–8). The MDF method with the FP collision operator is able to provide accurate predictions of the Knudsen paradox at a considerably lower computational cost compared to DSMC. Especially for large Knudsen numbers, complex velocity distributions were predicted at different locations in the channel (compare Fig. 7). It is virtually impossible to capture such effects with moment-based approaches like the R13 method.

The computational efficiency and stability of the time integration scheme is demonstrated in Fig. 15: velocity distributions of the flow around a cylinder in a channel with gravity are provided at different grid resolutions. The Knudsen number in these simulations was 0.4. At the bottom and top boundaries, periodic boundary conditions were applied and the left and right sides represent isothermal walls. For the estimation of macroscopic quantities at grid nodes, top-hat kernel functions were applied. Ten particles per grid cell were used and time averaging was applied to reduce the statistical error. The main flow characteristics are captured already at a very coarse resolution of 8×8 grid cells and the total flow rate is within 5.1% of the flow rate obtained with the 64×64 grid. This gives a clear indication of the accuracy and computational efficiency of the time integration scheme. Furthermore, the simulations illustrate that the MDF method can also be applied easily for more complex flow scenarios.

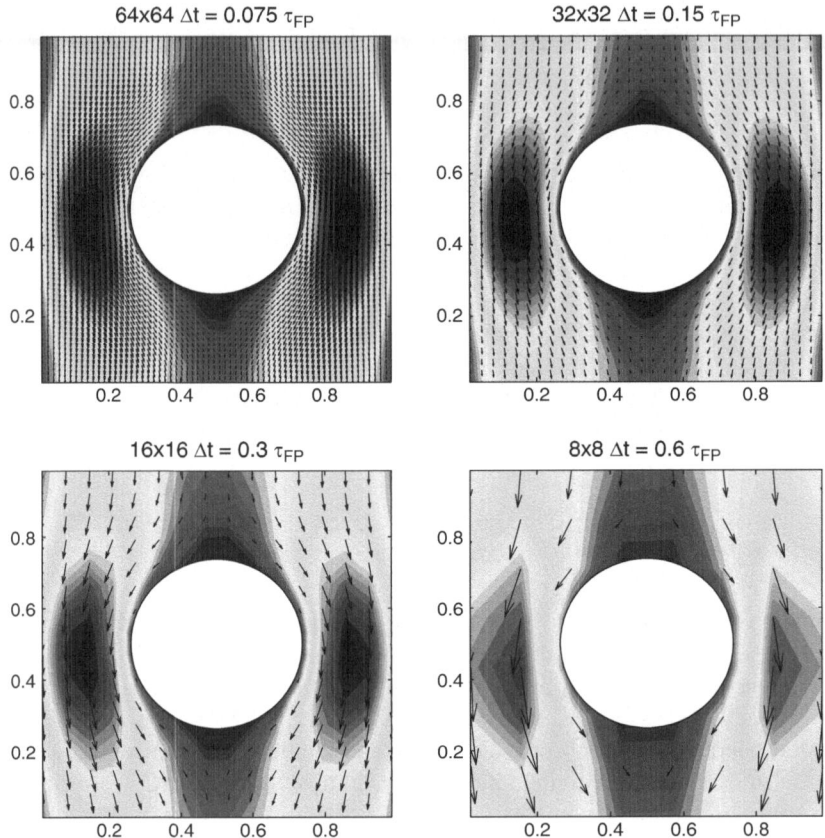

Fig. 15 Simulations of the flow around a cylinder in a channel with gravity. Simulations with different computational grids and time step sizes Δt are compared. All plots show velocity distributions and the colors represent velocity magnitudes ranging form Mach numbers equal to zero (*dark blue*) up to 0.3 (*dark red*)

5 Discussion

In this chapter, the basic structure of probability or mass density function (PDF/MDF) methods was presented. An efficient PDF (or MDF) method to solve Fokker–Planck (FP) equations was outlined. Such equations describe the evolution of multi-variate PDFs or MDFs that encapsulate probabilistic or statistical quantities relevant to certain problems. FP equations result from corresponding Kramers–Moyal expansions for continuous processes for which the theorem of Pawula is applicable. The coefficients in a FP equation are directly related to an equivalent set of temporal stochastic differential equations (SDEs) that determine the evolution of particles. These particles evolve in the multi-variate probability or property space and directly imply an efficient numerical solution algorithm for the FP equation. Macroscopic quantities are estimated from the particle ensemble on a computational grid with the help of kernel functions. Unresolved microscale physics is modeled by SDEs that may also depend on macroscopic quantities. The Lagrangian particle-based viewpoint facilitates the formulation of appropriate models in flow and transport problems. Furthermore, the particle ensemble provides not just means and covariances of relevant quantities, but complete distributions. This offers a higher level of closer compared to moment-based or deterministic methods that focus on the macroscale. These factors make PDF or MDF methods ideally suited for certain multi-scale problems. Several application examples were provided that illustrate these features.

For uncertainty quantification in subsurface flows, a joint PDF method was outlined that provides accurate results at a much lower computational cost compared to Monte Carlo (MC) simulations. The small-scale velocity fluctuations that determine advective transport are modeled with suitable stochastic processes that share certain key statistics with the real process observed in MC simulations. PSD effects, that also depend on small-scale structures, are modeled with a suitable dilution model. The outlined PDF method was validated for heterogeneous media with σ_Y^2 up to 2 and should be extended for spatially inhomogeneous cases including log-conductivity measurements.

For simulations of rarefied gas flows or flows in very small structures, an MDF method with a Fokker–Planck (FP) collision operator was outlined. If paired with an efficient and accurate integration scheme, the FP operator is computationally very efficient since large time steps can be used. Unlike BGK-operator or DSMC methods, collision timescales need not to be resolved. In statistically stationary cases, time averaging can be applied to reduce the number of computational particles. Recently, an improved FP operator was proposed [10] that resolves Prandtl number deficiencies of the linear FP operator outlined in this chapter. An interesting option for future work is to incorporated the FP-based MDF method into a hybrid simulation framework were moment equations for macroscopic quantities are solved. For large Knudsen numbers (Kn > 10), DSMC is accurate and computationally efficient and could be applied to determine unclosed higher-order moments like the molecular stress tensor. However, for intermediate Knudsen numbers, DSMC becomes expensive and a FP-based methodology seems advantageous. If the Knudsen number goes below unity, purely macroscopic or Eulerian methods like the R13 model, and for very low Kn, the Navier–Stokes equations are probably most efficient. The suggested framework is a unification in the sense that its components are identical for different Kn. Exceptions are the molecular stresses and heat fluxes, for which different Eulerian and Lagrangian methods are selected depending on Kn.

References

1. P. L. Bhatnagar, E. P. Gross, and M. Krook. A model for collision processes in gases. i. small amplitude processes in charged and neutral one-component systems. *Physical Review*, 94(3):511, 1954.
2. G. A. Bird. *Molecular gas dynamics and the direct simulation of gas flows*. Oxford engineering science series. Clarendon, Oxford, 1994.
3. Robert Brown and John Joseph Bennett. *The miscellaneous botanical works of Robert Brown*. Published for the Ray society by R. Hardwicke, London, 1866.
4. Elpidio Caroni and Virgilio Fiorotto. Analysis of concentration as sampled in natural aquifers. *Transport in Porous Media*, 59(1):19–45, 2005.
5. P. J. Colucci, F. A. Jaberi, P. Givi, and S. B. Pope. Filtered density function for large eddy simulation of turbulent reacting flows. *Physics of Fluids*, 10(2):499–515, 1998.
6. W. Dong. From stochastic processes to the hydrodynamic equations. *University of California Report No. UCRL-3353*, 1956.
7. A. Einstein. Über die von der molekularkinetischen Theorie der wärme geforderte Bewegung von in ruhenden Flüssigkeiten suspendierten Teilchen. *Annalen der Physik*, 322(8):549–560, 1905.
8. A. Fiori and G. Dagan. Concentration fluctuations in aquifer transport: A rigorous first-order solution and applications. *Journal of Contaminant Hydrology*, 45(1-2):139–163, 2000.
9. C. W. Gardiner. *Handbook of stochastic methods for physics, chemistry and the natural sciences*. Springer, Berlin, third edition, 2004.
10. M. Hossein Gorji and Patrick Jenny. A generalized stochastic solution algorithm for simulations of rarefied gas flows. In *Proceedings of the 2nd European Conference on Microfluidics*, 2010.
11. J. Janicka, W. Kolbe, and W. Kollmann. Closure of the transport-equation for the probability density-function of turbulent scalar fields. *Journal of Non-Equilibrium Thermodynamics*, 4(1):47–66, 1979.
12. P. Jenny, S. B. Pope, M. Muradoglu, and D. A. Caughey. A hybrid algorithm for the joint pdf equation of turbulent reactive flows. *Journal of Computational Physics*, 166(2):218–252, 2001.
13. P. Jenny, M. Torrilhon, and S. Heinz. A solution algorithm for the fluid dynamic equations based on a stochastic model for molecular motion. *Journal of Computational Physics*, 229(4):1077–1098, 2010.
14. A. G. Journel and Ch. Huijbregts. *Mining geostatistics*. Academic Press, London a.o., 1978.
15. A. Juneja and S. B. Pope. A dns study of turbulent mixing of two passive scalars. *Physics of Fluids*, 8(8):2161–2184, 1996.
16. P. Langevin. The theory of brownian movement. *Comptes Rendus Hébdomadaires des Séances de l'Academie des Sciences*, 146:530–533, 1908.
17. D. W. Meyer and P. Jenny. A mixing model for turbulent flows based on parameterized scalar profiles. *Physics of Fluids*, 18(3), 2006.
18. Daniel W. Meyer and Patrick Jenny. Micromixing models for turbulent flows. *Journal of Computational Physics*, 228(4):1275–1293, 2009.
19. Daniel W. Meyer, Patrick Jenny, and Hamdi A. Tchelepi. A joint velocity-concentration pdf method for tracer flow in heterogeneous porous media. *Water Resour. Res.*, 46(12):W12522, 2010.
20. Daniel W. Meyer and Hamdi A. Tchelepi. Particle-based transport model with Markovian velocity processes for tracer dispersion in highly heterogeneous porous media. *Water Resour. Res.*, 46(11):W11552, 2010.
21. Daniel Werner Meyer-Massetti. *On the modeling of molecular mixing in turbulent flows*. doctoral thesis, ETH, 2008.
22. S. B. Pope. A Monte-Carlo method for the pdf equations of turbulent reactive flow. *Combustion Science and Technology*, 25(5-6):159–174, 1981.

23. S. B. Pope. Pdf methods for turbulent reactive flows. *Progress in Energy and Combustion Science*, 11(2):119–192, 1985.
24. S. B. Pope. Lagrangian pdf methods for turbulent flows. *Annual Review of Fluid Mechanics*, 26:23–63, 1994.
25. H. Risken. *The Fokker-Planck equation: methods of solution and applications.* Springer-Verlag, Berlin; New York, 2nd edition, 1989.
26. P. Salandin and V. Fiorotto. Solute transport in highly heterogeneous aquifers. *Water Resources Research*, 34(5):949–961, 1998.
27. S. Subramaniam and S. B. Pope. A mixing model for turbulent reactive flows based on Euclidean minimum spanning trees. *Combustion and Flame*, 115(4):487–514, 1998.
28. M. Torrilhon and H. Struchtrup. Regularized 13-moment equations: shock structure calculations and comparison to Burnett models. *Journal of Fluid Mechanics*, 513:171–198, 2004.
29. Manav Tyagi. *Probability density function approach for modeling multi-phase flow in porous media.* doctoral thesis, ETH, 2010.
30. J. Villermaux and J. C. Devillon. Représentation de la coalescence et de la redispersion des domaines de ségrégation dans un fluide par un modèle d'interaction phénoménologique. In *Second International Symposium on Chemical Reaction Engineering*, pages 1–13, New York, 1972. Elsevier.

A Computational and Theoretical Investigation of the Accuracy of Quasicontinuum Methods

Brian Van Koten, Xingjie Helen Li, Mitchell Luskin, and Christoph Ortner

Abstract We give computational results to study the accuracy of several quasicontinuum methods for two benchmark problems – the stability of a Lomer dislocation pair under shear and the stability of a lattice to plastic slip under tensile loading. We find that our theoretical analysis of the accuracy near instabilities for one-dimensional model problems can successfully explain most of the computational results for these multi-dimensional benchmark problems. However, we also observe some clear discrepancies, which suggest the need for additional theoretical analysis and benchmark problems to more thoroughly understand the accuracy of quasicontinuum methods.

1 Introduction

Multiphysics model coupling has captured the excitement of the engineering research community for its potential to make possible the numerical simulation of heretofore computationally inaccessible multiscale problems. The capability to assess the accuracy of these multiphysics methods is crucial to both the verification of existing methods and the development of improved methods.

During the past several years, a theoretical basis has been developed for estimating the error of atomistic-to-continuum coupling methods [2, 3, 8–13, 16, 19, 21–28, 32, 34–36]. However, the accurate computation of lattice instabilities such as dislocation formation and movement, crack propagation, and plastic slip are primary

B. Van Koten · X. Helen Li · M. Luskin (✉)
School of Mathematics, 206 Church St. SE, University of Minnesota, Minneapolis, MN 55455, USA
e-mail: vank0068@umn.edu; lixxx835@umn.edu; luskin@umn.edu

C. Ortner
Mathematical Institute, St. Giles' 24–29, Oxford OX1 3LB, UK
e-mail: ortner@maths.ox.ac.uk

I.G. Graham et al. (eds.), *Numerical Analysis of Multiscale Problems*, Lecture Notes in Computational Science and Engineering 83, DOI 10.1007/978-3-642-22061-6_3, © Springer-Verlag Berlin Heidelberg 2012

goals of atomistic-to-continuum coupling methods; and a theoretical analysis of the accuracy of atomistic-to-coupling methods *up to the onset of lattice instability* of the atomistic energy has thus far only been achieved for one-dimensional model problems [5,11,12,14,34,35]. This theoretical analysis and corresponding numerical experiments have given a precise understanding of the varying accuracy of these methods for simple model problems, but it is not known to what degree these errors are significant for the multi-dimensional problems of scientific and technological interest.

We have developed a benchmark test code to study the atomistic-to-continuum coupling error for a face-centered cubic (FCC) crystal. Following the benchmark investigation of Miller and Tadmor [31], the displacement $U(x_1, x_2, x_3)$ of the atoms from their reference lattice positions is constrained by the atomistic analogue of continuum "plane strain" symmetry:

$$U = (U_1, U_2, 0) \quad \text{and} \quad U(x_1, x_2, x_3) = u(x_1, x_2)$$

where the crystal coordinates are given in terms of the basis vectors defined using Miller indices[1] by $e_1 = [1\,1\,0]$, $e_2 = [0\,0\,1]$, and $e_3 = [1\,\overline{1}\,0]$. We note that the computation of the atomistic energy and forces for an atomistic displacement with plane strain symmetry requires the summation of the interaction of each atom with neighboring atoms in three-dimensional space.

We have recomputed the benchmark study of Miller and Tadmor [31] for the Lomer dislocation dipole to more clearly separate the errors due to continuum modeling error, atomistic-to-continuum coupling error, and solution error. We allow a more general atomistic domain that is fully surrounded by a continuum region. We have also investigated plastic slip for tensile loading. Further investigation is underway to study the coarsening error in the continuum region and the far-field boundary condition error.

We have found that the *patch-test consistent* quasi-nonlocal (QNL) method [40] gave a more accurate critical shear strain than the popular "ghost force correction" quasicontinuum coupling method [30, 38] tested in [31] (see Table 1 and Fig. 7). This result confirms our earlier theoretical analysis of the dependence of the critical shear strain error on the accuracy of the atomistic-to-continuum coupling [14].

The motivation for the ghost force correction method (equation (12) below) was to correct the large coupling errors of the original quasicontinuum method (QCE) [41] by applying a "dead load" correction [30, 31, 38]. We have recast the ghost force correction method in a numerical analysis setting as an iterative method to solve the equilibrium equations for the force-based quasicontinuum (QCF) method [8,9], and we have proven for a one-dimensional model problem that the ghost force correction (QCE-QCF) method is inaccurate near lattice instabilities because it uses the *patch-test inconsistent* QCE method as a preconditioner for the QCF equilibrium equations [15]. The benchmark tests we present here confirm that the QCE-QCF method is not as accurate as the QNL method near lattice instabilities.

By contrast, we found that the $w^{1,\infty}$ error was smaller for the QNL method than for the QCE-QCF method far from lattice instabilities (see Figs. 7–9), which is contrary to expectations based on our analyses of a 1D model in [11, 13, 15, 34], where we have shown that the QCF method (the limit of the QCE-QCF iteration) is a more accurate approximation than the QNL method.

Another result of our numerical experiments that contradicts our 1D analysis [15] is that near a slip instability the QCE-QCF method has comparable accuracy to the QNL-QCF method, which uses the QNL energy as a preconditioner. To explain this, we note that our 1D analysis in [15] can be considered a good model for cleavage fracture, but not for the slip instabilities studied in the present paper. We are currently attempting to develop a 2D benchmark test for cleavage fracture to demonstrate that the QCE-QCF method may be less accurate than the QNL-QCF method near some types of lattice instabilities.

2 The Atomistic and Quasicontinuum Models

2.1 A Model for Plane Strain in the Face Centered Cubic Lattice

Let $\mathscr{L}$ be a face centered cubic (fcc) lattice [1] with cube side length a and nearest-neighbor distance $a/\sqrt{2}$. Our plane strain model is most easily derived by viewing the fcc lattice $\mathscr{L}$ as being generated by the primitive lattice vectors

$$a_1 = \frac{a}{2}(0,\ 1,\ 1), \quad a_2 = \frac{a}{2}(1,\ 0,\ 1), \quad \text{and} \quad a_3 = \frac{a}{2}(1,\ 1,\ 0). \tag{1}$$

The cubic supercells are then generated by the cubic lattice vectors

$$A_1 = a(1,\ 0,\ 0) = -a_1 + a_2 + a_3, \quad A_2 = a(0,\ 1,\ 0) = a_1 - a_2 + a_3,$$
$$A_3 = a(0,\ 0,\ 1) = a_1 + a_2 - a_3. \tag{2}$$

We let P be the orthogonal projection of $\mathscr{L}$ onto the plane with normal given by $a_1 - a_2 = \frac{a}{2}(-1,\ 1,\ 0)$ (or the $(1\bar{1}0)$ plane using Miller indices since we have that $A_1 - A_2 = -2(a_1 - a_2)$), and let $Q = I - P$ be the projection onto the line parallel to $a_1 - a_2 = \frac{a}{2}(-1,\ 1,\ 0)$ (or the $[1\bar{1}0]$ direction using Miller indices [1]). We observe that the projection of $\mathscr{L}$ onto the plane normal to $a_1 - a_2 = \frac{a}{2}(-1,\ 1,\ 0)$ (or the $(1\bar{1}0)$ plane using Miller indices) is a triangular lattice $\mathscr{T} := P\mathscr{L}$. Each point in the triangular lattice $\mathscr{T} := P\mathscr{L}$ is thus the projection of a column of atoms with spacing $a/\sqrt{2}$ parallel to $a_1 - a_2 = \frac{a}{2}(-1,\ 1,\ 0)$ (or the $[1\bar{1}0]$ direction using Miller indices) as depicted in Fig. 1.

Now let ω be a finite subset of the triangular lattice $\mathscr{T}$, and let Ω be the right cylinder over ω in the fcc lattice $\mathscr{L}$. We will call a displacement $U : \Omega \to \mathbb{R}^3$ periodic if

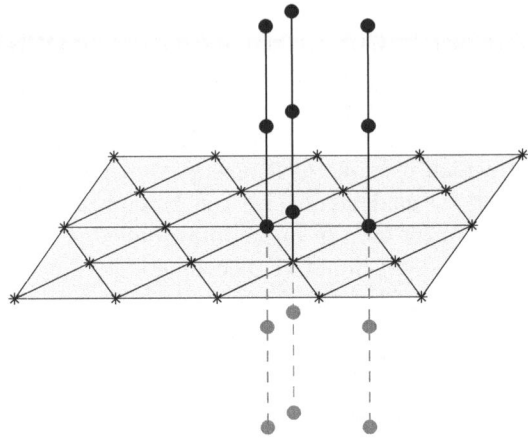

Fig. 1 Illustration of the FCC lattice and the projection onto the $(1\,\overline{1}\,0)$ plane. The grey shaded region represents the $(1\,\overline{1}\,0)$ plane, the asterisks are the points in the triangular lattice $\mathscr{T}$, and the disks are the atoms in the FCC lattice $\mathscr{L}$. The vertical lines emanating from the points in $\mathscr{T}$ represent the columns of atoms in the FCC lattice which project onto points of $\mathscr{T}$. The reader should imagine that there is a vertical column of atoms emanating from each point of $\mathscr{T}$

$$U(X) = U\left(X + n(a_1 - a_2)\right) \quad \text{for all } X \in \Omega \text{ and all } n \in \mathbb{Z}.$$

We note that the scale of the periodicity is $|a_1 - a_2| = a/\sqrt{2}$. Any periodic displacement U reduces to a displacement $u : \omega \to \mathbb{R}^3$ defined by

$$u(x) := U(X) \quad \text{for any } X \in \Omega \text{ with } PX = x.$$

We let $\mathscr{U}$ denote the space of displacements of ω.

We will now derive an *energy per period* $\mathscr{E}^a : \mathscr{U} \to \mathbb{R}$ and a corresponding force field on ω. Let ϕ be a pair potential. For example, we use the Morse potential in our numerical experiments below. We will adopt a fourth-nearest neighbor pair-interaction model. For this choice of the interaction model the quasi-nonlocal coupling method [40] can be applied. We note that, throughout, we understand nth nearest neighbours in terms of Euclidean distance in the reference configuration in $\mathbb{R}^3$ (i.e., in the 3D FCC lattice) [40].

Let $\mathscr{M}$ denote the $\mathscr{L}$-*interaction range*, by which we mean the set of all vectors pointing from an atom in $\mathscr{L}$ to one of its first, second, third, or fourth nearest-neighbors. Representative first, second, third, or fourth nearest-neighbor vectors are given in terms of the primitive lattice vectors (1) and cubic lattice vectors (2) by

$$a_1 = \left(0, \frac{a}{2}, \frac{a}{2}\right), \quad A_1 = (a, 0, 0), \quad a_2 + a_3 = \left(a, \frac{a}{2}, \frac{a}{2}\right), \quad 2a_1 = (0, a, a),$$
$$\tag{3}$$

with the full set of first, second, third, or fourth nearest-neighbor vectors given by symmetry.

For $X \in \Omega$, we let

$$\mathscr{M}_X := \{B \in \mathscr{M} : X + B \in \Omega\}.$$

We will often call directions $B \in \mathscr{M}_X$ bonds. We note that $\mathscr{M} \neq \mathscr{M}_X$ when X lies in a shallow layer along the boundary of Ω. We also note that we have $\mathscr{M}_{X_1} = \mathscr{M}_{X_2}$ if $PX_1 = PX_2 = x$, hence we can unambiguously use the notation $\mathscr{M}_x := \mathscr{M}_X$ where $PX = x$ for $x \in \omega$.

For any $B \in \mathscr{M}$ and any displacement $U(X)$, we define the difference operator ∂_B by

$$\partial_B U(X) := U(X + B) - U(X),$$

whenever $X,\ X + B \in \Omega$. We define the set of projections $b = PB$ of the bonds in $\mathscr{M}$ and $\mathscr{M}_X$ onto the plane $(1\,\bar{1}\,0)$ by

$$\mathscr{R} := P\mathscr{M} \qquad \text{and} \qquad \mathscr{R}_x := P\mathscr{M}_X.$$

For periodic displacements $U(X)$, we define

$$\partial_b u(x) := u(x + b) - u(x) = U(X + B) - U(X) = \partial_B U(X)$$

where $u(x) = U(X)$ for $x = PX$ and $b = PB$. It will also be convenient to use the notation

$$\partial_B u(x) := \partial_b u(x) = \partial_B U(X) \quad \text{for } PB = b.$$

The *energy per period* $\mathscr{E}^a : \mathscr{U} \to \mathbb{R}$ is given by

$$\mathscr{E}^a(u) := \tfrac{1}{2} \sum_{x \in \omega} \sum_{B \in \mathscr{M}_x} \phi(|B + \partial_B u(x)|)$$

where we note that the difference in the deformed positions of atoms at $X + B$ and X is $B + \partial_B u(x)$. The corresponding force at $x \in \omega$ is then given by

$$\mathscr{F}^a(u)(x) := -\frac{\partial}{\partial u(x)} \mathscr{E}^a(u)(x) = -\frac{\partial}{\partial u(x)} \sum_{B \in \mathscr{M}_x} \phi(|B + \partial_B u(x)|), \qquad (4)$$

where $\frac{\partial}{\partial u(x)}$ above denotes the partial derivative with respect to the displacement $u(x)$ of the atom with projected reference position $x \in \omega$. We define the first variation (or, gradient) of $\mathscr{E}^a(u)$ by $\delta \mathscr{E}^a(u)$, which is given for each $x \in \omega$ by

$$\delta \mathscr{E}^a(u)(x) := \frac{\partial}{\partial u(x)} \mathscr{E}^a(u)(x). \qquad (5)$$

We can then express (4) in operator notation by

$$\mathscr{F}^a(u) = -\delta \mathscr{E}^a(u).$$

We will now formulate a boundary value problem for the energy $\mathscr{E}^a$. We let $\Gamma \subset \omega$ denote the part of the "boundary" where the displacements are constrained. In the numerical experiments described below, Γ will be a shallow layer of atoms along all or some of the boundary of ω. Let the space of admissible displacements that are equal to $u_0 \in \mathscr{U}$ on the boundary Γ be denoted by

$$\text{Adm}(u_0) := \{u \in \mathscr{U} : u(x) = u_0(x) \text{ for all } x \in \Gamma\}.$$

We will study the minimization problem (or more precisely, the local minimization problem):

$$\text{Find } u \in \arg \text{ local min } \mathscr{E}^a(v). \tag{6}$$
$$\scriptstyle v \in \text{Adm}(u_0)$$

The Euler–Lagrange equation corresponding to problem (6) is

$$-\delta \mathscr{E}^a(u)(x) = 0 \qquad \text{for all } x \in \omega \setminus \Gamma,$$
$$u(x) = u_0(x) \qquad \text{for all } x \in \Gamma.$$

In our numerical experiments, we consider only initial displacements u_0 satisfying $u_0 : \omega \to (1\,\bar{1}\,0)$, where we recall that $(1\,\bar{1}\,0)$ is the plane with normal given by $a_1 - a_2 = \frac{a}{2}(-1, 1, 0)$. We will call any displacement $u : \omega \to (1\,\bar{1}\,0)$ a *plane displacement*, and we will let $\mathscr{V} \subset \mathscr{U}$ denote the space of plane displacements. For plane displacements, $\mathscr{E}^a$ can be interpreted as the energy of the two dimensional crystal ω computed using a pair potential which depends on the bond direction.

The restriction of the energy per period $\mathscr{E}^a(u) : \mathscr{U} \to \mathbb{R}$ to plane displacements $\mathscr{V}$ can then be given by

$$\mathscr{E}^a(u) = \frac{1}{2} \sum_{x \in \omega} \sum_{b \in \mathscr{R}_x} \phi_b(|b + \partial_b u(x)|), \quad u \in \mathscr{V}, \tag{7}$$

where $\phi_b : (0, \infty) \to \mathbb{R}$ is given by

$$\phi_b(r) := \sum_{\{B \in \mathscr{M}_x : PB = b\}} \phi\left(\left(r^2 + |QB|^2\right)^{1/2}\right)$$

where we recall that $Q = I - P$ is the projection onto the line parallel to $a_1 - a_2 = \frac{a}{2}(-1, 1, 0)$. Roughly speaking, $\phi_b(|b + \partial_b u(x)|)$ is the energy of the interaction of atom x with all of its neighbors in the column over $x + b$. For $x \in \omega$, we will call the inner sum

$$\mathcal{E}_x^a(u) := \tfrac{1}{2} \sum_{b \in \mathcal{R}_x} \phi_b(|b + \partial_b u(x)|)$$

the *atomistic energy* at x.

It will be convenient to establish coordinates adapted to the triangular lattice $\mathcal{T}$, which lies in the plane defined by the normal $a_1 - a_2 = \tfrac{a}{2}(-1, 1, 0)$ (or the $(1\,\overline{1}\,0)$ plane). This plane is spanned by the orthogonal vectors of length $a/\sqrt{2}$ in the $[1\,1\,0]$ direction and of length a in the $[0\,0\,1]$ direction given by

$$V_1 := \tfrac{1}{2}(A_1 + A_2) = a_3 = \tfrac{a}{2}(1, 1, 0),$$
$$V_2 := A_3 = a_1 + a_2 - a_3 = a(0, 0, 1).$$

Throughout the remainder of this paper, we will denote the coordinates of (x_1, x_2) in the $\{V_1, V_2\}$ basis by

$$\langle v_1, v_2 \rangle := v_1 V_1 + v_2 V_2.$$

The triangular lattice $\mathcal{T}$ is then generated by the basis vectors $\langle 1, 0 \rangle$ and $\langle 1/2, 1/2 \rangle$.

It can be checked that there are four distinct symmetry-related projected interaction potentials, $\phi_b(r)$, corresponding to interactions with bonds B such that

$b_1 = \langle 1, 0 \rangle \qquad PB = b_1 \text{ and } QB = n(a_1 - a_2) \text{ for } n = -1, 0, 1,$

$b_2 = \langle 2, 0 \rangle \qquad PB = b_2 \text{ and } QB = 0,$

$b_4 = \langle 1/2, 1/2 \rangle \quad PB = b_4 \text{ and } QB = n(a_1 - a_2) \text{ for } n = -3/2, -1/2,$

$$1/2, 3/2,$$

$b_5 = \langle 3/2, 1/2 \rangle \quad PB = b_5 \text{ and } QB = n(a_1 - a_2) \text{ for } n = -1/2, 1/2.$

where the planar bonds b_i are displayed in Fig. 2. The remaining interaction potentials, $\phi_b(r)$ for $b \in \mathcal{R}$, can be obtained by symmetry from $\phi_{b_1}(r) = \phi_{b_6}(r) = \phi_{b_7}(r)$, $\phi_{b_2}(r)$, $\phi_{b_4}(r)$, and $\phi_{b_5}(r)$. We note that the lattice constant a is contained in the definition given in the previous paragraph for the coordinates $\langle v_1, v_2 \rangle$.

Now observe that for $u \in \mathcal{V}$ we have $\delta\mathcal{E}^a(u) \in \mathcal{V}$ where the gradient $\delta\mathcal{E}^a(u)$ is defined by (5). Therefore, if a gradient-based minimization algorithm (such as the steepest descent method or the nonlinear conjugate gradient algorithm discussed below) is started from a plane displacement, it will terminate at a plane displacement. Thus, since we will only consider initial configurations which are plane displacements, we will be able to use the restriction of $\mathcal{E}^a$ to $\mathcal{V} \cap \text{Adm}(u_0)$ in our numerical experiments. This is important since $\mathcal{E}^a$ restricted to $\mathcal{V} \cap \text{Adm}(u_0)$ is the energy of a two dimensional crystal, and in that case the patch-test consistent atomistic to continuum coupling method developed in [37] can be used.

2.2 Energy-Based Quasicontinuum Approximations

We will construct a local approximation to the energy $\mathscr{E}^a$. Our methods follow [40]. First, we define an extrapolated difference operator $\bar{D}$. The difference operator $\bar{D}$ will approximate the vector pointing from an atom to one of its neighbors using only the vectors pointing from an atom to its nearest neighbors. The *nearest neighbors* are the atoms depicted in Fig. 2. Recall that $\mathscr{R} := P\mathscr{M}$; we thus have $b_i \in \mathscr{R}$ for the vectors b_i, $i = 1, \ldots, 7$, depicted in Fig. 2, and moreover, $\mathscr{R}$ is obtained by applying the lattice symmetries of $\mathscr{T}$ to $b_1, \ldots, b_7$. For these vectors, we define the operators

$$\bar{D}_{b_i} := \partial_{b_i} \text{ for } i \in \{1, 3, 4\},$$

$$\bar{D}_{b_2} := 2\partial_{b_1},$$

$$\bar{D}_{b_5} := \partial_{b_1} + \partial_{b_4},$$

$$\bar{D}_{b_6} := \partial_{b_3} + \partial_{b_5},$$

$$\bar{D}_{b_7} := 2\partial_{b_4},$$

and extend the definition by symmetries of $\mathscr{T}$ so that $\bar{D}_b$ is defined for all $b \in \mathscr{R}$.

Using the difference operator $\bar{D}$, we define the *Cauchy–Born energy* by

$$\mathscr{E}^{cb}(u) := \frac{1}{2} \sum_{x \in \omega} \sum_{b \in \mathscr{R}_x} \phi_b(|b + \bar{D}_b u(x)|).$$

We will call the inner sum

$$\mathscr{E}^{cb}_x(u) := \frac{1}{2} \sum_{b \in \mathscr{R}_x} \phi_b(|b + \bar{D}_b u(x)|)$$

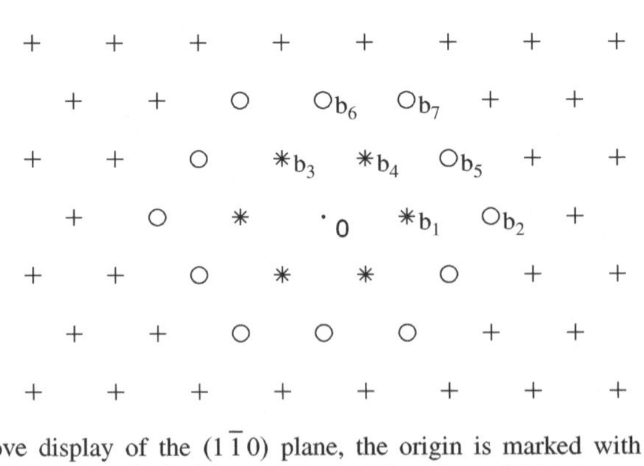

Fig. 2 In the above display of the $(1\bar{1}0)$ plane, the origin is marked with a dot, the nearest neighbors of the origin are marked with asterisks, and the other neighbors are marked with circles. The bond vector b_i is the vector pointing from the origin to the atom marked b_i

the *continuum energy* at x, so

$$\mathscr{E}^{cb}(u) = \sum_{x \in \omega} \mathscr{E}^{cb}_x(u).$$

The following remark gives a more detailed motivation for these definitions.

Remark 1 (The Cauchy–Born Approximation). We call $\mathscr{E}^{cb}(u)$ the Cauchy–Born energy since, if the displacement u is interpreted as a piecewise linear spline with respect to the canonical triangulation of $\mathscr{T}$, it can be rewritten as an integral over a stored energy density.

To make this precise, we observe that the continuum energy at $x \in \omega$ such that $x + b \in \omega$ for all $b \in \mathscr{R}$ is given for uniform displacements $u^F(x) := Fx - x$ where $F \in \mathbb{R}^{2 \times 2}$ by

$$\mathscr{E}^{cb}_x(u^F) = \tfrac{1}{2} \sum_{b \in \mathscr{R}} \phi_b(|Fb|).$$

We thus define the *Cauchy–Born stored energy density,* $W_{cb} : \mathbb{R}^{2 \times 2} \to \mathbb{R} \cup \{+\infty\}$, by

$$W_{cb}(F) = \frac{\tfrac{1}{2} \sum_{b \in \mathscr{R}} \phi_b(|Fb|)}{v},$$

where v is the area associated with each atom (in the 2D lattice). Moreover, for $u \in \mathscr{U}$, let $\bar{u}$ denote the piecewise affine interpolant of u with respect to the canonical triangulation of $\mathscr{T}$, let $\nabla \bar{u}$ denote the resulting displacement gradient, and let $\bar{\omega}$ denote the union of those elements.

For periodic boundary conditions, or by modifying the functional $\mathscr{E}^{cb}(u)$ at the boundary Γ of the domain ω, one can use Shapeev's bond density lemma [37] to prove that

$$\mathscr{E}^{cb}(u) = \int_{\bar{\omega}} W_{cb}(\nabla \bar{u}) \, dx.$$

We note that this modification of the functional $\mathscr{E}^{cb}(u)$ at the boundary does not affect the minimization of quasicontinuum energies whose set of admissible displacements, Adm(u_0), are constrained in the entire boundary of ω, such as for the Lomer Dislocation Dipole problem that we study. More detailed discussions and analyses of the Cauchy–Born approximation can be found in [4, 17, 18].

Now let $\omega := \mathscr{A} \cup \mathscr{C}$ be a partition of ω into an *atomistic region* $\mathscr{A}$ and a *continuum region* $\mathscr{C}$. We define the *energy based quasicontinuum* (QCE) energy by

$$\mathscr{E}^{qce}(u) := \sum_{x \in \mathscr{A}} \mathscr{E}^a_x(u) + \sum_{x \in \mathscr{C}} \mathscr{E}^{cb}_x(u). \tag{8}$$

Following [40], we also define the *quasi-nonlocal* (QNL) energy by

$$\mathscr{E}^{qnl}(u) := \frac{1}{2} \sum_{x \in \omega} \sum_{b \in \mathscr{R}_x} \chi_b(x) \phi_b(|b + \bar{D}_b u(x)|) + (1 - \chi_b(x)) \phi_b(|b + \partial_b u(x)|), \quad (9)$$

where

$$\chi_b(x) := \begin{cases} 1 & \text{if } x \in \mathscr{C} \text{ and } x + b \in \mathscr{C}, \\ 0 & \text{otherwise.} \end{cases}$$

The reason for the introduction of the QNL method was that the QCE method did not pass the *patch test* [11, 38], that is, the coupling mechanism defined in (8) results in non-zero forces (the "ghost forces") at the atomistic/continuum interface under homogeneous displacements (or, deformations). By contrast, it was shown in [40] that the QNL energy (9) does pass the patch test [11, 14, 40], as long as two atoms interact only when they share a common nearest neighbor. The authors of [40] consider only the case where the pair potential does not depend on the bond direction. Nonetheless, the argument which they give may be applied to show that the QNL energy (9) passes the patch test also in the present case.

The importance of passing the patch test lies in the fact that the "ghost forces" can be understood as a consistency error, which results in an $O(1)$ relative error in the displacement gradient [11]. Moreover, it was shown in [14], that they can result in a $O(1)$ relative error in the prediction of critical loads at which lattice instabilities occur.

To achieve an efficient quasicontinuum method, the positions of atoms in the continuum region are normally further constrained by piecewise linear inter-polation [30]. The development of an implementable and patch-test consistent quasicontinuum energy that allows coarsening by piecewise linear interpolation in the continuum region has been achieved so far only for two-dimensional problems with pair potential interactions [37]. The development of patch-test consistent quasicontinuum energies for many-body potentials such as the popular embedded atom method (EAM) or for three-dimensional problems is an open problem.

2.3   The Force Based Quasicontinuum Approximation

The force based quasicontinuum (QCF) approximation gives a patch-test consistent approximation for atomistic-to-continuum coupling even with coarsening in the continuum region [6, 8]. Force-based multiphysics coupling methods are generally popular because of their algorithmic simplicity. However, force-based coupling methods are known to give non-conservative force fields [6, 8], and our recent research described at the end of this section has discovered additional stability problems for QCF approximations and solution methods [15].

Instead of approximating the energy $\mathscr{E}^a$ by a local energy, we can approximate the forces $\mathscr{F}(u)$ directly. This leads to the *force based quasicontinuum* (QCF) approximation. As above, let $\omega := \mathscr{A} \cup \mathscr{C}$ be a partition of ω into an atomistic region $\mathscr{A}$ and a continuum region $\mathscr{C}$. Define $\mathscr{F}^{qcf} : \mathscr{V} \to \mathscr{V}$ by

$$\mathscr{F}^{qcf}(u)(x) := \begin{cases} -\delta\mathscr{E}^a(u)(x) & \text{if } x \in \mathscr{A}, \\ -\delta\mathscr{E}^{cb}(u)(x) & \text{if } x \in \mathscr{C}. \end{cases} \tag{10}$$

We consider the boundary value problem

$$\mathscr{F}^{qcf}(u)(x) = 0 \qquad \text{for all } x \in \omega \setminus \Gamma,$$
$$u(x) = u_0(x) \quad \text{for all } x \in \Gamma. \tag{11}$$

It was shown in [12, 13, 29], for one-dimensional model problems, that the solution of problem (11) approximates the solution of problem (6). It is also true that the QCF method passes the patch test [8].

In our numerical experiments we will solve (11) using an iterative method defined in [8, 9]. Let $u_0 \in \mathscr{V}$ be given, and let $\mathscr{E}^{qc}$ be either $\mathscr{E}^{qnl}$ or $\mathscr{E}^{qce}$. Then we let $u_{n+1} \in \text{Adm}(u_0) \cap \mathscr{V}$ be the solution of the problem:

$$u_{n+1} \in \arg\underset{v \in \text{Adm}(u_0) \cap \mathscr{V}}{\text{local min}} \left\{ \mathscr{E}^{qc}(v) - \sum_{x \in \omega} \left(\mathscr{F}^{qcf}(u_n)(x) + \delta\mathscr{E}^{qc}(u_n)(x) \right) \cdot v(x) \right\}. \tag{12}$$

The method (12) with $\mathscr{E}^{qc} = \mathscr{E}^{qce}$ is the popular *ghost force correction* method proposed in [30]. We have proposed the method $\mathscr{E}^{qc} = \mathscr{E}^{qnl}$ as a *ghost force correction* method with improved stability properties [15]. For $\mathscr{E}^{qc} = \mathscr{E}^{qce}$, we call the iteration (12) the QCF-QCE iteration; for $\mathscr{E}^{qc} = \mathscr{E}^{qnl}$, we call it the QCF-QNL iteration.

The Euler–Lagrange equation for problem (12) is

$$\delta\mathscr{E}^{qc}(u_{n+1})(x) = \mathscr{F}^{qcf}(u_n)(x) + \delta\mathscr{E}^{qc}(u_n)(x) \quad \text{for all } x \in \omega \setminus \Gamma,$$
$$u_{n+1}(x) = u_0(x) \qquad\qquad\qquad \text{for all } x \in \Gamma.$$

It is easy to see that, if the sequence of iterates converges, then its limit must be a solution of problem (11). Moreover, under certain technical conditions related to the stability of the preconditioner $\mathscr{E}^{qc}$, it can be shown that the sequence indeed converges [8, 9, 15].

A QCF solution method that can theoretically give inaccurate results is the use of a nonlinear conjugate gradient solver for the force-based quasicontinuum equations [6, 31, 39]. We have proven that the linearization of the force-based equilibrium equations is not positive definite [13], which implies that the conjugate gradient solution of this problem is unstable [15]. We have discovered from informal discussions with computational physicists that their conjugate gradient iterative solution of force-based multiphysics coupling methods sometimes oscillates rather than converges, a phenomenon partially explained by our theoretical analysis of this

instability. We are developing benchmark tests to further study the reliability of the
conjugate gradient solution of the force-based quasicontinuum equations.

We have shown theoretically and computationally for one-dimensional model
problems that the GMRES method is a reliable and efficient solver for the force-
based quasicontinuum equations [15]. We are thus also preparing multi-dimensional
benchmark tests to evaluate the reliability and efficiency of the GMRES solution of
the force-based quasicontinuum equations.

3 Numerical Experiments

3.1 The Lomer Dislocation Dipole: Analytical Model

Following the numerical experiments presented in [31], we consider a dipole of
Lomer dislocations, as depicted in Fig. 3. The Lomer dislocation has Burgers vector
$b = [1\,1\,0]$ in the $(0\,0\,1)$ plane [20]. The dipole should be a stable equilibrium if
the distance between the cores is large enough so that the Peach-Koehler elastic
attraction of the dislocations is weaker than the critical force needed to overcome
the Peierls energy barrier [31].

We will study the stability of the dipole when a shear is applied to the crystal. By
a shear, we mean a homogeneous deformation of the lattice $\mathscr{T}$ that takes the form

$$x \mapsto \sigma(\gamma)x \quad \text{where} \quad \sigma(\gamma) := \begin{pmatrix} 1 & \gamma \\ 0 & 1 \end{pmatrix}.$$

We call γ the *shear strain*. We say that the shear is *positive* if $\gamma > 0$ and *negative* if
$\gamma < 0$. More generally, we will apply a shear to the deformation $x + u(x)$ given by
the displacement $u(x)$ to obtain the sheared deformation

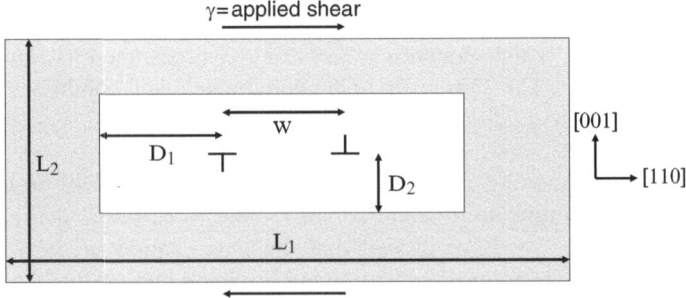

Fig. 3 The atomistic and continuum (*yellow*) domains for the Lomer dislocation dipole. The size
of the total computational domain is $L_1 = 75 \times a/\sqrt{2}$ and $L_2 = 60 \times a$ where a is the length of
the sides of the cubic supercell. The atomistic domain is given by $D_1 = k \times a/\sqrt{2}$ and $D_2 = k \times a$
for $k = 3, 4, 5$. The separation of the dipole used in the initial guess u_{elas} is $w = 10 \times a/\sqrt{2}$

$$x + u(x) \mapsto \sigma(\gamma) \, (x + u(x)) \, .$$

A positive shear applied to the top and bottom boundaries of the crystal changes the energy landscape to favor the movement of the dislocations apart. As γ increases, the shear force overwhelms both the Peierls force and the Peach-Koehler elastic attraction. Thus, the original equilibrium configuration of the dipole becomes unstable, and the dislocations move away from each other. We call the value of γ at which this instability occurs the *critical strain*. In our numerical experiments, we will simulate the process of slowly shearing the crystal until the critical strain is reached.

To understand the movement of the Lomer dipole theoretically, we model the local minima of the atomistic energy for displacements constrained on the boundary Γ to be $\sigma(\gamma)x - x$ and constrained to be in the energy well of a Lomer dipole with separation w:

$$\inf_{v \in \widetilde{\mathrm{Adm}}(\sigma(\gamma)x - x, w)} \mathscr{E}^a(v) \approx \tilde{\mathscr{E}}^a(w, \gamma) = \mathscr{E}^a_{\mathrm{misfit}}(w) + \mathscr{E}^a_{\mathrm{dipole\ attraction}}(w)$$

$$+ \, \mathscr{E}^a_{\mathrm{applied\ shear}}(w, \gamma) + \mathscr{E}^a_{\mathrm{boundary\ effect}}(w, \gamma)$$

where $\widetilde{\mathrm{Adm}}\,(\sigma(\gamma)x - x, w)$ is the set of admissible displacements roughly described above. Here the classical Peierls-Nabarro misfit energy [20], the classical Peach-Koehler interaction energy [20], and our modeling of the effect of the applied shear are given for $w \geq 0$ and $\gamma \geq 0$ by

$$\mathscr{E}^a_{\mathrm{misfit}}(w) = \frac{\mu b}{2\pi(1 - v)} \cos \frac{2\pi w}{b} \exp \left[-\frac{\pi d}{(1 - v)b} \right],$$

$$\mathscr{E}^a_{\mathrm{dipole\ attraction}}(w) = \frac{\mu b^2}{2\pi} \ln \frac{w}{L}, \tag{13}$$

$$\mathscr{E}^a_{\mathrm{applied\ shear}}(w, \gamma) = \beta_1 w + \beta_2 \gamma - \frac{\beta_3 \gamma \, w}{L},$$

where μ is a characteristic shear modulus, v is a characteristic Poisson's ratio, L is a characteristic length, $b = a/\sqrt{2}$ is the magnitude of the Burger's vector, $d = a/\sqrt{2}$ is the interatomic spacing between the $(1\,\bar{1}\,0)$ planes, and β_1, β_2 and β_3 are constants with $\beta_3 > 0$. Although our model for the misfit energy $\mathscr{E}^a_{\mathrm{misfit}}(w)$ and the dipole attraction $\mathscr{E}^a_{\mathrm{dipole\ attraction}}(w)$ are well-known approximations of the atomistic energy $\mathscr{E}^a(v)$ [20], our bilinear model for the shear energy $\mathscr{E}^a_{\mathrm{shear}}(w, \gamma)$ is not derived from the atomistic energy and our justification is based only on the qualitative behavior it predicts for the stability of the Lomer dipole.

After scaling the atomistic energy $\mathscr{E}^a(v)$, the dipole separation w, and the shear strain γ, and after neglecting the boundary effect $\mathscr{E}^a_{\mathrm{boundary\ effect}}(w, \gamma) = 0$, we can obtain from (13) the model

$$\tilde{\mathscr{E}}^a(w, \gamma) = \cos w + \beta \ln w - \gamma w \tag{14}$$

for some $\beta > 0$. We chose $\beta = 12$ in our numerical experiments. We then find that
the force conjugate to the dipole separation w is given by

$$-\frac{\partial \tilde{\mathscr{E}}^a(w, \gamma)}{\partial w} = \sin w - \frac{\beta}{w} + \gamma. \tag{15}$$

We can see from the force field displayed in Fig. 4 that the dipole separation w
becomes unstable at a critical shear strain γ and will tend to infinity under gradient
flow dynamics for the force field (15) (for example, consider the stability of the
equilibrium solution branch starting at $w = 1.5$ as γ is increased). The Lomer
dislocations similarly separated to the boundary at a critical shear strain γ in our
numerical experiments (see Fig. 5).

3.2 The Lomer Dislocation Dipole: Numerical Experiments

In out numerical experiments we use the Morse interatomic potential

$$\phi(r) := (1 - \exp(\alpha(r - 1)))^2 \quad \text{for } r > 0,$$

with

$$\alpha := 4.4.$$

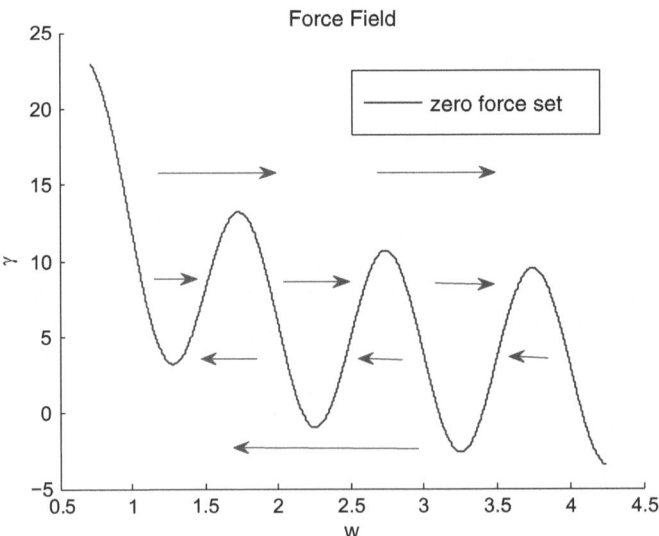

Fig. 4 The force field (15) for the Lomer dipole model (14) with $\beta = 12$

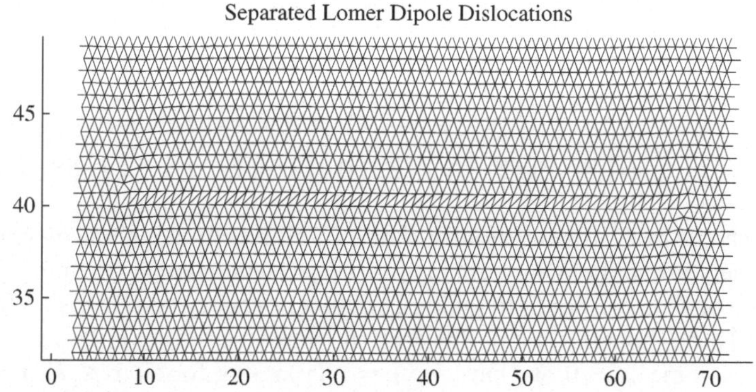

Separated Lomer Dipole Dislocations

Fig. 5 Separation of the Lomer dislocations to the boundary after continuation past a critical strain to $\gamma = 0.070$. Cartensian coordinates are used to label the axes. The top and bottom of the deformation of the reference domain ω are not displayed, but the entire width of the deformed reference domain is displayed. The triangular mesh was constructed from the reference lattice

We remark that we did not choose $\alpha = 4.4$ in order to model some specific material. Rather, we chose this value of α as small as possible while retaining a stable dipole. For smaller values of α we were unable to compute a stable dipole, whereas, if α is chosen too large then the atomistic, Cauchy–Born, and quasicontinuum models are almost identical since the first neighbor interactions, which would become dominant, are treated identically in all models.

We chose the lattice spacing a so that the FCC lattice $\mathscr{L}$ is a ground state for the atomistic energy defined by the Morse potential and using a cut-off radius that includes first, second, third, and fourth-nearest neighbor interactions (determined from the reference positions). Specifically, we let a be the minimizer of

$$\psi(r) := 12\phi\left(\sqrt{\frac{1}{2}}r\right) + 6\phi(r) + 24\phi\left(\sqrt{\frac{3}{2}}r\right) + 12\phi\left(\sqrt{2}r\right),$$

which is the energy per atom in the undeformed lattice with lattice spacing r since each atom in the $\mathscr{L}$ lattice has 12 nearest neighbors, 6 next-nearest neighbors, 24 third-nearest neighbors, and 12 fourth-nearest neighbors (see (3)). The minimum is attained when $a = 1.3338$.

We let ω be the rectangular domain consisting of 150 columns of atoms parallel to V_2 with each column containing 60 atoms defined in the $\langle v_1, v_2 \rangle$ coordinates by

$$\omega = \{\langle k, l \rangle : 0 \le k \le 74, \ 0 \le l \le 59\}$$
$$\cup \{\langle k + 1/2, l + 1/2 \rangle : 0 \le k \le 74, \ 0 \le l \le 59\}.$$

We take the boundary Γ of ω to be a layer of atoms four rows deep around the edges of ω, so that

$$\Gamma = \{\langle k, l \rangle : k = 0, 1, 73, 74, 0 \le l \le 59\}$$

$$\cup \{\langle k + 1/2, l + 1/2 \rangle : k = 0, 1, 73, 74, 0 \le l \le 59\}$$

$$\cup \{\langle k + 1/2, l + 1/2 \rangle : 0 \le k \le 74, l = 0, 1, 58, 59\}$$

$$\cup \{\langle k + 1/2, l + 1/2 \rangle : 0 \le k \le 74, l = 0, 1, 58, 59\}.$$

With this choice of Γ, for any atom $x \in \omega \setminus \Gamma$, all fourth-neighbors of x belong to ω. Thus, there are no boundary effects for the reference lattice when $a = 1.3338$. This choice of Γ corresponds to taking Dirichlet boundary conditions on the entire boundary. Throughout the rest of this section, we will consider only displacements u so that for some $\gamma > 0$, we have $u(x) = \sigma(\gamma)x - x$ for all $x \in \Gamma$. For such a displacement, we will call the shear strain γ the *shear on the boundary*.

We will now discuss how the stable Lomer dipoles displayed in Fig. 6 are computed. We first construct an initial guess u_{elas} for the displacement field corresponding to a Lomer dipole by using the displacement fields of the isotropic, linear elastic solution for the edge dislocations [20] at the left and right pole of the dipole, respectively. More precisely, the formula for these displacement fields is

$$u_1^* = \frac{b^*}{2\pi} \left[\tan^{-1} \frac{x_2 - x_2^*}{x_1 - x_1^*} + \frac{(x_1 - x_1^*)(x_2 - x_2*)}{2(1 - v)\left((x_1 - x_1^*)^2 + (x_2 - x_2^*)^2\right)} \right],$$

$$u_2^* = -\frac{b^*}{2\pi} \left[\frac{1 - 2v}{4(1 - v)} \ln \left((x_1 - x_1^*)^2 + (x_2 - x_2^*)^2 \right) \right.$$

$$\left. + \frac{(x_1 - x_1^*)^2 - (x_2 - x_2^*)^2}{4(1 - v)\left((x_1 - x_1^*)^2 + (x_2 - x_2^*)^2\right)} \right],$$

where $* = L, R$ refers to the left or the right edge dislocation, respectively, b^* is the corresponding Burgers vector, and x_i^*, $i = 1, 2$, are the components of the position vectors [20]. In our numerical experiments, we set the initial positions of the Lomer dislocations in the reference configuration to be (in the $\langle v_1, v_2 \rangle$ coordinates)

$$\langle v_1^L, v_2^L \rangle = \left\langle 32, 30 + \frac{1}{6} \right\rangle \quad \text{and} \quad \langle v_1^R, v_2^R \rangle = \left\langle 42, 30 + \frac{1}{3} \right\rangle.$$

We note that initial dislocation positions are placed between the center row of atoms at $v_2 = 30$ and the first row above the center $v_2 = 30 + 1/2$, with $v_2^L = 30 + 1/6$ and $v_2^R = 30 + 1/3$ placed symmetrically about $v_2 = 30 + 1/4$ (see Fig. 6).

The displacements u^L, u^R, are estimates of the displacement fields for isolated edge dislocations. By superimposing them, we obtain an estimate of the elastic displacement field for the dipole (without shear):

$$u_{elas} = u^L + u^R.$$

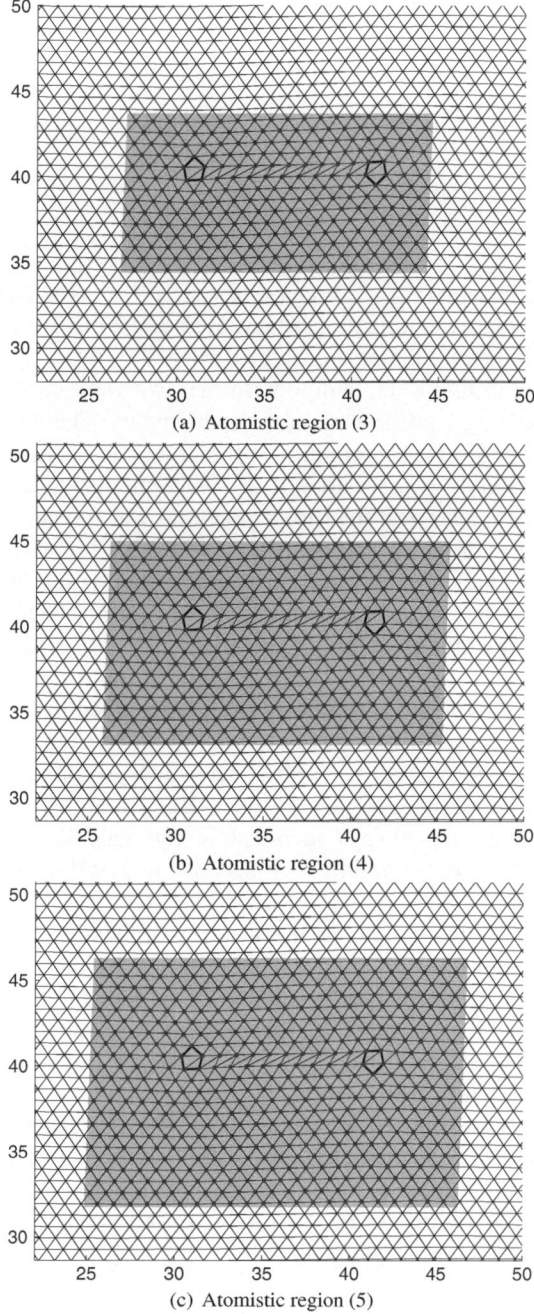

Fig. 6 The atomistic regions for the Lomer dislocation dipole. Not all of the continuum region is displayed. The mesh is the Delaunay triangulation [7] of the reference lattice (i.e., with the atoms as the nodes). The atomistic region is shaded grey. (**a**) Atomistic region (3) (**b**) Atomistic region (4) (**c**) Atomistic region (5)

Finally, we apply a shear deformation to the deformation field $x + u_{elas}$,

$$x + u_{elas}^0(x) = \sigma(\gamma)\,(x + u_{elas}(x))\,,$$

which yields an initial guess for the displacement field for the Lomer dipole under applied shear:

$$u_{elas}^0(x) = \sigma(\gamma)\,(x + u_{elas}(x)) - x.$$

We use the displacement $u_{elas}^0 = \sigma(\gamma_0)u_{elas}(x) - x$ as a starting guess for a preconditioned nonlinear conjugate gradient method (P-nCG), which is described in Sect. 4, to solve the minimization problem (6).

It is challenging to find a stable dipole. In fact, we were unable to find a stable dipole without applying a positive shear to the boundary. That is, we did not find a Lomer dipole that was a local minimum of the atomistic energy (7) subject to the boundary condition $u(x) = 0$ for all $x \in \Gamma$. We were also unable to find a stable dipole when the parameter α in the Morse potential was too small. Essentially, we found that as α decreased, the interval of shear strains for which a given dipole was stable became smaller. After some experimentation, we did find a stable Lomer dipole with $\alpha = 4.4$ and with shear on the boundary $\gamma_0 = 0.0375$.

Our atomistic and quasicontinuum models utilize a fourth-nearest neighbor cutoff calculated from the positions in the reference lattice, which is in fact the largest possible neighborhood for the QNL method applied to an FCC lattice [40]. Such a cutoff is acceptable for atomistic configurations that are "close" to the reference lattice, however, for the highly deformed positions near the dipole a cutoff in *deformed* coordinates ought to be used. Even though the dominant nearest-neighbour interactions are always included in our calculation, a more precise understanding of the error committed for second- to fourth-nearest neighbours is required.

Next we describe how we simulate the shearing of the dipole. For simplicity, we will explain how this is done for energy-based methods before we discuss the force-based method. Let u_0 be the displacement field of the stable Lomer dipole discussed above. Suppose that γ_0 is the shear on the boundary for u_0. Let γ_k be an increasing sequence of shear strains, and let $\delta\gamma_{k+1} := \gamma_{k+1} - \gamma_k$. $\mathscr{E}$ can be either the atomistic energy (7), the QCE energy (8), or the QNL energy (9).

We perform the following iteration starting with the displacement u_0. Suppose we have a dipole with displacement field u_k which solves the boundary value problem

$$-\delta\mathscr{E}(u_k)(x) = 0 \qquad\qquad \text{for all } x \in \omega \setminus \Gamma,$$
$$u_k(x) = \sigma(\gamma_k)x - x \quad \text{for all } x \in \Gamma.$$

We let

$$u_{k+1}^0(x) = \sigma(\delta\gamma_k)\,(x + u_k(x)) - x.$$

Observe that u_{k+1}^0 satisfies the boundary condition

$$u_{k+1}^0(x) = \sigma(\gamma_{k+1})x - x \quad \text{for all } x \in \Gamma.$$

We use the P-nCG algorithm to compute a new local minimizer u_{k+1} "near" u_{k+1}^0, which satisfies the new boundary condition $u_{k+1}(x) = \sigma(\gamma_{k+1})x - x$ for all $x \in \Gamma$. We call this process the *shear loading continuation*.

We terminate the shear loading continuation when we can no longer find an equilibrium $u_{k+1}(x)$ close to the initial guess $\sigma(\delta\gamma_{k+1})(x + u_k(x)) - x$. In practice, this means that we stop when the nCG algorithm returns a deformation in which the cores of the dislocations have moved apart. If the dislocations move apart at the $(k+1)$th step of the iteration, then we assume that the critical strain must be between γ_k and γ_{k+1}.

For the force based approximation (10), we use a similar iteration to find the critical strain. At each step in the iteration, we solve the boundary value problem

$$\begin{aligned}
\mathscr{F}^{qcf}(u_{k+1})(x) &= 0 && \text{for all } x \in \omega \setminus \Gamma, \\
u_{k+1}(x) &= \sigma(\gamma_{k+1})x - x && \text{for all } x \in \Gamma.
\end{aligned} \tag{16}$$

We solve (16) using the ghost force correction iteration defined in (12). We use the same initial guess

$$u_{k+1}^0(x) := \sigma(\delta\gamma_{k+1})(x + u_k(x)) - x$$

as in the energy-based case, and we solve each iteration of ghost force correction using the P-nCG algorithm.

We presented computational results for the deformation of a Lennard-Jones chain under tension in [9] that demonstrate the necessity of using a sufficiently small parameter step size to ensure that the computed solution remains in the domain of convergence of the *ghost force correction iteration* (defined below as QCE-QCF) method. These results exhibit fracture before the actual load limit if the parameter step size is too large. We thus conclude that the shear strain $\delta\gamma_{k+1}$ must be sufficiently small to ensure that our computed solution remains in the domain of convergence of the QCE-QCF method since it would otherwise predict instability for the Lomer dipole before that predicted by the QCF method.

We perform the shear loading iteration for each of the quasicontinuum methods using three different atomistic regions, as depicted in Fig. 6. The atomistic region (3) (resp., (4), (5)) is a box containing the dipole such that there are three (resp., four, five) rows of atoms between the continuum region and each of the atoms in the pentagons surrounding the cores of the dislocations, or more precisely, the atomistic region (k) is the box given in $\langle v_1, v_2 \rangle$ coordinates by $[32 - k, 43 + k] \times [30 - k, 30 + k]$ for $k = 3, 4, 5$. Throughout the remainder of this section, QNL-QCF (resp. QCE-QCF) will refer to the QCF method implemented using the ghost force correction iteration preconditioned by the QNL (resp. QCE) energy [15]. We will let

QNL-QCF(k) (resp. QCE-QCF(k), QNL(k), QCE(k)) refer to the QNL-QCF (resp. QCE-QCF, QNL, QCE) method with atomistic region (k) for k one of 3, 4, or 5.

We note that the atomistic regions studied in the benchmark tests given in [31] extended to the lateral boundaries of the computational domain ω, or, following the notation used in Fig. 3 these tests study the case $2D_1 + w = L_1$. Thus, the benchmark tests [31] do not present the capture of a Lomer dipole in an atomistic region fully surrounded by a continuuum region and do not study the effect of coupling error in the dipole plane (1 1 0).

In Table 1, we give the critical strains for each method in decreasing order. The columns γ^- and γ^+ are the shear strains at the last step before the dislocations moved apart, and at the first step for which the dislocations moved, respectively. The column labeled "Error" displays the relative error from the critical strain of the atomistic model, calculated by the formula

$$\text{Error} = \frac{\bar{\gamma}_{at} - \bar{\gamma}_{qc}}{\bar{\gamma}_{at}}, \qquad (17)$$

where $\bar{\gamma}_{at}$ is the mean of γ^- and γ^+ for the atomistic method, and $\bar{\gamma}_{qc}$ is the mean of γ^- and γ^+ for the quasicontinuum method. We wish to stress that if γ^+ for some method is equal to γ^- for another method, then no strong statement can be made comparing critical strains of the two methods. For example, our data do not show that QNL(5) has a significantly higher critical strain than QCE-QCF(5). Rather, our data suggest only that the critical strain of the QCE-QCF(5) method is in the interval $[0.038172, 0.038177]$ and that the critical strain of the QNL(5) method is in $[0.038177, 0.038181]$. Thus, the critical strains of the two methods are very close.

The reader will observe that the critical strain of the fully atomistic model is the highest. The critical strain predicted by the QCE method is the lowest, and is the farthest from the critical strain of the atomistic model. This is not surprising

Table 1 The critical strain for the Lomer dislocation dipole under shear. The Error is given by $(\bar{\gamma}_{at} - \bar{\gamma}_{qc})/\bar{\gamma}_{at}$ in percent which is defined precisely in the paragraph surrounding (17)

Method	Critical strain		Error
	γ^-	γ^+	
Atomistic	0.038190	0.038194	0.000 %
QNL(5)	0.038177	0.038181	0.034 %
QCE-QCF(5)	0.038172	0.038177	0.046 %
QNL-QCF(5)	0.038172	0.038177	0.046 %
QNL(4)	0.038164	0.038168	0.069 %
QCE-QCF(4)	0.038155	0.038159	0.092 %
QNL-QCF(4)	0.038155	0.038159	0.092 %
QNL(3)	0.038111	0.038116	0.206 %
QCE-QCF(3)	0.038081	0.038085	0.286 %
QNL-QCF(3)	0.038081	0.038085	0.286 %
QCE(3)	0.037750	0.037775	1.125 %

in light of the results on the stability of one dimensional chains obtained in [14]. It is surprising, however, that the QNL method predicts the critical strain of the atomistic model more accurately than the QNL-QCF force-based method, since we expected that the stability of the QNL-QCF method would be determined by the QNL preconditioner [15]. It also somewhat surprising that the QCE-QCF iteration predicted the critical strain as accurately as the QNL-QCF iteration. The analysis given in [15] suggests that the QCE-QCF iteration should lose stability at a lower strain than the QNL-QCF iteration.

Figure 7 consists of graphs showing the relative $w^{1,\infty}$ error versus the shear strain for various quasicontinuum methods. The $w^{1,\infty}$ norm is defined by

$$||u||_{w^{1,\infty}} := \max_{x \in \omega} \max_{b \in \mathscr{R}_x} \frac{|D_b u(x)|}{|b|},$$

and we define the *relative error* in $w^{1,\infty}$ by

$$\text{err}_{\text{rel}}(\mathsf{u}_{\text{qc}}, \mathsf{u}_{\text{a}}) := \frac{||\mathsf{u}_{\text{qc}} - \mathsf{u}_{\text{a}}||_{w^{1,\infty}}}{||\mathsf{u}_{\text{a}}||_{w^{1,\infty}}}. \tag{18}$$

Here u_{qc} denotes a solution of one of the quasicontinuum models, and u_a denotes a solution of the atomistic model with the same boundary conditions.

The reader will observe that the error of the QCE method is the greatest. This is largely due to the presence of ghost forces [10]. The error of the QNL method is the least. Again, this is surprising since various analyses [11, 13, 15, 29, 34] suggest that the force based method should be more accurate than the QNL method. However, we note that a similar effect was observed in one-dimensional numerical experiments in [29]: while for large atomistic regions the QCF method was considerably more accurate than the QNL method, for smaller atomistic regions the QNL method was clearly more accurate. The atomistic regions studied here in our numerical experiments for the Lomer dislocation dipole are very small.

3.3 Tensile Loading

In our next experiment, we examine the accuracy of various quasicontinuum methods for the computation of the resistance to slip under an applied tension. By a tension, we mean a homogeneous deformation of the lattice $\mathscr{T}$ which takes the form

$$x \mapsto \tau(\gamma)x \quad \text{where} \quad \tau(\gamma) := \begin{pmatrix} 1+\gamma & 0 \\ 0 & 1 \end{pmatrix}$$

when expressed in the basis of the coordinate vectors V_1 and V_2. We will let $\tau(\gamma)$ denote such a tension, and we will call γ the *tensile strain*. We call a tension *positive* if $\gamma > 0$ and *negative* if $\gamma < 0$.

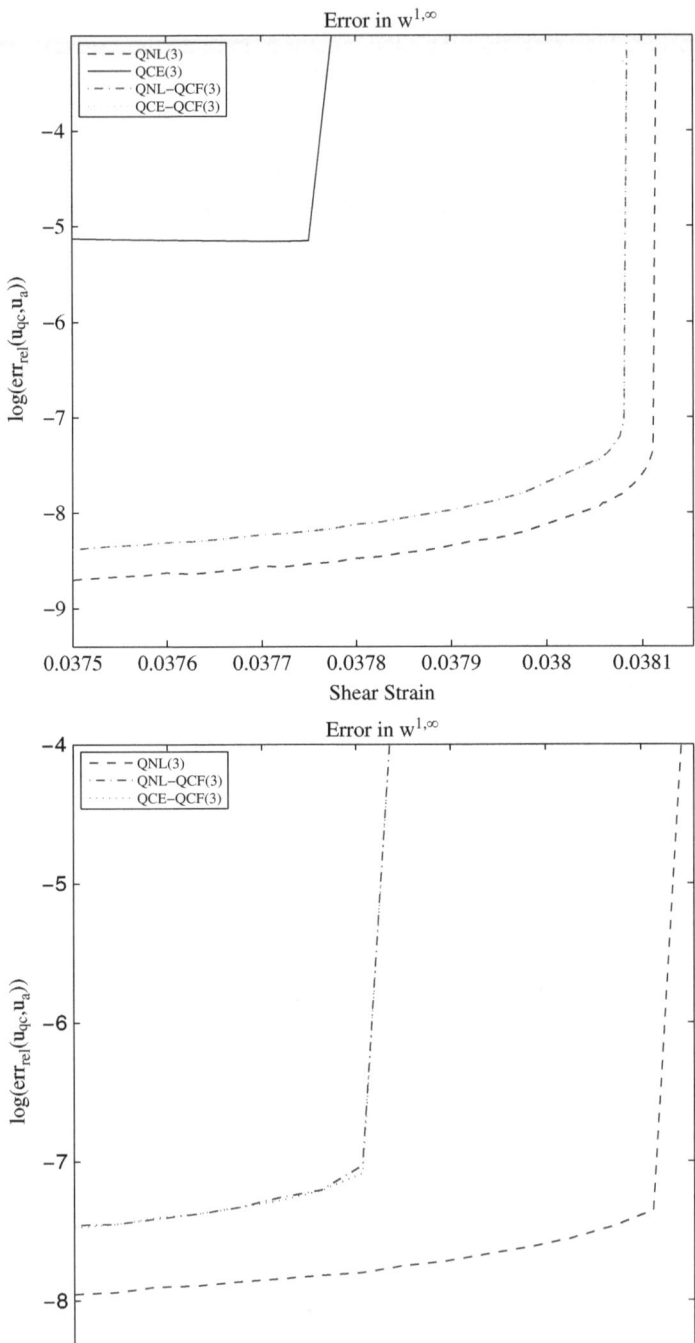

Fig. 7 The relative error $\mathrm{err}_{\mathrm{rel}}(u_{qc}, u_a) = \|u_{qc} - u_a\|_{w^{1,\infty}}/\|u_a\|_{w^{1,\infty}}$ in $w^{1,\infty}$ for the Lomer dislocation dipole

In our numerical experiments, we use the same pair potential and lattice spacing as for the Lomer dipole problem. We let ω be the rectangular domain consisting of 120 columns of atoms parallel to V_2 with each column containing 15 atoms defined in the $\langle v_1, v_2 \rangle$ coordinates by

$$\omega = \{\langle k, l \rangle : 0 \leq k \leq 59, \ 0 \leq l \leq 14\}$$
$$\cup \{\langle k + 1/2, l + 1/2 \rangle : 0 \leq k \leq 59, \ 0 \leq l \leq 14\}.$$

We will now let Γ consist of four columns of atoms parallel to V_2 at the left edge of ω, and four columns at the right edge of ω, that is,

$$\Gamma = \{\langle k, l \rangle : k = 0, \ 1, \ 58, \ 59, \ 0 \leq l \leq 14\}$$
$$\cup \{\langle k + 1/2, l + 1/2 \rangle : k = 0, \ 1, \ 58, \ 59, \ 0 \leq l \leq 14\}.$$

This corresponds to choosing Dirichlet boundary conditions on the sides of ω parallel to V_2, and free boundary conditions on the sides of ω parallel to V_1.

To find the critical strain for the crystal under tension, we perform an iteration similar to the shear loading iteration. Let γ_k be an increasing sequence of tensile strains with $\gamma_0 = 0$. We let u_0 be the displacement field of the undeformed reference lattice ω described above. Then we let u_{k+1} solve the problem

$$\mathcal{F}(u_{k+1})(x) = 0 \qquad \text{for all } x \in \omega \setminus \Gamma,$$
$$u_{k+1}(x) = \tau(\gamma_{k+1})x - x \quad \text{for all } x \in \Gamma. \tag{19}$$

Here $\mathcal{F}$ is the gradient of one of the energies (7), (8), (9), or else the force-based method (10). We solve (19) using the same methods discussed above for the shear loading iteration, except that in this case we start each step with the initial guess

$$u_{k+1}^0(x) := \tau\left(\frac{1+\gamma_{k+1}}{1+\gamma_k}\right)(x + u_k(x)) - x.$$

We call this iteration the *tensile loading continuation*. We stop the continuation when the P-nCG algorithm (or a ghost force correction iteration) returns a deformation in which a slip has occurred.

Our numerical experiments are designed to test how well the boundary between the atomistic and continuum regions resists slip under tensile loading. When we perform the tensile loading iteration for the fully atomistic model, we found that a slip tends to occur along the slip plane indicated in Fig. 8a when a critical tensile load is reached. Figure 8a shows the configuration of the crystal immediately after a typical slip has occurred. We note that the slip allows the crystal to accommodate an increased tensile strain with a decrease in the energy. We then chose an atomistic region $\mathscr{A}$ whose boundary is along that line, as depicted in Fig. 8b.

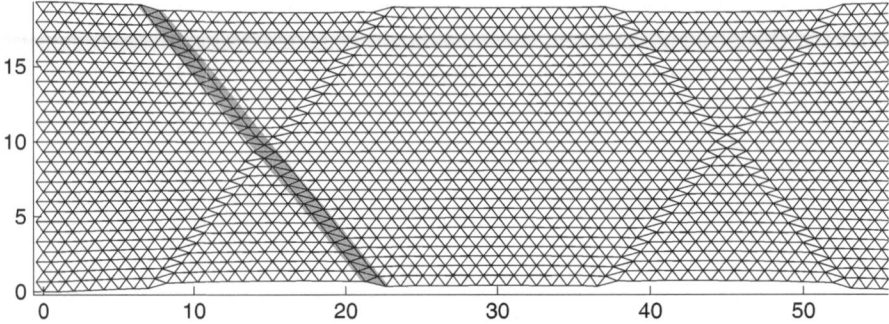

(a) The natural slip plane. We choose an atomistic region whose boundary is the plane highlighted in grey. The mesh is the Delaunay triangulation [7] of the reference lattice.

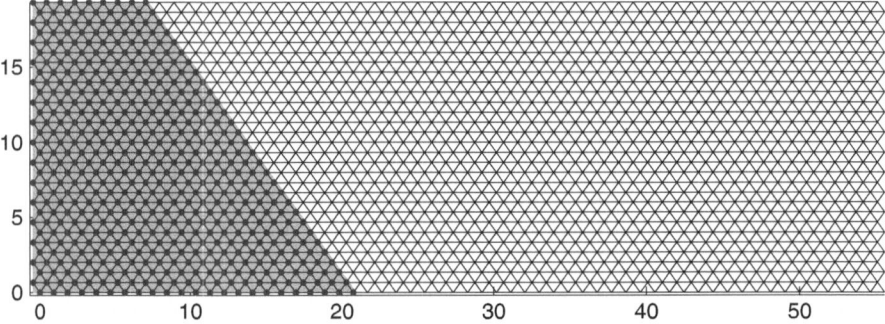

(b) The atomistic region $\mathscr{A}$ is shaded grey. All atoms in $\mathscr{A}$ are marked with a blue dot.

Fig. 8 The atomistic ($\mathscr{A}$) and continuum ($\mathscr{C}$) regions for the tensile loading experiment. (a) The natural slip plane. We choose an atomistic region whose boundary is the plane highlighted in *grey*. The mesh is the Delaunay triangulation [7] of the reference lattice. (b) The atomistic region $\mathscr{A}$ is shaded *grey*. All atoms in $\mathscr{A}$ are marked with a *blue dot*

Table 2 The critical strain under tension. The Error is given by $\left(\bar{\gamma}_{at} - \bar{\gamma}_{qc}\right) / \bar{\gamma}_{at}$ in percent, which is defined in (17)

Method	Critical strain		Error
	γ^-	γ^+	
Atomistic	0.081635	0.081649	0.000 %
QNL	0.081593	0.081607	0.052 %
QCE-QCF	0.081242	0.081270	0.473 %
QNL-QCF	0.081101	0.081130	0.645 %
QCE	0.063900	0.064200	21.548 %

We observe that the critical strains of the quasicontinuum models were lower than the critical strain of the atomistic model. Our results are summarized in Table 2. In Table 2, the columns γ^- and γ^+ are the tensile strains at the last step before slip is observed and at the first step for which slip is observed, respectively. The column labeled "Error" is the percent error from the critical strain of the atomistic model.

We observe that the critical strain of the QCE model is lower than the critical strain of any of the other models. This is in agreement with the one dimensional stability results established in [14]. However, we were surprised to find that the critical strain predicted by the QNL method is higher than the critical strain predicted by the QNL-QCF force based method, since we expected that the stability of the QNL-QCF method would be determined by the QNL preconditioner [15]. We were also surprised to observe that the critical strain of the QNL-QCF method is less than the critical strain of QCE-QCF. The one-dimensional analyses in [12, 15] suggest that the force based methods should have a comparable critical strain as QNL, and that the QCE-QCF iteration should lose stability at a lower strain than the QNL-QCF iteration.

Figure 9 consists of graphs showing the relative $w^{1,\infty}$ error versus the tensile strain for various quasicontinuum methods. We observe that the QCE model had the greatest error. This is primarily the result of the ghost forces which arise in the QCE model. The QNL model had the lowest error. Again, this is somewhat surprising in light of the one dimensional theory [11, 13, 15, 29, 34], which predicts that the force based methods should have lower error than the QNL method.

4 Conclusion

Materials scientists and engineers typically attempt to verify their multiphysics methods by benchmark tests. However, definitive conclusions from these benchmark tests are often not possible since they combine many sources of error (modeling error, coupling error, solution error, boundary condition error, etc.). Further supporting theory is needed to select a set of test problems that thoroughly samples the solution space.

We are developing an improved theoretical basis for benchmarking atomistic-to-continuum coupling methods based on the multiscale numerical analysis developed by us and others. Our goal is to be able to reliably predict the accuracy of atomistic-to-continuum coupling methods for general deformations and loads from numerical experiments for a small set of mechanics problems. This set of mechanics problems should sample the fundamental modes of material instability such as dislocation formation, slip, and fracture.

Our discussion of the benchmark tests presented in this paper give many examples of the predictive success of the theoretical analysis we have developed during the past few years, but we also describe several cases where our theoretical analysis seems to predict a different outcome than our computational experiments. This discrepancy between theory and computational experiment occurs when our theoretical analysis does not adequately model the computational problem and is motivation to develop more general theoretical analysis.

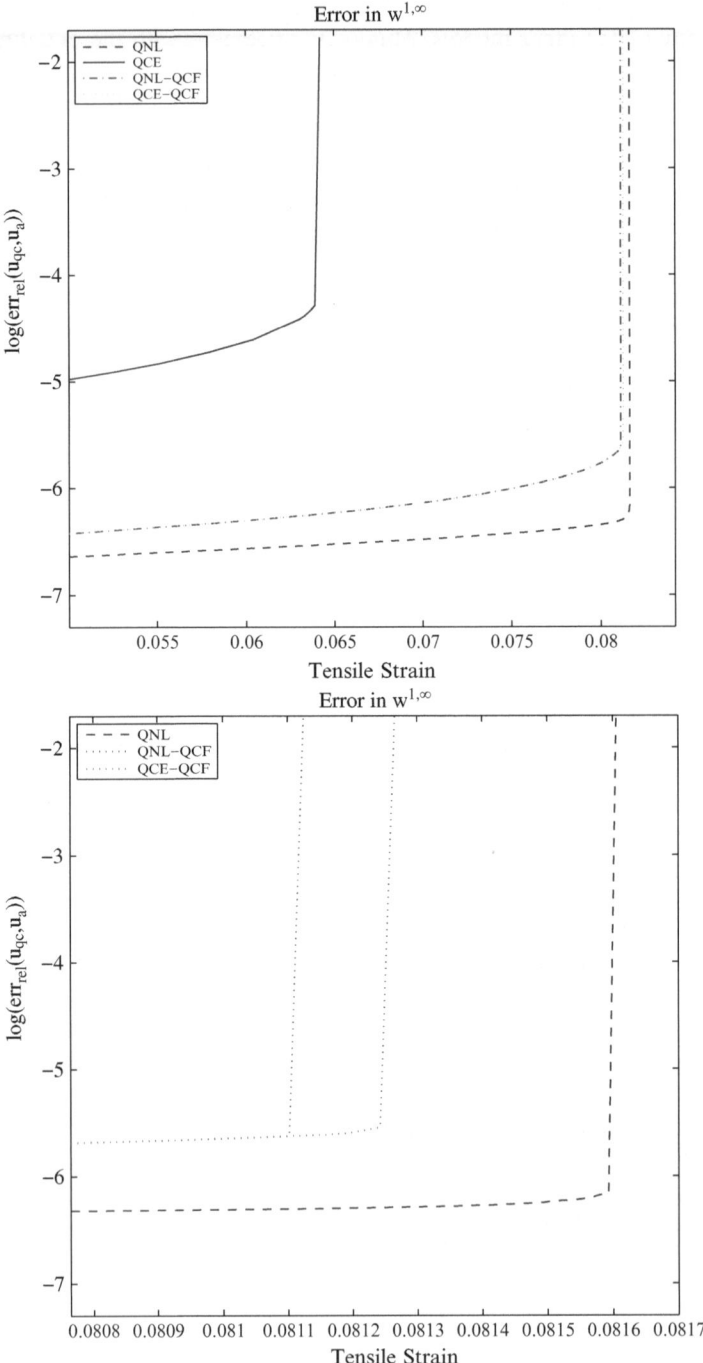

Fig. 9 The relative error $\mathrm{err}_{\mathrm{rel}}(\mathbf{u}_{\mathrm{qc}}, \mathbf{u}_{\mathrm{a}}) = \|\mathbf{u}_{\mathrm{qc}} - \mathbf{u}_{\mathrm{a}}\|_{w^{1,\infty}} / \|\mathbf{u}_{\mathrm{a}}\|_{w^{1,\infty}}$ in $w^{1,\infty}$ for tensile loading

Appendix: A Preconditioned Conjugate Gradient Algorithm

In this appendix, we describe the preconditioned nonlinear conjugate gradient optimization algorithm (P-nCG) and the corresponding linesearch method, which we used in the numerical experiments in this paper.

Let $\mathscr{E} : \mathbb{R}^N \to \mathbb{R} \cup \{+\infty\}$ be continuously differentiable in $\{u : \mathscr{E}(u) < \infty\}$. If $\mathscr{E}$ is an energy of the type discussed above, then the standard nonlinear conjugate gradient method [33, Sec. 5.2] is convergent, but very inefficient, due to the poor conditioning of the Hessian matrix at local minima.

Let $P \in \mathbb{R}^{N \times N}$ be a symmetric positive definite matrix, e.g., a discrete Laplacian or a modification thereof. A simple but considerably more efficient method is obtained if all inner products in the conjugate gradient algorithm are replaced by a P-inner product $(u, v)_P = u^T P v$, and all gradients $\delta\mathscr{E}(u)$ by P-gradients $\delta_P\mathscr{E}(u) = P^{-1}\delta\mathscr{E}(u)$. The P-gradient $\delta_P\mathscr{E}(u)$ "represents" the gradient $\delta\mathscr{E}(u)$ in the P-inner product $(v, w)_P := v^T P w$, since

$$(\delta_P\mathscr{E}(u), w)_P = (P^{-1}\delta\mathscr{E}(u))^T P w = \delta\mathscr{E}(u)^T w = (\delta\mathscr{E}(u), w).$$

In practice, we allow a new preconditioner to be computed at each step. A basic preconditioned conjugate gradient algorithm of Polak-Ribière type can be described as follows:

(0) Input: $u_0 \in \mathbb{R}^N$;
(1) Evaluate P_0; $g_0 = P_0^{-1}\delta\mathscr{E}(u_0)$; $s_0 = 0$;
(2) For $n = 1, 2, \ldots$ do:
(3)   Evaluate P_n;
(4) $g_n = P_n^{-1}\delta\mathscr{E}(u_n)$;
(5) $\beta_n = \max\{0, (g_n, g_n - g_{n-1})_{P_n}/(g_{n-1}, g_{n-1})_{P_{n-1}}\}$;
(6) $s_n = -g_n + \beta_n s_{n-1}$;
(7) $\alpha_n \leftarrow \text{LINESEARCH}$;
(8) $u_n = u_{n-1} + \alpha_n s_n$;

In the following we specify further crucial details of our implementation:

1. *Preconditioner:* The choice of preconditioner has the biggest influence on the efficiency of the optimization. We project all atoms onto a single plane and triangulate the resulting set of nodes. On this triangulation, we assemble the standard P1-finite element stiffness matrix, K_n, discretizing the negative Laplace operator. The preconditioner P_n is obtained by imposing homogeneous Dirichlet boundary conditions on the clamped nodes.
2. *LINESEARCH:* Our linesearch is implemented following Algorithms 3.5 and 3.6 in [33] closely. We use cubic interpolation from function and gradient values in the "Interpolate" step of Algorithm 3.6. We guarantee the strong Wolfe conditions, [33, (3.7)] with constants $c_1 = 10^{-4}$, $c_2 = 1/2$.

3. *Initial guess for α_n:* If the initial guess $\alpha_n^{(0)}$ for the new steplength α_n, which is passed to the LINESEARCH routine, is chosen well, then actual linesearch can be mostly avoided, which can result in considerable performance gains. Following [33, (3.60)], and extensive experimentation with alternative options, we choose $\alpha_n^{(0)} = 2(\mathscr{E}(u_{n-1}) - \mathscr{E}(u_{n-2}))/(g_n, s_n)_{P_n}$.

4. *Termination Criteria:* We terminate the iteration successfully if the following condition is satisfied:

$$\left(\|u_n - u_{n-1}\|_\infty \leq \mathrm{TOL}_u^\infty \quad \text{or} \quad \|u_n - u_{n-1}\|_{P_n} \leq \mathrm{TOL}_u^P \right)$$

$$\text{and} \quad \left(\|\delta\mathscr{E}(u_n)\|_\infty \leq \mathrm{TOL}_g^\infty \quad \text{or} \quad \|g_n\|_{P_n} \leq \mathrm{TOL}_g^P \right)$$

$$\text{and} \quad \left(\mathscr{E}(u_{n-1}) - \mathscr{E}(u_n) \leq \mathrm{TOL}_\mathscr{E} \right),$$

where $\|\cdot\|_\infty$ denotes the ℓ^∞-norm, and $\|\cdot\|_P$ denotes norm associated with the P-inner product. The tolerance parameters are adjusted for each problem. Typical choices are $\mathrm{TOL}_u^\infty = \mathrm{TOL}_u^P = 10^{-5}$, $\mathrm{TOL}_g^\infty = \mathrm{TOL}_g^P = 10^{-4}$, $\mathrm{TOL}_\mathscr{E} = 10^{-4}$.

We terminate the iteration unsuccessfully if a maximum number of iterations is reached, or if the LINESEARCH routine is unable to make any progress. Note also that, since P-nCG is a descent method, we have $\mathscr{E}(u_{n-1}) - \mathscr{E}(u_n) = |\mathscr{E}(u_{n-1}) - \mathscr{E}(u_n)|$.

5. *Robustness Checks:* In addition, our algorithm uses various minor modifications to increase its robustness. Since these do not significantly affect its performance, we have decided to not give any details.

The algorithm described above is both efficient and robust for most of the problems we consider. However, we wish to stress certain difficulties that arise in the presence of "meta-stable" states and particularly shallow local minimizers, which are difficult to distinguish numerically. In our experience, dislocations fall precisely into this category.

On some occasions, our algorithm would fail when a particularly low tolerance setting was used. The reason for the failure is usually that the directional derivative along the search direction is non-negative (up to numerical precision) and hence the linesearch fails or stagnates. Replacing the conjugate direction by a steepest descent direction (one of our robustness checks) resolves this problem only partially. We were able to overcome these difficulties mostly by tweaking the various optimization parameters.

Acknowledgements This work was supported in part by the National Science Foundation under DMS-0757355, DMS-0811039, the PIRE Grant OISE-0967140, the Institute for Mathematics and Its Applications, and the University of Minnesota Supercomputing Institute. This work was also supported by the Department of Energy under Award Number DE-SC0002085. CO was supported by the EPSRC grant EP/H003096/1 "Analysis of Atomistic-to-Continuum Coupling Methods."
We wish to thank Ellad Tadmor for helpful discussions.

References

1. N. Ashcroft and D. Mermin. *Solid State Physics*. Brooks Cole, 1976.
2. S. Badia, M. Parks, P. Bochev, M. Gunzburger, and R. Lehoucq. On atomistic-to-continuum coupling by blending. *Multiscale Model. Simul.*, 7(1):381–406, 2008.
3. X. Blanc, C. Le Bris, and F. Legoll. Analysis of a prototypical multiscale method coupling atomistic and continuum mechanics. *M2AN Math. Model. Numer. Anal.*, 39(4):797–826, 2005.
4. X. Blanc, C. Le Bris, and P.-L. Lions. From molecular models to continuum mechanics. *Arch. Ration. Mech. Anal.*, 164(4):341–381, 2002.
5. P. Bělík and M. Luskin. Sharp stability and optimal order error analysis of the quasi-nonlocal approximation of unconstrained linear and circular chains in 2-D. arXiv:1008.3716, 2010.
6. W. Curtin and R. Miller. Atomistic/continuum coupling in computational materials science. *Modell. Simul. Mater. Sci. Eng.*, 11(3):R33–R68, 2003.
7. M. de Berg, O. Cheong, M. van Kreveld, and M. Overmars. *Computational geometry*. Springer-Verlag, Berlin, third edition, 2008. Algorithms and applications.
8. M. Dobson and M. Luskin. Analysis of a force-based quasicontinuum approximation. *Mathematical Modelling and Numerical Analysis*, pages 113–139, 2008.
9. M. Dobson and M. Luskin. Iterative solution of the quasicontinuum equilibrium equations with continuation. *Journal of Scientific Computing*, 37:19–41, 2008.
10. M. Dobson and M. Luskin. An analysis of the effect of ghost force oscillation on the quasicontinuum error. *Mathematical Modelling and Numerical Analysis*, 43:591–604, 2009.
11. M. Dobson and M. Luskin. An optimal order error analysis of the one-dimensional quasicontinuum approximation. *SIAM. J. Numer. Anal.*, 47:2455–2475, 2009.
12. M. Dobson, M. Luskin, and C. Ortner. Sharp stability estimates for force-based quasicontinuum methods. *SIAM J. Multiscale Modeling & Simulation*, 8:782–802, 2010. arXiv:0907.3861.
13. M. Dobson, M. Luskin, and C. Ortner. Stability, instability, and error of the force-based quasicontinuum approximation. *Archive for Rational Mechanics and Analysis*, 197:179–202, 2010. arXiv:0903.0610.
14. M. Dobson, M. Luskin, and C. Ortner. Accuracy of quasicontinuum approximations near instabilities. *Journal of the Mechanics and Physics of Solids*, 58, 1741-1757 (2010). arXiv:0905.2914v2.
15. M. Dobson, M. Luskin, and C. Ortner. Iterative methods for the force-based quasicontinuum approximation: Analysis of a 1D model problem. *Computer Methods in Applied Mechanics and Engineering*, 200, 2697–2709 (2011). arXiv:0910.2013v3.
16. W. E, J. Lu, and J. Yang. Uniform accuracy of the quasicontinuum method. *Phys. Rev. B*, 74(21):214115, 2004.
17. W. E and P. Ming. Cauchy-Born rule and the stability of crystalline solids: static problems. *Arch. Ration. Mech. Anal.*, 183(2):241–297, 2007.
18. G. Friesecke and F. Theil. Validity and failure of the Cauchy-Born hypothesis in a two-dimensional mass-spring lattice. *J. Nonlinear Sci.*, 12(5):445–478, 2002.
19. M. Gunzburger and Y. Zhang. A quadrature-rule type approximation for the quasicontinuum method. *Multiscale Modeling and Simulation*, 8:571–590, 2010.
20. J. Hirth and J. Lothe. *Theory of Dislocations*. Krieger Publishing Company, 1992.
21. M. Iyer and V. Gavini. A field theoretical approach to the quasi-continuum method. *J. Mech. Phys. Solids* 59(8), 506–1535 (2011).
22. B. V. Koten and M. Luskin. Development and analysis of blended quasicontinuum approximations. arXiv:1008.2138v2, 2010.
23. X. H. Li and M. Luskin. A generalized quasinonlocal atomistic-to-continuum coupling method with finite-range interaction. *IMA Journal of Numerical Analysis*, (2011). doi: 10.1093/imanum/drq049.
24. X. H. Li and M. Luskin. A generalized quasi-nonlocal atomistic-to-continuum coupling method with finite range interaction. *International Journal for Multiscale Computational Engineering*, to appear. arXiv:1007.2336.

25. P. Lin. Theoretical and numerical analysis for the quasi-continuum approximation of a material particle model. *Math. Comp.*, 72(242):657–675, 2003.
26. P. Lin. Convergence analysis of a quasi-continuum approximation for a two-dimensional material without defects. *SIAM J. Numer. Anal.*, 45(1):313–332, 2007.
27. M. Luskin and C. Ortner. An analysis of node-based cluster summation rules in the quasicontinuum method. *SIAM. J. Numer. Anal.*, 47:3070–3086, 2009.
28. C. Makridakis, C. Ortner, and E. Süli. Stress-based atomistic/continuum coupling: A new variant of the quasicontinuum approximation. preprint, 2010.
29. C. Makridakis, C. Ortner, and E. Sli. A priori error analysis of two force-based atomistic/continuum hybdrid models of a periodic chain. OxMOS Report No. 28.
30. R. Miller and E. Tadmor. The Quasicontinuum Method: Overview, Applications and Current Directions. *Journal of Computer-Aided Materials Design*, 9:203–239, 2003.
31. R. Miller and E. Tadmor. Benchmarking multiscale methods. *Modelling and Simulation in Materials Science and Engineering*, 17:053001 (51pp), 2009.
32. P. Ming and J. Z. Yang. Analysis of a one-dimensional nonlocal quasi-continuum method. *Multiscale Model. Simul.*, 7(4):1838–1875, 2009.
33. J. Nocedal and S. J. Wright. *Numerical optimization*. Springer Series in Operations Research and Financial Engineering. Springer, New York, second edition, 2006.
34. C. Ortner. A priori and a posteriori analysis of the quasi-nonlocal quasicontinuum method in 1D, 2009. arXiv.org:0911.0671.
35. C. Ortner and E. Süli. Analysis of a quasicontinuum method in one dimension. *M2AN Math. Model. Numer. Anal.*, 42(1):57–91, 2008.
36. S. Prudhomme, H. Ben Dhia, P. T. Bauman, N. Elkhodja, and J. T. Oden. Computational analysis of modeling error for the coupling of particle and continuum models by the Arlequin method. *Comput. Methods Appl. Mech. Engrg.*, 197(41-42):3399–3409, 2008.
37. A. V. Shapeev. Consistent energy-based atomistic/continuum coupling for two-body potential: 1D and 2D case. preprint, 2010.
38. V. B. Shenoy, R. Miller, E. B. Tadmor, D. Rodney, R. Phillips, and M. Ortiz. An adaptive finite element approach to atomic-scale mechanics–the quasicontinuum method. *J. Mech. Phys. Solids*, 47(3):611–642, 1999.
39. L. E. Shilkrot, R. E. Miller, and W. A. Curtin. Multiscale plasticity modeling: Coupled atomistics and discrete dislocation mechanics. *J. Mech. Phys. Solids*, 52:755–787, 2004.
40. T. Shimokawa, J. Mortensen, J. Schiotz, and K. Jacobsen. Matching conditions in the quasicontinuum method: Removal of the error introduced at the interface between the coarse-grained and fully atomistic region. *Phys. Rev. B*, 69(21):214104, 2004.
41. E. B. Tadmor, M. Ortiz, and R. Phillips. Quasicontinuum Analysis of Defects in Solids. *Philosophical Magazine A*, 73(6):1529–1563, 1996.

Coarse-Grid Multiscale Model Reduction Techniques for Flows in Heterogeneous Media and Applications

Yalchin Efendiev and Juan Galvis

Abstract In this paper, we give an overview of our results [35, 38, 45, 46] from the point of view of coarse-grid multiscale model reduction by highlighting some common issues in coarse-scale approximations and two-level preconditioners. Reduced models discussed in this paper rely on coarse-grid spaces computed by solving local spectral problems. We define local spectral problems with a weight function computed with a choice of initial multiscale basis functions. We emphasize the importance of this initial choice of multiscale basis functions for both coarse-scale approximation and for preconditioners. In particular, we discuss various choices of initial basis functions and use some of them in our simulations. We show that a naive choice of initial basis functions, e.g., piecewise linear functions, can lead to a large dimensional spaces that are needed to achieve (1) a reasonable accuracy in the coarse-scale approximation or (2) contrast-independent condition number of preconditioned matrix within two-level additive Schwarz methods. While using a careful choice of initial spaces, we can achieve (1) and (2) with smaller dimensional coarse spaces.

1 Introduction

Many problems in applications occur in media that contain multiple scales and have a high contrast in the properties. For example, fractured reservoirs typically have large variations in their conductivities. These large variations bring additional small-scale parameters into the problem. Numerical discretization of flow problems in such heterogeneous media results to very large ill-conditioned systems of linear equations. Several approaches are proposed to solve such systems. Some

Y. Efendiev (✉) · J. Galvis
Texas A& M University, College Station, TX 77843-3368, USA
e-mail: efendiev@math.tamu.edu; jugal@math.tamu.edu

I.G. Graham et al. (eds.), *Numerical Analysis of Multiscale Problems*, Lecture Notes in Computational Science and Engineering 83, DOI 10.1007/978-3-642-22061-6_4, © Springer-Verlag Berlin Heidelberg 2012

approaches involve the solution on a coarse grid (e.g., [1–7, 15, 17–21, 25, 26, 30–34, 38–41, 43, 44, 48, 51, 52, 55, 56, 60–62, 66–69, 72, 75, 79]). In these approaches, coarse-grid properties, such as upscaled conductivities or multiscale basis functions, are constructed that represent the media or the solution on the coarse grid. For solving high-contrast problems on a fine grid, robust iterative methods that converge independent of the contrast and multiple scales are needed (e.g, [8, 28, 50, 63, 71, 73, 74, 76] and references therein). In this review paper, we will describe some of our work toward a design of robust preconditioners and coarse-grid solution techniques for multiscale high-contrast problems.

Multiscale methods attempt to compute the solution on a coarse grid. One of the commonly used approaches is upscaling methods (e.g., [16, 20, 30, 47, 58, 59, 79]) where coarse-grid conductivities are computed, and then the flow equation is solved on this coarse grid. Instead of coarse-grid conductivities, multiscale basis functions can be used to represent the solution on a coarse grid ([4, 14, 52, 56]). In the latter, multiscale basis functions are constructed on a coarse grid, and the approximation for the solution of fine-scale equation is sought on a finite dimensional space spanned by these basis functions. Multiscale basis functions are typically constructed by solving the local flow equation on a coarse grid subject to some boundary conditions. This shares similarities with upscaling methods where local problems are solved to compute effective properties. It is known that the boundary conditions of these local problems play an important role in the accuracy of the coarse-grid solution techniques. In particular, the use of artificial boundary conditions that do not contain correct small-scale information of the fine-scale solution leads to resonance errors [42, 49, 52]. To remedy this situation and improve subgrid capturing errors, various approaches are proposed. These approaches include the use of reduced boundary conditions [52, 56], oversampling methods [31, 40, 42, 52], limited global information and harmonic coordinates [39, 69], local-global approaches [24, 32, 57], and so on. All these approaches are intended to improve the subgrid accuracy.

To design robust iterative methods for the solution of the fine-scale problem, one can use coarse-grid multiscale solutions and additional local subdomain corrections (as in domain decomposition methods) to converge to a fine-scale solution in a few iterations. Domain decomposition methods use the solutions of local problems and a coarse problem in constructing preconditioners for the fine-scale system. The number of iterations required by domain decomposition preconditioners is adversely affected by the contrast in the media properties (e.g., [50] and the references therein). It is known that if high and low conductivity regions can be encompassed within coarse grid blocks such that the variation of the conductivity within each coarse region is bounded, several domain decomposition preconditioners result to a system with the condition number independent of the contrast (e.g., [63, 76]). Because of the complex geometry of fine-scale features (e.g., complex fracture geometry), it is very often *impossible* to separate low and high conductivity regions into different coarse grid blocks. Thus, it can require many iterations for iterative methods to converge.

In this paper, our goal is to improve the convergence rate of coarse-grid multiscale approximations as well as the performance of two-level domain decomposition iterative solvers by designing coarse spaces. This is very important for practical applications where one encounters small scales and high contrast in media properties.

We discuss constructing coarse spaces that are used in multiscale finite element methods (MsFEMs) for solving the problem on a coarse grid as well as in two-level preconditioners for iterative solutions. The construction of coarse spaces starts with an initial choice of multiscale basis functions that are supported in coarse regions sharing a common node. These basis functions are complemented using weighted local spectral problems that are defined in coarse blocks sharing a common node. The weight function in the local spectral problem is computed based on the initial choice of multiscale basis functions. Furthermore, we identify important eigenvalues (small eigenvalues in our case, see discussion below) and corresponding eigenvectors that represent important features of the solution. Coarse spaces are constructed by multiplying the selected eigenvectors with the initial multiscale basis functions. The estimate for the convergence of MsFEM and the condition number of two-level preconditioners discussed in the paper depends on the maximum of the inverse of the eigenvalues whose eigenvectors are not included in the coarse space. The maximum is taken over all coarse nodes. It is known that the number of iterations required by iterative methods, such as domain decomposition methods, is affected by the contrast in the media properties that are within each coarse-grid block ([50]). Our goal is to choose the initial multiscale space such that the eigenvalues of the local spectral problems increase rapidly.

The eigenvalues of the high-contrast spectral problem, depending on the weight function, can have multiple small eigenvalues, which are asymptotically vanishing as the contrast increases. The number of small eigenvalues depends on the choice of weight function, and consequently, on the choice of initial basis functions. The eigenvectors corresponding to the small eigenvalues contain important features of the local solution and need be included in the coarse space. With a naive choice of initial basis functions, one can get a large number of small eigenvalues, which are asymptotically vanishing as the contrast increases. The latter occurs when there are many isolated high-conductivity inclusions. By choosing initial basis functions appropriately, we can include all isolated inclusions into one basis function at each node. Consequently, the coarse space only includes basis functions corresponding to isolated channels (high-conductivity regions that connect the boundaries of the coarse block). We discuss the fact that the channels can not be removed with the help of initial multiscale basis functions.

Because the convergence of MsFEMs and the condition number of the pre-conditioned matrix depends on the inverse of the first eigenvalue, such that the corresponding eigenvector is not included in the coarse space, we need to include eigenvectors corresponding to all small eigenvalues, which are asymptotically vanishing as the contrast increases. Indeed, if the eigenvectors corresponding to the small, asymptotically vanishing as contrast increases, eigenvalues are not included in the coarse space, then the convergence of MsFEM may be reduced and the condition number of the preconditioned matrix will be extremely large. The initial multiscale coarse space is designed such that the corresponding local eigenvalue problem has fewer small eigenvalues which are asymptotically vanishing as the contrast increases.

We would like to mention that there have been many works where various multiscale methods for high-contrast problem and iterative solvers for multiscale high-contrast problems are investigated. We briefly discuss some of the contributions that are relevant to our work. There have been a number of works where multiscale basis functions are designed for high-contrast problems. This includes the work [18] where global multiscale spectral basis functions are constructed using the global spectral problem for Laplace equation. A spectral convergence has been proved. Further, the authors extend this approach and present basis localization in [70]. In this approach, they show that one can localize basis functions by using modified local problems in larger (than the target coarse-grid block) subdomains. The paper [27] designs multiscale basis functions for high-contrast problems where heterogeneities have special forms. The authors study local boundary conditions for basis functions and show that by solving carefully selected local problems one can achieve optimal convergence. In fact, if these basis functions are used as initial basis functions, then the coarse space dimension will be smaller compared to using reduced problems on the boundaries as well as standard multiscale basis functions. In [28], the authors propose a multiscale approach using an integral formulation of the high-contrast problem. The proposed approach uses contrast-independent problems on a coarse grid to construct an approximation of the fine-scale solution. Spectral basis functions are used in [9, 23, 51, 78] to approximate the solution on a coarse grid. However, these methods do not employ initial multiscale basis functions in designing local spectral problems. As we note in this paper, with a good choice of initial basis functions one can incorporate many small-scale features into one basis functions and achieve a dimension reduction. When complementing these spaces, one can obtain much smaller dimensional coarse spaces (depending on heterogeneities) if initial basis function space is appropriately chosen.

There have been many works on designing domain decomposition preconditioners that converge independent of contrast. In the context of domain decomposition methods we can consider overlapping and nonoverlapping methods. It was shown that nonoverlapping domain decomposition methods converge independently of the contrast (e.g. [29,63,65,76]) when conductivity variations within coarse regions are bounded. The final condition number estimate when using a two-level overlapping domain decomposition method involves the ratio H/δ, where H is the coarse-mesh size and δ is the size of the overlap region. The estimate with respect to the ratio H/δ can be improved with the help of the small overlap trick ([76]). When there are heterogeneities within coarse blocks, special local multiscale coarse spaces are needed. In a number of works [3, 50], the authors studied convergence of domain decomposition solvers when local heterogeneities have special forms, e.g., high-conductivity inclusions are strictly inside a coarse-grid block. Our works, that are summarized here, extend these methods to general heterogeneities by constructing special local multiscale coarse spaces. In [64], the authors consider enrichment of coarse spaces to obtain contrast-independent preconditioners. Their methods do not use initial basis functions in the construction of local spectral problems that is important for dimension reduction and one of the main ingredients of our theory.

Numerical results are presented. We show the results for the convergence of MsFEMs when the coarse space contains eigenvectors that correspond only to small, asymptotically vanishing (as contrast increases), eigenvalues as well as when additional eigenvectors are included. The convergence rate is observed as $H^{1+\beta}/\Lambda_*$, where $1/\Lambda_*$ is related to the inverse of the smallest eigenvalue. Furthermore, we present numerical results for two-level additive Schwarz preconditioners where we observe contrast-independent condition number of the preconditioned matrix as we increase the contrast. Using appropriate initial multiscale spaces we reduce the dimension of the coarse space that is needed for achieving contrast-independent bound for the preconditioner.

The paper is organized in the following way: in the next section, we present some background discussion on multiscale spaces. In Sect. 3 we present our strategy for complementing coarse spaces. In Sect. 4, we discuss the use of coarse spaces in domain decomposition preconditioners. Finally, numerical results are presented in Sect. 5.

2   Preliminaries. Multiscale Model Reduction: Solving Equations on a Coarse Grid with MsFEM

In this section, we will give a brief overview of MsFEM as a method for solving a problem on a coarse grid. MsFEMs consist of two major ingredients: (1) multiscale basis functions and (2) a global numerical formulation which couples these multiscale basis functions. Multiscale basis functions are designed to capture the fine-scale features of the solution. Important multiscale features of the solution need to be incorporated into these localized basis functions which contain information about the scales which are smaller (as well as larger) than the local numerical scale defined by the basis functions. In particular, we need to incorporate the features of the solution that can be localized and use additional basis functions to capture the information about the features that need to be separately included in the coarse space. A global formulation couples these basis functions to provide an accurate approximation of the solution. Many illustrations are done in two dimensions in this paper; however, the results hold in higher dimensions.

We consider the second order elliptic equation with heterogeneous coefficients

$$- \operatorname{div}(\kappa(x)\nabla u) = f \quad \text{in } D \tag{1}$$

subject to some boundary conditions. Here $\kappa(x)$ is a heterogeneous spatial field with multiple scales and high contrast (e.g., see Fig. 8 for an example of a permeability field). Let $\mathscr{T}^H$ be a usual conforming partition of D into finite elements (triangles,

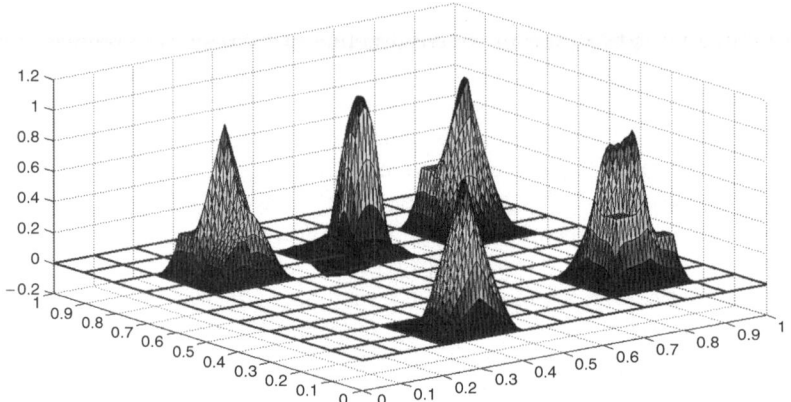

Fig. 1 Schematic description of coarse regions

quadrilaterals and etc.). We call this partition the coarse grid and assume that the coarse grid can be resolved via a finer grid (see Fig. 4 for illustration of a coarse and a fine grid). We denote by N_v the number of coarse nodes, by $\{y_i\}_{i=1}^{N_v}$ the vertices of the coarse mesh $\mathcal{T}^H$ and define the neighborhood of the node y_i by

$$\omega_i = \bigcup \{K_j \in \mathcal{T}^H; \ y_i \in \overline{K}_j\} \tag{2}$$

(see Fig. 1) and the neighborhood of the coarse element K by

$$\omega_K = \bigcup \{\omega_j \in \mathcal{T}^H; \ y_j \in \overline{K}\}. \tag{3}$$

Our objective is to seek multiscale basis functions for each node y_i. We denote the basis functions for the node i, χ_i^j, and assume that the basis functions are supported in ω_i. As in standard finite element methods, once multiscale basis functions are constructed (see Fig. 2 for illustration), we seek $u_0 = \sum_{ij} c_{ij} \chi_i^j$, where c_{ij} are determined from

$$a(u_0, v) = f(v), \quad \text{for all } v \in V_0, \tag{4}$$

$V_0 = \text{span}\{\chi_i^j\}$,

$$a(u, v) = \int_D \kappa(x) \nabla u(x) \nabla v(x) dx \quad \text{for all } u, v \in H_0^1(D) \tag{5}$$

and

$$f(v) = \int_D f(x) v(x) dx \quad \text{for all } v \in H_0^1(D).$$

Once c_{ij}'s are determined, one can define a fine-scale approximation of the solution by reconstructing via basis functions, $u_0 = \sum_{ij} c_{ij} \chi_i^j$.

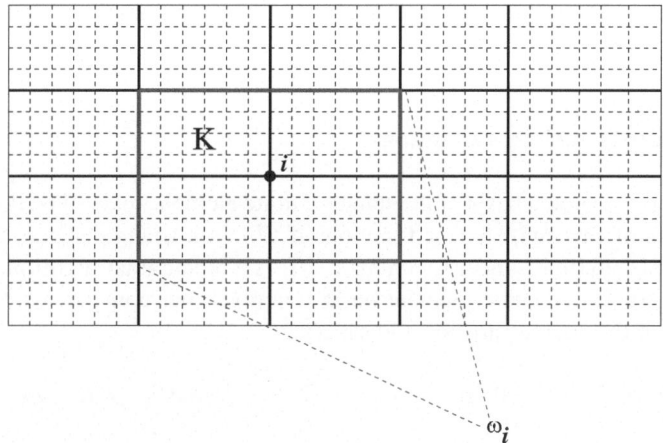

Fig. 2 Illustration of some multiscale basis functions

One can also view MsFEM in the discrete setting. Assume that the basis functions are defined on a fine grid as Φ_i with i varying from 1 to N_c, where N_c is the number of basis functions. Given coarse-scale basis functions, the coarse space is given by (for simplicity we keep the same notation as in the continuous formulation)

$$V_0 = \text{span}\{\Phi_i\}_{i=1}^{N_c}, \tag{6}$$

and the coarse matrix is given by $A_0 = R_0 A R_0^T$, where A is the fine-scale finite element matrix representation of the bi-linear form a in (5) and

$$R_0^T = [\Phi_1, \ldots, \Phi_{N_c}].$$

Here Φ_i's are discrete coarse-scale basis functions defined on a fine grid (i.e., vectors). Multiscale finite element solution is the finite element projection of the fine-scale solution into the space V_0. More precisely, multiscale solution u_0 is given by

$$A_0 u_0 = f_0,$$

where $f_0 = R_0^T b$.

2.1 Coarse Spaces

In this section, we will discuss some coarse spaces constructed to capture the fine-scale features of the solution. We will *first consider coarse spaces where there is only one basis function per coarse node y_i*. For this reason, we will use the notation

χ_i. Further, we discuss how these basis functions can be complemented so that our coarse-scale approximation converges to the fine-scale solution rapidly.

2.1.1 Linear Boundary Conditions

First, let χ_i^0 be the nodal basis of the standard finite element space W_H. For example, W_H consists of piecewise linear functions if $\mathscr{T}_H$ is a triangular partition or W_H consists of piecewise bi-linear functions if $\mathscr{T}_H$ is a rectangular partition. We define "standard" multiscale finite element basis functions that coincide with χ_i^0 on the boundaries of the coarse partition and satisfy:

$$\text{div}(\kappa \nabla \chi_i^{ms}) = 0 \text{ in } K \in \omega_i, \quad \chi_i^{ms} = \chi_i^0 \text{ in } \partial K, \ \forall \ K \in \omega_i, \tag{7}$$

where K is a coarse grid block within ω_i (see Fig. 2 for illustration of multiscale finite element basis functions).

Note that multiscale basis functions coincide with standard finite element basis functions on the boundaries of coarse grid blocks, and may be oscillatory in the interior of each coarse grid block depending on κ. Even though the choice of χ_i^0 can be quite arbitrary, our main assumption is that the basis functions satisfy the leading order homogeneous equations when the right hand side f is a smooth function (e.g., L_2 integrable).

Remark 1. Note that in the global formulation of MsFEM one can choose the test functions from W_H and arrive at the Petrov-Galerkin version of multiscale finite element method as introduced in [54] (see also [40]). Many other global variational formulations have been investigated in the literature.

Remark 2. We would like to remark that the MsFEM formulation allows one to take an advantage of scale separation. In particular, K in (7) can be chosen to be a volume smaller than the coarse grid. See [40] for discussions.

2.1.2 Oversampling Technique

Because of linear boundary conditions, the basis functions do not capture the fine-scale features of the solution along the boundaries. This can lead to large errors. When coefficients have a single physical scale ϵ (i.e., $\kappa(x) = \kappa(\frac{x}{\epsilon})$) it has been shown that the error (see [53]) is proportional to ϵ/H, and thus can be large when H is close to ϵ. To illustrate this concept, we depict in Fig. 3, the permeability field (left figure), fine-scale solution (middle figure), and multiscale solution (right figure) computed on a 3×3 coarse grid. As we can see from this figure, the multiscale solution with linear boundary conditions does not capture the fine-scale features of the solution along the boundaries of the coarse grid. In particular, multiscale solution differs from the fine-scale solution along the boundaries of coarse blocks. This can lead to large errors.

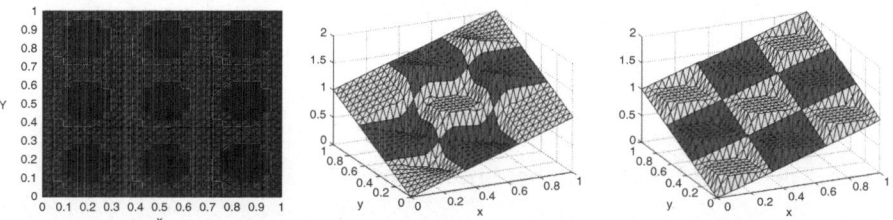

Fig. 3 *Left*: Permeability field. *Middle*: Fine-scale solution. *Right*: Coarse-scale solution with multiscale basis functions that have linear boundary conditions

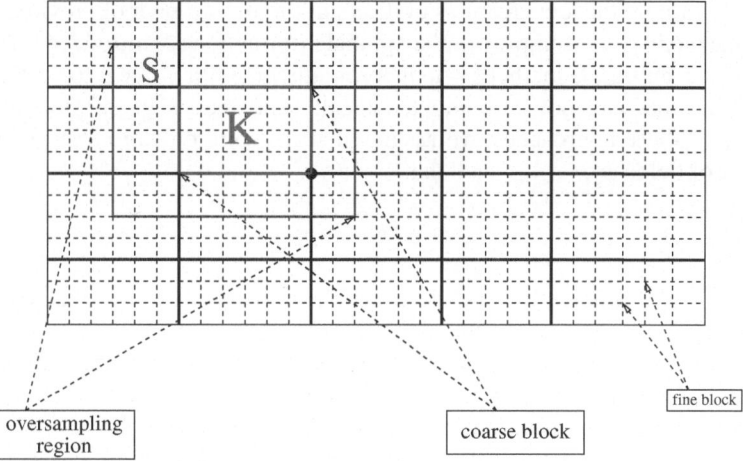

Fig. 4 Schematic description of oversampled regions

Motivated by such examples, Hou and Wu in [52] proposed an oversampling technique for multiscale finite element method. Specifically, let $\chi_i^{ovs,S}$ be the basis functions satisfying the homogeneous elliptic equation in the larger domain $S \supset K$ (see Fig. 4). We then form the actual basis χ_i^{ovs} by linear combination of $\chi_i^{ovs,S}$,

$$\chi_i^{ovs} = \sum_j \alpha_{ij} \chi_j^{ovs,S}.$$

The coefficients α_{ij} are determined by condition $\chi_i^{ovs}(y_j) = \delta_{ij}$, where y_j are nodal points. Other conditions can also be imposed (e.g., α_{ij} are determined based on homogenized parts of $\chi_i^{ovs,S}$). Note that this method is non-conforming. One can also multiply the oversampling test functions $\chi_i^{ovs,S}$ by linear basis functions to restrict them onto ω_i and thus obtain a conforming method. Numerical results and more discussions on oversampling methods can be found in [40].

Many other boundary conditions are introduced and analyzed in the literature. For example, reduced boundary conditions are found to be efficient in many porous media applications ([56]).

2.1.3 The Use of Limited Global Information

Previously, we discussed multiscale methods which employ local information in computing basis functions with the exception of energy minimizing basis functions. The accuracy of these approaches depends on local boundary conditions. Though effective in many cases, multiscale methods that only use local information may not accurately capture the local features of the solution. In a number of previous papers, multiscale methods that employ limited global information are introduced. The main idea of these multiscale methods is to incorporate some fine-scale information about the solution that can not be computed locally and that is given globally. More precisely, in these approaches, we assume that the solution can be represented by a number of fields $p_1,..., p_N$, such that

$$u \approx G(p_1, ..., p_N), \tag{8}$$

where G is a sufficiently smooth function, and $p_1,.., p_N$ are global fields. These fields typically contain the essential information about the heterogeneities at different scales and can also be local fields defined in larger domains. In the above assumption (8), p_i are solutions of homogeneous elliptic equations, $div(\kappa \nabla p_i) = 0$, with some prescribed boundary conditions. These global fields are used to construct multiscale basis functions (often multiple basis functions per coarse node). Finding $p_1, ..., p_N$, in general, can be a difficult task and we refer to Owhadi and Zhang [69] as well as to [39] where various choices of global information are proposed.

We will also consider energy minimizing basis functions (see [80]), where basis functions are obtained by minimizing the energy of the basis functions subject to a global constraint. More precisely, one can use the partition of unity functions $\{\chi_i^{emf}\}_{i=1}^{N_v}$, with N_v being the number of coarse nodes, that provide the least energy. This can be accomplished by solving

$$\min \sum_i \int_{\omega_i} \kappa |\nabla \chi_i^{emf}|^2 \tag{9}$$

subject to $\sum_i \chi_i^{emf} = 1$ with $\text{Supp}(\chi_i) \subset \omega_i$, $i = 1, \ldots, N_v$. Note that the restriction $\sum_i \chi_i^{emf} = 1$ is a global constraint though it is not tied to any particular global fields unlike the methods discussed previously. One can solve (9) following a procedure discussed in [80] or the preconditioned iteration in [77]. We note that the computation of these basis functions requires the solution of a global linear system and it is expensive compared to the local computation of multiscale finite element basis functions with linear boundary conditions χ_i^{ms}.

3 Complementing Coarse Spaces

3.1 Motivation

The coarse spaces discussed above often need to be complemented if more accurate coarse-scale solutions or more robust preconditioners are sought. For this reason, we seek additional basis functions that improve the accuracy of the approximation and we would like to keep the dimension of the coarse space as small as possible. We will consider complementing the coarse spaces described above by finding appropriate local fields in ω_i and by multiplying them with our multiscale functions; see [38] and also [10–14]. We start with the coarse space generated by one basis function per node mentioned above, e.g., χ_i^0 or χ_i^{ms} or χ_i^{emf} or so on, and further we complement this space by adding basis functions in each ω_i. *For simplicity from now on, unless otherwise is stated, we denote the initial basis function for node i by χ_i where χ_i can be computed with any of the above discussed methods.* We will emphasize that this initial choice of basis function is crucial for determining the dimension of the coarse space needed to obtain an accurate coarse-scale approximation and robust preconditioners.

In the procedure below, we will define local approximations of the solution in each ω_i, denoted by ψ_i^l (where i is the node and l is the number of the local approximant). Then, multiscale basis functions will be constructed as $\chi_i \psi_i^l$ (see Fig. 5 for illustration), and the space will be span($\chi_i \psi_i^l$). Furthermore, the coarse-scale solution is sought based on (4) with the coarse spaces defined by $V_0 = \text{span}(\chi_i \psi_i^l)$.

3.2 Coarse Space Dimension. Motivation for Dimension Reduction

We first motivate the choice of the coarse spaces based on our analysis presented in [38]. We briefly review the results of [38]. Consider the eigenvalue problem

$$div(\kappa \nabla \psi_l^{\omega_i}) = \lambda_l^{\omega_i} \widetilde{\kappa} \psi_l^{\omega_i}, \tag{10}$$

where $\lambda_l^{\omega_i}$ (or simply λ_l^i) and $\psi_l^{\omega_i}$ (or simply ψ_l^i) are eigenvalues and eigenvectors in ω_i and $\widetilde{\kappa}$ is defined by

$$\widetilde{\kappa} = \frac{1}{H^2} \kappa \sum_{j=1}^{N_v} |\nabla \chi_j|^2. \tag{11}$$

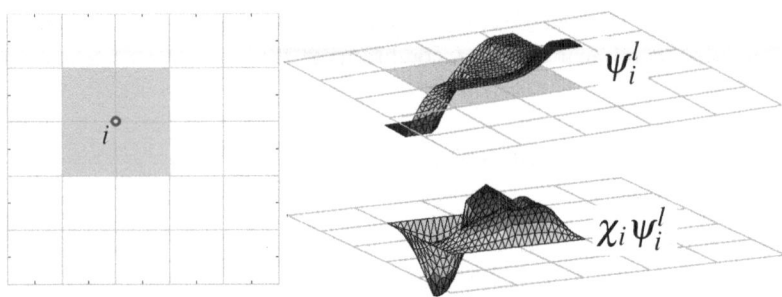

Fig. 5 Schematic description of basis function construction. *Left*: subdomain ω_i. *Right-Top*: Selected eigenvector ψ_i^{ℓ} with small eigenvalue. *Right-Bottom*: product $\chi_i \psi_i^{\ell}$ where χ_i is the initial basis function of node i

We recall that χ_i are initial multiscale basis functions, e.g., $\chi_i = \chi_i^0$ or $\chi_i = \chi_i^{ms}$ or $\chi_i = \chi_i^{emf}$, and N_v is the number of the coarse nodes. One can choose other multiscale basis functions, for example, multiscale basis functions that employ limited global information. The eigenvalue problem considered above is solved with zero Neumann boundary condition and it is understood in a discrete setting (see Appendix for the details of eigenvalue problem setup in the discrete setting). Note that the eigenvalues of (10) depend on the initial basis functions χ_i. Assume eigenvalues are given by

$$0 = \lambda_1^{\omega_i} \leq \lambda_2^{\omega_i} \leq \dots.$$

Basis functions are computed by selecting a number of eigenvalues (starting with small ones) and multiplying corresponding eigenvectors by initial multiscale basis functions χ_i (see Fig. 5 for the illustration). Thus, multiscale space is defined for each i as the span of $\chi_i \psi_l^{\omega_i}$, $l = 1, \dots, L_i$, where L_i is the number of selected eigenvectors.

We note that the dimension of the coarse space depends on the choice of $\widetilde{\kappa}$ and thus it is important to have a good choice of $\widetilde{\kappa}$ when solving the local eigenvalue problem. In this paper, we design a good choice for $\widetilde{\kappa}$ via initial multiscale basis functions. The essential ingredient in designing $\widetilde{\kappa}$ is to guarantee that there are fewer small eigenvalues of (10) which are asymptotically vanishing as the contrast increases. With an initial choice of multiscale basis functions that contain many small-scale localizable features of the solution, one can reduce the dimension of the coarse space as we show in this paper. In particular, we note that the eigenvectors corresponding to small, asymptotically vanishing, eigenvalues represent the local features of the solution that are not captured by the initial multiscale basis functions. This gives a natural way to complement initial coarse spaces and emphasizes the importance of the initial multiscale spaces. If initial basis functions are not chosen carefully, it can result to large dimensional coarse spaces as discussed in the paper. Thus, the use of advanced multiscale techniques in constructing initial basis helps to reduce the dimension of the coarse space needed to achieve contrast-independent two-level domain decomposition preconditioners and more accurate coarse-grid solutions.

In [38], we show that under some conditions the convergence rate is inversely related to the smallest eigenvalue whose eigenvector is not included in the coarse space. This is also observed in our numerical results. More precisely, we show that the convergence rate in the energy norm $\int_D \kappa(x)|\nabla(u - u_0)|^2$ is proportional to

$$\max_i \frac{H^{1+\beta}}{\lambda_{L_i+1}^{\omega_i}}, \tag{12}$$

where H is the coarse mesh size and $\beta \geq 0$ is related to the smoothness of the coefficients. For example, for smooth coefficients one can have $\beta = 1$ and thus recover classical error estimates. We note that these results assume that the right hand side f is a smooth function. We refer to [22] for including the effects of rough source terms in our multiscale simulations. We see from here that one needs to reach larger, O(1), eigenvalues as fast as possible. More precisely, with a choice of initial basis functions χ_i, we need to get rapid increase of eigenvalues. We note that (12) also is a main factor in the condition number of the preconditioned matrix that is computed using two-level additive Schwarz (see discussions later and [35, 45, 46]). Thus, our goal is to choose initial basis functions such that eigenvalues increase rapidly and more precisely, we have fewer asymptotically small eigenvalues.

3.3   Eigenvalue Behavior

In this section, we will discuss how various choices of initial multiscale basis functions can affect the eigenvalue behavior. We start with piecewise linear basis functions, i.e., χ_i are linear basis functions. In this case, $\widetilde{\kappa}$ and κ have similar high-contrast regions because $\nabla \chi_i$ are piecewise constant functions over coarse-grid regions. This eigenvalue problem is the same as the one considered in [35, 45, 46]. In particular, if there are m separated inclusions and channels, then one can observe m small eigenvalues which are asymptotically vanishing as the contrast increases. These small eigenvalues represent high-conductivity regions of the weight function (this can be seen from the corresponding Rayleigh-Quotient). We recall that the inclusions are isolated high-conductivity regions within a coarse block, while channels are isolated high-conductivity regions that connect the boundaries of the coarse blocks (ω_i in this case). In fact, these small eigenvalues are inversely related to the high-conductivity values of the coefficients. We assume throughout that the coefficient takes values 1 and η, where η is large. The corresponding eigenvectors have constant values (in the asymptotic sense when the contrast increase to infinity) within each high-contrast region.

In Fig. 6, we depict a region with three high-conductivity inclusions and one high-conductivity channel. The first four eigenvalues are small and are given by $[0, 0.0745e-5, 0.0918e-5, 0.2416e-5]$ for $\eta = 1e+8$. Here we used piecewise linear functions χ_j to compute $\widetilde{\kappa}$ in (11). We depict eigenvectors corresponding to

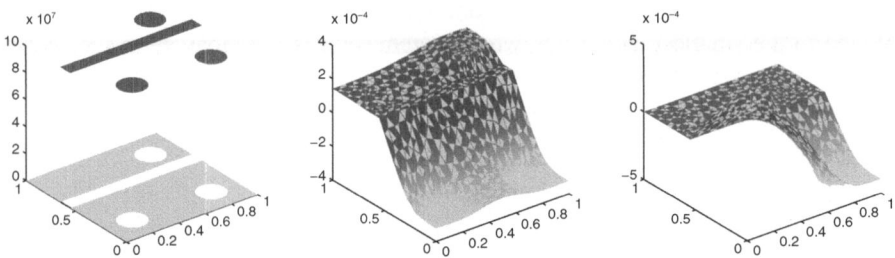

Fig. 6 Illustration of eigenvectors in a unit square coarse block. *Left*: Permeability field κ. *Middle*: Second eigenvector. *Right*: Third eigenvector. The first four eigenvalues using $\widetilde{\kappa} = \kappa$ are $[0\ 0.0745e - 5\ 0.0918e - 5\ 0.2416e - 5]$. The fifth eigenvalue is 9.88

the second and third smallest eigenvalues. The eigenvectors corresponding to the small, asymptotically vanishing, eigenvalues represent all possible constants within high-conductivity regions. The fifth eigenvalue is 9.88 and thus, there is a large gap in the spectrum.

The choice of initial multiscale basis functions, and consequently the choice of $\widetilde{\kappa}$ can result in large coarse dimensional spaces that are needed to eliminate small eigenvalues. Note that without including eigenvectors that correspond to small eigenvalues, one expects very large errors that are proportional to the contrast. Thus, it is essential that we reduce the number of small eigenvalues, which are asymptotically vanishing as the contrast increases.

Note that in many cases, one can have a very large number of small isolated high-conductivity regions within the domain. To illustrate this, in Fig. 7, we schematically depict a coarse region with many small isolated high-conductivity inclusions. In this case, the dimension of the coarse space is very large. For example, if the size of inclusions is of order ϵ, then, we may need a coarse space with the dimension that can scale as $\frac{1}{\epsilon^\gamma}$ (for some $\gamma > 0$ depending density of the inclusions) when there are many inclusions. However, it turns out that one can encapsulate the effects of isolated inclusions in a single basis function per coarse node that will ultimately reduce the dimension of the coarse space. This is done via the initial choice of partition of unity functions resulting to a new weight matrix $\widetilde{\kappa}$. This is discussed in the next section.

3.4 Reduced Dimensional Coarse Spaces

In this section, we will discuss how one can reduce the dimension of the coarse space that is spanned by eigenvectors corresponding to asymptotically small eigenvalues. In particular, we will show that one can reach $O(1)$ eigenvalues with fewer basis functions if initial multiscale basis functions are chosen appropriately.

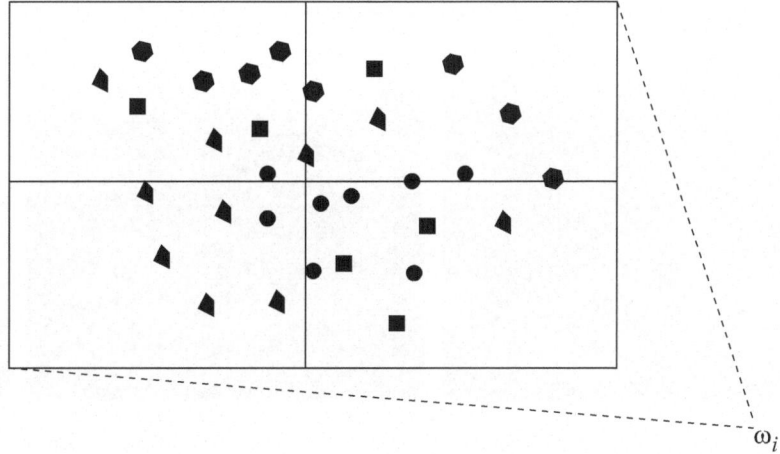

Fig. 7 Schematic description of a coarse region with many isolated inclusions. This will lead to a large dimensional coarse space unless the initial multiscale space is chosen properly

As we mentioned, if the partition of unity functions χ_j are piecewise linear polynomials then $\widetilde{\kappa}$ and κ have the same high-contrast structure. We are interested in partition of unity functions that can "eliminate" isolated high-conductivity inclusions. This can be achieved by minimizing the number of high-conductivity components in $\widetilde{\kappa}$. We note that the high-conductivity regions of the weight $\widetilde{\kappa}$ determine the number of small eigenvalues. By choosing multiscale finite element basis functions or energy minimizing basis functions as defined above (see Sect. 2.1), we can eliminate all isolated high-conductivity inclusions, while preserving the channels. The elimination of isolated high-conductivity inclusions is understood in a sense that their fine-scale features are incorporated into a single basis function. Indeed, the energy of local solutions remains bounded in the high-conducting regions that are isolated. This can be observed in our numerical experiments. In Fig. 8 we depict the coarse grid and κ (left plot) and $\widetilde{\kappa}$ (right picture) using multiscale basis functions on the coarse grid. One can observe that isolated inclusions are removed in $\widetilde{\kappa}$, and consequently, the coarse space contains only long channels that connect boundaries of the coarse grid.

We are interested in the partition of unity functions that can "eliminate" isolated high-conductivity inclusions and thus reduce the dimension of the coarse space. In particular, the effects of isolated inclusions are represented by a coarse-scale basis function in each ω_i. This can be achieved by minimizing high-conductivity components in $\widetilde{\kappa}$. In particular, by choosing multiscale finite element basis functions or energy minimizing basis functions (as discussed in Sect. 2.1), we can eliminate all isolated (i.e., not touching the boundaries of the coarse grids) high-conductivity inclusions, while preserving the channels. To remove the degrees of freedoms associated with isolated inclusions that intersect only a single boundary segment of the coarse-grid block, we can use oversampling, energy minimizing basis functions,

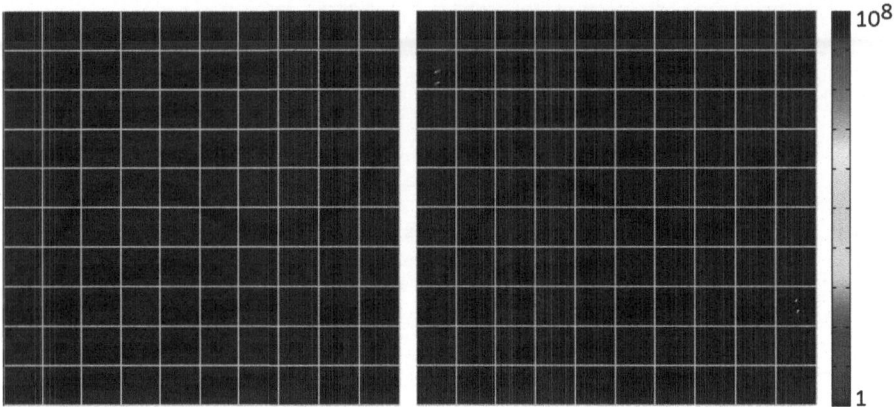

Fig. 8 *Left*: Coarse mesh and original coefficient. Here $\eta = 10^8$. *Right*: Coarse mesh and coefficient $\widetilde{\kappa}$ computed as in (11) using (linear) multiscale basis functions

multiscale basis functions with limited global information or basis functions proposed in [27]. In our numerical results, we will use energy minimizing basis functions.

We note that one can not remove the channels with the partition of unity functions. Indeed, because the partition of unity functions have "unit" gradient flow within coarse-grid blocks, there will be a non-zero gradient within channels, and thus, $\widetilde{\kappa}$ will remain high. Thus, an optimal initial partition of unity functions will be those that can eliminate all isolated inclusions. As we show numerically, multiscale finite element basis functions can achieve that. For example, for the permeability field depicted in Fig. 6, we can remove all three inclusions and $\widetilde{\kappa}$ will contain only one channel. In fact, more sophisticated multiscale basis functions can further reduce the dimension by eliminating the inclusions that touch boundaries.

4   The Use of Coarse Spaces in Domain Decomposition Preconditioners

In this section, we describe how the coarse spaces proposed earlier can be used in two-level domain decomposition preconditioners. In particular, we show that one can obtain preconditioners that yield a contrast-independent condition number, and thus they are optimal in terms of physical parameters. The details of these results can be found in [35, 45, 46]. See [36] for extension of the results to multilevel methods.

Next, we briefly describe the two-level domain decomposition setting that we use. We denote by $\{D_i'\}_{i=1}^N$ the overlapping decomposition obtained from the original nonoverlapping decomposition $\{D_i\}_{i=1}^N$ by enlarging each subdomain D_i to

$$D_i' = D_i \cup \{x \in D, \text{dist}(x, D_i) < \delta_i\}, \quad i = 1, \dots, N, \tag{13}$$

where dist is some distance function and let $V_0^i(D_i')$ be the set of finite element functions with support in D_i'. We also denote by $R_i^T : V_0^i(D_i') \to V^h$ the extension by zero operator. Using the coarse triangulation $\mathcal{T}^H$, we denote the basis functions $\{\Phi_i\}_{i=1}^{N_c}$ (N_c is the number of coarse basis functions), and $A_0 = R_0 A R_0^T$, $R_0^T = [\Phi_1, \ldots, \Phi_{N_c}]$ (see Sect. 2).

We can solve the fine-scale linear system iteratively with the preconditioned conjugate gradient (PCG) method. Any other suitable iterative scheme can be used as well. We use the two-level additive preconditioner of the form

$$B^{-1} = R_0^T A_0^{-1} R_0 + \sum_{i=1}^{N} R_i^T A_i^{-1} R_i, \qquad (14)$$

where the local matrices are defined by

$$v A_i w = a(v, w) \quad \text{for all } v, w \in V_0^i(D_i'), \qquad (15)$$

$i = 1, \ldots, N$. See [63, 76] and references therein. The application of the preconditioner involves solving a coarse-scale system and solving local problems in each iteration. In domain decomposition methods, our main goal is to reduce the number of iterations in the iterative procedure. The appropriate construction of the coarse space V_0 plays a key role in obtaining robust iterative domain decomposition methods.

In the general setting of domain decomposition methods, the overlapping subdomains $\{D_i'\}$ and the coarse triangulation $\mathcal{T}^H$ are not related. The two partitions of unity used here can be chosen independently of each other. Both partitions of unity are needed to construct a contrast-independent domain decomposition method. In the numerical experiments, we will assume that the overlapping subdomains $\{D_i'\}$ coincide with the coarse vertex neighborhoods $\{\omega_i\}$ of $\mathcal{T}^H$. In this case $\delta \asymp H$, where $\delta = \max_{1 \leq i \leq N} \delta_i$ is the overlapping parameter.

Next, we present main results from [35, 45, 46]. We define the coarse interpolation $I_0 : V^h(D) \to V_0$ by

$$I_0 v = \sum_{i=1}^{N_v} \sum_{\ell=1}^{L_i} \left(\int_{\omega_i} \widetilde{\kappa} v \psi_\ell^{\omega_i} \right) I^h(\chi_i \psi_\ell^{\omega_i}), \qquad (16)$$

where I^h is the fine-scale nodal value interpolation (in previous discussions, we have simply used $\chi_i \psi_\ell^{\omega_i}$ for fine-scale interpolation). The following weighted L^2 approximation and weighted H^1 stability properties are valid (see [35, 45, 46] for the proof).

Lemma 1. *For all coarse element K we have*

$$\int_K \widetilde{\kappa}(v - I_0 v)^2 \leq \frac{1}{\lambda_{K,L+1}} \int_{\omega_K} \kappa |\nabla v|^2 \qquad (17)$$

$$\int_K \kappa |\nabla I_0 v|^2 \preceq \max\{1, \frac{1}{\lambda_{K,L+1}}\} \int_{\omega_K} \kappa |\nabla v|^2, \qquad (18)$$

where $\lambda_{K,L+1} = \min_{y_i \in K} \lambda_{L_i+1}^{\omega_i}$ and ω_K is in (3).

Using Lemma 1, the condition number of the preconditioned operator $B^{-1}A$ with B^{-1} defined in (14) using the coarse space V_0 in (A.6) can be estimated. See [35, 45, 46, 63, 76].

Lemma 2. *The condition number of the preconditioned operator $B^{-1}A$ with B^{-1} defined in (14) using the coarse space V_0 defined in (A.6) and (A.5) is of order*

$$cond(B^{-1}A) \preceq C_0^2 \preceq \max\{1, \frac{1}{\lambda_{L+1}}\} \leq 1 + \frac{1}{\lambda_{L+1}},$$

where $\lambda_{L+1} = \min_{1 \leq i \leq N_v} \lambda_{L_i+1}^{\omega_i}$.

Remark 3. The computations of the eigenvectors and coarse spaces can be expensive, in general; however, the proposed coarse spaces can reduce the number of iterations and CPU time significantly if the conductivity has high contrast or the problem is solved multiple times. One can consider the cost of computing eigenvectors (and coarse spaces) as a pre-processing step. Eigenvalue computations can be done consecutively and one can stop once the next eigenvalue exceeds a certain threshold. For example, for problems without high-contrast, one will stop after the computation of the first eigenvalue and eigenvector.

We note that the computations of the eigenvectors can be carried out hierarchically and this allows reducing the computational cost in solving the eigenvalue problem, see [36, 38]. We have shown that with such inexpensive hierarchical computations, we still have optimal preconditioners in terms of the contrast.

5 Numerical Results

In this section, we present representative numerical results for coarse-scale approximation and for the two level additive preconditioner (14) with the local spectral multiscale coarse spaces as discussed above. The equation $-\text{div}(\kappa \nabla u) = 1$ is solved with boundary conditions $u = x + y$ on ∂D. For the coarse-scale approximation, we will vary the dimension of the coarse spaces and investigate the convergence rate as a function of coarse-mesh size, minimum eigenvalue, and the contrast. For preconditioning results, we will investigate the behavior of the condition number as we increase the contrast for various choices of coarse spaces. We will show that, using multiscale basis functions as the initial coarse space, one can achieve contrast-independent results with a small dimensional coarse space. In our simulations, we run the Preconditioned Conjugate Gradient (PCG) until the ℓ_2 norm of the residual is reduced by a factor of 10^{10}. We take $D = [0, 1] \times [0, 1]$ that is divided into 10×10

Fig. 9 Coefficients κ. The numerical results are presented in Tables 1–5

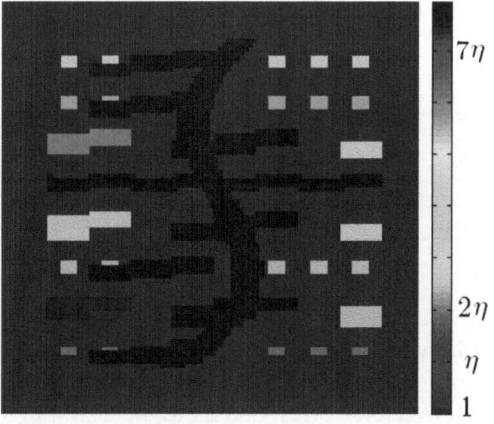

equal square subdomains. Inside each subdomain we use a fine-scale triangulation, where triangular elements constructed from 10×10 squares are used.

We consider the scalar coefficient $\kappa(x)$ depicted in Fig. 9 that corresponds to a background one and high conductivity channels and inclusions. The values of conductivities vary within each region. In particular, we choose the high conductivities between η and 10η randomly for each region.

First, we present the results for a two-level preconditioner. We implement a two-level additive preconditioner with the following coarse spaces: multiscale functions with linear boundary conditions (MS); energy minimizing functions (EMF); spectral coarse spaces with $\widetilde{\kappa} = \kappa$ where piecewise linear partition of unity functions are used as an initial space (LSM $\widetilde{\kappa} = \kappa$); spectral coarse spaces with $\widetilde{\kappa}$ defined by (11) where multiscale finite element basis functions with linear boundary conditions (χ^{ms}) are used as an initial space (LSM $\widetilde{\kappa}$ with MS); spectral coarse spaces with $\widetilde{\kappa}$ defined by (11) where energy minimizing basis functions (χ^{emf}) are used as an initial space (LSM $\widetilde{\kappa}$ with EMF). In Table 1, we show the number of PCG iterations and estimated condition numbers. We also show the dimensions of the coarse spaces. First, we note that the standard coarse space with one basis per coarse node has the dimension 81×81. The smallest dimension can be achieved by using energy minimizing basis functions as an initial coarse space. Similarly, more sophisticated multiscale basis functions, e.g., those proposed in [27], can also remove isolated inclusions that intersect one of the boundary segments of a coarse block and the computations of these basis functions are cheaper. We observe that the number of iterations does not change as the contrast increases when spectral coarse spaces are used. On the contrary, when using multiscale basis functions (one basis per coarse node), the condition number of the preconditioned matrix increases as the contrast increases.

In our next numerical tests, the accuracy of MsFEMs is investigated (see also [38] for more details). In particular, we test the accuracy of MsFEMs when coarse spaces include eigenvectors corresponding to small, asymptotically vanishing (as contrast increases), eigenvalues as well as the cases when additional eigenvectors

Table 1 Number of iterations until convergence and estimated condition number for the PCG and different values of the contrast η with the coefficient depicted in Fig. 9. We set the tolerance to 1e-10. Here $H = 1/10$ with $h = 1/100$

η	MS	EMF	LSM $(\widetilde{\kappa} = \kappa)$	LSM $(\widetilde{\kappa}, \text{MS})$	LSM $(\widetilde{\kappa}, \text{EMF})$
10^3	97(1.92e+03)	71(1.01e+03)	32(8.79e+00)	34(1.01e+01)	32(9.63e+00)
10^5	149(1.92e+05)	97(9.93e+04)	33(8.83e+00)	35(1.02e+01)	34(9.73e+00)
10^7	210(1.92e+07)	116(9.93e+06)	35(8.83e+00)	47(2.34e+01)	41(1.17e+01)
Dim	81	81	165	113	113

are included in the coarse space (see [38] for more discussions). We choose the following notation: LSM+n indicates that the coarse spaces that include eigenvectors corresponding to small, asymptotically vanishing, eigenvalues and n additional eigenvectors corresponding to the next n eigenvalues (in an increasing order). E.g., LSM+0 indicates the coarse space that only includes eigenvectors corresponding to small, asymptotically vanishing eigenvalues, while LSM+1 indicates the coarse space that includes eigenvectors corresponding to small, asymptotically vanishing eigenvalues, plus one more eigenvector in each coarse region that corresponds to the next largest eigenvalue. We consider two different contrasts and two different coarse mesh sizes. We study the convergence of the methods LSM (when $\widetilde{\kappa} = \kappa$) and $\widetilde{\text{LSM}}$ (when multiscale basis functions, either χ^{ms} or χ^{emf}, are used to construct $\widetilde{\kappa}$).

First, in Fig. 10, we compare coarse-scale approximations of the solution for various spaces on 10×10 coarse grid. In the top left figure, the fine-scale solution is depicted. In the top middle figure, the solution computed with multiscale basis functions with linear boundary conditions (χ^{ms}) is plotted. In the top right, we depict the solution computed with energy minimizing basis functions (χ^{emf}). The figures in the second row correspond to coarse-scale approximations computed using spectral basis functions where we use the eigenvectors that correspond to asymptotically small eigenvalues as the contrast increases. Here, the bottom left figure corresponds to the case where the initial space consists of piecewise linear functions, the bottom middle figure corresponds to the case where the initial space consists of multiscale basis functions with linear boundary conditions, and the bottom right figure corresponds to the case where the initial space consists of energy minimizing multiscale basis functions. We see that the approximations with one basis function per node are not accurate (note that we do not use any special global information to construct boundary conditions). In Fig. 11, we depict the coarse-scale approximations with local spectral basis functions when two additional eigenvectors are taken in the coarse space. The latter has better accuracy as one can see comparing Figs. 10 and 11. The approximation errors are presented in Tables 2–5 that we will discuss next.

In all numerical results, the errors in the energy norm ($|\cdot|_A^2$), H^1 norm ($|\cdot|_{H^1}^2$), L^2-weighted norm ($|\cdot|_{L^2}^2$), maximum of the first non-chosen eigenvalue $\lambda_{L+1}^{min} H^2 = \min_i \lambda_{L_i+1} H^2$ and the mean of first non-chosen eigenvalues $\{\lambda_{L_i+1} H^2\}$ (mean($\lambda_{L_i+1} H^2$)) are presented. The last two columns have been scaled by a factor of H^2 to properly compare with the case of the coefficient $\widetilde{\kappa}$ defined

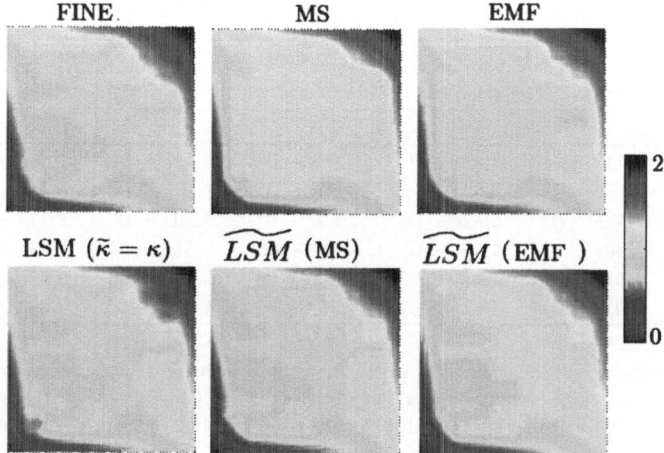

Fig. 10 The results for coarse-scale approximations. Coarse space contains only the eigenvectors that correspond to asymptotically small eigenvalues

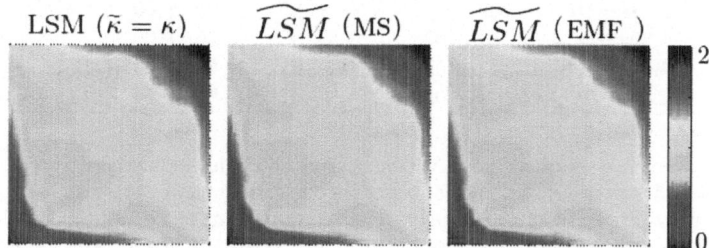

Fig. 11 The results for coarse-scale approximations. Coarse space contains the eigenvectors that correspond to asymptotically small eigenvalues and two additional eigenvectors for each coarse node

in (11). We present the convergence as we increase the number of additional eigenvectors.

In Tables 2 and 3, we present the numerical results for LSM when the initial coarse space consists of piecewise linear basis functions for the contrasts $\eta = 10^4$ and $\eta = 10^6$, respectively. The dimension of the coarse space is 165. We observe that the convergence is robust with respect to the contrast; however, the dimension of the coarse space can be reduced without sacrificing the accuracy. Moreover, the error reduces as the eigenvalue increases.

In Tables 4 and 5, we present the numerical results when the initial coarse space consists of multiscale basis functions with linear boundary conditions for the contrasts $\eta = 10^4$ and $\eta = 10^6$, respectively. The dimension of the coarse space is reduced to 113. We observe that the convergence is robust with respect to the contrast and the error reduces as the minimum eigenvalue increases. Moreover, we observe smaller errors compared to the case with piecewise linear initial

Table 2 Convergence results for LSM with the increasing dimension of the coarse space. LSM+n indicates that the coarse spaces include eigenvectors corresponding to small, asymptotically vanishing eigenvalues, and n additional eigenvectors corresponding to the next n eigenvalues. Here, $\eta = 10^4$. The coefficient is depicted in Fig. 9

$H = 1/10$	$\lvert \cdot \rvert_A^2$	$\lvert \cdot \rvert_{H^1}^2$	$\lvert \cdot \rvert_{L^2}^2$	$\lambda^{min} H^2$	mean$(\lambda_{L_i+1})H^2$
LSM+0	$1.43e-01$	$1.42e-01$	$6.17e-04$	$5.42e-01$	$2.54e+00$
LSM+1	$5.60e-02$	$5.60e-02$	$9.36e-05$	$1.85e+00$	$6.42e+00$
LSM+2	$5.32e-02$	$5.32e-02$	$8.59e-05$	$3.20e+00$	$9.37e+00$
LSM+3	$4.70e-02$	$4.70e-02$	$6.53e-05$	$6.61e+00$	$1.23e+01$
LSM+4	$4.01e-02$	$4.01e-02$	$4.80e-05$	$9.45e+00$	$1.62e+01$
LSM+5	$2.93e-02$	$2.93e-02$	$2.74e-05$	$1.12e+01$	$1.89e+01$

Table 3 Convergence results for LSM with increasing dimension of the coarse space. LSM+n indicates that the coarse spaces include eigenvectors corresponding to small, asymptotically vanishing eigenvalues, and n additional eigenvectors corresponding to the next n eigenvalues. Here, $\eta = 10^6$. The coefficient is depicted in Fig. 9

$H = 1/10$	$\lvert \cdot \rvert_A^2$	$\lvert \cdot \rvert_{H^1}^2$	$\lvert \cdot \rvert_{L^2}^2$	$\lambda^{min} H^2$	mean$(\lambda_{L_i+1})H^2$
LSM+0	$1.43e-01$	$1.43e-01$	$6.17e-04$	$5.41e-01$	$2.54e+00$
LSM+1	$5.60e-02$	$5.60e-02$	$9.36e-05$	$1.85e+00$	$6.42e+00$
LSM+2	$5.33e-02$	$5.33e-02$	$8.60e-05$	$3.20e+00$	$9.37e+00$
LSM+3	$4.70e-02$	$4.70e-02$	$6.54e-05$	$6.61e+00$	$1.23e+01$
LSM+4	$4.01e-02$	$4.01e-02$	$4.80e-05$	$9.45e+00$	$1.62e+01$
LSM+5	$2.93e-02$	$2.93e-02$	$2.74e-05$	$1.12e+01$	$1.89e+01$

Table 4 Convergence results for $\widetilde{LSM}$ with increasing dimension of the coarse space. $\widetilde{LSM}+n$ indicates that the coarse spaces include eigenvectors corresponding to small, asymptotically vanishing eigenvalues and n additional eigenvectors corresponding to the next n eigenvalues. Here, $\eta = 10^4$. The initial coarse space is spanned by multiscale basis functions with piecewise linear boundary conditions (χ^{ms}). The coefficient is depicted in Fig. 9

$H = 1/10$	$\lvert \cdot \rvert_A^2$	$\lvert \cdot \rvert_{H^1}^2$	$\lvert \cdot \rvert_{L^2}^2$	$\widetilde{\lambda}^{min}$	mean$(\widetilde{\lambda}_{L_i+1})$
$\widetilde{LSM}+0$	$1.28e-01$	$1.28e-01$	$7.83e-04$	$1.42e-01$	$8.72e-01$
$\widetilde{LSM}+1$	$4.06e-02$	$4.06e-02$	$6.04e-05$	$8.53e-01$	$1.75e+00$
$\widetilde{LSM}+2$	$3.80e-02$	$3.79e-02$	$4.76e-05$	$1.42e+00$	$2.85e+00$
$\widetilde{LSM}+3$	$3.16e-02$	$3.16e-02$	$3.54e-05$	$2.35e+00$	$4.16e+00$
$\widetilde{LSM}+4$	$2.50e-02$	$2.49e-02$	$2.87e-05$	$3.34e+00$	$4.93e+00$
$\widetilde{LSM}+5$	$2.06e-02$	$2.05e-02$	$1.95e-05$	$4.10e+00$	$5.70e+00$

conditions. We repeat the numerical results for coarse-scale approximations when the initial coarse space consists of energy minimizing basis functions for the contrasts $\eta = 10^4$ and $\eta = 10^6$ in Tables 6 and 7. When using energy minimizing basis functions, we observe even smaller errors compared to the case with the initial coarse space consisting of multiscale basis functions with piecewise linear boundary conditions.

In our next numerical results, we present convergence rates when the coarse mesh is chosen to be $H = 1/20$. We will present this only for the case of multiscale basis functions with linear boundary conditions. In Tables 8 and 9, numerical results for

Table 5 Convergence results for $\widetilde{LSM}$ with increasing dimension of the coarse space. $\widetilde{LSM} + n$ indicates that the coarse spaces include eigenvectors corresponding to small, asymptotically vanishing eigenvalues and n additional eigenvectors corresponding to the next n eigenvalues. Here, $\eta = 10^6$. The initial coarse space is spanned by multiscale basis functions with piecewise linear boundary conditions (χ^{ms}). The coefficient is depicted in Fig. 9

$H = 1/10$	$\|\cdot\|_A^2$	$\|\cdot\|_{H^1}^2$	$\|\cdot\|_{L^2}^2$	$\widetilde{\lambda}^{min}$	mean($\widetilde{\lambda}_{L_i+1}$)
$\widetilde{LSM}+0$	$1.44e-01$	$1.39e-01$	$8.10e-04$	$1.42e-01$	$8.72e-01$
$\widetilde{LSM}+1$	$4.52e-02$	$4.26e-02$	$6.43e-05$	$8.53e-01$	$1.75e+00$
$\widetilde{LSM}+2$	$4.00e-02$	$3.97e-02$	$4.92e-05$	$1.42e+00$	$2.85e+00$
$\widetilde{LSM}+3$	$3.36e-02$	$3.34e-02$	$3.80e-05$	$2.35e+00$	$4.16e+00$
$\widetilde{LSM}+4$	$2.69e-02$	$2.67e-02$	$3.27e-05$	$3.34e+00$	$4.93e+00$
$\widetilde{LSM}+5$	$2.23e-02$	$2.21e-02$	$2.16e-05$	$4.10e+00$	$5.70e+00$

Table 6 Convergence results for $\widetilde{LSM}$ with energy minimizing functions (χ^{emf}) as initial coarse space and increasing dimension of the resulting coarse space. $\widetilde{LSM} + n$ indicates that the coarse spaces include eigenvectors corresponding to small, asymptotically vanishing eigenvalues and n additional eigenvectors corresponding to the next n eigenvalues. Here, $\eta = 10^4$. The coefficient is depicted in Fig. 9

$H = 1/10$	$\|\cdot\|_A^2$	$\|\cdot\|_{H^1}^2$	$\|\cdot\|_{L^2}^2$	$\widetilde{\lambda}^{min}$	mean($\widetilde{\lambda}_{L_i+1}$)
$\widetilde{LSM}+0$	$1.28e-01$	$1.28e-01$	$7.83e-04$	$1.46e-01$	$9.22e-01$
$\widetilde{LSM}+1$	$3.85e-02$	$3.85e-02$	$5.73e-05$	$8.70e-01$	$1.94e+00$
$\widetilde{LSM}+2$	$3.52e-02$	$3.52e-02$	$4.45e-05$	$1.01e+00$	$2.70e+00$
$\widetilde{LSM}+3$	$2.94e-02$	$2.94e-02$	$3.21e-05$	$1.93e+00$	$3.62e+00$
$\widetilde{LSM}+4$	$1.96e-02$	$1.96e-02$	$1.84e-05$	$2.76e+00$	$4.35e+00$
$\widetilde{LSM}+5$	$1.17e-02$	$1.17e-02$	$7.72e-06$	$3.25e+00$	$5.37e+00$

Table 7 Convergence results for $\widetilde{LSM}$ with energy minimizing functions (χ^{emf}) as initial coarse space and increasing dimension of the resulting coarse space. $\widetilde{LSM} + n$ indicates that the coarse spaces include eigenvectors corresponding to small, asymptotically vanishing eigenvalues and n additional eigenvectors corresponding to the next n eigenvalues. Here, $\eta = 10^6$. The coefficient is depicted in Fig. 9

$H = 1/10$	$\|\cdot\|_A^2$	$\|\cdot\|_{H^1}^2$	$\|\cdot\|_{L^2}^2$	$\widetilde{\lambda}^{min}$	mean($\widetilde{\lambda}_{L_i+1}$)
$\widetilde{LSM}+0$	$1.22e-01$	$1.21e-01$	$7.41e-04$	$1.45e-01$	$9.22e-01$
$\widetilde{LSM}+1$	$4.20e-02$	$4.15e-02$	$6.32e-05$	$8.70e-01$	$1.94e+00$
$\widetilde{LSM}+2$	$3.76e-02$	$3.74e-02$	$4.74e-05$	$1.01e+00$	$2.70e+00$
$\widetilde{LSM}+3$	$3.16e-02$	$3.15e-02$	$3.52e-05$	$1.93e+00$	$3.62e+00$
$\widetilde{LSM}+4$	$2.16e-02$	$2.16e-02$	$2.21e-05$	$2.76e+00$	$4.35e+00$
$\widetilde{LSM}+5$	$1.34e-02$	$1.34e-02$	$9.98e-06$	$3.25e+00$	$5.37e+00$

the contrasts $\eta = 10^4$ and $\eta = 10^6$ are presented. We observe that the error is smaller when compared to the case $H = 1/10$. We compute the parameter β (see (12)) and the corresponding β's are $\beta = 0.3474, 0.1335, 1.1116, 1.8323, 2.6273, 3.0011$ for the case $\eta = 10^4$ and $\beta = 0.5117, 0.2855, 1.1863, 1.9202, 2.7302, 3.1174$ for the case $\eta = 10^6$.

Table 8 Convergence results for $\widetilde{LSM}$ with the increasing dimension of the coarse space. $\widetilde{LSM} + n$ indicates that the coarse spaces include eigenvectors corresponding to small, asymptotically vanishing eigenvalues and n additional eigenvectors corresponding to the next n eigenvalues. Here, $\eta = 10^4$. The initial coarse space is spanned by multiscale basis functions with piecewise linear, boundary conditions (χ^{ms}). The coefficient is depicted in Fig. 9

$H = 1/20$	$\lvert\cdot\rvert^2_A$	$\lvert\cdot\rvert^2_{H^1}$	$\lvert\cdot\rvert^2_{L^2}$	$\widetilde{\lambda}^{min}$	mean$(\widetilde{\lambda}_{L_i+1})$
$\widetilde{LSM}+0$	$1.20e-01$	$1.20e-01$	$1.79e-04$	$1.19e-01$	$8.20e-01$
$\widetilde{LSM}+1$	$3.45e-02$	$3.45e-02$	$2.00e-05$	$9.15e-01$	$1.54e+00$
$\widetilde{LSM}+2$	$2.10e-02$	$2.10e-02$	$7.68e-06$	$1.19e+00$	$2.25e+00$
$\widetilde{LSM}+3$	$1.10e-02$	$1.10e-02$	$2.36e-06$	$1.90e+00$	$3.40e+00$
$\widetilde{LSM}+4$	$4.93e-03$	$4.92e-03$	$7.80e-07$	$2.73e+00$	$4.14e+00$
$\widetilde{LSM}+5$	$2.82e-03$	$2.82e-03$	$3.44e-07$	$3.74e+00$	$5.00e+00$

Table 9 Convergence results for $\widetilde{LSM}$ with the increasing dimension of the coarse space. $\widetilde{LSM} + n$ indicates that the coarse spaces include eigenvectors corresponding to small, asymptotically vanishing eigenvalues and n additional eigenvectors corresponding to the next n eigenvalues. Here, $\eta = 10^6$. The initial coarse space is spanned by multiscale basis functions with piecewise linear, boundary conditions (χ^{ms}). The coefficient is depicted in Fig. 9

$H = 1/20$	$\lvert\cdot\rvert^2_A$	$\lvert\cdot\rvert^2_{H^1}$	$\lvert\cdot\rvert^2_{L^2}$	$\widetilde{\lambda}^{min}$	mean$(\widetilde{\lambda}_{L_i+1})$
$\widetilde{LSM}+0$	$1.21e-01$	$1.21e-01$	$1.79e-04$	$1.19e-01$	$8.20e-01$
$\widetilde{LSM}+1$	$3.45e-02$	$3.45e-02$	$2.01e-05$	$9.15e-01$	$1.54e+00$
$\widetilde{LSM}+2$	$2.10e-02$	$2.10e-02$	$7.69e-06$	$1.19e+00$	$2.25e+00$
$\widetilde{LSM}+3$	$1.10e-02$	$1.10e-02$	$2.36e-06$	$1.90e+00$	$3.40e+00$
$\widetilde{LSM}+4$	$4.95e-03$	$4.94e-03$	$7.83e-07$	$2.73e+00$	$4.14e+00$
$\widetilde{LSM}+5$	$2.82e-03$	$2.82e-03$	$3.44e-07$	$3.74e+00$	$5.00e+00$

6 Conclusions

In this paper, we discuss coarse-scale spaces for MsFEMs and two-level domain decomposition methods. The coarse space construction starts with an initial choice of a multiscale space. This space is complemented based on a local spectral problem. In this local spectral problem, the initial multiscale space is used to construct a suitable weight function. We show that a careful choice of initial spaces is important and can lead to a substantial dimension reduction for coarse spaces when the media contain many isolated inclusions. We discuss the convergence of coarse-scale approximation and contrast-independent estimates for the condition number of the preconditioned matrix using these coarse spaces. We present numerical results when the initial coarse space is chosen to be piecewise linear where one needs a large dimensional coarse space. We show that one can use smaller dimensional coarse spaces without sacrificing the convergence. This is accomplished via a careful choice of the initial multiscale space. We also present the estimates for the condition number of the preconditioned matrix that shows that the condition number is independent of the contrast when using smaller dimensional coarse spaces that are discussed in the paper. The extension of these results to more complex problems, such as Stokes-Brinkman equations, is presented in [37].

Acknowledgements Yalchin Efendiev's research is partially supported by NSF (0724704, 0811180, 0934837) and DOE.

We would like to thank Ivan Graham, Tom Hou, and Rob Scheichl for organizing LMS-EPSRC Durham Symposium on Numerical Analysis of Multiscale Problems and for their many contributions that inspired our work.

Appendix: Discrete Formulation of Eigenvalue Problem and Basis Construction

For any ω_i, we define the Neumann matrix A^{ω_i} by

$$vA^{\omega_i}w = \int_{\omega_i} \kappa \nabla v \nabla w \quad \text{for all } v, w \in V^h(\Omega), \quad i = 1, \dots, N, \qquad \text{(A.1)}$$

and the mass matrix of same dimension M^{ω_i} by

$$vM^{\omega_i}w = \int_{\omega_i} \widetilde{\kappa} vw \quad \text{for all } v, w \in V^h(\Omega). \qquad \text{(A.2)}$$

We consider the finite dimensional symmetric eigenvalue problem

$$A^{\omega_i}\phi = \lambda M^{\omega_i}\phi \qquad \text{(A.3)}$$

and denote its eigenvalues and eigenvectors by $\{\lambda_\ell^{\omega_i}\}$ and $\{\psi_\ell^{\omega_i}\}$, respectively. Note that the eigenvectors $\{\psi_\ell^{\Omega}\}$ form an orthonormal basis of $V^h(\Omega)$ with respect to the M^{ω_i} inner product. Note that $\lambda_1^{\omega_i} = 0$. We order eigenvalues as

$$\lambda_1^{\omega_i} \leq \lambda_2^{\omega_i} \leq \cdots \leq \lambda_j^{\omega_i} \leq \dots. \qquad \text{(A.4)}$$

The eigenvalue problem above corresponds to the approximation of the eigenvalue problem

$$-\text{div}(\kappa \nabla v) = \lambda \widetilde{\kappa} v$$

in Ω with homogeneous Neumann boundary condition, where $\psi_\ell^{\omega_i}$ denotes the ℓth eigenvector of the Neumann matrix associated to the neighborhood of y_i.

We choose the basis functions that span the eigenfunctions corresponding to small, asymptotically vanishing, eigenvalues in the way described below. Let $\{\chi_i\}_{i=1}^{N_v}$ be a partition of unity subordinated to the covering $\{\omega_i\}$. Define the set of coarse basis functions

$$\Phi_{i,\ell} = \chi_i \psi_\ell^{\omega_i} \quad \text{for } 1 \leq i \leq N_v \text{ and } 1 \leq \ell \leq L_i, \qquad \text{(A.5)}$$

where L_i is the number of eigenvalues that will be chosen for the node i. Denote by V_0, as before, the *local spectral multiscale* space

$$V_0 = \text{span}\{\Phi_{i,\ell} : 1 \leq i \leq N_v \text{ and } 1 \leq \ell \leq L_i\}. \qquad (A.6)$$

References

1. J. E. Aarnes, *On the use of a mixed multiscale finite element method for greater flexibility and increased speed or improved accuracy in reservoir simulation*, SIAM MMS, 2 (2004), 421–439.
2. J. E. Aarnes, Y. Efendiev, and L. Jiang, *Analysis of multiscale finite element methods using global information for two-phase flow simulations*, SIAM MMS, 2008.
3. J. Aarnes and T. Hou, *Multiscale domain decomposition methods for elliptic problems with high aspect ratios*, Acta Math. Appl. Sin. Engl. Ser., 18(1):63–76, 2002.
4. J. E. Aarnes, S. Krogstad, and K.-A. Lie, *A hierarchical multiscale method for two-phase flow based upon mixed finite elements and nonuniform grids*, Multiscale Model. Simul. 5(2) (2006), 337–363.
5. A. Abdulle and B. Engquist, *Finite element heterogeneous multiscale methods with near optimal computational complexity*, Multiscale Model. Simul. 6 (2007), no. 4, 10591084.
6. G. Allaire and R. Brizzi, *A multiscale finite element method for numerical homogenization*, SIAM MMS, 4(3), 2005, 790–812.
7. T. Arbogast, *Implementation of a locally conservative numerical subgrid upscaling scheme for two-phase Darcy flow*, Comput. Geosci., 6 (2002), 453–481.
8. T. Arbogast, G. Pencheva, M. F. Wheeler, and I. Yotov, *A multiscale mortar mixed finite element method*, SIAM J. Multiscale Modeling and Simulation, 6(1), 2007, 319–346.
9. I. Babuška and R. Lipton, *Optimal Local approximation spaces for generalized finite element methods with application to multiscale problems*, Multiscale Model. Simul. 9(1), 373–406 (2011)
10. I. Babuška, Ivo, V. Nistor and N. Tarfulea, *Generalized finite element method for second-order elliptic operators with Dirichlet boundary conditions*, J. Comput. Appl. Math. 218 (2008), no 1, 175–183.
11. I. Babuška, U. Banerjee, and J. E. Osborn, *Survey of meshless and generalized finite element methods: A unified approach*, Acta Numerica, 1–125, 2003.
12. I. Babuška, G. Caloz, and E. Osborn, *Special finite element methods for a class of second order elliptic problems with rough coefficients*, SIAM J. Numer. Anal., 31:945–981, 1994.
13. I. Babuška and J. M. Melenk, *The partition of unity method*, Internat. J. Numer. Methods Engrg., 40:727–758, 1997.
14. I. Babuška and E. Osborn, *Generalized finite element methods: Their performance and their relation to mixed methods*, SIAM J. Numer. Anal., 20:510–536, 1983.
15. M. Balhoff, S. Thomas, and M. Wheeler, *Mortar coupling and upscaling of pore-scale models*, Computational Geosciences, vol. 12 (September, 2007), no. 1, pp. 15–27.
16. A. Bensoussan, J. L. Lions, and G. Papanicolaou, *Asymptotic analysis for periodic structures*, Volume 5 of Studies in Mathematics and Its Applications, North-Holland Publ., 1978.
17. L. Berlyand and A. Novikov, *Error of the network approximation for densely packed composites with irregular geometry*, SIAM Journal on Mathematical Analysis, 34(2) (2002). 385–408.
18. L. Berlyand and H. Owhadi, *Flux norm approach to finite dimensional homogenization approximations with non-separated scales and high contrast*, Archives for Rational Mechanics and Analysis (2010, Volume 198, Number 2, 677–721).

19. L. Borcea and G.C. Papanicolaou, *Network approximation for transport properties of high contrast materials*, SIAM Journal on Applied Mathematics, vol. 58, no. 2, 1998, 501–539.
20. A. Bourgeat and A. Piatnitski, *Approximations of effective coefficients in stochastic homogenization*, Ann. Inst. H. Poincare Probab. Statist. 40 (2004), no. 2, 153–165.
21. A. Brandt, *Multiscale solvers and systematic upscaling in computational physics*, Computer Physics Communication, 169 (2005) 438–441.
22. V. Calo, Y. Efendiev, and J. Galvis, *A note on variational multiscale methods for high-contrast heterogeneous flows with rough source terms*, article in press (available online February 02-2011), Advances in Water Resources.
23. T. Chartier, R. Falgout, V.E. Henson, J. Jones, T. Manteuffel, S. McCormick, J. Ruge, and P.S. Vassilevski, *Spectral element agglomerate AMGe*, in Domain Decomposition Methods in Science and Engineering XVI, Lecture Notes in Computational Science and Engineering, Springer-Verlag, Berlin Heidelberg **55**(2007), 515–524.
24. Y. Chen and L.J. Durlofsky, *Adaptive local-global upscaling for general flow scenarios in heterogeneous formations*, Transport in Porous Media, 62 (2006), 157–185.
25. Z. Chen and T.Y. Hou, *A mixed multiscale finite element method for elliptic problems with oscillating coefficients*, Math. Comp., 72 (2002), 541–576.
26. Z. Chen and T. Savchuk, *Analysis of the multiscale finite element method for nonlinear and random homogenization problems*, SIAM J. Numer. Anal., 46 (2008), 260–279.
27. C.C. Chu, I.G. Graham, and T.Y. Hou, *A new multiscale finite element method for high-contrast elliptic interface problem*, Math. Comp., 79 , 1915–1955, 2010.
28. E. Chung and Y. Efendiev, *Reduced-contrast approximations for high-contrast multiscale flow problems*, Multiscale Model. Simul. 8 (2010), no. 4, 1128"1153.
29. M. Dryja, *Multilevel Methods for Elliptic Problems with Discontinuous Coefficients in Three Dimensions*, Seventh International Conference of Domain Decomposition Methods in Scientific and Engineering Computing, by David E. Keyes and Jinchao Xu, vol. 180, 1994, 43–47.
30. L.J. Durlofsky, *Numerical calculation of equivalent grid block permeability tensors for heterogeneous porous media*, Water Resour. Res., 27 (1991), 699–708.
31. L.J. Durlofsky, *Coarse scale models of two-phase flow in heterogeneous reservoirs: Volume averaged equations and their relation to existing upscaling techniques*, Comp. Geosciences 2 (1998), 73–92.
32. L.J. Durlofsky, Y. Efendiev, and V. Ginting, *An adaptive local-global multiscale finite volume element method for two-phase flow simulations*, Advances in Water Resources, 30 (2007), 576–588.
33. J. Eberhard, S. Attinger, and G. Wittum, *Coarse graining for upscaling of flow in heterogeneous porous media*, Multiscale Model. Simul. 2 (2004), no. 2, 269"301.
34. J. Eberhard and G. Wittum, *A coarsening multigrid method for flow in heterogeneous porous media*. Multiscale methods in science and engineering, 111"132, Lect. Notes Comput. Sci. Eng., 44, Springer, Berlin, 2005.
35. Y. Efendiev and J. Galvis, *A domain decomposition preconditioner for multiscale high-contrast problems*, in Domain Decomposition Methods in Science and Engineering XIX, Huang, Y.; Kornhuber, R.; Widlund, O.; Xu, J. (Eds.), Volume 78 of Lecture Notes in Computational Science and Engineering, Springer-Verlag, 2011, Part 2, 189–196.
36. Y. Efendiev, J. Galvis and P. Vassielvski, *Spectral element agglomerate algebraic multigrid methods for elliptic problems with high-Contrast coefficients*, in Domain Decomposition Methods in Science and Engineering XIX, Huang, Y.; Kornhuber, R.; Widlund, O.; Xu, J. (Eds.), Volume 78 of Lecture Notes in Computational Science and Engineering, Springer-Verlag, 2011, Part 3, 407–414.
37. Y. Efendiev, J. Galvis, R. Lazarov and J. Willems, *Robust domain decomposition preconditioners for abstract symmetric positive definite bilinear forms*, submitted.
38. Y. Efendiev, J. Galvis, and X. H. Wu, *Multiscale finite element methods for high-contrast problems using local spectral basis functions*, Journal of Computational Physics. Volume 230, Issue 4, 20 February 2011, Pages 937–955.

39. Y. Efendiev, V. Ginting, T. Hou, and R. Ewing, *Accurate multiscale finite element methods for two-phase flow simulations*, J. Comp. Physics, 220 (1), 155–174, 2006.
40. Y. Efendiev and T. Hou, *Multiscale finite element methods. Theory and applications*, Springer, 2009.
41. Y. Efendiev, T. Hou, and V. Ginting, *Multiscale finite element methods for nonlinear problems and their applications*, Comm. Math. Sci., 2 (2004), 553–589.
42. Y. Efendiev, T. Y. Hou, and X. H. Wu, *Convergence of a nonconforming multiscale finite element method*, SIAM J. Num. Anal., 37 (2000), 888–910.
43. W. E and B. Engquist, *The heterogeneous multiscale methods*, Commun. Math. Sci. 1 (2003), no. 1, 87–132.
44. W. E, P. Ming, and P.W. Zhang, *Analysis of the heterogeneous multiscale method for elliptic homogenization problems*, J. AMS., 2005, Vol. 18, 121–156.
45. J. Galvis and Y. Efendiev, *Domain decomposition preconditioners for multiscale flows in high contrast media*, SIAM MMS, Volume 8, Issue 4, 1461–1483 (2010).
46. J. Galvis and Y. Efendiev, *Domain decomposition preconditioners for multiscale flows in high-contrast media: Reduced dimension coarse spaces*, SIAM MMS, Volume 8, Issue 5, 1621–1644 (2010).
47. M. Gerritsen and L.J. Durlofsky, *Modeling of fluid flow in oil reservoirs*, Annual Reviews in Fluid Mechanics, 37 (2005), pp. 211–238.
48. A. Gloria, *Analytical framework for the numerical homogenization of elliptic monotone operators and quasiconvex energies*, SIAM MMS, 5 (2006), No. 3, pp 996–1043.
49. A. Gloria, *Reduction of the resonance error. Part 1: Approximation of homogenized coefficients*, Math. Models Methods Appl. Sci. (M3AS), Issue: 1(2011) pp. 1–30 DOI: 10.1142/S0218202511005507.
50. I. G. Graham, P. O. Lechner, and R. Scheichl, *Domain decomposition for multiscale PDEs*, Numer. Math., 106(4):589–626, 2007.
51. U. Hetmaniuk and R. Lehoucq, *Special finite element methods based on component mode synthesis techniques*, Sandia National Laboratories, Technical report SAND 2009–0115J. Accepted for publication in ESAIM: Mathematical Modelling and Numerical Analysis.
52. T.Y. Hou and X.H. Wu, *A multiscale finite element method for elliptic problems in composite materials and porous media*, Journal of Computational Physics, 134 (1997), 169–189.
53. T.Y. Hou, X.H. Wu, and Z. Cai, *Convergence of a Multiscale Finite Element Method for Elliptic Problems With Rapidly Oscillating Coefficients*, Math. Comput., 68 (1999), 913–943.
54. T.Y. Hou, X.H. Wu, and Y. Zhang, *Removing the cell resonance error in the multiscale finite element method via a Petrov-Galerkin formulation*, Communications in Mathematical Sciences, 2(2) (2004), 185–205.
55. T. Hughes, G. Feijoo, L. Mazzei, and J. Quincy, *The variational multiscale method - a paradigm for computational mechanics*, Comput. Methods Appl. Mech. Engrg, 166 (1998), 3–24.
56. P. Jenny, S.H. Lee, and H. Tchelepi, *Multi-scale finite volume method for elliptic problems in subsurface flow simulation*, J. Comput. Phys., 187 (2003), 47–67.
57. L. Jiang, Y. Efendiev and I. Mishev, *Mixed multiscale finite element methods using approximate global information based on partial upscaling*, Computational Geosciences 14 (2010), 319–341.
58. V. Jikov, S. Kozlov, and O. Oleinik. *Homogenization of differential operators and integral functionals,* Springer-Verlag, 1994, Translated from Russian.
59. J. V. Lambers, M. G. Gerritsen, and B. T. Mallison, *Accurate Local Upscaling with Variable Compact Multi-point Transmissibility Calculations*, Computational Geosciences 12, Special Issue on Multiscale Methods for Flow and Transport in Heterogeneous Porous Media (2008), p. 399–416.
60. I. Lunati and P. Jenny, *Multi-scale finite-volume method for compressible multi-phase flow in porous media*, J. Comp. Phys., 216:616–636, 2006.
61. I. Lunati and P. Jenny, *Multiscale finite-volume method for compressible multiphase flow in porous media*, Journal of Computational Physics, 216, (2006), 616–636.

62. I. Lunati and P. Jenny, *The Multiscale Finite Volume Method: A flexible tool to model physically complex flow in porous media*, 10th European Conference on the Mathematics of Oil Recovery Amsterdam, The Netherlands, 4 7 September, 2006.

63. T. P. A. Mathew, *Domain decomposition methods for the numerical solution of partial differential equations*, volume 61 of *Lecture Notes in Computational Science and Engineering*, Springer-Verlag, Berlin, 2008.

64. F. Nataf, H. Xiang, V. Dolean, N. Spillane, *A coarse space construction based on local Dirichlet to Neumann maps*, accepted for publication in SIAM J. Sci Comput., 2011.

65. S.V. Nepomnyaschikh, *Mesh theorems on traces, normalizations of function traces and their inversion*, Soviet J. Numer. Anal. Math. Modelling, 6(2):151–168, 1991.

66. J. Nolen, G. Papanicolaou, and O. Pironneau. *A framework for adaptive multiscale method for elliptic problems*. SIAM MMS., 7:171–196, 2008.

67. J.M. Nordbotten, *Adaptive variational multiscale methods for multiphase flow in porous media*, SIAM MMS, 7(3), doi:10.1137/080724745.

68. M. Ohlberger, *A posterior error estimates for the heterogenoeous mulitscale finite element method for elliptic homogenization problems* SIAM Multiscale Mod. Simul. pages 88–114 vol. 4 num. 1–2005.

69. H. Owhadi and L. Zhang, *Metric based up-scaling*, Comm. Pure and Applied Math., vol. LX:675–723, 2007.

70. H. Owhadi and L. Zhang, *Localized bases for finite dimensional homogenization approximations with non-separated scales and high-contrast*, accepted SIAM MMS, Available at Caltech ACM Tech Report No 2010–04. arXiv:1011.0986

71. C. Pechstein and R. Scheichl, *Analysis of FETI methods for multiscale PDEs*, Numerische Mathematik 111(2):293–333, 2008.

72. M. Peszynska, M. F. Wheeler, I. Yotov, *Mortar upscaling for multiphase flow in porous media*, Comp. Geosciences (6), pp. 73–100, 2002.

73. M. Sarkis, *Nonstandard coarse spaces and Schwarz methods for elliptic problems with discontinuous coefficients using non-conforming elements*, Numer. Math., 77(3), 383–406, 1997.

74. M. Sarkis, *Partition of unity coarse spaces: enhanced versions, discontinuous coefficients and applications to elasticity*, Domain decomposition methods in science and engineering, Natl. Auton. Univ. Mex., Mexico, 149–158, 2003.

75. D. Svyatskiy, D. Moulton and K. Lipnikov, *Multilevel Multiscale Mimetic (M3) method for two-phase flows in porous media*, Journal of Computational Physics, vol. 227(14), 6727–6753, 2008.

76. A. Toselli and O. Widlund. *Domain decomposition methods—algorithms and theory*, volume 34 of Springer Series in Computational Mathematics, Springer-Verlag, Berlin, 2005.

77. J. Van lent, R. Scheichl, and I.G. Graham, *Energy-minimizing coarse spaces for two-level Schwarz methods for multiscale PDEs*. Numer. Linear Algebra Appl. 16 (2009), no. 10, 775–799.

78. P.S. Vassilevski, *Coarse Spaces by Algebraic Multigrid: Multigrid Convergence and Upscaling Error Estimates*, (2010) (to appear). Available as Lawrence Livermore National Laboratory Technical Report LLNL-PROC-432896, May 21, 2010.

79. X.H. Wu, Y. Efendiev, and T.Y. Hou, *Analysis of upscaling absolute permeability*, Discrete and Continuous Dynamical Systems, Series B, 2 (2002), 185–204.

80. J. Xu and L. Zikatanov, *On an energy minimizing basis for algebraic multigrid methods*, Comput. Visual Sci., 7:121–127, 2004.

Fast Algorithms for High Frequency Wave Propagation

Björn Engquist and Lexing Ying

Abstract High frequency wave propagation is computationally challenging due to the very large number of unknowns that are needed in direct numerical approximations. We will present new fast algorithms for the solution of the linear systems, which follow from discretization of the Helmholtz equation and its related integral equation formulation. For the Helmholtz equation we present a new type of preconditioner, which, together with the GMRES iterative method, results in a near optimal computational complexity. The cost of the preconditioner scales essentially linearly with the number of unknowns and the number of iterations is independent of frequency. In the integral equation case, a directional fast multilevel technique also results in a near optimal computational complexity.

1 Introduction

There is currently an increasing interest in simulation of high frequency wave propagation because of several driving forces behind this. An obvious reason is the importance of wireless communication. Another is seismic exploration for natural resources. Further applications include ultra sound medical imaging, acoustics, and nondestructive testing of elastic structures. The underlying wave phenomena are well described by the scalar wave equation, Maxwell's equations, and the equations of linear elasticity for acoustic, electromagnetic, and elastic waves respectively. We will focus our presentation in this paper on the scalar wave equation,

$$\frac{\partial u(x,t)}{\partial t} = c^2(x)\Delta u(x,t), \quad u : D \times \mathbb{R}^+ \to \mathbb{R}, \quad D \subset \mathbb{R}^d \tag{1}$$

B. Engquist (✉) · L. Ying
Department of Mathematics and ICES, The University of Texas at Austin, Austin, TX, USA
e-mail: engquist@math.utexas.edu; lexing@math.utexas.edu

I.G. Graham et al. (eds.), *Numerical Analysis of Multiscale Problems*, Lecture Notes in Computational Science and Engineering 83, DOI 10.1007/978-3-642-22061-6_5, © Springer-Verlag Berlin Heidelberg 2012

and in particular on the corresponding frequency formulation,

$$\Delta u(x) + k^2(x)u(x) = f(x), \quad k^2(x) = \frac{\omega^2}{c^2(x)}, \quad x \in D \subset \mathbb{R}^d. \qquad (2)$$

We will discuss the core algorithms in the solution of the related linear systems of equations and these algorithms are relevant also for electromagnetic and elastic wave propagation.

Simulations of high frequency wave propagation are computationally challenging. High frequency means that the important wavelengths in the solution are short compared to the size of the computational domain. From the classical Shannon sampling theorem we know that one needs at least two points per wavelength just in order to represent the solution, [55]. Typical engineering simulations are done with 5 to 20 points per wavelength in all dimensions due to numerical dispersion. Similar requirements on the number of unknowns are also needed in finite element and discontinuous Galerkin simulations.

This means that a 3D computation in the time domain requires at least $O(\omega^4)$ unknowns and thus at least $O(\omega^4)$ flops. Due to error propagation these estimates should actually be slightly higher in order to guarantee a given accuracy, [43].

One option to meet this challenge is to use one of the many high frequency analytic approximations instead of the full wave equation. Classical examples are geometrical optics, geometrical theory of diffraction, physical optics and Gaussian beams. For a discussion of computations based on this class of techniques, see the survey in [25]. We will consider numerical approximations of the full wave equation, which contain all physical wave phenomena. Asymptotic models are in our case only used as guidelines in the design and in analysis.

Even if all wave phenomena are kept, it is common to consider simplifications in terms of the number of dimensions. A standard technique is to study the frequency domain problem (2). In order to analyze a broad band signal, a superposition of solutions to equations with different ω is applied. Our original problem, which required at least (ω^4) unknowns is thus reduced to one with at least $O(\omega^3)$. For piecewise constant materials we can reformulate the wave equation as an integral equation over the boundary and interface sets. The dimension is thus reduced further and the requirement is $O(\omega^2)$ unknowns.

The choice of technique for the discretization of the wave equation in terms of the selection of difference stencils or finite elements is important, but is outside the scope of this presentation. The discretization method influences to some extent the number of unknowns required for a given accuracy. The properties that we need in our discussion of solutions of the linear algebraic systems resulting from the discretization are common to many numerical methods.

The linear system of equations that follows from the discretization of the Helmholtz equation or the corresponding integral equation can of course be solved by Gaussian elimination. The system from an approximation of the Helmholtz equation in 3D can be solved by a multifrontal type Gaussian elimination with nested dissection in $O(N^2)$ operations for N unknowns. The system from an integral equation formulation is dense and Gaussian elimination requires $O(N^3)$ flops for 3D problems. (Note that N is typically much smaller for the integral equation based methods.) It is natural to look at iterative methods for solving the linear systems to reduce the computational cost. We will consider standard iterative techniques as, for example, GMRES and QMR and focus on the crucial preconditioning in the Helmholtz case and matrix vector multiply in the integral equation case.

The key to a successful iterative method for the Helmholtz equation (2) is an efficient preconditioner. There are many excellent preconditioning techniques for standard elliptic positive definite problems, see e.g. the text [53]. The challenges related to the high frequency Helmholtz equation are (on the algebraic side) that the resulting linear system is not positive definite and (on the analytic side) that sharp signals may propagate over long distances. The latter means that a preconditioner approximating the solution with a slight error in the phase velocity creates a large pointwise error even for quite short distances. This reduces the efficiency of the preconditioning and serves as an explanation why methods based only on independent approximations of wave propagation but not on the given discretization have been disappointing. One such example is the use of multigrid, which otherwise is highly successful for regular elliptic problems. See Sect. 2.1 and the article by Ernst and Gander [33] in this volume for further discussions.

There are other techniques that start from the given discretization and simplify or compress it as, for example, in many different incomplete LU decompositions. The reason for success of the techniques presented here is that they have enough accuracy over reasonable distances and that they focus on highly compressible quantities. For the latter property the sweeping nature of the ordering of elimination and compression is essential, see Sect. 2.2 for details. Other orderings of elimination, for example the ones in [48, 59, 60], do not result the same optimal computational complexity for the Helmholtz equation.

In Sects. 2.3 and 2.4 we present two sweeping preconditioners, one based on hierarchical matrix representation for the compression and the other on restricting the domain of interaction between sets of unknowns by perfectly matched layers. Section 2.5 extends the technique to the 3D case. The theory is developed for constant coefficients and two dimensions in the first approach but all 2D and 3D numerical examples indicate that both techniques for variable coefficients generate methods, which scale essentially linearly with the number of unknowns independent of frequency.

Section 3 deals with linear systems from discretzations of boundary integral formulations. Again iterative methods supply the outer loop and our challenge is to speed up the matrix vector multiplication, which otherwise for N unknowns requires $O(N^2)$ operations, since we are dealing with dense matrices. The fast multipole method (FMM), of Greengard and Rokhlin [39] reduces the computational cost to $O(N)$ for standard elliptic problems. As for the Helmholtz equation discussed above sharp signals are carried by a solution over long distances. The interaction between sets of unknowns at large distances is not automatically of low rank, independent of frequency. The original FMM thus fails to give the desired improvement in computational complexity. Again the choice of sets for which the interaction matrices can be compressed is crucial.

In Sect. 3 we present a hierarchical decomposition of the unknowns which is the basis for compression that reduces the computational cost to $O(N \log N)$. The decomposition is not just in far and near fields but also in different directions, see Sect. 3.2. An approximation theorem, together with complexity estimates for the algorithm gives the theoretical foundation for the overall computational cost. The approximation theorem is constructive but we only use it as a guideline for how the hierarchical decomposition should be done. The practical compression is done by singular value decomposition (SVD) of the relevant interaction sub matrices. If the SVD would be applied to the full interaction sub matrices the cost would be high but it is enough to apply it to a random selection of rows and columns of these matrices. The directional 2D algorithm is given in Sect. 3.3 and in Sect. 3.4 the 3D case is discussed.

2   Sweeping Preconditioner for the Helmholtz Equation

Let the domain of interest be the unit box $D = (0, 1)^d$ with $d = 2, 3$. The time-independent wave field $u(x)$ for $x \in D$ satisfies

$$\Delta u(x) + \frac{\omega^2}{c^2(x)} u(x) = f(x), \quad x \in D, \tag{3}$$

where ω is the time frequency, $c(x)$ is the velocity field, and $f(x)$ is the external force supported in D. Commonly used boundary conditions are approximations of the Sommerfeld condition which guarantees that the wave field generated by $f(x)$ propagates out of the domain if $c(x)$ is constant outside a sufficiently large ball. By appropriately rescaling the system, we can assume conveniently that the mean of $c(x)$ is equal to 1 and $\lambda = \frac{2\pi}{\omega}$ is the (typical) wavelength.

2.1   *Background*

Efficient and accurate numerical solution of the Helmholtz problem is one of the urgent problems in computational mathematics. This is, however, a very difficult problem due to two main reasons. Firstly as we mentioned already, in a typical engineering application, the Helmholtz equation is discretized with at least 5 to 20 points per wavelength. Therefore, the number of samples n in each dimension is proportional to ω, the total number of samples N is $n^d = O(\omega^d)$, and the discrete system of the Helmholtz equation is of size $O(\omega^d) \times O(\omega^d)$. In the high frequency range where ω is large, this is an enormous system. Secondly, as the discrete system is highly indefinite and has a very oscillatory Green's function due to the wave nature of the Helmholtz equation, most of the modern multiscale techniques developed for elliptic or parabolic problems are no longer effective.

The most efficient direct methods for solving the discretized Helmholtz systems are the multifrontal methods with nested dissection [23, 36, 46]. The multifrontal methods exploit the locality of the discrete operator and construct an LDL^t factorization based on a hierarchical partitioning of the domain. Their computational costs depend quite strongly on the dimensionality. In 2D, for a problem with $N = n \times n$ unknowns, a multifrontal method takes $O(N^{3/2})$ steps and $O(N \log N)$ storage space. The prefactor is usually rather small, making the multifrontal methods effectively the default choice for the 2D Helmholtz problem. In 3D, for a problem with $N = n \times n \times n$ unknowns, a multifrontal method takes $O(n^6) = O(N^2)$ steps and $O(n^4) = O(N^{4/3})$ storage space, which can be very costly for large scale 3D problems.

There has been a surge of developments in iterative methods for solving the Helmholtz equation. The following discussion is by no means complete and more details can be found in [30].

Standard multigrid methods do not work very well for the Helmholtz equation for several reasons. The most important one is that the oscillations on the scale of the wavelength cannot be carried on the coarse grids. Several methods have been proposed to address this issue [9, 24, 34, 45, 47, 58]. Among them is the wave-ray method that performs quite well for the constant coefficient case. Others methods including [3, 22, 57] leverage the idea of domain decomposition. These methods are typically quite suitable for parallel implementation, as the computation in each subdomain can essentially be done independently. However, convergence rates of these methods are usually quite slow [30]. Another class of methods [2, 31, 32, 44] that attracts a lot of attention recently seeks to precondition the Helmholtz operator with a shifted Laplacian operator $\Delta - \frac{\omega^2}{c^2(x)}(\alpha + i\beta)$ with $\alpha > 0$ to improve the spectral property of the discrete Helmholtz system. Since the shifted Laplacian operator is elliptic, standard algorithms such as multigrid can be used for its inversion. These methods offer quite significant improvements for the convergence rate, but the reported number of iterations typically still grow linearly with respect to ω. Another class of preconditioners [4, 35, 50] is based on incomplete LU (ILU) decomposition, i.e., generating only a small portion of the entries of the LU factorization of the discrete Helmholtz operator and applying this ILU decomposition as a preconditioner. Recent approaches based on ILUT (incomplete LU factorization with thresholding) and ARMS (algebraic recursive multilevel solver) have been reported in [50]. Though these ILU preconditioners bring down the number of iterations quite significantly, the number of iterations still scales typically linearly in ω.

In this section, we describe the preconditioners recently proposed in [28, 29]. What distinguish these preconditioners from previous developments is that the number of iterations is essentially independent of the frequency ω. We describe these preconditioners in 2D first and then comment on the differences in 3D at the end.

2.2 Sweeping Factorization

In 2D, the computational domain is $D = (0, 1)^2$ with Sommerfeld boundary condition specified at infinity. One standard way of approximating the Sommerfeld boundary condition at finite distance is to use the perfectly matched layer (PML) [5, 15, 42]. We introduce

$$\sigma(t) = \begin{cases} \frac{C}{\eta} \cdot \left(\frac{t-\eta}{\eta}\right)^2 & t \in [0, \eta] \\ 0 & t \in [\eta, 1-\eta] , \\ \frac{C}{\eta} \cdot \left(\frac{t-1+\eta}{\eta}\right)^2 & t \in [1-\eta, 1] \end{cases} \tag{4}$$

and

$$s_1(x_1) = \left(1 + 1\frac{\sigma(x_1)}{\omega}\right)^{-1}, \quad s_2(x_2) = \left(1 + 1\frac{\sigma(x_2)}{\omega}\right)^{-1},$$

where $1 = \sqrt{-1}$, η is typically around one wavelength, and C is an appropriate positive constant independent of ω. Computationally, the PML method replaces ∂_1 and ∂_2 with $s_1(x_1)\partial_1$ and $s_2(x_2)\partial_2$, respectively. This effectively provides a damping layer of width η and the resulting equation becomes

$$\left((s_1\partial_1)(s_1\partial_1) + (s_2\partial_2)(s_2\partial_2) + \frac{\omega^2}{c^2(x)}\right) u = f, \quad x \in D,$$

$$u = 0, \quad x \in \partial D.$$

If we assume without loss of generality that $f(x)$ is supported inside $[\eta, 1-\eta]^2$ (away from the PML), dividing the above equation by $s_1(x_1)s_2(x_2)$ results

$$\left(\partial_1\left(\frac{s_1}{s_2}\partial_1\right) + \partial_2\left(\frac{s_2}{s_1}\partial_2\right) + \frac{\omega^2}{s_1 s_2 c^2(x)}\right) u = f,$$

which is complex symmetric. We discretize the domain $[0, 1]^2$ with a Cartesian grid with spacing $h = 1/(n + 1)$, where the number of points n in each dimension is proportional to the wave number ω, since a constant number of points is required for each wavelength. The set of all interior points of this grid is denoted by

$$P = \{p_{i,j} = (ih, jh) : 1 \leq i, j \leq n\}$$

(see Fig. 1 (left)) and the total number of points in P is $N = n^2$.

We denote by $u_{i,j}$, $f_{i,j}$, and $c_{i,j}$ the values of $u(x)$, $f(x)$, and $c(x)$ at point $p_{i,j} = (ih, jh)$, respectively. The 5-point stencil finite difference method gives the equation at points in P using central differences. The resulting equation at $x_{i,j} = (ih, jh)$ is

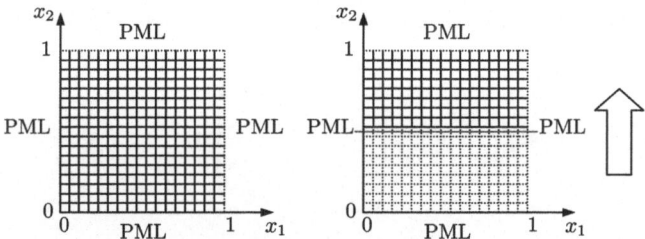

Fig. 1 *Left*: Discretization grid in 2D. *Right*: Sweeping order in 2D. The dotted grid indicates the part that has already been eliminated

$$\frac{1}{h^2}\left(\frac{s_1}{s_2}\right)_{i-\frac{1}{2},j} u_{i-1,j} + \frac{1}{h^2}\left(\frac{s_1}{s_2}\right)_{i+\frac{1}{2},j} u_{i+1,j} + \frac{1}{h^2}\left(\frac{s_2}{s_1}\right)_{i,j-\frac{1}{2}} u_{i,j-1}$$

$$+ \frac{1}{h^2}\left(\frac{s_2}{s_1}\right)_{i,j+\frac{1}{2}} u_{i,j+1} + \left(\frac{\omega^2}{(s_1 s_2)_{i,j}\cdot c_{i,j}^2} - (\cdots)\right) u_{i,j} = f_{i,j} \quad (5)$$

with $u_{i',j'}$ equal to zero for index (i', j') that violates $1 \le i', j' \le n$. Here $(\cdots)$ stands for the sum of the first four coefficients. We order both $u_{i,j}$ and $f_{i,j}$ row by row starting from the first row $j = 1$, define the vectors

$$u = (u_{1,1}, u_{2,1}, \ldots, u_{n,1}, \ldots, u_{1,n}, u_{2,n}, \ldots, u_{n,n})^t,$$
$$f = (f_{1,1}, f_{2,1}, \ldots, f_{n,1}, \ldots, f_{1,n}, f_{2,n}, \ldots, f_{n,n})^t,$$

and denote the discrete system of (5) by $Au = f$. We further introduce its block version by defining P_m to be set of the indices in the mth row, i.e., $P_m = \{p_{1,m}, p_{2,m}, \ldots, p_{n,m}\}$ and introducing

$$u_m = (u_{1,m}, u_{2,m}, \ldots, u_{n,m})^t, \quad f_m = (f_{1,m}, f_{2,m}, \ldots, f_{n,m})^t.$$

Clearly $u = (u_1^t, u_2^t, \ldots, u_n^t)^t$, $f = (f_1^t, f_2^t, \ldots, f_n^t)^t$, and the system $Au = f$ takes the following block tridiagonal form

$$\begin{pmatrix} A_{1,1} & A_{1,2} & & \\ A_{2,1} & A_{2,2} & \ddots & \\ & \ddots & \ddots & A_{n-1,n} \\ & & A_{n,n-1} & A_{n,n} \end{pmatrix} \begin{pmatrix} u_1 \\ u_2 \\ \vdots \\ u_n \end{pmatrix} = \begin{pmatrix} f_1 \\ f_2 \\ \vdots \\ f_n \end{pmatrix} \quad (6)$$

where $A_{m,m}$ are tridiagonal and $A_{m,m-1} = A_{m-1,m}^t$ are diagonal matrices.

We then introduce the concept of *sweeping factorization* of the matrix A, which is essentially a block tridiagonal LDL^t factorization (or Thomas algorithm [17]) that eliminates the unknowns layer by layer, starting from the absorbing layer next to $x_2 = 0$ (see [28] for details). The result of this process is a factorization

$$A = L_1 \cdots L_{n-1} \begin{pmatrix} S_1 & & & \\ & S_2 & & \\ & & \ddots & \\ & & & S_n \end{pmatrix} L_{n-1}^t \cdots L_1^t, \tag{7}$$

where $S_1 = A_{1,1}$, $S_m = A_{m,m} - A_{m,m-1} S_{m-1}^{-1} A_{m-1,m}$ for $m = 2, \ldots, n$, and L_k is given by $L_k(P_{k+1}, P_k) = A_{k+1,k} S_k^{-1}$, $L_k(P_i, P_i) = I$ ($1 \le i \le n$), and zero otherwise. This process is illustrated graphically in Fig. 1 (right). Inverting this factorization for A gives the following formula for u:

$$u = (L_1^t)^{-1} \cdots (L_{n-1}^t)^{-1} \begin{pmatrix} S_1^{-1} & & & \\ & S_2^{-1} & & \\ & & \ddots & \\ & & & S_n^{-1} \end{pmatrix} L_{n-1}^{-1} \cdots L_1^{-1} f. \tag{8}$$

By introducing $T_m = S_m^{-1}$, the construction of the sweeping factorization is summarized in the following algorithm.

Algorithm 1 Construction of the sweeping factorization of A.

1: $S_1 = A_{1,1}$ and $T_1 = S_1^{-1}$.
2: **for** $m = 2, \ldots, n$ **do**
3:     $S_m = A_{m,m} - A_{m,m-1} T_{m-1} A_{m-1,m}$ and $T_m = S_m^{-1}$.
4: **end for**

Since S_m and T_m are in general dense matrices of size $n \times n$, the cost of the construction algorithm is of order $O(n^4) = O(N^2)$. Once the factorization is ready, the computation of $u = A^{-1} f$ in (8) is given as follows. The cost of computing u

---

Algorithm 2 Computation of $u = A^{-1} f$ using the sweeping factorization of A.

1: **for** $m = 1, \ldots, n$ **do**
2:     $u_m = f_m$
3: **end for**
4: **for** $m = 1, \ldots, n-1$ **do**
5:     $u_{m+1} = u_{m+1} - A_{m+1,m} (T_m u_m)$
6: **end for**
7: **for** $m = 1, \ldots, n$ **do**
8:     $u_m = T_m u_m$
9: **end for**
10: **for** $m = n-1, \ldots, 1$ **do**
11:     $u_m = u_m - T_m (A_{m,m+1} u_{m+1})$
12: **end for**

---

with Algorithm 2 is of order $O(n^3) = O(N^{3/2})$. Obviously the product $T_m u_m$ in the second and the third loops only needs to be carried out once for each m. However, we prefer to write the algorithm this way to simplify the presentation.

Both Algorithms 1 and 2 are about $O(N^{1/2})$ times more expensive compared to the multifrontal method. A natural question is that whether it is possible to make Algorithms 1 and 2 more efficient by approximating T_m accurately and efficiently. To do that, we consider the physical meaning of the Schur complement matrices T_m of the sweeping factorization. Let us restrict to the top-left $m \times m$ block of the factorization (7).

$$
\begin{pmatrix}
A_{1,1} & A_{1,2} & & \\
A_{2,1} & A_{2,2} & \ddots & \\
& \ddots & \ddots & A_{m-1,m} \\
& & A_{m,m-1} & A_{m,m}
\end{pmatrix}
= L_1 \cdots L_{m-1}
\begin{pmatrix}
S_1 & & & \\
& S_2 & & \\
& & \ddots & \\
& & & S_m
\end{pmatrix}
L_{m-1}^t \cdots L_1^t, \quad (9)
$$

where the L_k matrices are redefined to their restrictions to the top-left $m \times m$ blocks. The matrix on the left is in fact the discrete Helmholtz equation restricted to the half space below $x_2 = (m+1)h$ and with zero boundary condition on this line. Inverting the factorization (9) gives

$$
\begin{pmatrix}
A_{1,1} & A_{1,2} & & \\
A_{2,1} & A_{2,2} & \ddots & \\
& \ddots & \ddots & A_{m-1,m} \\
& & A_{m,m-1} & A_{m,m}
\end{pmatrix}^{-1}
= (L_1^t)^{-1} \cdots (L_{m-1}^t)^{-1}
\begin{pmatrix}
S_1^{-1} & & & \\
& S_2^{-1} & & \\
& & \ddots & \\
& & & S_m^{-1}
\end{pmatrix}
$$

$$
L_{m-1}^{-1} \cdots L_1^{-1}.
$$

The matrix on the left side is an approximation of the discrete half-space Green's function of the Helmholtz operator with zero boundary condition. On the right side, due to the definition of the matrices $L_1, \ldots, L_{m-1}$, the (m, m)th block of the product is exactly equal to S_m^{-1}. Therefore, we reach the following essential observation: $T_m = S_m^{-1}$ is the discrete half-space Green function of the Helmholtz operator with zero boundary at $x_2 = (m+1)h$, restricted to the points on $x_2 = mh$. Based on this observation, we propose two different approaches that approximate T_m effectively. When combined with Algorithms 1 and 2, they result in efficient preconditioners.

2.3 Hierarchical Matrix Algebra Approach

The main observation of the first approach is that T_m and S_m are highly compressible since their off-diagonal blocks are numerically low-rank. The following theorem

[28] proves this for constant velocity field $c(x) = 1$ for the continuous half-space Green's function.

Theorem 3. *Let*

$$Y = \{p_{i,m} = (ih, mh), i = 1, \ldots, \ell\} \quad and$$

$$X = \{p_{i,m} = (ih, mh), i = \ell + 1, \ldots, n\},$$

for $1 < \ell < n$ and G be the (continuous) half-space Green's function of the Helmholtz operator for the domain $(-\infty, \infty) \times (-\infty, (m + 1)h)$ with zero boundary condition on $x_2 = (m + 1)h$ and Sommerfeld condition elsewhere. Then $(G(x, y))_{x \in X, y \in Y}$ is numerically low-rank. More precisely, for any $\epsilon > 0$, there exist $R = O(\log \omega |\log \epsilon|^2)$, functions $\{\alpha_r(x)\}_{1 \leq r \leq R}$ for $x \in X$, and functions $\{\beta_r(y)\}_{1 \leq r \leq R}$ for $y \in Y$ such that

$$\left| G(x, y) - \sum_{r=1}^{R} \alpha_r(x)\beta_r(y) \right| \leq \epsilon \quad for \quad x \in X, y \in Y.$$

For a fixed $\epsilon > 0$, Theorem 3 says that the rank R grows logarithmically with respect to ω (and thus to n). Numerical experiments confirm the result of Theorem 3. For the constant coefficient case $c(x) = 1$ with $\frac{\omega}{2\pi} = 32$ ($n = 256$), Fig. 2 (left) shows the numerical ranks of the off-diagonal blocks of T_m at $m = 128$. For each off-diagonal block, the singular values of this block are calculated and the value in each block indicates the number of singular values that are greater than 10^{-6}. For non-constant velocity fields $c(x)$, the rank estimate would depend on the variations in $c(x)$ and numerical results suggest that the off-diagonal blocks of T_m and S_m still admit this low-rank property for a wide class of $c(x)$. An example for the non-constant velocity field is given in Fig. 2 (right). We would like to emphasize

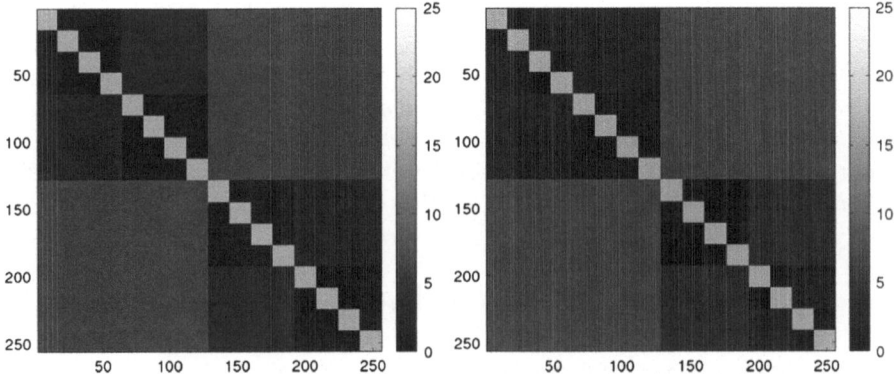

Fig. 2 Numerical ranks of off-diagonal blocks of T_m. *Left*: Constant coefficient case with PML boundary condition. *Right*: Non-constant coefficient case with PML boundary condition

that both the Sommerfeld boundary condition and the layer-by-layer sweeping order starting from a PML layer are essential for the low-rankness result. Other elimination patterns, such as the ones in the multifrontal algorithm or in [48] do not result efficient solution methods for the Helmholtz equation.

Since T_m and S_m are highly compressible with numerically low-rank off-diagonal blocks, it is natural to represent these matrices using the hierarchical matrix framework proposed by Hackbusch et al [7, 38, 40]. For each m, we construct a hierarchical decomposition of the grid points of P_m through bisection. At level 0 (the top level), there is only one group $J_1^0 = P_m$. At level ℓ, there are 2^ℓ groups J_i^ℓ for $i = 1, \ldots, 2^\ell$ given by

$$J_i^\ell = \{p_{t,m} : (i-1) \cdot n/2^\ell + 1 \le t \le i \cdot n/2^\ell\}.$$

The bisection is stopped when each group J_i^ℓ contains only a small number of points. Hence, the number of total levels L of bisection is equal to $\log_2 n - O(1)$ (see Fig. 3 (left)). Two groups J_i^ℓ and $J_{i'}^\ell$ are called *well-separated* if the distance between them is greater than or equal to their width. It is clear that if J_i^ℓ and $J_{i'}^\ell$ are well-separated then $T_m(J_i^\ell, J_{i'}^\ell)$ (the restriction of T_m to J_i^ℓ and $J_{i'}^\ell$) is numerical low-rank and stored in a low-rank factorized form (the same is also true for S_m). It is not difficult to see that the storage cost of matrices T_m and S_m is of order $O(Rn \log n)$. In our implementation, the rank R of off-diagonal blocks is chosen a priori to be a fixed number, which turns out to be sufficient for the purpose of constructing a preconditioner. All basic operations of the standard linear algebra have their hierarchical version in this hierarchical matrix framework:

- hmatvec(G, f): matrix vector multiplication of matrix G with vector f. Its cost is $O(Rn \log n)$ where R is the rank of the low-rank approximation.
- hadd(G, H) and hsub(G, H): matrix addition and subtraction. Both of them take $O(R^2 n \log n)$ steps.

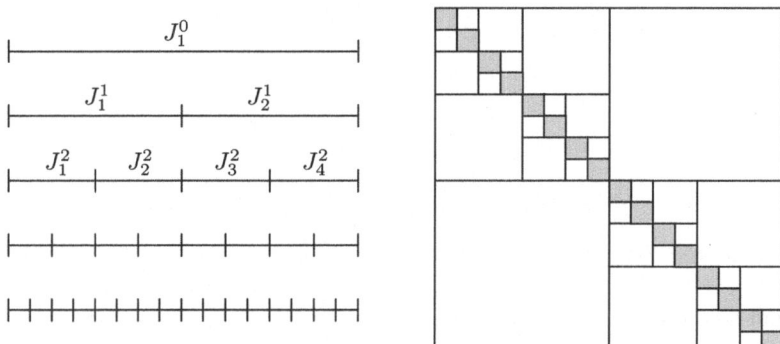

Fig. 3 Hierarchical matrix representation. *Left*: Hierarchical partitioning of the index set J for each layer. *Right*: Induced partitioning of the matrix T_m in the weakly admissible case. Off-diagonal blocks (*in white*) are stored in low-rank factorized form. Diagonal blocks (*in gray*) are stored densely

- hmul(G, H) and hinv(G): matrix multiplication and matrix inversion. For both, the computational cost is $O(R^2 n \log^2 n)$.
- hdiagmul(G, D) and hdiagmul(D, G): multiplication of a hierarchical matrix G with a diagonal matrix D on the right and on the left. These procedures take $O(Rn \log n)$ steps each.

We can replace the operations in Algorithms 1 and 2 with their hierarchical versions. Let us denote the approximations of S_m and T_m in the hierarchical matrix representation by $\widetilde{S}_m$ and $\widetilde{T}_m$, respectively.

Algorithm 4 Construction of the approximate sweeping factorization of A in the hierarchical matrix framework.

1: $\widetilde{S}_1 = A_{1,1}$ and $\widetilde{T}_1 = \text{hinv}(\widetilde{S}_1)$.
2: **for** $m = 2, \ldots, n$ **do**
3:    $\widetilde{S}_m = \text{hsub}(A_{m,m}, \text{hdiagmul}(A_{m,m-1}, \text{hdiagmul}(\widetilde{T}_{m-1}, A_{m-1,m})))$ and $\widetilde{T}_m = \text{hinv}(\widetilde{S}_m)$.
4: **end for**

The cost of Algorithm 4 is $O(R^2 n^2 \log^2 n) = O(R^2 N \log^2 N)$. Once the factorization is ready, the computation of $u \approx A^{-1} f$ takes the following steps.

Algorithm 5 Computation of $u \approx A^{-1} f$ using the approximate sweeping factorization of A in the hierarchical matrix framework.

1: **for** $m = 1, \ldots, n$ **do**
2:    $u_m = f_m$
3: **end for**
4: **for** $m = 1, \ldots, n-1$ **do**
5:    $u_{m+1} = u_{m+1} - A_{m+1,m} \cdot \text{hmatvec}(\widetilde{T}_m, u_m)$
6: **end for**
7: **for** $m = 1, \ldots, n$ **do**
8:    $u_m = \text{hmatvec}(\widetilde{T}_m, u_m)$
9: **end for**
10: **for** $m = n-1, \ldots, 1$ **do**
11:    $u_m = u_m - \text{hmatvec}(\widetilde{T}_m, A_{m,m+1} u_{m+1})$
12: **end for**

The cost of Algorithm 5 is $O(Rn^2 \log n) = O(RN \log N)$. Algorithm 5 defines an operator

$$M : f = (f_1^t, f_2^t, \ldots, f_n^t)^t \to u = (u_1^t, u_2^t, \ldots, u_n^t)^t,$$

which is an approximate inverse of the discrete Helmholtz operator A. When the threshold $\epsilon > 0$ is set to be sufficiently small, M can be used directly as the inverse of A and u can be taken as the solution. However, a small $\epsilon > 0$ value means that the rank R of the low-rank factorized form needs to be fairly large, thus resulting large storage and computation cost. On the other hand, when R is kept rather small, Algorithms 4 and 5 are more efficient both in terms of storage and time. Though the resulting M is not accurate enough as the direct inverse of A, it serves as an excellent preconditioner and we call it the *sweeping preconditioner*. We can then combine it with iterative solvers such as GMRES and TFQMR [53, 54] by solving the preconditioned system

$$MAu = Af.$$

Since the cost of applying M to any vector is $O(RN \log N)$, the total cost of the iterative solver is $O(N_I RN \log N)$, where N_I is the number of iterations. The numerical results demonstrate that N_I is in practice very small, thus resulting in an algorithm with almost linear complexity.

In the presentation of the sweeping preconditioner, we choose the sweeping direction to be in the positive x_2 direction. It is clear that sweeping along the other three directions (i.e., the positive x_1, the negative x_1, and the negative x_2) also gives a slightly different sweeping preconditioner. Due to the variations in the velocity field (more precisely the existence of turning rays), a carefully selected sweeping direction can often result in significantly fewer GMRES iterations than the other directions do.

We illustrate the effectiveness of this preconditioner with a numerical example. The velocity field $c(x)$ is randomly generated and is tested with two external forces: (1) a Gaussian point source located at $(x_1, x_2) = (0.5, 0.125)$ and (2) a Gaussian wave packet at $(x_1, x_2) = (0.125, 0.125)$ and pointing in the $(1, 1)$ direction. We perform tests for $\frac{\omega}{2\pi} = 16, 32, \ldots, 256$ and discretize the equation with $q = 8$ points per wavelength. Recall that R is the rank of the off-diagonal blocks in the hierarchical matrix and we fix it to be 2. In all tests, the sweeping direction is bottom-up from $x_2 = 0$ to $x_2 = 1$. The numerical results are reported in Table 1 and from the numbers we see clearly the frequency-independent iteration number and linear scaling of the algorithm.

2.4 Moving PML Approach

Let us recall that the main task in making Algorithms 1 and 2 more efficient is to find efficient approximation of T_m. In the second approach, we approximate T_m in a different way. Since $T_m : \mathbb{C}^n \to \mathbb{C}^n$ maps an external force $g_m \in \mathbb{C}^n$ loaded only on the mth layer to the solution $v_m \in \mathbb{C}^n$ restricted to the same layer, its domain of interest is a neighborhood of $x_2 = mh$. We recall that the main idea of the PML approximation is to approximate an infinite domain problem with a finite domain problem that wraps the domain of interest with efficient damping layers. Therefore it is natural to replace the half-space Helmholtz problem associated with T_m with an approximate local subproblem with a PML close to $x_2 = mh$. So the central idea is to *push the PML from $x_2 = 0$ to a location that is a few buffer layers away from $x_2 = mh$ when approximating T_m*. We call this approach the *moving PML* method, since these new PMLs do not exist in the original problem as they are only introduced in order to approximate T_m efficiently. The purpose of keeping a few extra buffer layers is that the resulting approximation is more accurate. In practice

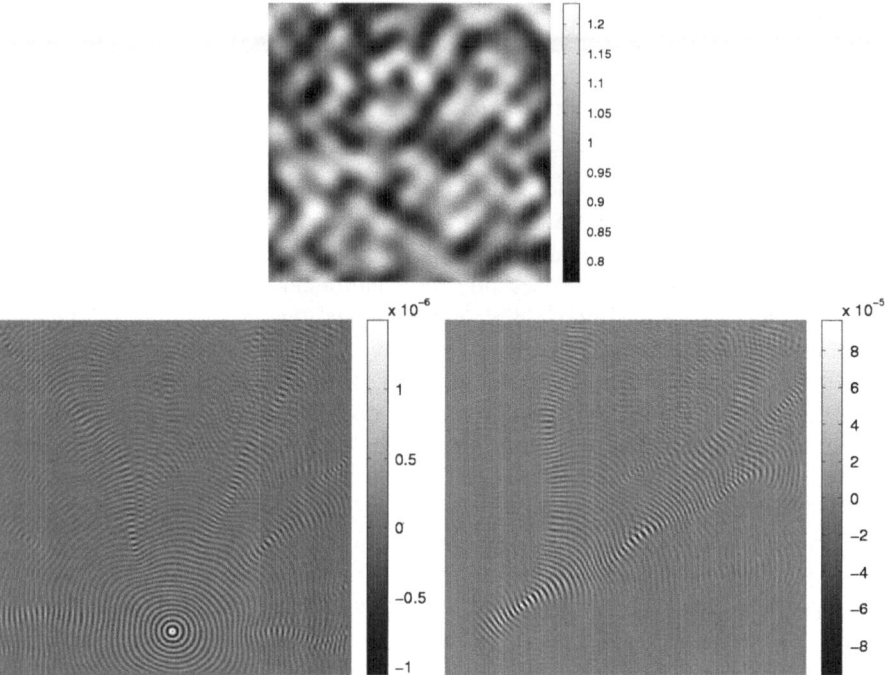

Table 1 Results for different ω. *Top*: Velocity field $c(x)$. *Middle*: Solutions for two external forces with $\omega/(2\pi) = 64$. *Bottom*: Results for different ω

$\omega/(2\pi)$	q	$N = n^2$	R	T_{setup}	Test 1		Test 2	
					N_{iter}	T_{solve}	N_{iter}	T_{solve}
16	8	128^2	2	6.50e−01	2	5.00e−02	2	5.00e−02
32	8	256^2	2	5.10e+00	2	2.50e−01	3	3.00e−01
64	8	512^2	2	3.48e+01	3	1.49e+00	3	1.48e+00
128	8	1024^2	2	2.16e+02	4	8.99e+00	3	7.37e+00
256	8	2048^2	2	1.26e+03	5	4.64e+01	3	3.25e+01

it is often reasonable to even move the PML right next to $x_2 = mh$ in order to gain more efficiency. In fact, numerical results show that the extra buffer layers provide little improvement on the approximation accuracy and hence the moving PML is indeed placed right next to $x_2 = mh$ in the following discussion.

To make this precise, let us assume that the width η of the PML is an integer multiple of h and let $b = \eta/h$ be the number of grid points in the PML layer in the x_2 direction. Define $s_2^m(x_2) = \left(1 + i\frac{\sigma(x_2 - (m-b)h)}{\omega}\right)^{-1}$ and introduce an auxiliary problem on the domain $D_m = [0, 1] \times [(m - b)h, (m + 1)h]$:

$$\left((s_1 \partial_1)(s_1 \partial_1) + (s_2^m \partial_2)(s_2^m \partial_2) + \frac{\omega^2}{c^2(x)} \right) u = f, \ x \in D_m, \tag{10}$$

$$u = 0, \ x \in \partial D_m.$$

This equation is discretized with the subgrid

$$G_m = \{ p_{i,j}, 1 \le i \le n, m - b + 1 \le j \le m \}$$

of the original grid P and the resulting $bn \times bn$ discrete Helmholtz operator is denoted by H_m. Following the main idea mentioned above, the operator $\widetilde{T}_m : g_m \to v_m$ defined by

$$\begin{pmatrix} * \\ \vdots \\ * \\ v_m \end{pmatrix} = H_m^{-1} \begin{pmatrix} 0 \\ \vdots \\ 0 \\ g_m \end{pmatrix} \tag{11}$$

is an approximation to the matrix T_m. Here each 0 stands for a zero vector of length n and each $*$ for some vector of the same length. Notice that applying $\widetilde{T}_m$ to an arbitrary vector g_m involves solving a linear system with matrix H_m, which comes from the *local* 5-point stencil on the narrow grid G_m that contains only b layers. Let us introduce a new ordering for G_m that iterates through the x_2 direction first and denote the permutation matrix induced from this new ordering by P_m. Now the matrix $P_m H_m P_m^t$ is a banded matrix with only $b - 1$ lower diagonals and $b - 1$ upper diagonals. Clearly the LU factorization $P_m H_m P_m^t = L_m U_m$ takes $O(n)$ steps since b is a constant and the application of $\widetilde{T}_m$ following (11) can be done also in $O(n)$ steps once the factorization is ready.

Let us incorporate the moving PML technique into Algorithms 1 and 2. The computation at the first b layers needs to be handled differently, since it does not make sense to introduce moving PMLs for these initial layers. Similarly, the last b layers require some special treatment. Let us call the first b layers the *front* part and the last b layers the *end* part. Define

$$u_F = (u_1^t, \dots, u_b^t)^t, \quad f_F = (f_1^t, \dots, f_b^t)^t$$

$$u_E = (u_{n-b+1}^t, \dots, u_n^t)^t, \quad f_E = (f_{n-b+1}^t, \dots, f_n^t)^t.$$

Then we can rewrite $Au = f$ as

$$\begin{pmatrix} A_{F,F} & A_{F,b+1} & & \\ A_{b+1,F} & A_{b+1,b+1} & \ddots & \\ & \ddots & \ddots & A_{n-b,E} \\ & & A_{E,n-b} & A_{E,E} \end{pmatrix} \begin{pmatrix} u_F \\ u_{b+1} \\ \vdots \\ u_E \end{pmatrix} = \begin{pmatrix} f_F \\ f_{b+1} \\ \vdots \\ f_E \end{pmatrix}.$$

The construction of the approximate sweeping factorization of A takes the following steps.

Algorithm 6 Construction of the approximate sweeping factorization of A with moving PML.

1: **for** $m = F, b + 1, \ldots, n - b, E$ **do**
2:     If $m = F$, let G_F be the subgrid of the first b layers, $H_F = A_{F,F}$, and P_F be the permutation matrix induced by the new ordering (x_2 first) of G_F. Construct the LU factorization $L_F U_F = P_F H_F P_F^t$. For other values of m, let $G_m = \{p_{i,j}, 1 \le i \le n, m - b + 1 \le j \le m\}$, H_m be the discrete system of (10) on G_m, and P_m be the permutation induced by the new ordering of G_m. Construct the LU factorization $L_m U_m = P_m H_m P_m^t$.
3: **end for**

The cost of Algorithm 6 is $O(b^3 n^2) = O(b^3 N)$. Notice that as T_m are approximated directly there is no need to compute S_m. Algorithm 2 takes the following form now under this moving PML approximation.

Algorithm 7 Computation of $u \approx A^{-1} f$ using the sweeping factorization of A with moving PML.

1: **for** $m = F, b + 1, \ldots, n - b, E$ **do**
2:     $u_m = f_m$
3: **end for**
4: **for** $m = F, b + 1, \ldots, n - b$ **do**
5:     $u_{m+1} = u_{m+1} - A_{m+1,m}(\widetilde{T}_m u_m)$. For $m = F$, The application $\widetilde{T}_F u_F$ is done as $P_F^t U_F^{-1} L_F^{-1} P_F u_F$. For other values of m, $\widetilde{T}_m u_m$ is done by forming the vector $(0, \ldots, 0, u_m^t)^t$, applying $P_m^t U_m^{-1} L_m^{-1} P_m$ to it, and extracting the value on the last layer.
6: **end for**
7: **for** $m = F, b + 1, \ldots, n - b, E$ **do**
8:     $u_m = \widetilde{T}_m u_m$. See step 5 for the application of $\widetilde{T}_m$.
9: **end for**
10: **for** $m = n - b, \ldots, b + 1, F$ **do**
11:     $u_m = u_m - \widetilde{T}_m (A_{m,m+1} u_{m+1})$. See step 5 for the application of $\widetilde{T}_m$.
12: **end for**

The cost of Algorithm 7 is $O(b^2 n^2) = O(b^2 N)$, which is essentially linear as b is a fixed constant. Algorithm 7 defines an operator

$$M : f = (f_F^t, f_{b+1}^t, \ldots, f_E^t)^t \to u = (u_F^t, u_{b+1}^t, \ldots, u_E^t)^t,$$

which is an approximate inverse of the discrete Helmholtz operator A. In practice, instead of generating the sweeping factorization of the original matrix A, it is more effective to generate the factorization for the matrix A_α associated with the modified Helmholtz equation

$$\Delta u(x) + \frac{(\omega + 1\alpha)^2}{c^2(x)} u(x) = f(x), \tag{12}$$

where the damping parameter α is an $O(1)$ positive constant. We would like to emphasize that (12) is very different from the equation used in the shifted Laplacian approach (for example [31, 32, 44]), since in the shifted Laplacian formulation the damping parameter is $O(\omega)$ while here it is $O(1)$. We denote by $M_\alpha : f \rightarrow u$ the operator defined by Algorithm 7 with this modified equation. Since α is small, A_α and M_α are close to A and M, respectively. Therefore, we propose to solve the preconditioner system

$$M_\alpha A u = M_\alpha f$$

using the GMRES solver [53, 54]. As the cost of applying M_α to any vector is $O(n^2) = O(N)$, the total cost of the iterative solver scales like $O(N_I N)$, where N_I is the number of iterations. The numerical results demonstrate that N_I depends at most logarithmically on N, thus providing a solver with almost linear complexity.

We illustrate this second sweeping preconditioner with the same randomly generated velocity field and the results are reported in Table 2. Compared with the first approach, the number of iterations is still essentially independent of ω but it does grow due to the introduction of the damping parameter α. On the other hand, the setup cost is much lower.

2.5 3D Case

In 3D, the computational domain is $D = (0, 1)^3$. By introducing also $s_3(x_3) = \left(1 + 1\frac{\sigma(x_3)}{\omega}\right)^{-1}$, we can write down the equation with the PML condition

$$\left(\partial_1 \left(\frac{s_1}{s_2 s_3} \partial_1\right) + \partial_2 \left(\frac{s_2}{s_1 s_3} \partial_2\right) + \partial_3 \left(\frac{s_3}{s_1 s_2} \partial_3\right) + \frac{\omega^2}{s_1 s_2 s_3 c^2(x)}\right) u = f.$$

The domain $[0, 1]^3$ is discretized with a Cartesian grid with spacing $h = 1/(n + 1)$ and the interior points of this grid are

$$P = \{p_{i,j,k} = (ih, jh, kh) : 1 \leq i, j, k \leq n\}.$$

We denote by $u_{i,j,k}$, $f_{i,j,k}$, and $c_{i,j,k}$ the values of $u(x)$, $f(x)$, and $c(x)$ at point $p_{i,j,k} = (ih, jh, kh)$, respectively. The 7-point stencil finite difference method gives the equation at points in P using central differences. We order $u_{i,j,k}$ and $f_{i,j,k}$ by going through the dimensions in order and define the vectors

$$u = (u_{1,1,1}, u_{2,1,1}, \ldots, u_{n,1,1}, \ldots, u_{1,n,n}, u_{2,n,n}, \ldots, u_{n,n,n})^t,$$
$$f = (f_{1,1,1}, f_{2,1,1}, \ldots, f_{n,1,1}, \ldots, f_{1,n,n}, f_{2,n,n}, \ldots, f_{n,n,n})^t.$$

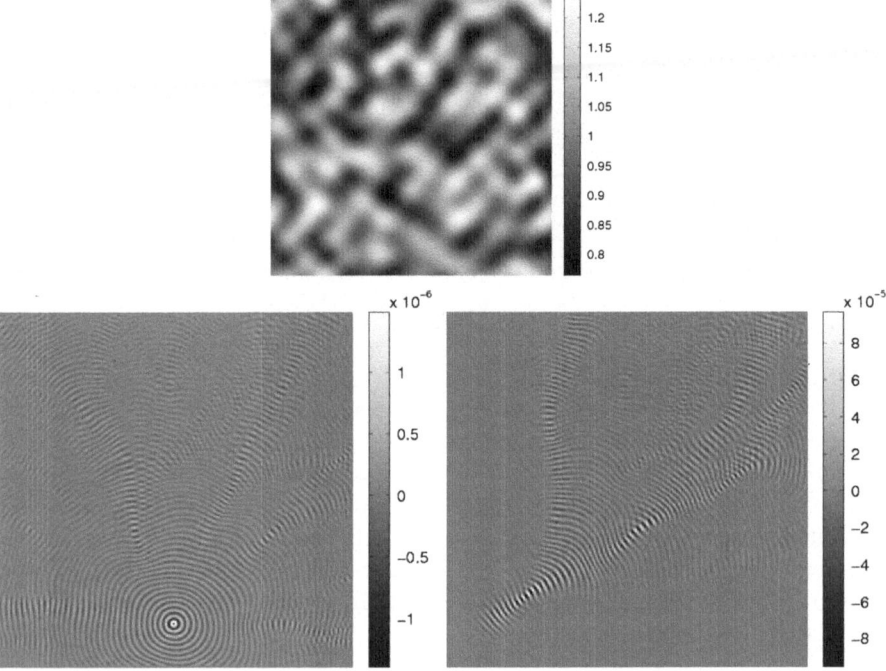

Table 2 Results with varying ω. *Top*: Velocity field $c(x)$. *Middle*: Solutions for two external forces with $\omega/(2\pi) = 64$. *Bottom*: Results for different ω

$\omega/(2\pi)$	q	$N = n^2$	T_{setup}	Test 1		Test 2	
				N_{iter}	T_{solve}	N_{iter}	T_{solve}
16	8	127^2	3.56e−01	18	4.91e−01	19	5.03e−01
32	8	255^2	9.31e−01	18	2.42e+00	19	2.61e+00
64	8	511^2	3.76e+00	17	8.66e+00	23	1.24e+01
128	8	1023^2	1.60e+01	19	3.90e+01	22	4.80e+01
256	8	2047^2	6.82e+01	17	1.54e+02	17	1.48e+02

The whole system then takes the form $Au = f$. We further introduce a block version by defining P_m to be the indices in the mth row $P_m = \{p_{1,1,m}, p_{2,1,m}, \ldots, p_{n,n,m}\}$ and denoting

$$u_m = (u_{1,1,m}, u_{2,1,m}, \ldots, u_{n,n,m})^t, \quad f_m = (f_{1,1,m}, f_{2,1,m}, \ldots, f_{n,n,m})^t.$$

Then $u = (u_1^t, u_2^t, \ldots, u_n^t)^t$, $f = (f_1^t, f_2^t, \ldots, f_n^t)^t$, and we get the same block tridiagonal system (6) but each block is now of size $n^2 \times n^2$. The sweeping factorization takes the same form as the 2D one given in (7).

2.5.1 Hierarchical Matrix Representation Approach

Recall that the main idea of our first approach is to approximate T_m using the hierarchical matrix representation. At the mth layer for fixed m, we build a hierarchical structure for the grid points in P_m through bisections in both x_1 and x_2 directions. At the top level (level 0), the only group is $J_{11}^0 = P_m$. At level ℓ, there are $2^\ell \times 2^\ell$ index groups $J_{ij}^\ell, i, j = 1, \ldots, 2^\ell$

$$J_{ij}^\ell = \{p_{s,t,m} : (i-1) \cdot n/2^\ell + 1 \leq s \leq i \cdot n/2^\ell, (j-1) \cdot n/2^\ell + 1 \leq t \leq j \cdot n/2^\ell\}.$$

The bisection is stopped when each set J_{ij}^ℓ contains only a small number of indices. This hierarchical partition is illustrated in Fig. 4 (left)). Two index sets J_{ij}^ℓ and $J_{i'j'}^\ell$ on the same level ℓ are considered well-separated from each other if $\max(|i - i'|, |j - j'|) > 1$. When J_{ij}^ℓ and $J_{i'j'}^\ell$ are well-separated, the numerical rank of their interaction $T_m(J_{ij}^\ell, J_{i'j'}^\ell)$ or $S_m(J_{ij}^\ell, J_{i'j'}^\ell)$ is of order $O(n/2^\ell)$. Since the number of indices in J_{ij}^ℓ and $J_{i'j'}^\ell$ is equal to $(n/2^\ell)^2$, this numerical rank scales like the square root of the number of indices in each set and hence it is still favorable to store $G(J_{ij}^\ell, J_{i'j'}^\ell)$ in a factorized form. In principle, the rank R of the factorized form should scale like $O(n/2^\ell)$. However, as the construction cost of the approximate sweeping factorization scales like $O(R^2 n^3 \log^2 n) = O(R^2 N \log^2 N)$, following this scaling can be rather costly in practice. Instead, we choose R to be a rather small constant since our goal is only to construct a preconditioner.

We illustrate the effectiveness of this 3D preconditioner with an example. The velocity field is randomly generated and is tested with two external forces: (1) a Gaussian point source located at $(x_1, x_2, x_3) = (0.5, 0.5, 0.25)$ and (2) a Gaussian wave packet at $(x_1, x_2, x_3) = (0.5, 0.25, 0.25)$ and pointing in the $(0, 1, 1)$ direction. We perform tests for $\frac{\omega}{2\pi} = 5, 10, 20$ and discretize the equation with $q = 8$ points per wavelength. Recall that R is the rank of the factorized form of the hierarchical matrix representation. It is clear from the previous discussion that the value of R

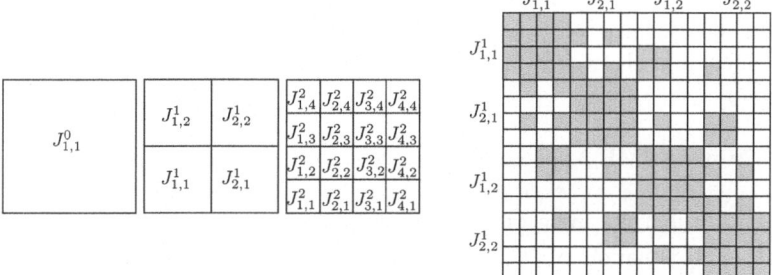

Fig. 4 Hierarchical matrix representation. *Left*: hierarchical decomposition of the index set for each layer. *Right*: Induced partitioning of the matrix T_m in the strongly admissible case. Blocks in white are stored in low-rank factorized form. Blocks in gray are stored densely

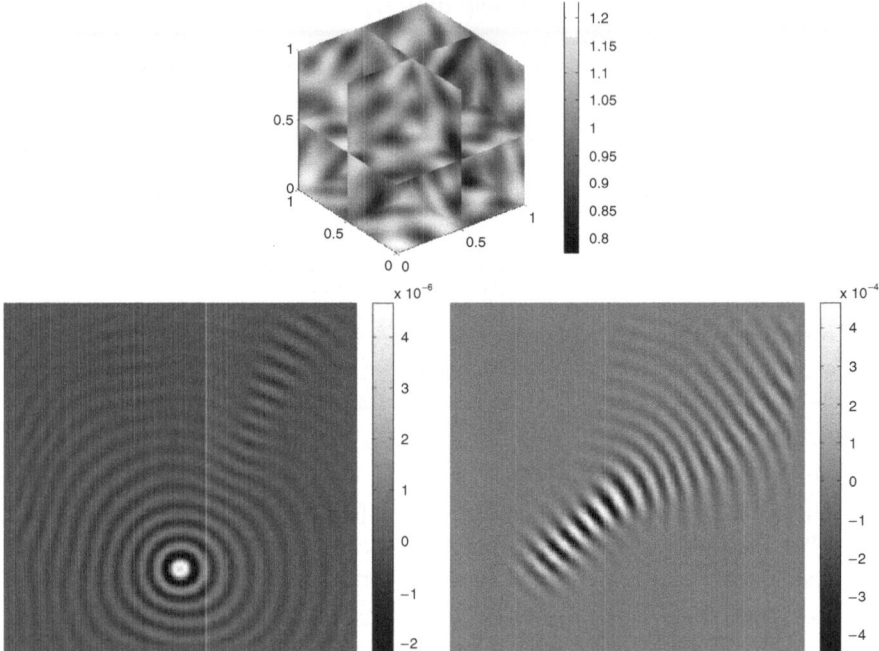

Table 3 Results for different ω. *Top*: Velocity field $c(x)$ in cross-section view. *Middle*: Solutions for two external forces with $\omega/(2\pi) = 20$ on a plane near $x_1 = 0.5$. *Bottom*: Results for different ω

$\omega/(2\pi)$	q	$N = n^3$	R	T_{setup}	Test 1		Test 2	
					N_{iter}	T_{solve}	N_{iter}	T_{solve}
5	8	40^3	2	9.00e+01	3	7.20e−01	3	7.20e−01
10	8	80^3	3	2.37e+03	4	1.22e+01	3	9.90e+00
20	8	160^3	4	4.74e+04	4	1.37e+02	3	1.07e+02

should grow with ω (and n). Here, we choose $R = 2, 3, 4$ for $\omega = 5, 10, 20$, respectively. The sweeping direction is bottom-up from $x_3 = 0$ to $x_3 = 1$. The numbers reported in Table 3 demonstrate that the number of iterations is essentially independent of ω. However, the setup cost is rather high.

2.5.2 Moving PML Approach

Let us recall that the main idea of our second approach is to use moving PMLs for the approximation of $T_m : \mathbb{C}^{n^2} \to \mathbb{C}^{n^2}$, which maps an external force $g_m \in \mathbb{C}^{n^2}$ loaded only on the mth layer to the solution $v_m \in \mathbb{C}^{n^2}$ restricted to the same layer. Following the idea of pushing the PML near $x_3 = mh$, we define $s_3^m(x_3) = \left(1 + i\frac{\sigma(x_3 - (m-b)h)}{\omega}\right)^{-1}$ and introduce an auxiliary problem on the domain

$D_m = [0, 1] \times [0, 1] \times [(m - b)h, (m + 1)h]$:

$$\left((s_1 \partial_1)(s_1 \partial_1) + (s_2 \partial_2)(s_2 \partial_2) + (s_3^m \partial_3)(s_3^m \partial_3) + \frac{\omega^2}{c^2(x)} \right) u = f, \quad x \in D_m, \quad (13)$$

$$u = 0, \quad x \in \partial D_m.$$

This equation is then discretized with the subgrid

$$G_m = \{ p_{i,j,k}, 1 \leq i, j \leq n, m - b + 1 \leq k \leq m \}$$

of the original grid P, with the resulting $bn^2 \times bn^2$ discrete Helmholtz operator denoted by H_m. The operator $\widetilde{T}_m : g_m \in \mathbb{C}^{n^2} \to v_m \in \mathbb{C}^{n^2}$ defined in (11) based on solving H_m is then an approximation of T_m. Since H_m comes from the 7-point stencil with b layers, it can be viewed as a quasi-2D problem that can be solved efficiently using the multifrontal method [23, 36, 46]. The construction phase of the multifrontal method takes $O(b^3 n^3)$ steps and applying to an arbitrary vector takes $O(b^2 n^2 \log n)$ steps. As a result, the cost of Algorithm 6 in 3D is $O(b^3 n^4) = O(b^3 N^{4/3})$ and the one for Algorithm 7 is $O(b^2 n^3 \log n) = O(b^2 N \log N)$.

Though the complexity of Algorithm 6 is slightly higher than linear, it can be improved by building the inverses of H_m under the hierarchical matrix framework used in [28] or using the moving PML idea once again to the solution of H_m. Either one of these two choices gives strictly linear complexity and they are indeed of significant theoretical interest. However, we observe that, for many practical problems that are not extremely large, the current version is at least equally competitive since the efficiency of the multifrontal implementation has been highly optimized due to its simple structure.

For the reason mentioned in the 2D case, we apply Algorithms 6 and 7 to the discrete operator A_α of the modified system $\Delta u(x) + \frac{(\omega + i\alpha)^2}{c^2(x)} u(x) = f(x)$, where α is an $O(1)$ positive constant. We denote by $M_\alpha : f \to u$ the operator defined by Algorithm 7 for this modified equation. Since A_α is close to A when α is small, we propose to solve the preconditioner system $M_\alpha A u = M_\alpha f$ using the GMRES solver [53, 54]. Because the cost of applying M_α to any vector is $O(N \log N)$, the total cost of the GMRES solver is $O(N_I N \log N)$, where N_I is the number of iterations required. The numerical results demonstrate that N_I is essentially independent of the number of unknowns N, thus resulting an algorithm with almost linear complexity.

We illustrate the effectiveness of this 3D preconditioner with a numerical example again with randomly generated velocity field. The results are reported in Table 4 and we see clearly that the number of iterations is independent of ω, the computational cost follows linear scaling, and finally the setup cost is significantly improved compared to the approach based the hierarchical matrix framework.

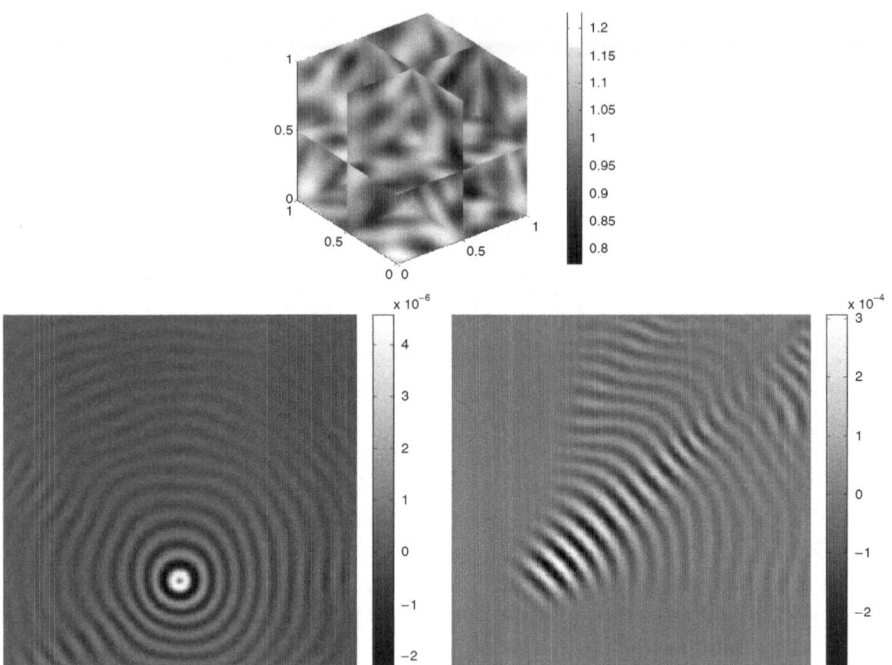

Table 4 Results for different ω. *Top*: Velocity field $c(x)$ in cross-section view. *Middle*: Solutions for two external forces with $\omega/(2\pi) = 16$ on a plane near $x_1 = 0.5$. *Bottom*: Results for different ω

$\omega/(2\pi)$	q	$N = n^3$	T_{setup}	Test 1		Test 2	
				N_{iter}	T_{solve}	N_{iter}	T_{solve}
5	8	39^3	4.85e+00	12	5.26e+00	12	5.44e+00
10	8	79^3	6.69e+01	11	5.10e+01	13	5.99e+01
20	8	159^3	8.42e+02	11	5.58e+02	13	6.28e+02

3 Fast Directional Algorithm for Integral Equation Formulation

In this section, we focus on the integral equation formulation of the exterior scattering problem of the Helmholtz equation, which is arguably the simplest case of the Helmholtz equation with piecewise constant materials. Suppose that the scatterer D is a smooth object with $O(1)$ diameter in the unit ball $B(0, 1)$ and let $u^{inc}(x)$ to be the incoming field. The scattered field $u(x)$ satisfies the Helmholtz equation with the Dirichlet boundary condition:

$$\Delta u(x) + \omega^2 u(x) = 0 \quad \text{for} \quad x \in \mathbb{R}^d \setminus \bar{D}$$

$$u(x) = -u^{inc}(x) \quad \text{for} \quad x \in \partial D \tag{14}$$

$$\lim_{|x| \to \infty} |x|^{(d-1)/2} \left(\left(\frac{x}{|x|}, \nabla u(x) \right) - i\omega u(x) \right) = 0.$$

The last condition is the Sommerfeld radiation condition and it guarantees that the scattered field $u(x)$ propagates to infinity. The wavelength λ is equal to $\frac{2\pi}{\omega}$ and the scatterer D is about ω wavelengths across.

One efficient way of solving this problem is to reformulate it with a boundary integral equation (BIE) [16] such as for $x \in \partial D$

$$\frac{1}{2}\phi(x) + \int_{\partial D} \left(\frac{\partial G(x, y)}{\partial n(y)} - i\eta \cdot G(x, y) \right) \phi(y) dy = -u^{inc}(x), \quad x \in \partial D, \tag{15}$$

where $\phi(x)$ for $x \in \partial D$ is the unknown distribution on ∂D, $n(y)$ is the exterior normal of ∂D at y, $\eta = O(\omega)$ is a constant as a function of space, and the kernel

$$G(x, y) = \begin{cases} \frac{1}{4} H_0^{(1)}(\omega|x - y|), & \text{2D} \\ \frac{e^{i\omega|x-y|}}{|x-y|}, & \text{3D} \end{cases}$$

is the fundamental solution of the free-space Helmholtz equation. The main advantages of the BIE approach is that it reduces an unbounded d-dimensional problem (14) to a bounded $(d - 1)$-dimensional problem (15). In addition, the condition number is dramatically improved. Once $\phi(x)$ is solved from (15), the scattered field $u(x)$ can be simply computed by

$$u(x) = \int_{\partial D} \left(\frac{\partial G(x, y)}{\partial n(y)} - i\eta \cdot G(x, y) \right) \phi(y) dy, \quad \forall x \in \mathbb{R}^2 \setminus \bar{D}.$$

3.1 Background

The boundary formulation (15) is often discretized with Galerkin, collocation, or Nyström methods, see [16] and the references therein for details. In a typical discretization, one uses a constant number of degrees of freedom for each wavelength, hence the number of unknowns N of the discrete system scales like $O(1/\lambda^{d-1}) = O(\omega^{d-1})$.

One obvious disadvantage of the boundary formulation (15) is that the discrete system is dense since the integral operator is global. Direct solvers are clearly too expensive, especially for high frequency problems with large ω. Therefore, iterative algorithms such as GMRES [54] become the natural tool for solving the resulting discrete system. At each step of the iterative solver, one then needs to evaluate the N-body problem of the high frequency Helmholtz kernel

$$u_i = \sum_{j=1}^{N} G(p_i, p_j) \cdot f_j, \tag{16}$$

where $\{p_i\}_{1 \le i \le N} \subset B(0, 1)$ are the appropriate quadrature or nodal points, $\{f_i\}_{1 \le i \le N}$ are the sources at $\{p_i\}_{1 \le i \le N}$, and $\{u_i\}_{1 \le i \le N}$ are called the potentials. A similar problem can be formulated for the kernel $\frac{\partial G(x,y)}{\partial n(y)}$ as well.

The direct computation of (16) takes $O(N^2) = O(\omega^{2(d-1)})$ operations. This can be quite time consuming when ω is large. Various fast algorithms have been proposed to reduce this complexity in the past two decades. Among them, the most popular approach is the high frequency fast multipole method (HF-FMM) developed by Rokhlin et al. [12, 13, 51, 52], which has an optimal $O(N \log N)$ complexity and has been widely used. Other algorithms using related techniques can be found in [14, 18, 56]. In [49], Michielssen and Boag proposed a multilevel multiplication algorithm to bring the overall complexity down to $O(N \log^2 N)$. Another approach for speeding up the computation of (16) is to exploit the translation-invariant property of the kernel and use the fast Fourier transform to perform the non-adjacent computation in the Fourier domain [6, 10]. Though quite efficient for many situations, the asymptotic complexity of this approach is not optimal. A different way to accelerate the N-body computation is to discretize the integral equation (15) under the Galerkin framework with local Fourier bases or wavelet packets. Then the stiffness matrix becomes approximately sparse under these bases since most of the entries are close to zero and can be safely discarded [1, 8, 11, 19–21, 37, 41].

In the rest of this section, we describe a directional algorithm for evaluating (16) recently proposed in [26, 27]. Not only this algorithm has the optimal $O(N \log N)$ complexity, our implementation is also comparable to the most efficient implementation of the HF-FMM in literature [12, 13]. We focus the presentation of the algorithm on the 2D case first and then comment on the difference in the 3D case.

3.2 Directional Low-Rank Property

Let us focus on the 2D case first. The main observation is a low-rank property of the Helmholtz kernel.

Definition 1. Let $f(x, y)$ be a function for $x \in X$ and $y \in Y$. We say that $f(x, y)$ has a t-term ϵ-expansion for X and Y if there exist functions $\{\alpha_s(x)\}_{1 \le s \le t}$ and $\{\beta_s(y)\}_{1 \le s \le t}$ such that

$$\left| f(x, y) - \sum_{s=1}^{t} \alpha_s(x) \beta_s(y) \right| \le \epsilon, \quad \forall x \in X, \forall y \in Y.$$

Since the two sets of functions $\{\alpha_s(x)\}_{1 \le s \le t}$ and $\{\beta_s(y)\}_{1 \le s \le t}$ depend only on x and y respectively, the above expansion is called *separated*. Let us consider

$$Y = B(0, w\lambda) \quad \text{and} \quad X = \{x : \theta(x, \ell) \le 1/w, |x| \ge w^2\lambda\}, \qquad (17)$$

where $w \ge 2$ and ℓ is a given unit vector, and $\theta(x, \ell)$ is the spanning angle between vectors x and ℓ (see Fig. 5 for an illustration).

The following theorem from [27] serves as the theoretical foundation of our approach.

Theorem 8. *For any $\epsilon > 0$, there exists a number t_ϵ independent of w such that*

$$G(x, y) = \frac{1}{4} H_0^{(1)}(\omega|x - y|)$$

has a t_ϵ-term ϵ-expansion for $x \in X = \{x : \theta(x, \ell) \le 1/w, |x| \ge w^2\lambda\}$ and $y \in Y = B(0, w\lambda)$ for any ℓ.

The actual expansion is computed using a randomized procedure presented in [26, 27] and we refer to these papers for detailed description. The end result of this randomized procedure is that one can find points $\{e_t\}_{1 \le t \le t_\epsilon}$ in Y, points $\{c_s\}_{1 \le s \le t_\epsilon}$ in X, and a matrix $D = (d_{ts})_{1 \le t, s \le t_\epsilon}$ such that

$$\left| G(x, y) - \sum_{t=1}^{t_\epsilon} G(x, e_t) \left(\sum_{s=1}^{t_\epsilon} d_{ts} G(c_s, y) \right) \right| = O(\epsilon) \qquad (18)$$

for any $x \in X$ and $y \in Y$. In order to represent such a low rank approximation, one only needs to store $\{c_s\}_{1 \le s \le t_\epsilon}$, $\{e_t\}_{1 \le t \le t_\epsilon}$, and the matrix $D = (d_{ts})_{1 \le s, t \le t_\epsilon}$ since the kernels $G(x, e_t)$ and $G(c_s, y)$ can be evaluated on the fly. Since all of them together takes at most $O(t_\epsilon^2)$ storage space, this approximation is extremely efficient storage-wise.

3.3 Directional Algorithm

In order to use the low-rank property to speed up the calculation of (16), we construct an adaptive quadtree that contains the points $\{p_i\}_{1 \le i \le N}$. Starting from the domain $[-1, 1]^2$ that contains the whole scatterer, we subdivide the domain until that

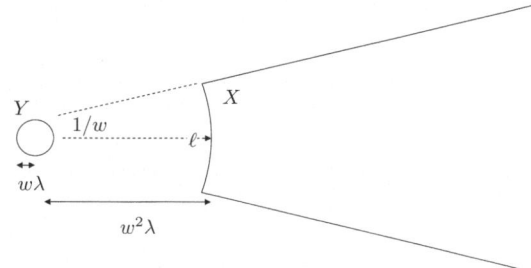

Fig. 5 Two sets Y and X that satisfy the directional parabolic separation condition

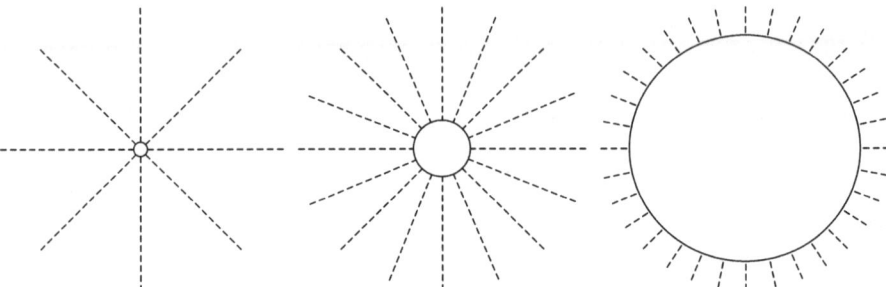

Fig. 6 The far field is partitioned into wedges. From left to right, the width of the box B (not shown) is $w\lambda = \lambda, 2\lambda, 4\lambda$. The radii of the far field boundaries are λ, 4λ, and 16λ, respectively

Fig. 7 B is a square with width $w \geq 2$. For any fixed ℓ, there exists ℓ' such that $W^{B,\ell}$ is contained in $W^{B',\ell'}$ where B' is any one of B's children

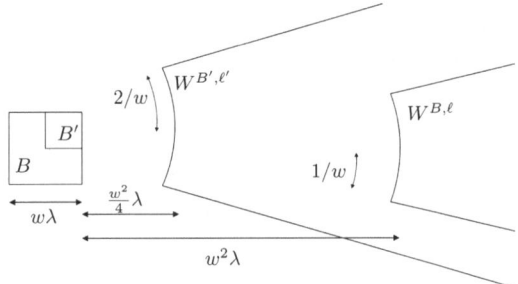

each leaf box contains only a small number of points. Since the scatterer boundary is in fact discretized with $O(1)$ points per wavelength, we can assume without loss of generality that leaf boxes have width λ in order to simplify the discussion.

We use B to denote a square in the quadtree and $w\lambda$ for its width. The near field N^B is the union of the squares A that satisfy dist$(A, B) \leq w^2\lambda$, the far field F^B is the complement of N^B, and the interaction list I^B contains the squares in $N^P \backslash N^B$ on B's level, where P is B's parent square. Notice that the far field of a square B in the high frequency regime is pushed away in order to be compatible with the directional parabolic separation condition. In order to take advantage of the directional separated approximations discussed in the previous section, the far field F^B is further partitioned into a group of directional wedges, each belonging to a cone with spanning angle $O(1/w)$. We denote the set of the wedges of B by $\{W^{B,\ell}\}$ and Fig. 6 illustrates the angular decompositions of F^B for $w\lambda = \lambda, 2\lambda$, and 4λ. We would like to emphasize that the wedges from adjacent levels also enjoy a natural nested property: For each direction ℓ of box B, there exists a direction ℓ' such that $W^{B,\ell}$ is contained in $W^{B',\ell'}$ where B' is any one of B's children (see Fig. 7).

Let us consider the interaction between a square B of width $w\lambda$ and one of its wedges $W^{B,\ell}$ using the separated approximations (18) with $Y = B$ and $X = W^{B,\ell}$. We decorate the quantities $\{e_t\}_{1 \leq t \leq t_\epsilon}$, $\{c_s\}_{1 \leq s \leq t_\epsilon}$, and $D = (d_{ts})_{1 \leq t, s \leq t_\epsilon}$ of (18) with superscript $()^{B,\ell}$ to denote their dependence on B and ℓ.

We first consider the potentials in $W^{B,\ell}$ generated by the points in B. Applying (18) to $y = p_i$ for each $p_i \in B$ and summing them up with weight f_i gives the following estimate:

$$\left| \sum_{p_i \in B} G(x, p_i) f_i - \sum_{t=1}^{t_\epsilon} G(x, e_t^{B,\ell}) \left(\sum_{s=1}^{t_\epsilon} d_{ts}^{B,\ell} \left(\sum_{p_i \in B} G(c_s^{B,\ell}, p_i) f_i \right) \right) \right|$$

$$\leq \left(\sum_{p_i \in B} |f_i| \right) \cdot \epsilon.$$

This implies that we can place a set of sources

$$\left\{ f_t^{B,\ell} := \sum_{s=1}^{t_\epsilon} d_{ts}^{B,\ell} \left(\sum_{p_i \in B} G(c_s^{B,\ell}, p_i) f_i \right) \right\}_{1 \leq t \leq t_\epsilon} \tag{19}$$

at points $\{e_t^{B,\ell}\}_{1 \leq t \leq t_\epsilon}$ to approximate the potential generated by the sources $\{f_i\}$ located at points $\{p_i \in B\}$. Conceptually, the sources $\{f_t^{B,\ell}\}_{1 \leq t \leq r_\epsilon}$ encode the potential in $W^{B,\ell}$ generated by the points in B and we call them the *directional equivalent sources* of B in direction ℓ. It is clear from (19) that $\{f_t^{B,\ell}\}_{1 \leq t \leq t_\epsilon}$ can be computed simply from kernel evaluations and a matrix multiplication with $D^{B,\ell} = (d_{ts}^{B,\ell})_{1 \leq t,s \leq t_\epsilon}$.

Let us now reverse the situation and consider the potential in B generated by the points $\{p_i \in W^{B,\ell}\}$. Since $G(x, y) = G(y, x)$, summing over (18) with $x = p_i \in W^{B,\ell}$ and weights f_i leads us to the estimate

$$\left| \sum_{p_i \in W^{B,\ell}} G(y, p_i) f_i - \sum_{s=1}^{t_\epsilon} G(y, c_s^{B,\ell}) \left(\sum_{t=1}^{t_\epsilon} d_{ts}^{B,\ell} \left(\sum_{p_i \in W^{B,\ell}} G(e_t^{B,\ell}, p_i) f_i \right) \right) \right|$$

$$\leq \left(\sum_{p_i \in W^{B,\ell}} |f_i| \right) \cdot \epsilon.$$

This means that once the potentials

$$\left\{ u_t^{B,\ell} := \sum_{p_i \in W^{B,\ell}} G(e_t^{B,\ell}, p_i) f_i \right\}_{1 \leq t \leq t_\epsilon} \tag{20}$$

at points $\{e_t^{B,\ell}\}_{1 \leq t \leq t_\epsilon}$ are given, we can then approximate at any $y \in B$ the potential generated by the points in $W^{B,\ell}$, through a matrix multiplication with $D^{B,\ell} = (d_{ts}^{B,\ell})$ and kernel evaluations. Conceptually, the potentials $\{u_t^{B,\ell}\}_{1 \leq t \leq r_\epsilon}$ encode the potential in B generated by the points in $W^{B,\ell}$ and we call them the *directional check potentials* of B in direction ℓ.

The directional equivalent sources $\{f_t^{B,\ell}\}_{1 \le t \le t_\epsilon}$ will play the role of the multipole expansion in classical FMM and the directional check potentials $\{u_t^{B,\ell}\}_{1 \le t \le t_\epsilon}$ will play the role of the local expansion in classical FMM. However, it is important to notice that these quantities now depend on ℓ and thus vary from one wedge to another.

Although (19) and (20) defines the equivalent sources and check potentials, they do not result in efficient computation when either B or $W^{B,\ell}$ contains a large number of points. This problem is remedied by the translation operators to be introduced now. We follow the convention in [39, 51] to name these translation operators the M2M, L2L, and M2L translations, even though no multipole or local expansions are involved in our algorithm.

For a square B and a direction ℓ, the *M2M translation* constructs the directional equivalent sources $\{f_t^{B,\ell}\}_t$ from the equivalent sources of B's children. Suppose that the children squares have already computed their directional equivalent sources. Let B' be one of B's child boxes. Due to the nested structure between the wedges of B and B' (i.e, for each ℓ, there exists a direction ℓ' such that $W^{B,\ell}$ is contained in $W^{B',\ell'}$ for each child B' of B, see Fig. 7), the M2M translation takes the equivalent sources $\{f_t^{B',\ell'}\}_{t,B'}$ of the children squares as the true source and performs the following step similar to (19)

$$f_t^{B,\ell} \Leftarrow \sum_s d_{ts}^{B,\ell} \left(\sum_{B'} \sum_{t'} G(c_s^{B,\ell}, e_{t'}^{B',\ell'}) f_{t'}^{B',\ell'} \right). \tag{21}$$

For a square B and a direction ℓ, the *L2L translation* constructs the check potentials of B's children from the directional check potentials $\{u_t^{B,\ell}\}_t$ of B. If $W^{B',\ell'}$ is the wedge of B' that contains $W^{B,\ell}$, we construct the check potentials $\{u_t^{B',\ell'}\}_t$ as follows:

$$u_t^{B',\ell'} \Leftarrow u_t^{B',\ell'} + \sum_s G(e_t^{B',\ell'}, c_s^{B,\ell}) \left(\sum_{t'} d_{t's}^{B,\ell} u_{t'}^{B,\ell} \right). \tag{22}$$

Finally, the *M2L translation* is applied to all pairs of squares A and B that are in each other's interaction list. Suppose that B is in the wedge $W^{A,\ell'}$ of A and A is in the wedge $W^{B,\ell}$ of B. The implementation of the M2L translation contains only one step:

$$u_t^{B,\ell} \Leftarrow u_t^{B,\ell} + \sum_{t'} G(e_t^{B,\ell}, e_{t'}^{A,\ell'}) f_{t'}^{A,\ell'}. \tag{23}$$

We would like to emphasize that all three operators take $O(1)$ steps and involve only kernel evaluation and matrix-vector multiplication with precomputed matrices. Therefore, they are simple to implement and highly efficient.

With all these preparations, we are ready to give the overall structure of our new algorithm.

1. Partition the domain recursively into an adaptive quadtree until each leaf box has width $O(\lambda)$. Since the boundary is discretized with a constant number of points per wavelength, the number of points in each leaf box is $O(1)$.
2. Travel up in the quadtree. For every such square B and each ℓ, construct the directional equivalent sources $\{f_t^{B,\ell}\}_t$ using (19) if B is a leaf box or the directional M2M translation (21) if B is not. We skip the squares with width greater than $\sqrt{\omega}\lambda$ since their interaction lists are empty.
3. Travel down in the quadtree. For every such square B and for each direction ℓ, perform the following two steps:

   (a) For each square A that is in $W^{B,\ell}$ and also in B's interaction list, perform the directional M2L translation (23).
   (b) Perform the directional L2L translation (22) to transform $\{u_t^{B,\ell}\}_t$ to the incoming check potentials for B's children.

   Again, we skip the squares with width greater than $\sqrt{\omega}\lambda$.
4. Nearby interaction. For each leaf square B and for each $p_i \in B$, we add to u_i the nearby interaction from the points $p_j \in N^B$.

The following theorem summarizes the complexity of the proposed algorithm.

Theorem 9. *Let ∂D be a piecewise smooth boundary curve in $B(0,1)$. Suppose that the points $\{p_i\}_{1 \le i \le N}$ are $N = O(\omega)$ samples of ∂D with a constant number of points per wavelength. For any prescribed accuracy the proposed algorithm has a computational complexity $O(\omega \log \omega) = O(N \log N)$.*

We illustrate our 2D algorithm with a numerical example. The scatterer is a kite-shaped object and the results are summarized in Table 5. Here N is the number of points, K is the size of the problem in terms of the wavelength, ϵ is the prescribed error threshold such that the final error is to be bounded by a small constant multiple of ϵ, T_a is the running time of our algorithm in seconds, T_d is the estimated running time of the direct evaluation in seconds, T_d/T_a is the speedup factor, and ϵ_a is the relative error of our algorithm. Here ϵ_a is estimated by comparing the results of our algorithm with the results of direct calculation at 200 randomly selected points. These numbers demonstrate clearly that our algorithm scales like $O(N \log N)$ in terms of the number of points. Furthermore, the error seems to grow only slightly as we increase the number of points, indicating that the separated approximations are stable.

3.4 3D Case

Let us now turn to the 3D case. For X and Y defined as Fig. 5 in 3D, we have proved the following theorem in [26].

Theorem 10. *For any $\epsilon > 0$, there exists a number t_ϵ that is independent of w such that*

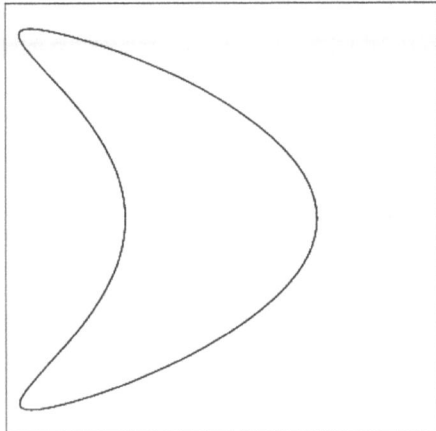 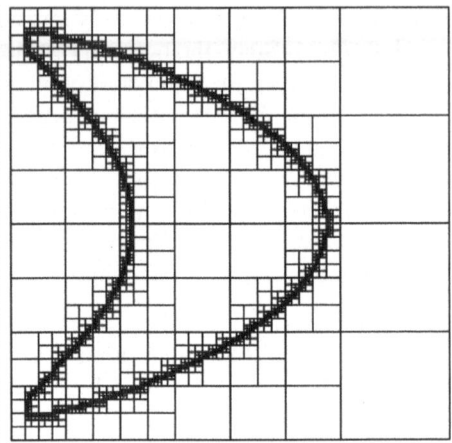

Table 5 Results of a kite-shaped model

(K, ϵ)	N	T_a(sec)	T_d(sec)	T_d/T_a	ϵ_a
(2048,1e−4)	1.13e+5	4.00e+1	8.11e+3	2.03e+2	1.08e−4
(8192,1e−4)	4.53e+5	1.77e+2	1.30e+5	7.36e+2	1.33e−4
(32768,1e−4)	1.81e+6	8.04e+2	2.09e+6	2.60e+3	1.41e−4
(2048,1e−6)	1.13e+5	6.10e+1	8.11e+3	1.33e+2	9.35e−7
(8192,1e−6)	4.53e+5	2.72e+2	1.30e+5	4.78e+2	9.15e−7
(32768,1e−6)	1.81e+6	1.24e+3	2.10e+6	1.70e+3	8.80e−7
(2048,1e−8)	1.13e+5	9.20e+1	8.16e+3	8.87e+1	1.45e−8
(8192,1e−8)	4.53e+5	4.05e+2	1.30e+5	3.22e+2	1.31e−8
(32768,1e−8)	1.81e+6	1.80e+3	2.11e+6	1.17e+3	1.52e−8

$$G(x, y) = \frac{e^{\imath \omega |x-y|}}{|x - y|}$$

has a t_ϵ-term ϵ-expansion for $x \in X = \{x : \theta(x, \ell) \le 1/w, |x| \ge w^2 \lambda\}$ and $y \in Y = B(0, w\lambda)$ for any ℓ.

Based on this theorem, an expansion in 3D similar to (18) can be constructed using the randomized procedure in [26, 27]. Starting from the domain $[-1, 1]^3$ that contains the whole scatterer, we subdivide the domain until that each of the leaf boxes contains only a small number of points. In order to take advantage of the directional separated approximations, the far field F^B for a box B of width $w\lambda$ is further partitioned into $O(w^2)$ directional wedges, each belonging to a cone with spanning angle $O(1/w)$. The translations are defined in exactly the same way and the algorithm proceeds in the same fashion. The following theorem summarizes the complexity analysis in 3D.

Theorem 11. *Let ∂D be a piecewise smooth boundary surface in $B(0, 1)$. Suppose that the points $\{p_i\}_{1 \le i \le N}$ are $N = O(\omega^2)$ samples of ∂D with a constant number*

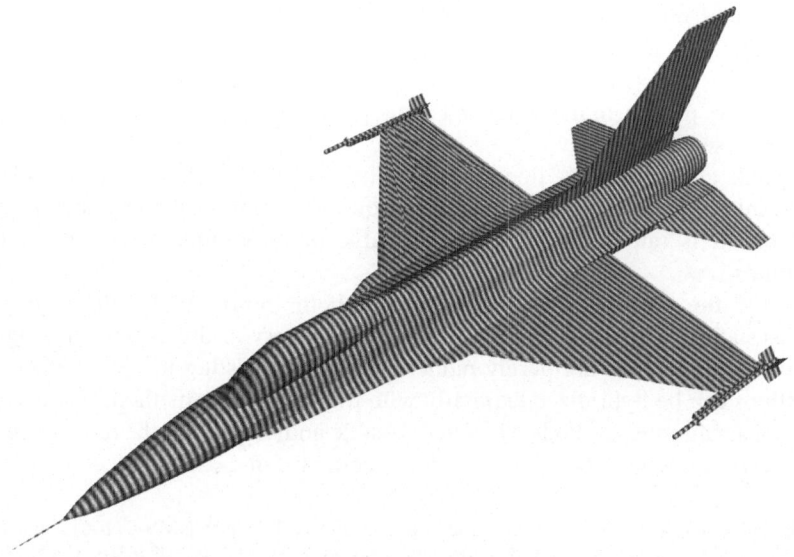

Table 6 Results of the F16 model with the Helmholtz kernel

(K, ϵ)	N	T_a(sec)	T_d(sec)	T_d/T_a	ϵ_a
(32, 1e−4)	1.87e+5	5.00e+1	4.17e+3	8.34e+1	6.13e−4
(64, 1e−4)	7.46e+5	2.27e+2	6.58e+4	2.90e+2	6.69e−4
(128,1e−4)	2.98e+6	1.04e+3	1.03e+6	9.87e+2	6.89e−4
(256,1e−4)	1.19e+7	5.04e+3	1.64e+7	3.25e+3	7.63e−4
(32, 1e−6)	1.87e+5	1.18e+2	4.06e+3	3.44e+1	2.72e−6
(64, 1e−6)	7.46e+5	6.12e+2	6.56e+4	1.07e+2	3.30e−6
(128,1e−6)	2.98e+6	3.07e+3	1.06e+6	3.45e+2	4.16e−6
(32, 1e−8)	1.87e+5	2.38e+2	4.07e+3	1.71e+1	6.34e−8
(64, 1e−8)	7.46e+5	1.29e+3	6.64e+4	5.14e+1	8.10e−8
(128,1e−8)	2.98e+6	6.42e+3	1.06e+6	1.64e+2	6.55e−8

of points per wavelength. Then for any prescribed accuracy, the proposed algorithm has a computational complexity $O(\omega^2 \log \omega) = O(N \log N)$.

We illustrate this 3D algorithm with a numerical example. The scatterer is a fighter jet and we report the results in Table 6. These numbers demonstrate clearly that our algorithm scales exactly like $O(N \log N)$ in terms of the number of points.

4   Conclusions and Future Work

In this paper, we presented the sweeping preconditioners for solving high frequency Helmholtz equation and a directional multilevel algorithm for its integral equation formulation. Both algorithms achieve essentially linear complexity. There are several future research directions along this line. First, we would like to extend these techniques to the Maxwell equations for electromagnetic scattering and the equations for linear elastic waves. In both cases we expect the overall structure of the algorithm to remain the same and some progress has already been made in the Maxwell case.

Second, although with linear complexity, these algorithms might still take significant running time due to the high frequency nature of the physical problem. Therefore, it is important to develop scalable parallel implementations of these algorithms.

Finally, the methods proposed here are both motivated by the asymptotic theories such as geometric optics and geometric theory of diffraction, although the algorithms themselves are purely numerical. It is interesting to see whether these algorithms can be hybridized explicitly with the asymptotic methods, for example, applying asymptotic methods wherever possible and only using the more expensive numerical techniques when the asymptotic methods fail.

Acknowledgements B. E. is partially supported by the NSF grants DMS-0714612 and DMS-1016577. L. Y. is partially supported by the NSF CAREER award DMS-0846501, the NSF grant DMS-1016577, and an Alfred P. Sloan Fellowship. The authors thank the reviewer for detailed comments and suggestions.

References

1. A. Averbuch, E. Braverman, R. Coifman, M. Israeli, and A. Sidi. Efficient computation of oscillatory integrals via adaptive multiscale local Fourier bases. *Appl. Comput. Harmon. Anal.*, 9(1):19–53, 2000.
2. A. Bayliss, C. I. Goldstein, and E. Turkel. An iterative method for the Helmholtz equation. *Journal of Computational Physics*, 49(3):443–457, 1983.
3. J.-D. Benamou and B. Desprès. A domain decomposition method for the Helmholtz equation and related optimal control problems. *J. Comput. Phys.*, 136(1):68–82, 1997.
4. M. Benzi, J. C. Haws, and M. Tůma. Preconditioning highly indefinite and nonsymmetric matrices. *SIAM J. Sci. Comput.*, 22(4):1333–1353 (electronic), 2000.
5. J.-P. Berenger. A perfectly matched layer for the absorption of electromagnetic waves. *J. Comput. Phys.*, 114(2):185–200, 1994.
6. E. Bleszynski, M. Bleszynski, and T. Jaroszewicz. AIM: Adaptive integral method for solving large-scale electromagnetic scattering and radiation problems. *Radio Science*, 31:1225–1252, 1996.
7. S. Börm, L. Grasedyck, and W. Hackbusch. Hierarchical matrices, 2006. Max-Planck-Institute Lecture Notes.
8. B. Bradie, R. Coifman, and A. Grossmann. Fast numerical computations of oscillatory integrals related to acoustic scattering. I. *Appl. Comput. Harmon. Anal.*, 1(1):94–99, 1993.
9. A. Brandt and I. Livshits. Wave-ray multigrid method for standing wave equations. *Electron. Trans. Numer. Anal.*, 6(Dec.):162–181 (electronic), 1997. Special issue on multilevel methods (Copper Mountain, CO, 1997).
10. O.P. Bruno and L. A. Kunyansky. A fast, high-order algorithm for the solution of surface scattering problems: basic implementation, tests, and applications. *J. Comput. Phys.*, 169(1):80–110, 2001.

11. F. X. Canning. Sparse approximation for solving integral equations with oscillatory kernels. *SIAM J. Sci. Statist. Comput.*, 13(1):71–87, 1992.
12. H. Cheng, W. Crutchfield, Z. Gimbutas, L. Greengard, J. Huang, V. Rokhlin, N. Yarvin, and J. Zhao. Remarks on the implementation of the wideband FMM for the Helmholtz equation in two dimensions. In *Inverse problems, multi-scale analysis and effective medium theory*, volume 408 of *Contemp. Math.*, pages 99–110. Amer. Math. Soc., Providence, RI, 2006.
13. H. Cheng, W. Y. Crutchfield, Z. Gimbutas, L. F. Greengard, J. F. Ethridge, J. Huang, V. Rokhlin, N. Yarvin, and J. Zhao. A wideband fast multipole method for the Helmholtz equation in three dimensions. *J. Comput. Phys.*, 216(1):300–325, 2006.
14. W. Chew, E. Michielssen, J. M. Song, and J. M. Jin, editors. *Fast and efficient algorithms in computational electromagnetics*. Artech House, Inc., Norwood, MA, USA, 2001.
15. W. C. Chew and W. H. Weedon. A 3-d perfectly matched medium from modified Maxwell's equations with stretched coordinates. *Microwave Opt. Tech. Lett*, 7:599–604, 1994.
16. D. Colton and R. Kress. *Inverse acoustic and electromagnetic scattering theory*, volume 93 of *Applied Mathematical Sciences*. Springer-Verlag, Berlin, second edition, 1998.
17. S. D. Conte. *Elementary numerical analysis: An algorithmic approach*. McGraw-Hill Book Co., New York, 1965.
18. E. Darve. The fast multipole method: numerical implementation. *J. Comput. Phys.*, 160(1): 195–240, 2000.
19. L. Demanet and L. Ying. Scattering in flatland: efficient representations via wave atoms. *Found. Comput. Math.*, 10(5):569–613, 2010.
20. H. Deng and H. Ling. Fast solution of electromagnetic integral equations using adaptive wavelet packet transform. *Antennas and Propagation, IEEE Transactions on*, 47(4):674–682, Apr 1999.
21. H. Deng and H. Ling. On a class of predefined wavelet packet bases for efficient representation of electromagnetic integral equations. *Antennas and Propagation, IEEE Transactions on*, 47(12):1772–1779, Dec 1999.
22. B. Després. Domain decomposition method and the Helmholtz problem. In *Mathematical and numerical aspects of wave propagation phenomena (Strasbourg, 1991)*, pages 44–52. SIAM, Philadelphia, PA, 1991.
23. J. Duff and J. Reid. The multifrontal solution of indefinite sparse symmetric linear equations. *ACM Trans. Math. Software*, 9:302–325, 1983.
24. H. C. Elman, O. G. Ernst, and D. P. O'Leary. A multigrid method enhanced by Krylov subspace iteration for discrete Helmholtz equations. *SIAM J. Sci. Comput.*, 23(4):1291–1315 (electronic), 2001.
25. B. Engquist and O. Runborg. Computational high frequency wave propagation. *Acta Numer.*, 12:181–266, 2003.
26. B. Engquist and L. Ying. Fast directional multilevel algorithms for oscillatory kernels. *SIAM J. Sci. Comput.*, 29(4):1710–1737 (electronic), 2007.
27. B. Engquist and L. Ying. A fast directional algorithm for high frequency acoustic scattering in two dimensions. *Commun. Math. Sci.*, 7(2):327–345, 2009.
28. B. Engquist and L. Ying. Sweeping preconditioner for the Helmholtz equation: hierarchical matrix representation. *Comm. Pure Appl. Math.*, 64(5):697–735, May 2011.
29. B. Engquist and L. Ying. Sweeping preconditioner for the Helmholtz equation: moving perfectly matched layers. Preprint, 2010.
30. Y. A. Erlangga. Advances in iterative methods and preconditioners for the Helmholtz equation. *Arch. Comput. Methods Eng.*, 15(1):37–66, 2008.
31. Y. A. Erlangga, C. W. Oosterlee, and C. Vuik. A novel multigrid based preconditioner for heterogeneous Helmholtz problems. *SIAM J. Sci. Comput.*, 27(4):1471–1492 (electronic), 2006.
32. Y. A. Erlangga, C. Vuik, and C. W. Oosterlee. On a class of preconditioners for solving the Helmholtz equation. *Appl. Numer. Math.*, 50(3-4):409–425, 2004.
33. O. G. Ernst and M. J. Gander. Why it is difficult to solve Helmholtz problems with classical iterative methods. In *this volume*, page 325, Springer, 2011.

34. J. Fish and Y. Qu. Global-basis two-level method for indefinite systems. I. Convergence studies. *Internat. J. Numer. Methods Engrg.*, 49(3):439–460, 2000.
35. M. J. Gander and F. Nataf. An incomplete LU preconditioner for problems in acoustics. *J. Comput. Acoust.*, 13(3):455–476, 2005.
36. J. George. Nested dissection of a regular finite element mesh. *SIAM J. Numer. Anal.*, 10:345–363, 1973.
37. W. Golik. Wavelet packets for fast solution of electromagnetic integral equations. *Antennas and Propagation, IEEE Transactions on*, 46(5):618–624, May 1998.
38. L. Grasedyck and W. Hackbusch. Construction and arithmetics of $\mathcal{H}$-matrices. *Computing*, 70(4):295–334, 2003.
39. L. Greengard and V. Rokhlin. A fast algorithm for particle simulations. *J. Comput. Phys.*, 73(2):325–348, 1987.
40. W. Hackbusch. A sparse matrix arithmetic based on $\mathcal{H}$-matrices. Part I: Introduction to $\mathcal{H}$-matrices. *Computing*, 62:89–108, 1999.
41. D. Huybrechs and S. Vandewalle. A two-dimensional wavelet-packet transform for matrix compression of integral equations with highly oscillatory kernel. *J. Comput. Appl. Math.*, 197(1):218–232, 2006.
42. S. Johnson. Notes on perfectly matched layers. Technical Report, Massachusetts Institute of Technology, 2010.
43. H.-O. Kreiss and J. Oliger. Comparison of accurate methods for the integration of hyperbolic equations. *Tellus*, 24:199–215, 1972.
44. A. Laird and M. Giles. Preconditioned iterative solution of the 2D Helmholtz equation. Technical Report, NA 02-12, Computing Lab, Oxford University, 2002.
45. B. Lee, T. A. Manteuffel, S. F. McCormick, and J. Ruge. First-order system least-squares for the Helmholtz equation. *SIAM J. Sci. Comput.*, 21(5):1927–1949 (electronic), 2000. Iterative methods for solving systems of algebraic equations (Copper Mountain, CO, 1998).
46. J. Liu. The multifrontal method for sparse matrix solution: Theory and practice. *SIAM Rev.*, 34:82–109, 1992.
47. I. Livshits and A. Brandt. Accuracy properties of the wave-ray multigrid algorithm for Helmholtz equations. *SIAM J. Sci. Comput.*, 28(4):1228–1251 (electronic), 2006.
48. P.-G. Martinsson. A fast direct solver for a class of elliptic partial differential equations. *Journal of Scientific Computing*, 38:316–330, 2009.
49. E. Michielssen and A. Boag. A multilevel matrix decomposition algorithm for analyzing scattering from large structures. *IEEE Transactions on Antennas and Propagation*, 44(8):1086–1093, 1996.
50. D. Osei-Kuffuor and Y. Saad. Preconditioning Helmholtz linear systems. *Appl. Numer. Math.*, 60(4):420–431, 2010.
51. V. Rokhlin. Rapid solution of integral equations of scattering theory in two dimensions. *J. Comput. Phys.*, 86(2):414–439, 1990.
52. V. Rokhlin. Diagonal forms of translation operators for the Helmholtz equation in three dimensions. *Appl. Comput. Harmon. Anal.*, 1(1):82–93, 1993.
53. Y. Saad. *Iterative methods for sparse linear systems.* Society for Industrial and Applied Mathematics, Philadelphia, PA, second edition, 2003.
54. Y. Saad and M. H. Schultz. GMRES: a generalized minimal residual algorithm for solving nonsymmetric linear systems. *SIAM J. Sci. Statist. Comput.*, 7(3):856–869, 1986.
55. C. E. Shannon. Communication in the presence of noise. *Proc. I.R.E.*, 37:10–21, 1949.
56. J. M. Song and W. C. Chew. Multilevel fast-multipole algorithm for solving combined field integral equations of electromagnetic scattering. *Microwave Opt. Tech. Lett.*, 10(1):15–19, 1995.
57. R. F. Susan-Resiga and H. M. Atassi. A domain decomposition method for the exterior Helmholtz problem. *J. Comput. Phys.*, 147(2):388–401, 1998.

58. P. Vaněk, J. Mandel, and M. Brezina. Two-level algebraic multigrid for the Helmholtz problem. In *Domain decomposition methods, 10 (Boulder, CO, 1997)*, volume 218 of *Contemp. Math.*, pages 349–356. Amer. Math. Soc., Providence, RI, 1998.
59. S. Wang, M. V. de Hoop, and J. Xia. Acoustic inverse scattering via Helmholtz operator factorization and optimization. *J. Comput. Phys.*, 229(22):8445–8462, 2010.
60. J. Xia, S. Chandrasekaran, M. Gu, and X. S. Li. Superfast multifrontal method for large structured linear systems of equations. *SIAM J. Matrix Anal. Appl.*, 31(3):1382–1411, 2009.

Uncertainty Quantification for Subsurface Flow Problems Using Coarse-Scale Models

Louis J. Durlofsky and Yuguang Chen

Abstract The multiscale nature of geological formations can have a strong impact on subsurface flow processes. In an attempt to characterize these formations at all relevant length scales, highly resolved property models are typically constructed. This high degree of detail greatly complicates flow simulations and uncertainty quantification. To address this issue, a variety of computational upscaling (numerical homogenization) procedures have been developed. In this chapter, a number of the existing approaches are described. These include single-phase parameter upscaling (the computation of coarse-scale permeability or transmissibility) and two-phase parameter upscaling (the computation of coarse-scale relative permeability curves) procedures. Methods that range from purely local to fully global are considered. Emphasis is placed on the performance of these techniques for uncertainty quantification, where many realizations of the geological model are considered. Along these lines, an ensemble-level upscaling approach is described, in which the goal is to provide coarse models that capture ensemble flow statistics (such as the cumulative distribution function for oil production) consistent with those of the underlying fine-scale models rather than agreement on a realization-by-realization basis. Numerical results highlighting the relative advantages and limitations of the various methods are presented. In particular, the ensemble-level upscaling approach is shown to provide accurate statistical predictions at an acceptable computational cost.

L.J. Durlofsky (✉)
Department of Energy Resources Engineering, Stanford University, Stanford, CA 94305, USA
e-mail: lou@stanford.edu

Y. Chen
Chevron Energy Technology Company, PO Box 6019, San Ramon, CA 94583, USA
e-mail: ychen@chevron.com

I.G. Graham et al. (eds.), *Numerical Analysis of Multiscale Problems*, Lecture Notes in Computational Science and Engineering 83, DOI 10.1007/978-3-642-22061-6_6, © Springer-Verlag Berlin Heidelberg 2012

1   Introduction

Subsurface formations such as oil reservoirs and aquifers typically display heterogeneous properties that vary over a range of length scales. Precise geological characterization of these formations is challenging, however, as field measurements are very sparse spatially. Property models are generally constructed through application of a variety of geological and geostatistical modeling techniques. Because of the stochastic nature of the characterization problem, multiple realizations of the formation are typically generated, with each realization assigned an appropriate probability. Then, by performing flow simulations for each of these models, statistical measures of performance (e.g., cumulative distribution function for oil production or aquifer discharge rate) can be computed.

The computational cost of the flow simulations required by this approach can be prohibitive. Because geological features at many length scales can impact flow and transport, models that resolve all relevant scales of heterogeneity must be constructed. And, because there can be many uncertain geological properties and parameters, a large number of geological realizations must be generated. Flow simulation on even a single highly-detailed geological model can represent a computational challenge. Simulation of hundreds or thousands of these models for uncertainty quantification in practical field cases may not currently be realizable.

This general problem can be addressed in different ways. Specialized sampling procedures could be applied to select appropriate models for flow simulation. Existing sampling approaches, however, still require that a large number of flow simulations be performed. Alternatively, the geological models can be coarsened or upscaled prior to flow simulation. This entails the application of numerical homogenization procedures to provide coarse-scale models that can be simulated much more efficiently. These procedures, and their application and performance for ensembles of geological models, will be the topic of this article.

There have been a wide variety of upscaling approaches that have been developed in recent years. Some of these procedures are discussed in detail in the following sections. Upscaling methods can be generally classified in terms of the types of coarse-scale quantities computed (single-phase only versus single-phase and two-phase flow functions) and the spatial domain over which the coarse-scale parameters are computed (e.g., local versus global). Generally speaking, the accuracy of the coarse model improves when larger spatial domains are used for the numerical homogenization and when two-phase flow parameters are computed in addition to single-phase parameters (for, e.g., oil-water problems). These extra computations lead to additional expense, which must be balanced against the enhanced accuracy provided.

Reviews discussing single-phase numerical upscaling procedures are presented in, e.g., [35, 37, 62, 67]. Local single-phase upscaling techniques include those described by [28, 60, 68, 71]. Many of these approaches differ in terms of the boundary conditions applied, the size of the domain used for the numerical homogenization computations, and the particular parameters computed (coarse-block permeability or transmissibility). Techniques that use global fine-scale information include the methods described in [41, 56, 69, 72, 74], while approaches that incorporate approximate (e.g., coarse-scale) global effects are presented in [9, 11, 17, 36, 49]. Early approaches that apply two-phase parameter upscaling appear in [51, 64]. A number of two-phase upscaling techniques are discussed by [7, 23, 25]. More recent approaches include those of [12, 14, 33, 40, 44, 55, 65, 66]. These approaches vary in terms of the boundary conditions and detailed upscaling calculations, the size of the computational domain, and the form of the coarse-scale equations. We note that, in many cases, the coarse-scale models generated using these procedures can be used for a variety of flow problems.

Upscaling methods are related to multiscale finite element (MsFEM) and finite-volume-based (MsFVM) techniques. Multiscale techniques have been studied extensively for subsurface flow problems and our discussion of these approaches will be brief. A comprehensive presentation is provided by [31]. Specific techniques include the original MsFEM approach of [42], the MsFVM implementations of [38, 39, 45, 46, 53, 70], and the mixed finite element approaches of [1, 3, 5, 21]. Both upscaling and MsFEM/MsFVM approaches seek to construct coarse-scale operators that capture the effects of subgrid property variation. In the case of upscaling procedures, most (or all, depending on the specific approach) of the required numerical homogenization computations are performed in a preprocessing step. Multiscale techniques, by contrast, typically include both fine and coarse-scale computations during the course of the simulation. In addition, the goal of most multiscale procedures is to reproduce, as closely as possible, the fine-scale solution. Thus these approaches often include some type of reconstruction step. Upscaling techniques, by contrast, seek to provide a coarse-scale rather than a fine-scale solution. For example, for subsurface flow problems, an ideal upscaling method will give accurate boundary fluxes or injection and production rates, but it will not provide state variables at the fine-scale level. Coarse-grid state variables should, however, approximate (volume) averages of corresponding fine-scale quantities. There are some techniques that effectively combine multiscale and upscaling ideas; see, for example [53, 57].

Just as the boundary conditions applied for the computation of upscaled parameters can affect coarse-grid accuracy, the boundary conditions used in the computation of the multiscale basis can also impact solution accuracy. Approaches for addressing this issue include the use of fine-scale information outside the target coarse-grid block (this approach is referred to as oversampling; see [42]), the use of reduced boundary conditions [46], the use of global [1, 32] and approximate global [30, 47] single-phase flow information, and correction functions [38]. The impact of local boundary conditions in heterogeneous multiscale methods is considered in [73].

The need to quantify the impact of geological and other uncertainties on multiscale flow problems adds an additional set of challenges. As noted earlier, uncertainty is generally handled through use of multiple realizations, which increases computational requirements. Within the context of upscaling procedures, this problem has been addressed using ensemble-level upscaling [13, 20], where the intent is for the coarse models to capture key fine-scale flow quantities in a statistical sense rather than in a deterministic (realization-by-realization) manner. Uncertainty quantification has also been treated within a multiscale context – see, e.g., [2].

This chapter proceeds as follows. We first present, in Sect. 2, the fine and coarse-scale governing equations as well as the basic finite volume discretization procedure typically used in reservoir simulation. Single and two-phase upscaling techniques are described in Sects. 3 and 4. Then, in Sect. 5, ensemble-level upscaling approaches are presented. Numerical results in Sects. 3, 4 and 5 illustrate the performance of a wide variety of numerical upscaling procedures. We conclude with a summary and some suggestions for future research directions.

2 Governing Equations and Finite Volume Discretization

In this section we present the equations describing the flow of two immiscible fluids (here taken to be oil and water) in a porous medium. These equations are provided for both the fully-resolved (fine-scale) system and the coarse-scale model. The finite volume discretization technique commonly used for reservoir simulation is briefly discussed.

2.1 Oil-Water Flow Equations

We consider a two-phase, two-component system containing oil and water. The oil component exists only in the oleic (or oil) phase and the water component exists only in the aqueous (or water) phase; i.e., there is no mass transfer between phases. The continuum (Darcy-scale) equations describing the flow of oil and water through porous formations are derived by combining expressions for mass conservation with Darcy's law. Using the subscript j to designate component or phase ($j = o$ for oil and w for water), these equations can be written as:

$$\frac{\partial}{\partial t}\left(\phi \rho_j S_j\right) - \nabla \cdot \left[\rho_j \lambda_j \mathbf{k} \left(\nabla p_j - \rho_j g \mathbf{i}_z\right)\right] + q_j^w = 0. \tag{1}$$

Here $\mathbf{k}$ is the absolute permeability tensor (essentially a flow conductivity) of the rock, $\lambda_j = k_{rj}/\mu_j$ is the phase mobility, with k_{rj} the relative permeability to phase j and μ_j the phase viscosity, p_j is phase pressure, ρ_j is the phase density, g is gravitational acceleration, $\mathbf{i}_z$ is the unit vector pointing in the downward-vertical (gravity) direction, t is time, ϕ is porosity (volume fraction of the pore space), S_j is saturation (phase j volume fraction within the pore space) and q_j^w is the source/sink term (positive for production). The relative permeabilities $k_{rj}(S_j)$, and thus the mobilities λ_j, capture the phase interference effect; i.e., the impact of saturation on flow rate. The general oil-water model is completed by incorporating the saturation constraint ($S_o + S_w = 1$) and by specifying a capillary pressure relationship $p_c(S_w) = p_o - p_w$. In practice, the relative permeability curves and the capillary pressure curves are determined empirically through laboratory measurements (though pore and grain-scale modeling, such as the methodology described in [58], is emerging as an alternative to time consuming experimental procedures). These functions typically vary with rock type (and thus with spatial location), though in the simulations presented here only a single set of curves will be used.

In many oil-water systems, after a short early-transient period, the effects of compressibility are very small and can be neglected. In addition, even at the fine-grid block scale, the effects of capillary pressure are often negligible, which means we can take $p_o = p_w = p$. With these simplifications, and the further assumption that the rock is incompressible (which means ϕ does not vary in time), (1) can be rearranged into the following form:

$$\nabla \cdot \{\lambda \mathbf{k} \left[\nabla p + g(\lambda_o \rho_o + \lambda_w \rho_w) \mathbf{i}_z \right] \} = R, \tag{2}$$

$$\phi \frac{\partial S}{\partial t} + \nabla \cdot [\mathbf{u} f - \mathbf{k} \cdot \mathbf{i}_z \lambda_o g(\rho_w - \rho_o)] = -R_w. \tag{3}$$

Note that we now use S in place of S_w. In (2) and (3), $\lambda(S) = \lambda_o + \lambda_w$, $\mathbf{u} = -\lambda \mathbf{k}[\nabla p + g(\lambda_o \rho_o + \lambda_w \rho_w) \mathbf{i}_z]$ is the total (oil + water) Darcy velocity, $f(S) = \lambda_w/(\lambda_o + \lambda_w)$ is the so-called Buckley-Leverett flux function, $R_w = q_w/\rho_w$ and $R = q_o/\rho_o + q_w/\rho_w$. Equation (2), often referred to as the pressure equation, is elliptic, while (3), the saturation equation, is hyperbolic. For equal density fluids, or in the absence of gravitational effects, (3) reduces to $\phi \partial S/\partial t + \nabla \cdot (\mathbf{u} f) = -R_w$.

2.2 Coarse-Scale Flow Equations

In any type of upscaling or numerical homogenization procedure, a key first step is the determination of the form of the coarse-scale equations. In general, the coarse-scale equations can include terms, functional dependencies, and levels of anisotropy that are not present in the fine-scale description. We now describe the types of coarse-scale models that have been used for subsurface flow simulation. The computation of the parameters appearing in these equations will be the subject of Sects. 3 and 4, while the statistical assignment of some of these parameters will be considered in Sect. 5.

We first discuss the coarse-scale single-phase flow equation. For this purpose we consider the steady-state incompressible flow of a single component in a single phase. Taking $\mu = 1$ and neglecting gravitational effects and the source term, (1) (and (2)) reduces to:

$$\nabla \cdot [\mathbf{k} \nabla p] = 0. \tag{4}$$

Under the assumption of length scale separation, in which $\mathbf{k}$ is assumed to vary over a scale $\mathbf{y}$ that is small compared to the global length scale $\mathbf{x}$, the form of the coarse-scale equation is [8]:

$$\nabla \cdot \left[\mathbf{k}^* \nabla p^c \right] = 0, \tag{5}$$

where $\mathbf{k}^*$ is the upscaled or effective permeability tensor and the superscript c designates a coarse-scale quantity.

For two-phase flow problems, the general form of the coarse-scale equation is often assumed to be the same as the fine-scale model. For example, for incompressible two-phase flow in the absence of gravitational effects and source terms, the coarse model is of the form:

$$\nabla \cdot \left[\lambda^* \mathbf{k}^* \nabla p^c \right] = 0, \tag{6}$$

$$\phi^* \frac{\partial S^c}{\partial t} + \nabla \cdot \left[\mathbf{u}^c f^* \right] = 0. \tag{7}$$

The upscaled parameters $\mathbf{k}^*(\mathbf{x})$ and $\phi^*(\mathbf{x})$ and functions $\lambda^*(\mathbf{x}, S^c)$ and $f^*(\mathbf{x}, S^c)$ are computed in a preprocessing step. Note that, even if the fine-scale functions $f(S)$ and $\lambda(S)$ do not vary from block to block (i.e., they are not functions of $\mathbf{x}$ or $\mathbf{y}$), the coarse-scale functions are, in general, dependent on $\mathbf{x}$ and thus must be computed for each coarse block.

Alternate upscaled representations have also been devised. Coarse-scale models involving nonlocal effects were developed and applied by, e.g., [34, 43]. A related model, which includes only local terms and may thus be more amenable for computation, is the generalized convection-diffusion (GCD) representation, described by [12, 33]. This model contains, in addition to convective subgrid effects (as are captured in f^* in (7)), a diffusive term. This term allows the model to properly represent subgrid effects driven by short-length-scale permeability variations, as this type of permeability correlation structure leads to diffusive rather than convective corrections.

In the GCD model, rather than apply (7), the coarse-scale saturation equation is given by:

$$\phi^* \frac{\partial S^c}{\partial t} + \nabla \cdot [\mathbf{u}^c f] + \nabla \cdot \mathbf{m} - \nabla \cdot [\mathbf{D} \nabla S^c] = 0, \tag{8}$$

where $\mathbf{m}(\mathbf{x}, S^c)$ is a correction to the convective flux and $\mathbf{D}(\mathbf{x}, S^c) \nabla S^c$ is a diffusive flux ($\mathbf{D}$ is the coarse-scale diffusion tensor). Equation (8) could also be used for systems in which capillary pressure effects are important. The general GCD model additionally includes a ∇S^c dependency in the coarse-scale pressure equation, though a variant that uses an equation of the same form as (6) has been applied successfully [12]. In the case of the GCD model (with (6) for the coarse-scale pressure equation), the coarse-scale functions $\mathbf{k}^*(\mathbf{x})$, $\phi^*(\mathbf{x})$, $\lambda^*(\mathbf{x}, S^c)$, $\mathbf{m}(\mathbf{x}, S^c)$ and $\mathbf{D}(\mathbf{x}, S^c)$ must be computed in a preprocessing step.

2.3 Finite Volume Discretization

We now briefly describe the finite volume representation typically applied for (1). See [6] or [37] for more details. Analogous procedures are used to discretize the coarse-scale flow equations. For simplicity we assume the model domain in an $x - y - z$ coordinate system is partitioned into N uniform grid blocks of dimensions $\Delta x \times \Delta y \times \Delta z$. Permeability $\mathbf{k}$ and porosity ϕ are taken to be constant in each grid block, though they are discontinuous (with potentially large jumps) from block to block. Neglecting capillary pressure and considering horizontal flow in the x-direction (in which gravity does not act), the flow term in (1), for grid block i, is represented in a fully-implicit formulation as:

$$\frac{\partial}{\partial x}\left[k\rho_j\lambda_j\left(\frac{\partial p}{\partial x}\right)\right] \approx \left\{(T_j)_{i-1/2}^{n+1}\left[p_{i-1}^{n+1} - p_i^{n+1}\right] + (T_j)_{i+1/2}^{n+1}\left[p_{i+1}^{n+1} - p_i^{n+1}\right]\right\}\frac{1}{V},$$

(9)

where subscript j denotes phase and i denotes grid block, superscript $n+1$ specifies the next time step, and $V = \Delta x \Delta y \Delta z$ is the grid block volume. The transmissibility $(T_j)_{i-1/2}^{n+1}$ relates the mass flow rate of phase j to the pressure difference between grid blocks $i - 1$ and i and is given by:

$$(T_j)_{i-1/2}^{n+1} = \left(\frac{kA}{\Delta x}\right)_{i-1/2}(\rho_j\lambda_j)_{i-1/2}^{n+1},$$

(10)

where $A = \Delta y \Delta z$ is the area of the interface between blocks $i - 1$ and i. The interface permeability $k_{i-1/2}$ is computed as the harmonic mean of k_i and k_{i-1} and the $\rho_j\lambda_j$ term is upstream weighted. The transmissibility $(T_j)_{i+1/2}^{n+1}$ is defined similarly, as are the transmissibilities in the y and z-directions. Complications arise with full-tensor permeabilities, as discussed in Sect. 3.3. The accumulation term in (1) is represented using a first-order implicit (backward Euler) method.

In a typical reservoir simulation, grid blocks may be tens of meters on a side in x and y while wellbore diameters are 0.25 m or less. This size disparity means that the wellbore pressure, which is a quantity commonly specified (or computed) in flow simulations, differs from the pressure of the grid block containing the well (the well block). To capture this effect, the source/sink term in (1) is typically modeled using a well index representation. This approach incorporates the analytical solution for radial flow (in which $p \sim \log r$, where r is radial distance from the wellbore) into the numerical representation for q_j. Specifically, a well equation of the following form is introduced:

$$\left(q_j^w\right)_i^{n+1} = W_i\left(\rho_j\lambda_j\right)_i^{n+1}(p_i^{n+1} - p_i^w).$$

(11)

Here $(q_j^w)_i^{n+1}$ is the flow rate of phase j (in units of mass/time) from block i into the well (or vice versa) at time $n + 1$, p_i^{n+1} is grid-block pressure at time $n + 1$, p_i^w is the wellbore pressure for the well located in grid block i, and W_i is the well index. For a vertical well that fully penetrates block i, W_i is given by [59]:

$$W_i = \left[\frac{2\pi k \Delta z}{\log (r_0/r_w)} \right]_i, \tag{12}$$

where r_w is the wellbore radius and $r_0 \approx 0.2\Delta x$ for square grid blocks and isotropic permeability fields. See [59] for further discussion and for expressions for W_i for more general cases.

The discretized representation of (1) entails $2N$ nonlinear algebraic equations which must be solved to determine pressure and saturation in each grid block at time $n + 1$. Solution of this system is accomplished using Newton's method.

3   Upscaling of Single-Phase Flow Parameters

We now describe the numerical computation of the upscaled functions that appear in the coarse-scale models presented in Sect. 2.2. We first consider upscaled single-phase flow parameters. Our descriptions here and in Sect. 4 are for two-dimensional systems, though the techniques considered are equally valid for three-dimensional cases.

3.1   Local Permeability and Transmissibility Upscaling

The coarse-scale single-phase flow equation is given by (5). The upscaled permeability $\mathbf{k}^*$ can be computed using a variety of numerical homogenization procedures (analytical approaches can also be used for idealized cases). Figure 1a depicts the global flow domain, where the finer lines represent the fine-scale grid and the heavier lines the coarse-scale grid. In local permeability upscaling procedures, the fine-scale pressure equation (4) is solved on the fine-scale domain corresponding to the target coarse block, subject to specified boundary conditions. Common choices for these boundary condition are constant pressure on inlet and outlet boundaries and no flow on all other boundaries (this is the standard procedure), or periodicity [28, 60]. Equation (4) must be solved twice in two-dimensional problems and three times in three-dimensional problems, with pressure driven in each coordinate direction. The upscaled permeability $\mathbf{k}^*$ can then be computed in various ways (see discussion in [75]); an accurate and robust approach is through the inversion of the coarse-scale Darcy's law [68, 71]:

$$\bar{\mathbf{u}} = -\mathbf{k}^*\overline{\nabla p}, \tag{13}$$

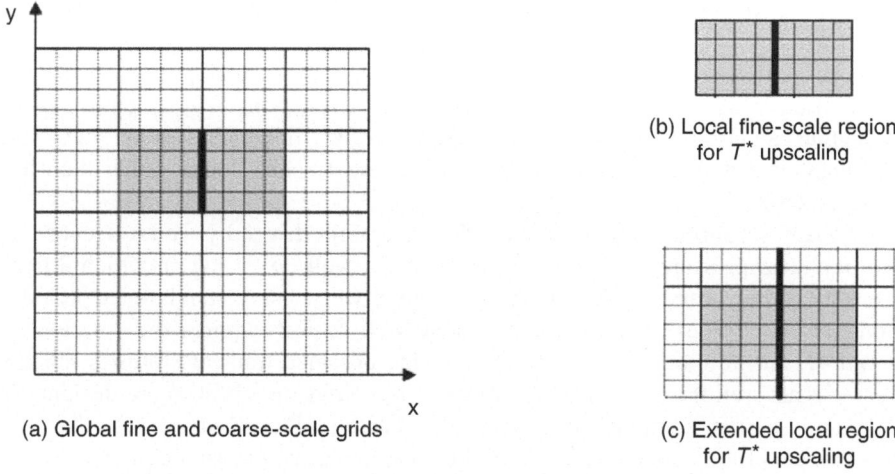

Fig. 1 Schematic showing global fine and coarse-scale grids and local regions to compute upscaled parameters. (**a**) Global fine and coarse-scale grids. (**b**) Local fine-scale region for T^* upscaling. (**c**) Extended local region for T^* upscaling

where $\bar{\mathbf{u}}$ represents the averaged fine-scale velocities over the coarse-scale block and $\overline{\nabla p}$ is the averaged fine-scale pressure gradient over the coarse block. For Cartesian grids, $\overline{\nabla p} = \nabla \bar{p}$, as shown in [75].

Once $\mathbf{k}^*$ is computed for all coarse blocks, coarse-scale transmissibility, as appears in the coarse analog of (9), can be computed by harmonically averaging appropriate components of $\mathbf{k}^*$ in adjacent blocks (more general multipoint flux techniques are required if full-tensor effects are important – see, e.g., [4]). An alternate procedure, however, is to compute upscaled transmissibility (T^*) directly. This approach avoids the second (harmonic) averaging and has been shown to provide improved coarse-model accuracy for highly heterogeneous systems [e.g., 17, 63].

Figure 1b illustrates the local fine-scale region used to directly compute T^*. Note that T^* is defined at the interface between the two coarse blocks. Local boundary conditions with constant pressure at the left and right boundaries and no flow specified on the upper and lower boundaries (for flow in the x-direction) are often assumed. Then the upscaled T^* (in the x direction) is determined via:

$$(T_x^*)_{i+1/2} = \frac{\bar{q}_{i+1/2}}{\bar{p}_i - \bar{p}_{i+1}},\qquad(14)$$

where $\bar{q}$ designates the integrated fine-scale flow rate through the interface and $\bar{p}_i$ and $\bar{p}_{i+1}$ are the volume averages of the fine-scale pressures over the two coarse blocks. An analogous problem with flow driven in the y-direction is solved to determine T_y^*.

To reduce the effect of the assumed local boundary conditions, some neighboring fine-scale cells can be included in the computation, as shown in Fig. 1c. This approach, referred to as extended local upscaling [68], is analogous to oversampling, used in the computation of multiscale basis functions [42]. Extended local upscaling can also be applied for the calculation of $\mathbf{k}^*$, in which case the fine-scale computational domain is centered around the target coarse block rather than the coarse interface as in Fig. 1c.

In many subsurface flow problems, flow is mainly driven by wells. Because of the nonlinear pressure field in the vicinity of the well ($p \sim \log r$), which differs from the slowly varying "linear" flow assumed when using standard or periodic boundary conditions, upscaling in the near-well region requires the solution of a well-driven flow problem. In addition to the upscaled transmissibilities linking coarse-scale well blocks to adjacent blocks (these transmissibilities are designated T_w^*), an important additional quantity is the coarse-scale well index W_i^*. This can be computed following the solution of the local fine-scale pressure equation with a (well-driven) source term using:

$$W_i^* = \frac{q_i^w}{p_i^w - \overline{p}_i}. \tag{15}$$

Here i designates the coarse block in which the well is located, q_i^w is the flow rate into or out of the well, p_i^w is wellbore pressure and $\overline{p}_i$ is the volume-averaged fine-scale pressure over coarse block i. For details on near-well upscaling, refer to [16, 27, 29, 54].

We note that specialized treatments for well-driven flow have also been developed for multiscale finite element and finite volume methods. Such implementations include those discussed in [1, 22, 45, 48, 50, 70].

3.2 Global and Quasi-global Approaches

The local procedures described above are efficient, as the large-scale global flow problem is decomposed into a series of small-scale, local flow problems (which can be easily solved in parallel). As a result of assumptions regarding local boundary conditions, they may however lose accuracy for highly heterogeneous permeability fields that lack scale separation. In recent years, techniques that incorporate global flow into upscaling calculations have been shown to improve the accuracy of coarse-scale models.

This can be achieved by directly computing the upscaled properties from global fine-scale (single-phase) flow solutions. Such global (single-phase flow) upscaling procedures [41, 56, 69, 72, 74] entail the computation of $\mathbf{k}^*$, T^* and near-well parameters. Global upscaling eliminates the need for (assumed) local boundary conditions, though it requires one or more global fine-scale solutions. It should be kept in mind, however, that the global steady-state single-phase flow problem need

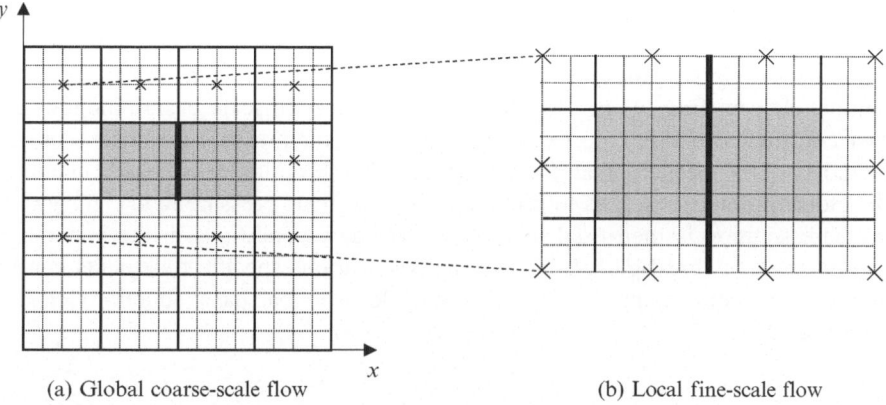

(a) Global coarse-scale flow   (b) Local fine-scale flow

Fig. 2 Schematic showing global coarse-scale grid and local fine-scale region. Local boundary conditions are determined from global coarse-scale pressures (represented by ×'s). (**a**) Global coarse-scale flow. (**b**) Local fine-scale flow

be solved only once (or perhaps a few times), and that the resulting parameters can then be applied to coarse-scale time-dependent multiphase flow simulations. Thus, global single-phase upscaling is viable in practice in many cases.

Quasi-global approaches also incorporate global flow effects, though in this case this information is only approximate. In local-global upscaling [e.g. 11, 17, 36, 49], for example, the fine-scale boundary conditions used for the local upscaling computations are determined from global coarse-scale solutions. The basic procedure is illustrated in Fig. 2. First, a local single-phase upscaling is performed and the global coarse-scale pressure equation is solved. Global coarse-scale pressures (represented by ×'s in Fig. 2a) are then interpolated to provide boundary conditions for extended local upscaling computations (as depicted in Fig. 2b). The approach thus incorporates (approximate) global flow effects while avoiding the solution of the global fine-scale problem. Iteration and thresholding can be introduced to improve accuracy and efficiency; for detailed algorithms refer to [11, 17].

We note that there are two basic ways in which global flow information (either exact or approximate) can be used in upscaling calculations. The first approach entails the use of "generic" global flow information, in which flow is driven by, e.g., large-scale pressure differences in each coordinate direction. Approaches of this type include those of [17, 36, 72]. Generic flow information can be expected to provide reasonable accuracy for many different flow scenarios, though for some particular problems it may not provide the best accuracy achievable. The second approach addresses this issue by using global information based on the specific flow scenario [e.g., 11, 41, 49, 56, 74]. In this case the coarse-scale properties are adapted to the specific global problem and can thus be expected to provide improved accuracy relative to those computed from generic flows. Some or all of these "adaptive" properties may, however, need to be recomputed when global flow conditions change considerably. Thus there is a potential tradeoff between accuracy and efficiency with these two basic approaches.

The local-global procedure has also been applied within a multiscale finite volume element framework [30]. In this approach, multiscale basis functions are first constructed using, e.g., oversampling and linear pressure conditions on the boundary of the local (oversampled) regions. Using this initial set of basis functions, the global problem is solved to provide the coarse-scale pressure field. This global pressure solution is then used to provide boundary conditions in the generation of a new set of (local-global) basis functions. Various specific treatments can be applied, and conforming or nonconforming local-global basis functions can be constructed.

3.3 Accounting for Full-Tensor Effects in Coarse Models

It is well known that fine-scale heterogeneity can lead to large-scale permeability anisotropy, as can be seen immediately through consideration of layered systems. It is also evident that, if the layering is skewed relative to the coordinate system, the coarse-scale permeability will be a full tensor. Numerical effects similar to those that appear with full-tensor permeabilities also arise from grid nonorthogonality. In many practical cases, full-tensor and grid nonorthogonality effects are relatively small and can be neglected or treated approximately. When these effects are important, methods must be applied that are suitable for use with heterogeneous full-tensor permeability fields. Within the context of permeability upscaling, one can proceed by computing $\mathbf{k}^*$ for each coarse block and then applying a multipoint flux approximation (MPFA) for discretization [e.g., 4].

Alternatively, coarse-scale transmissibilities that capture subgrid and full-tensor effects can be directly computed. Such an approach was developed for local upscaling in [52, 61]. We note that MsFEM and MsFVM techniques also employ a coarse-scale operator that captures full-tensor effects. In addition to these local approaches, global and local-global transmissibility upscaling procedures that represent full-tensor effects have also been developed and applied [9, 10].

In some cases it is possible to account approximately for full-tensor effects using a nonlinear two-point flux approximation. Specifically, in [19], it was shown that the use of a specialized two-point flux approximation in conjunction with global or local-global upscaling procedures can approximately capture full-tensor effects in coarse-scale simulations.

3.4 Numerical Results for Single-Phase Flow

We now present results for single-phase flow problems. We first illustrate the performance of several different upscaling techniques. Then, results for the local-global multiscale finite volume element method are provided.

3.4.1 Performance of Various Upscaling Techniques

The performance of various single-phase upscaling techniques will now be assessed for a challenging case. The example presented here is from [18]. We consider a two-dimensional synthetic channelized system with highly variable permeability. A total of 100 realizations, conditioned to seismic and well permeability data, are generated to account for uncertainty in the permeability field. Three realizations are shown in Fig. 3. One injection well and one production well penetrate high permeability blocks in all realizations, as indicated in Fig. 3. The fine-scale model (of dimensions 100×100) is uniformly coarsened to 20×20. We consider a unit dimensionless pressure difference between the injector and producer, and compare the total flow rate Q between the fine and coarse-scale models.

Displayed in Fig. 4 are cross-plots for the fine and coarse-scale flow rates for the 100 realizations. The range of flow rate values reflects the uncertainty in the permeability field. A perfect upscaling procedure would result in all points falling on the $45°$ line. Figure 4a shows coarse-scale flow rates for models generated using local $\mathbf{k}^*$ upscaling without near-well upscaling (in this case, $\mathbf{k}^*$ is computed for the well block using a local approach and (12) is applied using coarse dimensions and $(k_{xx}^* k_{yy}^*)^{1/2}$ in place of k). The reference fine-scale flow rates are significantly underestimated by these coarse models, demonstrating the inadequacy of the local $\mathbf{k}^*$ approach for this case. This underestimation is due to the inability of purely local upscaling procedures to capture the large-scale permeability connectivity that strongly impacts global flow and to the inaccurate near-well treatment. The coarse-scale results are considerably improved through use of extended local T^* and near-well upscaling (displayed in Fig. 4b). Further improvement is achieved using adaptive local-global upscaling (which includes near-well upscaling), where the upscaled properties are adapted to the specific well configuration, as is evident in Fig. 4c. These results demonstrate the advantage of incorporating global flow in the upscaling calculations.

When considering large numbers of realizations, realization-by-realization agreement may not be as important as agreement in key statistical quantities such

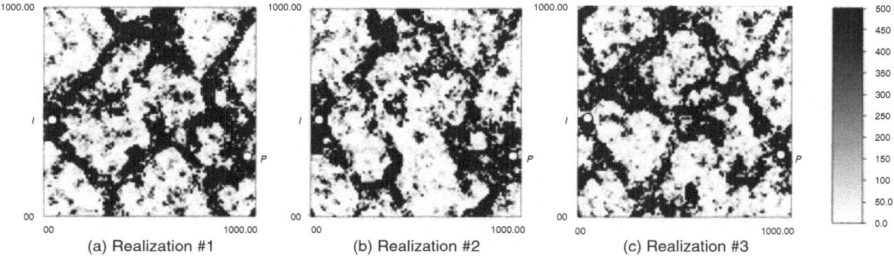

(a) Realization #1 (b) Realization #2 (c) Realization #3

Fig. 3 Three realizations of the permeability distribution (of dimensions 100×100) for the channelized reservoir conditioned to synthetic seismic and well data. Locations of injection $(3, 48)$ and production $(98, 28)$ wells are indicated as I and P (figure modified from [11]). (**a**) Realization #1. (**b**) Realization #2. (**c**) Realization #3

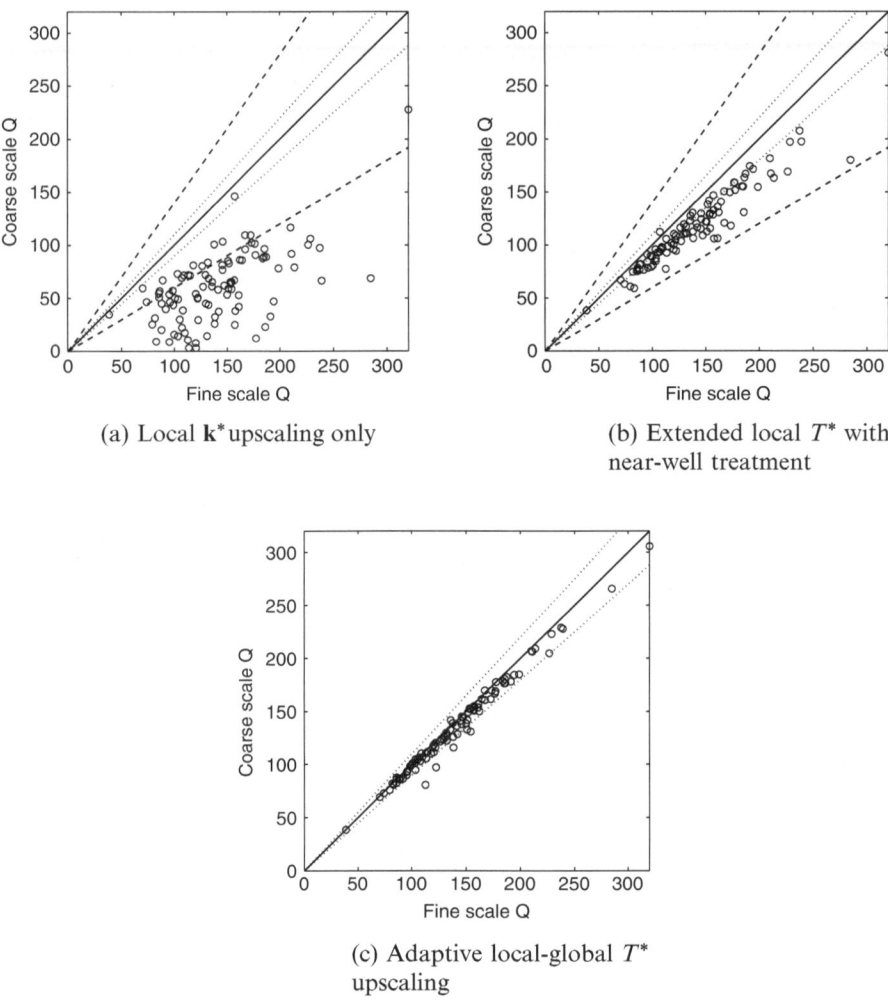

(a) Local **k***upscaling only

(b) Extended local T^* with near-well treatment

(c) Adaptive local-global T^* upscaling

Fig. 4 Well flow rates for fine and coarse-scale models for 100 realizations. Dotted lines represent 10% error and dashed lines 40% error (from [18]). (**a**) Local **k*** upscaling only. (**b**) Extended local T^* with near-well treatment. (**c**) Adaptive local-global T^* upscaling

as the cumulative distribution function (CDF) for flow rate over the ensemble, as shown in Fig. 5. The fine-scale CDF is represented by the thick solid curve. The CDF that results from the use of local **k*** upscaling is shifted significantly to the left, indicating a clear bias (underestimation) in flow rate. The use of extended local T^* and near-well upscaling provides coarse-scale models with much better accuracy – the dot-dashed CDF curve is relatively close to the fine-scale CDF. Consistent with the cross-plot shown in Fig. 4c, the use of adaptive local-global T^* upscaling (thin solid curve) leads to a very accurate CDF. This demonstrates the

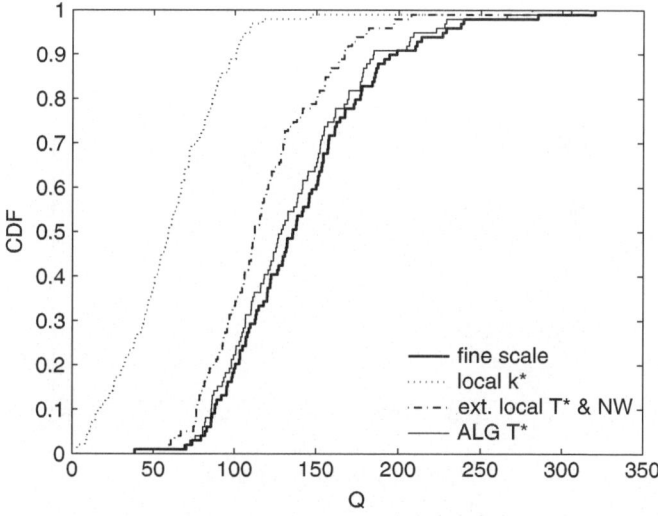

Fig. 5 Comparison of CDF of well flow rates between fine and coarse-scale models

importance of using appropriate coarse-scale models in uncertainty quantification and the potential bias that can arise through use of inaccurate upscaling procedures.

3.4.2 Results for Local-Global Multiscale Finite Volume Element Method

We now present a single set of results using the local-global multiscale finite volume element method developed in [30]. These simulations involve 50 two-dimensional channelized models taken from [24]. Although we do not show results for this case using the upscaling techniques considered in the previous section, we note that the application of standard upscaling procedures for this system results in significant error. Local-global upscaling can, however, provide accurate coarse models (as demonstrated in [17]).

For this problem, pressure is fixed at the left and right boundaries and no-flow conditions are imposed at the upper and lower boundaries. Flow is single phase and incompressible. The total flow rate through the system Q is then computed by integrating the flux over the inlet (or outlet) boundary. The fine-scale model is of dimensions 220×60 and the coarse-scale model is of dimensions 22×6.

Results for flow rates for both the standard (upper) and local-global (lower) multiscale finite volume element method are shown in Fig. 6. For the standard method, we use no oversampling and apply linear pressure boundary conditions for the construction of the basis functions (better accuracy can be achieved using oversampling or reduced boundary conditions). Both sets of results are presented as cross plots of the coarse-scale flow rates against the corresponding fine-scale rates. The improvement achieved through the local-global updating of the basis

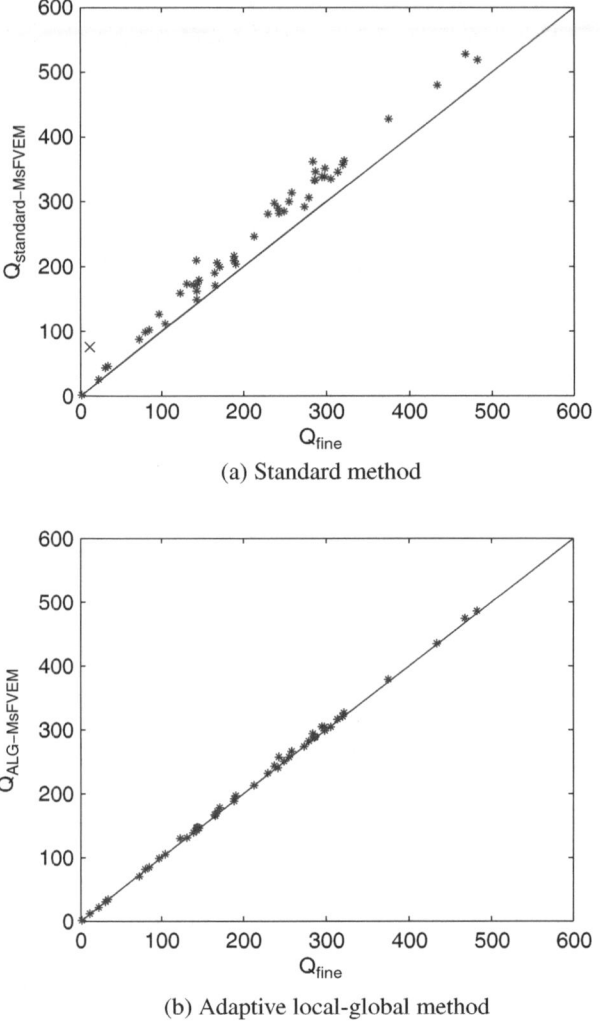

(a) Standard method

(b) Adaptive local-global method

Fig. 6 Comparison of global flow rates for side to side flow for (**a**) standard and (**b**) local-global multiscale finite volume element method (from [30])

functions is evident. The average relative error in Q for the standard multiscale finite volume element method (with the high-error point, designated by the $\times$ in the upper figure, removed) is about 18%. For the local-global method, the average error (based on all of the points) is reduced to about 2%. This example illustrates that, consistent with our observations for upscaling procedures, the use of approximate global information can also act to enhance the accuracy of multiscale methods.

4 Upscaling of Two-Phase Flow Functions

In many cases, reasonably accurate coarse-scale models for two-phase or multiphase flow can be constructed with only the single-phase flow parameters upscaled (assuming these single-phase flow parameters are sufficiently accurate). This is often the case when the level of coarsening is moderate (relative to the permeability correlation structure) and when the coarse grid is constructed to resolve key high-permeability features. Some examples demonstrating accurate two-phase simulation results using only single-phase parameter upscaling are presented in [19, 68, 74]. In other cases, however, such as when a high degree of coarsening is applied or the underlying permeability field is highly heterogeneous, the numerical homogenization of two-phase flow parameters is also required. Even when two-phase parameter upscaling is introduced, it is still essential to use a sufficiently accurate single-phase upscaling procedure, so all of the methods described in the previous section remain applicable. In this section, we focus on two-phase upscaling for coarse-scale models described by (6) and (7). Reviews and comparative studies of some of the earlier two-phase upscaling procedures are presented in [7, 23, 25].

The calculation of the upscaled two-phase flow functions can be somewhat time consuming. However, it should be noted that the resulting coarse model can generally be used for many flow simulations. For example, if a computational optimization procedure is to be applied to optimize well rates, the same coarse model can be used for many if not all of the required simulations (and these optimizations may require hundreds or thousands of flow simulations). In such cases the overhead associated with the upscaling will not be of major concern.

4.1 Numerical Procedures for Computing Upscaled Functions

In local two-phase upscaling, the fine-scale two-phase flow problem is solved on a local domain (some amount of border region is commonly used). This entails the solution of both the pressure and saturation equations (2) and (3) subject to pressure and saturation boundary conditions.

In analogy to the commonly used boundary conditions for single-phase upscaling, an intuitive choice for local boundary conditions for the two-phase problem is to prescribe constant pressures along the inlet and outlet of the local region and no flow through the boundaries parallel to the flow direction. Saturation at the inlet is specified to be one. These "standard" local boundary conditions have been shown to overestimate the local fine-scale fluxes, often leading to coarse-scale models that predict early breakthrough of the injected fluid relative to the reference fine-scale global solution.

In an attempt to correct this bias, "effective flux" boundary conditions (EFBCs) were proposed [66]. These boundary conditions view the local region as an inclusion immersed in a global domain, and specify the inlet and outlet local fluxes based on the local fine-scale permeabilities and a global background (or effective) permeability. Thus, EFBCs consider global effects, in a very approximate sense. Although EFBCs have been shown to reduce the bias associated with standard boundary conditions, there are still cases, such as those characterized by very small vertical permeability correlation lengths, where error persists [12]. For such cases it is useful to incorporate more accurate global effects in the two-phase upscaling calculations, as will be illustrated below.

Following the local fine-scale flow solution of (2) and (3) subject to appropriate boundary conditions, the upscaled two-phase functions λ^* and f^* (as appear in (6) and (7)) are computed to preserve the average fine-scale total flow rate and fractional flow of water. The upscaled total mobility function $\lambda^*(S^c)$ is computed to satisfy:

$$\lambda^*(S^c)\mathbf{k}^*\nabla p^c \approx \overline{\lambda \mathbf{k}\nabla p} = -\overline{\mathbf{u}}, \tag{16}$$

where the overbar represents volume averaging and $\overline{\mathbf{u}}$ designates the averaged fine-scale total velocity. Note that the upscaled two-phase functions $\lambda^*(S^c)$ and $f^*(S^c)$ are both treated as directional quantities. The x-component in the above equation gives $\lambda_x^*(S^c)k_x^*\Delta\overline{p}/\Delta x^c = \overline{u}_x$, where $\Delta\overline{p}$ represents the difference of the averaged fine-scale pressures (of opposite sign to ∇p^c). Therefore $\lambda_x^*(S^c)$ can be computed as:

$$\lambda_x^*(S^c) = \frac{\overline{u}_x}{k_x^*\Delta\overline{p}/\Delta x^c} = \frac{\overline{u}_x\Delta y^c h}{(k_x^*\Delta\overline{p}/\Delta x^c)\Delta y^c h} = \frac{\overline{q}_x}{T_x^*\Delta\overline{p}}, \tag{17}$$

where Δx^c and Δy^c designate the dimensions of a coarse block, h is the model thickness (Δz), $\overline{q}_x$ is the total volumetric flow rate in the x-direction, and k_x^* and T_x^* are coarse-scale permeability and transmissibility in the x-direction (note permeability is here taken as a diagonal tensor).

The upscaled fractional flow function $f^*(S^c)$ is computed to preserve the averaged water flux $\overline{\mathbf{u}f}$ in the volume-averaged saturation equation. We then have

$$\mathbf{u}^c f^*(S^c) = \overline{\mathbf{u}f}. \tag{18}$$

The directional fractional flow function in the x-direction is thus given by:

$$f_x^*(S^c) = \frac{\overline{u_x f}}{\overline{u}_x}. \tag{19}$$

The coarse-scale saturation S^c can be computed either as the pore-volume average saturation in the upstream coarse block or over the fine-grid cells along the block interface. This computation should be consistent with the numerical scheme used for the coarse-scale saturation equation.

The upscaled functions λ_y^* and f_y^* are computed analogously, with the local flow imposed in the y-direction. The calculation is repeated for each coarse block (or coarse-block interface) in each flow direction. Note that after λ^* and f^* are computed, the upscaled relative permeabilities k_{rj}^* can be readily determined through the relationships $\lambda^* = k_{ro}^*/\mu_o + k_{rw}^*/\mu_w$ and $f^* = (k_{rw}^*/\mu_w)/\lambda^*$. For more detailed discussion of the computation of the upscaled two-phase flow functions, refer to [12, 14].

4.2 Global Methods and Near-Well Treatment

As is evident from the above discussion, the choice of local boundary conditions can impact the accuracy of the coarse-scale two-phase flow parameters. To address this issue, global and local-global two-phase upscaling approaches have been developed.

Global two-phase upscaling methods are expensive as they entail the solution of the full fine-scale problem. Following this solution, coarse-scale functions are computed as described above, using (17) and (19). As discussed below, in some settings the use of global methods may be appropriate, particularly when the upscaled functions can be reused for many simulations.

Local-global approaches, which incorporate approximate global information into the determination of the local boundary conditions, also exist for two-phase upscaling problems [14, 15]. These methods are more involved than those used for single-phase upscaling as boundary information evolves in time and is required for both pressure and saturation. The approach in [14] considers generic global flows, i.e., global flow in the x and y-directions (for a two-dimensional system). Global coarse information is incorporated by interpolating the coarse-scale solutions to determine the local fine-scale boundary conditions.

More specifically, for a target coarse domain containing $n_x \times n_y$ fine cells, the fine-scale fluxes and saturations (q_{x-}^f and S_{x-}^f) at the inlet are computed from the coarse-scale flux and saturation (q_{x-}^c and S_{x-}^c) using (for flow in the x-direction):

$$\left(q_i^f\right)_{x-} = \left(\frac{\tau_{max} - \tau_i}{\tau_{max} - \tau_{min}}\right)_{x-} q_{x-}^c, \quad 1 \leq i \leq n_y,$$

$$\left(S_i^f\right)_{x-} = \left(\frac{\tau_{max} - \tau_i}{\tau_{max} - \tau_{min}}\right)_{x-} S_{x-}^c, \quad 1 \leq i \leq n_y, \tag{20}$$

where τ_{max} and τ_{min} represent the maximum and minimum values of time of flight (τ) along the fine-scale boundary $x-$ and i is the fine-block index along the local boundary. Note that τ is computed from the global single-phase fine-scale velocity field (time of flight is inversely proportional to the velocity along streamlines). Fluxes at the outlet boundary are prescribed analogously. To incorporate two-phase flow effects in the global flow information, the local boundary conditions are updated in accordance with the coarse-scale solutions. A criterion was introduced such that the change of (averaged) local fine-scale saturation is approximately equal to that of the global coarse-scale solution. For more detailed discussion, refer to [14].

As noted earlier, the upscaled properties can also be derived from a specific global flow, in which case the resulting coarse model can provide enhanced accuracy. Such an approach was applied in [15] to systems with strong full-tensor effects and was shown to adequately capture the two-phase flow induced by the full-tensor anisotropy. This approach shares some similarities with the nonlinear two-point flux approximation for single-phase flow, discussed in Sect. 3.3.

We note that for alternate coarse-scale descriptions, such as the GCD model discussed in Sect. 2.2, the coarse-scale functions appearing in (8) must also be numerically computed. Therefore, similar issues with local boundary conditions and the dependency of the upscaled functions on global flow also arise. Although global and local-global two-phase upscaling approaches could be applied to the GCD model, such approaches have not yet been investigated.

Important near-well effects occur in multiphase flow simulations, and these can be difficult to capture in coarse models in the absence of specialized treatments. Challenging examples include models with large contrasts in phase mobilities (e.g., high-mobility water injected into a low-mobility oil) or cases where dissolved gas is liberated from the oil phase in the vicinity of production wells. To accurately model such cases in coarse-scale simulations, two-phase near-well upscaling procedures have been devised [e.g., 44, 55].

In the approach described in [55], coarse-scale well-block parameters, including a parameter (R_s) that quantifies the volume of gas dissolved in the oil phase as a function of pressure, are computed for oil-gas models. Boundary conditions for the "local well model" are determined using a local-global technique. Because of the nonlinear nature of the equations and the complex interaction between two-phase flow, phase behavior and near-well heterogeneity, direct averaging of the form described in (16)–(19) does not guarantee accurate results. Rather, coarse-scale parameters are computed in this case through use of a gradient-based optimization procedure, where the difference between coarse-model phase flow rates and the corresponding (integrated) fine-model flow rates are minimized over the local well region. This minimization is accomplished by adjusting the coarse functions (e.g., λ_o^*).

An adjoint procedure is used to provide the gradients. Using this approach, the augmented cost function J_A, which we seek to minimize, is given by:

$$J_A = \sum_{n=1}^{N} L^n\left(\mathbf{p}^n, \mathbf{u}\right) + \sum_{n=1}^{N} (\boldsymbol{\lambda}^T)^n \mathbf{g}^n\left(\mathbf{p}^n, \mathbf{p}^{n-1}, \mathbf{u}\right). \tag{21}$$

Here n designates time step and N is the total number of time steps, L^n quantifies the mismatch between fine and coarse models at time step n, $\mathbf{p}^n$ represents all of the coarse-model states (pressure and saturation) at time step n, $\mathbf{u}$ represents tabular values for the upscaled functions we wish to compute, $\boldsymbol{\lambda}^n$ represents the vector of Lagrange multipliers at time step n, and $\mathbf{g}^n(\mathbf{p}^n, \mathbf{p}^{n-1}, \mathbf{u})$ designates the dynamic system (coarse-scale reservoir flow equations). The adjoint procedure provides $\partial J_A / \partial \mathbf{u}$, which is then used to minimize the cost function J_A using a gradient-based technique. For details on the computations and the performance of the overall procedure, see [55].

4.3   Numerical Results Using Two-Phase Upscaling

We now present flow results for coarse-scale models generated using various two-phase upscaling approaches. We first show results for individual realizations and then present comparisons of ensemble statistics of the flow response for multiple realizations. These examples are from [12,14], where coarse-scale simulation results using different procedures, for a variety of fine-scale geological models, are shown.

4.3.1   Flow Results for Individual Realizations

We consider log-normal permeability distributions, generated using sequential Gaussian simulation [26]. The permeability field is characterized by dimensionless correlation lengths l_x and l_y (dimensionless correlation length is defined as the correlation length divided by the model length in the corresponding direction) and variance of $\log k$ (designated as σ^2). Two such permeability fields are shown in Fig. 7. The field in Fig. 7a is characterized by $l_x = 0.5$ and $l_y = 0.05$, while that in Fig. 7b is characterized by $l_x = 0.4$ and $l_y = 0.01$.

For both cases, the fine-scale model is of dimensions 100×100. These models are uniformly coarsened to provide coarse models that are of dimensions 10×10. Flow is driven by specifying different pressures along the left and right boundaries of the models, with no-flow conditions on the top and bottom. Water is injected at the left boundary and oil and water are produced at the right boundary. The fine-scale

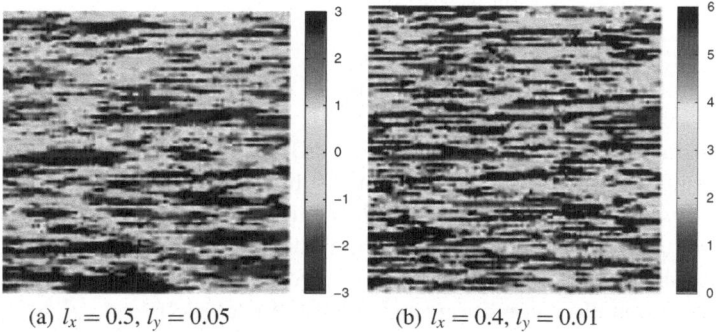

| (a) $l_x = 0.5$, $l_y = 0.05$ | (b) $l_x = 0.4$, $l_y = 0.01$ |

Fig. 7 Fine-scale permeability fields, of dimensions 100×100 with different correlation lengths, shown in log scale. For the permeability field in (a), the mean of $\log k$ is 0, and for that in (b), the mean of $\log k$ is 3.0. The variance (σ^2) of $\log k$ is 4.0 for both fields. (**a**) $l_x = 0.5$, $l_y = 0.05$. (**b**) $l_x = 0.4$, $l_y = 0.01$

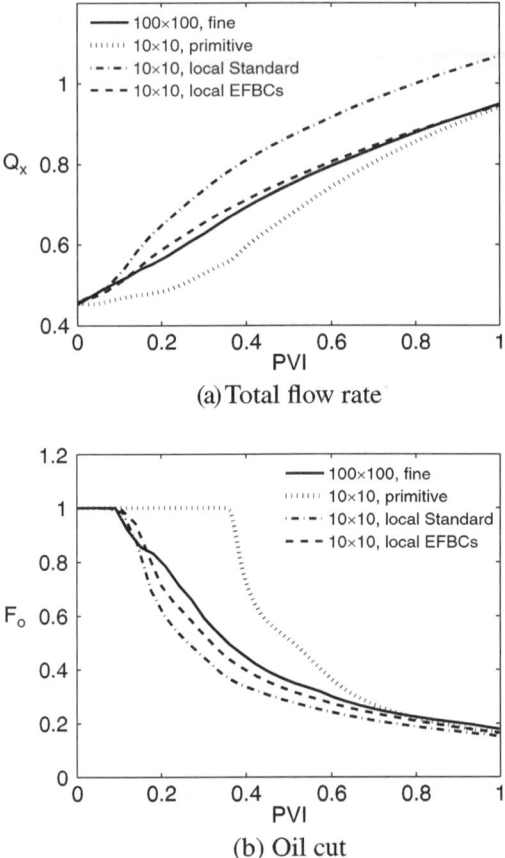

(a) Total flow rate

(b) Oil cut

Fig. 8 Flow results using local two-phase upscaling with standard boundary conditions and EFBCs for a log normal permeability field ($l_x = 0.5$, $l_y = 0.05$, $\sigma = 2.0$) and $M = 5$ (figure modified from [12]). (**a**) Total flow rate. (**b**) Oil cut

relative permeability functions are specified as $k_{rw} = S^2$ and $k_{ro} = (1 - S)^2$; the oil-water viscosity ratio (μ_o/μ_w) is 5.

We generate coarse models using several different procedures and compare simulation results with the fine-scale reference solutions. The primitive coarse model employs only the upscaled single-phase flow parameters, without any upscaled two-phase flow functions (in this case we simply take $k_{rj}^* = k_{rj}$). Coarse models are also generated through application of local two-phase upscaling procedures using both standard boundary conditions and EFBCs. Local-global two-phase upscaling is additionally considered [14]. A global single-phase upscaling procedure [19] was used to generate T^* for all of the coarse models. This approach was employed to ensure accuracy in the single-phase flow parameters.

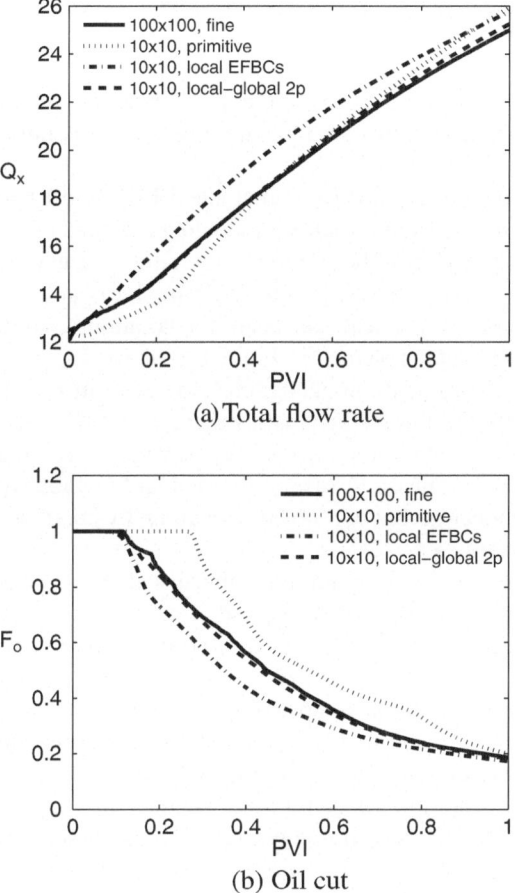

Fig. 9 Flow results using various coarse models for a log normal permeability field ($l_x = 0.4$, $l_y = 0.01$, and $\sigma = 2.0$), $M = 5$ (from [14]). (**a**) Total flow rate. (**b**) Oil cut

Flow results are presented in Figs. 8 and 9. These results display the total flow rate (Q_x) and the oil cut (F_o) as a function of pore volume injected (PVI), defined as $(1/V_p) \int_0^t Q(\tau) d\tau$, where V_p is the total pore volume. Note that PVI corresponds to a dimensionless time. Here total flow rate can be viewed as either the water injection rate or the total (oil+water) production rate, as the system is incompressible. Total flow rate increases with time as more mobile water replaces less mobile oil within the model. Oil cut is the fraction of oil in the produced fluid – it is 1 at early time and then decreases once the injected water appears at the right edge of the model. Note that Q_x is greater in Fig. 9a than in Fig. 8a as a result of the higher permeability values in Fig. 7b.

Shown in Fig. 8 are the results for the case $l_x = 0.5$ and $l_y = 0.05$. The primitive model shows substantial error, particularly for F_o. The coarse model based on k_{rj}^*

generated using standard boundary conditions overcorrects the primitive model – Q_x is now too large and F_o is generally underestimated relative to the fine model. The coarse model based on k_{rj}^* computed using EFBCs is accurate for both Q_x and F_o, which demonstrates the impact that local boundary conditions can have on global flow results. These results are typical for permeability fields in this parameter range.

It was observed in [12], however, that the EFBC model loses accuracy for $l_y \leq 0.02$. There it was shown that accurate coarse models for very small l_y can be generated through use of either a global upscaling procedure or via application of the GCD model (as noted earlier, the GCD model is able to capture diffusive effects driven by short-length-scale permeability variations). These observations are consistent with the results presented in Fig. 9 for the case $l_x = 0.4$ and $l_y = 0.01$. For this example, the primitive model again shows error in F_o. The EFBC model is less accurate for both Q_x and F_o than in the previous case (Fig. 8). More specifically, it overpredicts the flow rate of the injected water, which leads to an overestimate of Q_x and to an underestimate of F_o (use of the standard method would result in even larger errors). The local-global two-phase upscaling, by contrast, provides accurate results for both quantities.

The time required for the computation of the upscaled two-phase flow parameters can be considerable. Timings are reported, for example, in [12], where two-dimensional models containing 100×100 fine cells were coarsened to 10×10 coarse models. In that work directional λ^* and f^* were computed for all coarse grid blocks using local two-phase upscaling methods. Overall speedups for these coarse models relative to the fine-scale models (overall speedup accounts for the time required for the upscaling calculations and the coarse-model simulations) of 4-10 were achieved. Runtime speedups were significantly larger, however, ranging from 400 to 900. Thus, as noted above, the overhead associated with two-phase upscaling is much less of a concern if the coarse model is to be run many times.

4.3.2   Flow Results for Multiple Realizations

The next set of results involves flow simulation on multiple permeability realizations, as would be required for uncertainty quantification. We generate 100 realizations of the fine-scale permeability field with $l_x = 0.4$, $l_y = 0.01$ and $\sigma = 2$ (one realization is shown in Fig. 7b). Results for the fine and primitive coarse-scale models are shown in Fig. 10 as the lighter gray curves. The black solid curve shows the P50 (median) response, and the black dashed curves display the so-called P10 and P90 results. These curves are computed such that 10% and 90% of the responses fall below the corresponding curves. Figure 10a displays the fine-scale predictions, while Fig. 10b is for the primitive coarse model. Note that time is nondimensionalized here and in subsequent two-dimensional simulations using system variables (i.e., not in terms of PVI). We next compute these key statistical quantities (P50, P10 and P90 curves) for coarse-scale models generated using different approaches.

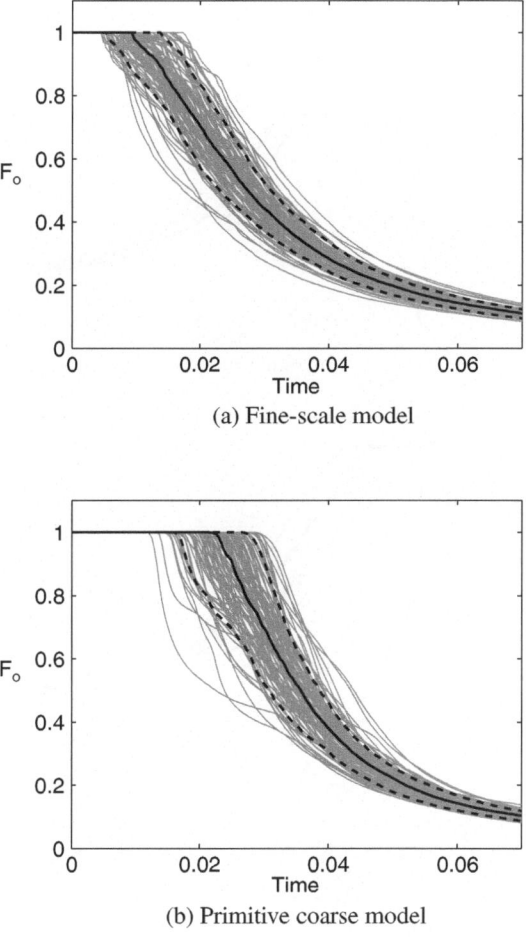

(a) Fine-scale model

(b) Primitive coarse model

Fig. 10 Oil cut for 100 realizations (*represented by gray curves*) for log normal permeability fields ($l_x = 0.4, l_y = 0.01, \sigma = 2.0$) and $M = 5$. *Black curves* represent P50 (*solid curve*) and P10-P90 interval (*dashed curves*) (from [14]). (**a**) Fine-scale model. (**b**) Primitive coarse model

The comparisons between results for coarse-scale models and the reference fine-scale results are shown in Fig. 11. Here, the solid curves correspond to the fine-scale results and the dot-dash curves to the coarse-scale models. The thick curves represent P50 results and the thin curves the P10 (lower curves) and P90 (upper curves) flow responses. Figure 11a compares the fine and primitive coarse models. The primitive coarse model shows large errors, with the P10-P90 intervals predicted by the fine and coarse models overlapping very little. Displayed in Fig. 11b are the results using local EFBC upscaling. These results are better than those using the primitive model, though there is a clear bias toward underprediction of F_o (consistent with the results in Fig. 9). The results using local-global two-phase

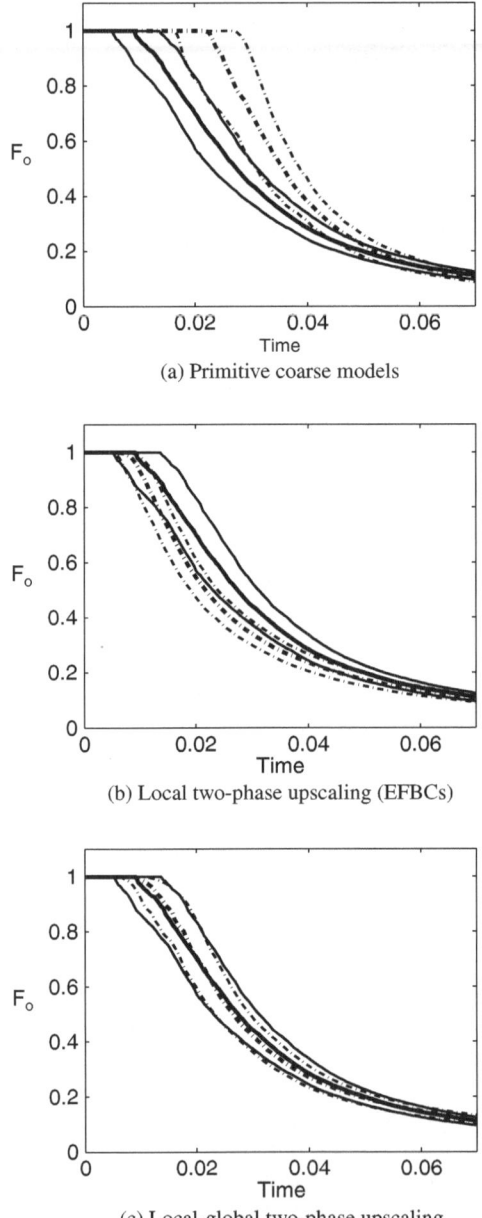

(a) Primitive coarse models

(b) Local two-phase upscaling (EFBCs)

(c) Local-global two-phase upscaling

Fig. 11 Comparison of P50 (*thick curves*) and P10-P90 interval (*thin curves*) for oil cut between fine-scale (*solid curves*) and coarse-scale (*dot-dash curves*) models for log normal permeability fields ($l_x = 0.4$, $l_y = 0.01$, $\sigma = 2.0$) and $M = 5$ (from [14]). (**a**) Primitive coarse models. (**b**) Local two-phase upscaling (EFBCs). (**c**) Local-global two-phase upscaling

blocks in the remaining $(N_r - N_c)$ models through statistical simulation. It will also be important to preserve the spatial correlation structure of the coarse-scale $\mathbf{k}^*$ or T^* fields, which can be accomplished using the geostatistical simulation described below.

5.1.1 Parameterization, Attributes and Clustering

The procedures described here provide, in the general case, k_{rw}^* and k_{ro}^* curves in each coordinate direction for every coarse block. The description below is for k_{rj}^* in the x-direction; analogous procedures are applied for k_{rj}^* in the other coordinate directions. We apply a simple one-parameter representation of the upscaled k_{rj}^* curves. We characterize any upscaled curve based on the area between the k_{rj}^* curve and the original fine-scale $k_{rj}(S)$ curve; i.e., we compute

$$\delta k_{rj} = \frac{1}{N} \sum_{i=1}^{N} \left| k_{rj}^*(S_i^c) - k_{rj}(S_i) \right|, \quad j = o, w, \tag{22}$$

where N is the number of (equally spaced) saturation values at which the two curves are evaluated. This approach is illustrated in Fig. 12 for k_{ro}^* and k_{rw}^* for two different grid blocks. Although the underlying $k_{rj}(S)$ curves are the same in all fine-grid blocks in this example, the k_{rj}^* curves (and thus δk_{rj}) differ for the two coarse blocks as a result of the different underlying fine-scale permeability fields. This parameterization does not uniquely define the k_{rj}^* curve, and more complex parameterizations could be used (see [13, 20] for discussion of other approaches). However, this single-parameter representation appears adequate for the EnLU procedure.

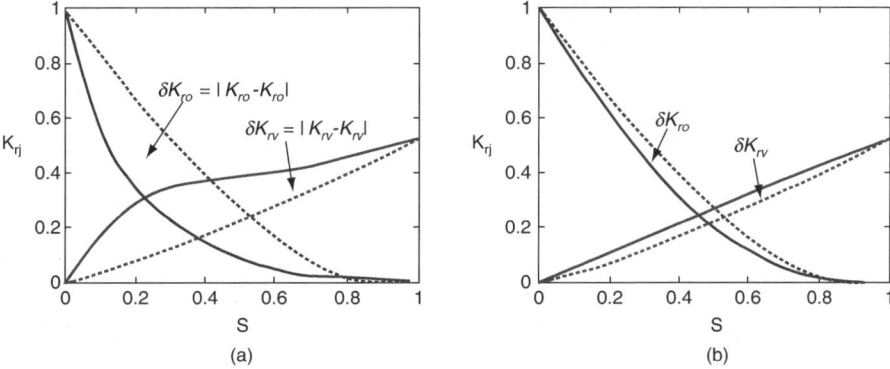

Fig. 12 Parameterization of upscaled relative permeabilities. *Dashed curves* represent fine-scale k_{rj} and *solid curves* represent upscaled k_{rj}^*: (**a**) k_{rj}^* in x for one coarse block and (**b**) k_{rj}^* in x for another coarse block (from [20])

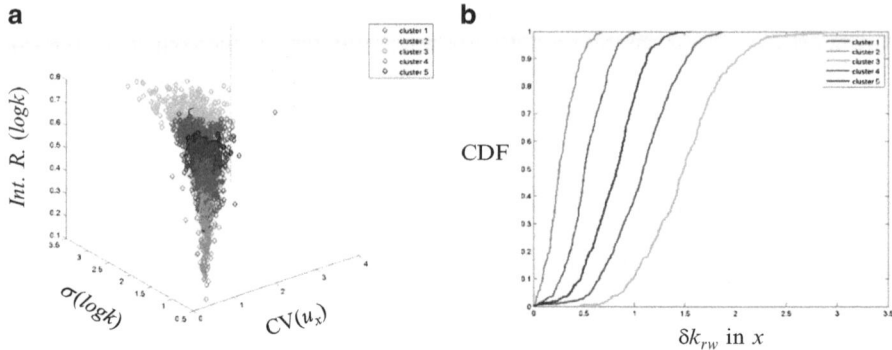

Fig. 13 Clustering and resultant CDFs using attributes $CV(u_x)$, $\sigma(\log k)$ and integral range of autocorrelation function of $\log k$: (**a**) 5 clusters and (**b**) calibration CDFs for the 5 clusters (from [20])

The attributes used in EnLU should, to the extent possible, be fast to compute, correlate with δk_{rj}, and be statistically independent of one another. In our earlier work we investigated the use of various sets of attributes. In [13], the average and standard deviation of single-phase fine-scale Darcy velocities within the coarse block region, $\bar{u}_x$ and $\sigma(u_x)$, available from the solution of (4) (which is solved during the calculation of $\mathbf{k}^*$ or T^*), were used as the coarse-block attributes. The quantity $\sigma(u_x)$ provides a measure of subgrid velocity fluctuations, which strongly impact the coarse-scale two-phase flow functions. In subsequent work [20], improved results were achieved using the coefficient of variation (CV) of the fine-scale single-phase Darcy velocity ($\sigma(u_x)/\bar{u}_x$), the integral spatial range of the autocorrelation function of the fine-scale $\log k$, and the standard deviation of the fine-scale $\log k$, designated as $\sigma(\log k)$. These quantities are computed over the target coarse block and involve the fine-scale single-phase velocity field and the underlying permeability field.

Clusters are then formed by plotting all of the training data points in the space defined by the attributes and applying a K-means clustering algorithm. The δk_{rj} values of the points are not used in forming the clusters. The CDF for δk_{rj} corresponding to each cluster can now be easily generated.

The resulting clusters and their associated CDFs (for δk_{rw} in x) using the three attributes are shown in Fig. 13. It is evident that the clusters are distinct in attribute space and that the CDFs show clear separation, which are both desirable properties. This demonstrates that the selected attributes are indeed meaningful for use in simulating upscaled relative permeabilities.

5.1.2 Stochastic Simulation of Coarse-Block Relative Permeabilities

Simulation of upscaled relative permeability (we designate the statistically simulated function as $\delta\tilde{k}_{rj}$) by sampling from the target cluster CDF can be expressed as follows:

$$\delta \tilde{k}_{rj} = F_i^{-1} \left(U(u\,[0,1]) \right), \quad j = o, w, \tag{23}$$

where F_i^{-1} is the inverse CDF of target cluster i and U denotes the CDF of a uniform random function between 0 and 1 (represented by $u\,[0,1]$). This approach is referred to as EnLU with random drawing.

As discussed in [20], application of (23) does not directly incorporate any information regarding the spatial correlation structure of δk_{rj} into $\delta \tilde{k}_{rj}$. Such information might be available, for example, from the training data; i.e., the N_c realizations for which we compute k_{rj}^* using flow-based two-phase upscaling. Alternatively, one might seek to constrain $\delta \tilde{k}_{rj}$ to honor the correlation structure of the underlying fine-scale permeability field. This can be accomplished most readily in the case of Gaussian permeability models, which are often used to represent permeability in subsurface formations. Such models are fully characterized by the two-point autocorrelation function. In geological contexts, this information is typically described in terms of the variogram.

In the simulation of $\delta \tilde{k}_{rj}$, information regarding spatial correlation can be introduced by replacing the uniform random function in (23) by a multi-Gaussian random function g. We then have:

$$\delta \tilde{k}_{rj} = F_i^{-1} \left(G(g) \right), \quad j = o, w, \tag{24}$$

where G is the CDF of g. Standard geostatistical algorithms (e.g., sequential Gaussian simulation [26]) can be used to generate g. The use of the back-transform F_i^{-1} assures that the calibration CDF from the target cluster is honored. Once $\delta \tilde{k}_{rj}$ is determined through statistical simulation, the full $k_{rj}^*(S^c)$ curve is determined by searching the target cluster for the nearest δk_{rj} and then assigning the associated curve to the block.

The use of additional information in the simulation of $\delta \tilde{k}_{rj}$, including cross-correlation between different variables, is discussed in [20]. In that work, an example is presented that demonstrates improved accuracy in the upscaled model (as measured by the correlation coefficient between models constructed using flow-based δk_{rj} and statistically simulated $\delta \tilde{k}_{rj}$) when spatial correlation information is included in $\delta \tilde{k}_{rj}$. We have not, however, observed significant differences in ensemble-level flow quantities using (24) in place of (23). It is nonetheless possible that the realization-by-realization agreement between fine and coarse-scale simulations (at least for some simulation quantities) may be improved through use of (24), though this issue requires further study.

5.2 Numerical Results Using Ensemble-Level Upscaling

In this section we present results for both two and three-dimensional systems using EnLU. Various core upscaling procedures are applied, and different variants of EnLU are used.

5.2.1 Results for 2D Systems

The first example is from [13]. This case involves 100 realizations of a model characterized by dimensionless correlation lengths $l_x = 0.4$ and $l_y = 0.01$ and variance of log permeability (σ^2) of 4. This is the same set of realizations described in Sect. 4.3 (one realization is displayed in Fig. 7b). Recall that the fine-scale models contain 100×100 grid blocks. The problem set up is as described in Sect. 4.3 (water is injected at the left boundary and production is at the right boundary).

Coarse-scale results (on 10×10 grids) for this example are presented for three different upscaling approaches. These include the use of primitive models (in which fine-scale k_{rj} curves are applied in the coarse-scale model), models in which all upscaled functions are computed numerically using a flow-based procedure, and models in which EnLU is applied. In the numerically upscaled models, the coarse-scale relative permeability functions are determined from global solutions (i.e., global two-phase flow upscaling). This is the most expensive approach, as it entails the fine-scale solution of the two-phase flow equations. In many practical settings this global approach would not be appropriate, but it is useful for current purposes as it acts to minimize upscaling error. In addition this approach is, as we will see, reasonable for use with EnLU because only a small fraction of realizations need to be simulated. Specifically, in this case, 10% of the realizations are numerically upscaled, while for the remaining 90% the k_{rj}^* curves are assigned using EnLU with random drawing, as described above. In all cases, the upscaled single-phase flow parameters (T^*) are computed using global single-phase flow simulations (recall that this entails the global solution of (4)).

The ensemble results for oil cut (P50, P10-P90 interval) for the primitive coarse models were presented earlier (in Fig. 11a). Results for the other procedures are shown in Fig. 14. As shown in the previous section, the use of the primitive models (Fig. 11a) clearly results in a significant shift to the right for all three curves, with F_o significantly overpredicted. The use of numerically simulated k_{rj}^* curves (Fig. 14a) provides results of excellent accuracy, as would be expected. The EnLU results, shown in Fig. 14b, are of essentially the same accuracy as the results with the numerically simulated k_{rj}^*. This demonstrates the applicability of EnLU for this problem.

It is of interest to consider other measures of the level of agreement provided by EnLU. Shown in Fig. 15 is a cross-plot of the time at which $F_o = 0.8$ for each coarse model against the corresponding fine model (for all 100 realizations). The bias associated with the primitive models (open circles) is evident, as is the accuracy of the numerically computed k_{rj}^* (solid circles). The EnLU results (crosses) display more scatter than the models that use numerically computed k_{rj}^*, though the bias evident in the results from the primitive models is essentially eliminated. The CDF for the time at which $F_o = 0.8$ is shown for the various models in Fig. 16. Here again we see essential agreement in results using EnLU with the reference fine-scale results, along with the bias in the primitive model results. These results demonstrate that, as expected, there is a loss of accuracy in the realization-by-realization agreement for models constructed using EnLU, but

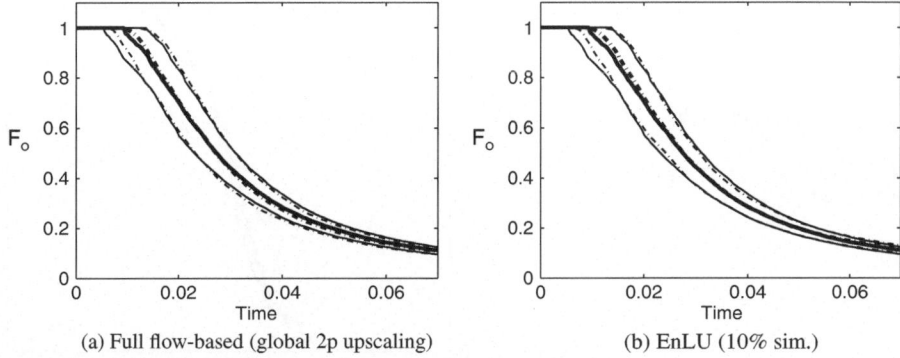

(a) Full flow-based (global 2p upscaling)   (b) EnLU (10% sim.)

Fig. 14 Comparison of P50 (*thick curves*) and P10-P90 interval (*thin curves*) for oil cut between fine-scale (*solid curves*) and coarse-scale (*dot-dash curves*) models for a log normal permeability field ($l_x = 0.4$, $l_y = 0.01$, $\sigma = 2.0$) and $M = 5$ (from [13]). (**a**) Full flow-based (global 2p upscaling). (**b**) EnLU (10% sim.)

Fig. 15 Realization-by-realization comparison between fine-scale and coarse-scale models for time at which $F_o = 0.8$ for log normal permeability fields ($l_x = 0.4$, $l_y = 0.01$, $\sigma = 2.0$) and $M = 5$ (figure modified from [13])

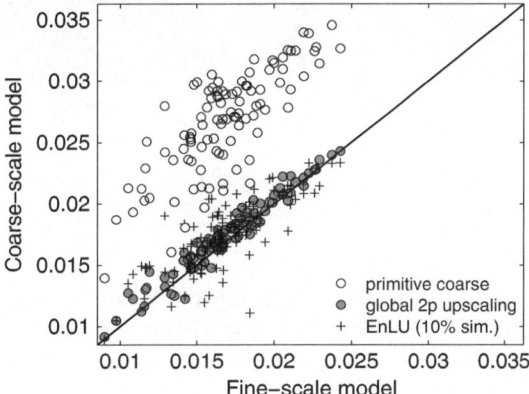

the statistical response of the EnLU models agrees closely with that of their fine-scale counterparts.

5.2.2 Results for 3D Systems with Well-Driven Flow

We next consider a three-dimensional example taken from [20]. In this case the fine and coarse models contain $55 \times 55 \times 25$ and $11 \times 11 \times 5$ grid blocks respectively. The fine-scale models are characterized by $l_x = l_y = 0.5$, $l_z = 0.025$ and $\sigma^2 = 3$. Five realizations of the permeability field are shown in Fig. 17a. Flow is driven by a water injection well and four production wells (arranged in a five-spot pattern as shown in Fig. 17b). For this case, $\mu_o/\mu_w = 10$.

Fig. 16 Comparison of CDFs between fine-scale and coarse-scale models for time at which $F_o = 0.8$ for log normal permeability fields ($l_x = 0.4$, $l_y = 0.01$, $\sigma = 2.0$) and $M = 5$ (figure modified from [13])

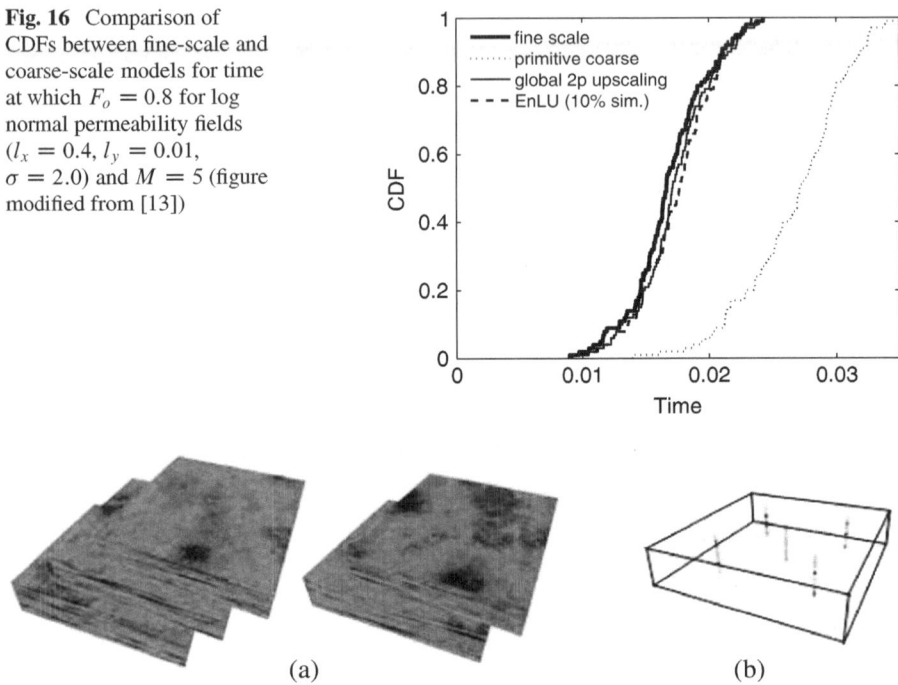

(a)   (b)

Fig. 17 (a) Log normal permeability fields ($l_x = 0.5$, $l_y = 0.5$, $l_z = 0.025$, $\sigma = 1.73$) and (b) well locations – one injector (*in the center*) and four producers

The upscaled permeability $\mathbf{k}^*$ for this example is computed using a local method and the k_{rj}^* are determined through application of EFBCs. The use of local $\mathbf{k}^*$ upscaling is adequate in this case because the permeability fields are Gaussian and σ is not that large. Single-phase near-well upscaling (T_w^* and W_i^*) is applied for all of the wells, though two-phase near-well upscaling is applied only for the injection well (it was shown in [44] that, for this type of oil-water model, the impact of well-block k_{rj}^* is greatest for injection wells). In the models for which EnLU is applied, the upscaled functions are computed for 3% of the models and are assigned to the other 97% through application of (24). See [20] for more details.

Results for well oil cut (P10, P50, P90) for two of the production wells are shown in Fig. 18 for the primitive models, the models with numerically computed k_{rj}^* and the models using EnLU. The bias in the primitive model results is again apparent. The upscaled models using both the numerically computed (Fig. 18b) and statistically simulated (Fig. 18c) k_{rj}^* curves provide very accurate results. CDFs for field oil production rate (oil production summed over the four production wells) are presented in Fig. 19. The primitive models again show substantial bias, which is essentially eliminated using the other two approaches. There is some inaccuracy in the results using the numerically computed k_{rj}^* for CDF > 0.7, which also appears in the EnLU results. This is due to the limitations of the underlying (local) upscaling

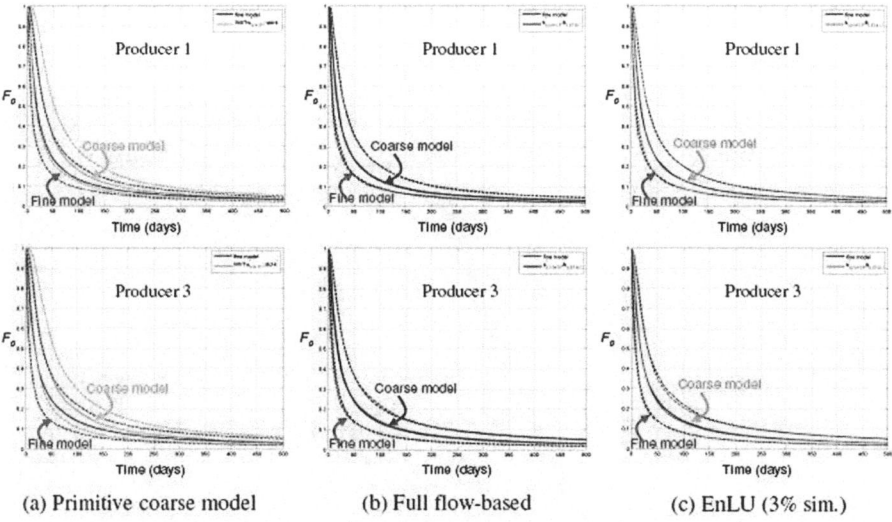

Fig. 18 Comparison of fine and coarse-scale P50, P10 and P90 predictions of well oil cut (producers 1 and 3) for log-normal permeability fields ($l_x = 0.5$, $l_y = 0.5$, $l_z = 0.025$, $\sigma = 1.73$) with $M = 10$ (from [20]). (**a**) Primitive coarse model. (**b**) Full flow-based. (**c**) EnLU (3% sim.)

procedures used for this example. Because EnLU uses training data generated by the core upscaling method, it inherits the limitations associated with that approach.

6 Concluding Remarks

In this chapter we described a variety of numerical homogenization procedures applicable for use in subsurface flow simulation. Techniques for the computational upscaling of both single-phase and two-phase flow parameters were discussed. A new methodology, ensemble-level upscaling, in which the aim is to provide coarse models suitable for computing key flow statistics rather than realization-by-realization agreement, was also presented. Numerical results for many of the methods discussed illustrated the relative advantages and limitations of the various procedures.

There are many useful directions for future research on upscaling for subsurface flow modeling and uncertainty quantification. The core computational upscaling methodologies are well developed for single-phase and two-phase flow, though there has been much less work addressing three-phase systems, compositional models (which may contain many components) and thermal models (required to simulate steam injection, for example). Future effort should be directed toward the development of upscaling procedures for these practically important yet challenging processes. It will also be of interest to devise coarse-scale modeling

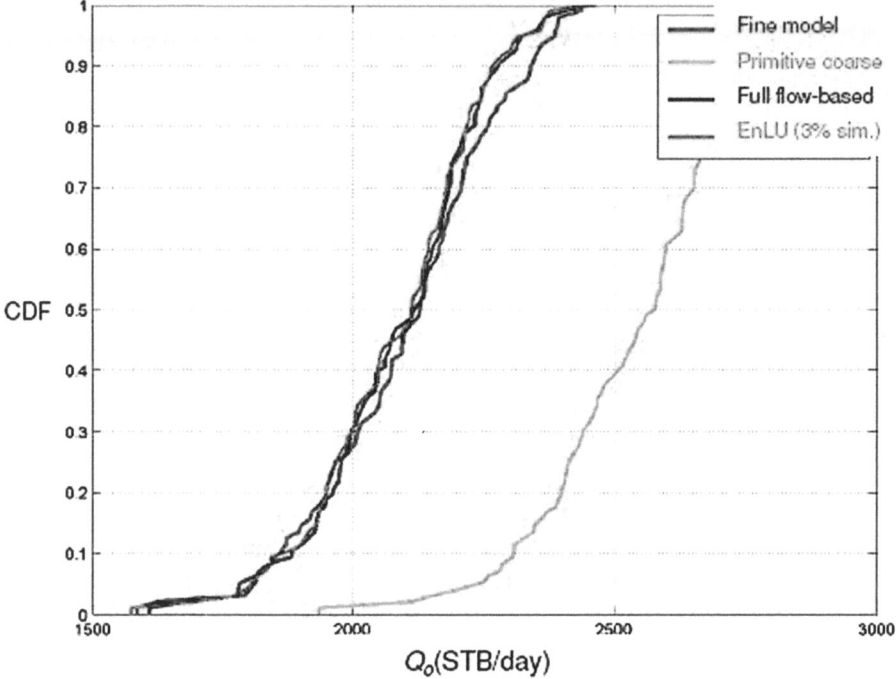

Fig. 19 Comparison of CDFs of field oil production rate between fine and coarse-scale models at $t = 100$ days for log-normal permeability fields ($l_x = 0.5$, $l_y = 0.5$, $l_z = 0.025$, $\sigma = 1.73$) with $M = 10$ (from [20])

capabilities for problems that involve coupled flow and geomechanical effects, as this coupling can be important in both oil recovery and geological carbon storage operations.

It is now recognized that uncertainty quantification for oil production should account for uncertainty in geological "style" (e.g., depositional setting, structural geological model) in addition to considering multiple realizations for a particular style. In the ensemble-level upscaling procedure discussed here, all realizations were sampled from the same ensemble, so different geological styles were not considered. It will be of interest to extend the EnLU procedures to handle realizations generated for a variety of geological styles. It will also be useful to combine EnLU approaches with data assimilation procedures to enable the construction of coarse models that are consistent with production data and other field measurements. This will enable more realistic predictions for future production and will provide models that are well suited for use in computational optimization procedures.

References

1. Aarnes JE (2004) On the use of a mixed multiscale finite element method for greater flexibility and increased speed or improved accuracy in reservoir simulation. Multiscale Modeling and Simulation 2:421–439
2. Aarnes JE, Efendiev Y (2007) Mixed multiscale finite element methods for stochastic porous media flows. SIAM Journal on Scientific Computing 30(5):2319–2339
3. Aarnes JE, Krogstad S, Lie KA (2006) A hierarchical multiscale method for two-phase flow based upon mixed finite elements and nonuniform coarse grids. Multiscale Modeling and Simulation 5(2):337–363
4. Aavatsmark I (2002) An introduction to multipoint flux approximations for quadrilateral grids. Computational Geosciences 6:405–432
5. Arbogast T (2002) Implementation of a locally conservative numerical subgrid upscaling scheme for two-phase Darcy flow. Computational Geosciences 6:453–481
6. Aziz K, Settari A (1986) Petroleum Reservoir Simulation. Elsevier, New York
7. Barker JW, Thibeau S (1997) A critical review of the use of pseudorelative permeabilities for upscaling. SPE Reservoir Engineering 12:138–143
8. Bourgeat A (1984) Homogenized behavior of two-phase flows in naturally fractured reservoirs with uniform fractures distribution. Computer Methods in Applied Mechanics and Engineering 47:205–216
9. Chen T, Gerritsen MG, Durlofsky LJ, Lambers JV (2009) Adaptive local-global VCMP methods for coarse-scale reservoir modeling. Paper SPE 118994 presented at the SPE Reservoir Simulation Symposium, The Woodlands, Texas, February 2–4
10. Chen T, Gerritsen MG, Lambers JV, Durlofsky LJ (2010) Global variable compact multipoint methods for accurate upscaling with full-tensor effects. Computational Geosciences 14:65–81
11. Chen Y, Durlofsky LJ (2006) Adaptive local-global upscaling for general flow scenarios in heterogeneous formations. Transport in Porous Media 62:157–185
12. Chen Y, Durlofsky LJ (2006) Efficient incorporation of global effects in upscaled models of two-phase flow and transport in heterogeneous formations. Multiscale Modeling and Simulation 5:445–475
13. Chen Y, Durlofsky LJ (2008) Ensemble-level upscaling for efficient estimation of fine-scale production statistics. SPE Journal 13:400–411
14. Chen Y, Li Y (2009) Local-global two-phase upscaling of flow and transport in heterogeneous formations. Multiscale Modeling and Simulation 8:125–153
15. Chen Y, Li Y (2010) Incorporation of global effects in two-phase upscaling for modeling flow and transport with full-tensor anisotropy. In: Proceedings of the 12th European Conference on the Mathematics of Oil Recovery, Oxford, UK
16. Chen Y, Wu XH (2008) Upscaled modeling of well singularity for simulating flow in heterogeneous formations. Computational Geosciences 12:29–45
17. Chen Y, Durlofsky LJ, Gerritsen M, Wen XH (2003) A coupled local-global upscaling approach for simulating flow in highly heterogeneous formations. Advances in Water Resources 26:1041–1060
18. Chen Y, Durlofsky LJ, Wen XH (2004) Robust coarse scale modeling of flow and transport in heterogeneous reservoirs. In: Proceedings of the 9th European Conference on the Mathematics of Oil Recovery, Cannes, France
19. Chen Y, Mallison BT, Durlofsky LJ (2008) Nonlinear two-point flux approximation for modeling full-tensor effects in subsurface flow simulations. Computational Geosciences 12:317–335
20. Chen Y, Park K, Durlofsky LJ (2011) Statistical assignment of upscaled flow functions for an ensemble of geological models. Computational Geosciences 15:35–51
21. Chen Z, Hou TY (2003) A mixed finite element method for elliptic problems with rapidly oscillating coefficients. Mathematics of Computation 72:541–576

22. Chen Z, Yue X (2003) Numerical homogenization of well singularities in the flow transport through heterogeneous porous media. Multiscale Modeling and Simulation 1:260–303
23. Christie MA (2001) Flow in porous media – scale up of multiphase flow. Current Opinion in Colloid & Interface Science 6:236–241
24. Christie MA, Blunt MJ (2001) Tenth SPE comparative solution project: a comparison of upscaling techniques. SPE Reservoir Evaluation & Engineering 4:308–317
25. Darman NH, Pickup GE, Sorbie KS (2002) A comparison of two-phase dynamic upscaling methods based on fluid potentials. Computational Geosciences 6:5–27
26. Deutsch CV, Journel AG (1998) GSLIB: Geostatistical Software Library and User's Guide, 2nd edition. Oxford University Press, New York
27. Ding Y (1995) Scaling-up in the vicinity of wells in heterogeneous field. Paper SPE 29137 presented at the SPE Symposium on Reservoir Simulation, San Antonio, Texas, February 12–15
28. Durlofsky LJ (1991) Numerical calculation of equivalent grid block permeability tensors for heterogeneous porous media. Water Resources Research 27:699–708
29. Durlofsky LJ, Milliken WJ, Bernath A (2000) Scaleup in the near-well region. SPE Journal 5:110–117
30. Durlofsky LJ, Efendiev YR, Ginting V (2007) An adaptive local-global multiscale finite volume element method for two-phase flow simulations. Advances in Water Resources 30: 576–588
31. Efendiev Y, Hou TY (2009) Multiscale Finite Element Methods: Theory and Applications. Springer, New York
32. Efendiev Y, Ginting V, Hou T, Ewing R (2006) Accurate multiscale finite element methods for two-phase flow simulations. Journal of Computational Physics 220(1):155–174
33. Efendiev YR, Durlofsky LJ (2003) A generalized convection-diffusion model for subgrid transport in porous media. Multiscale Modeling and Simulation 1:504–526
34. Efendiev YR, Durlofsky LJ, Lee SH (2000) Modeling of subgrid effects in coarse scale simulations of transport in heterogeneous porous media. Water Resources Research 36: 2031–2041
35. Farmer CL (2002) Upscaling: a review. International Journal for Numerical Methods in Fluids 40:63–78
36. Gerritsen M, Lambers J (2008) Integration of local-global upscaling and grid adaptivity for simulation of subsurface flow in heterogeneous formations. Computational Geosciences 12:193–218
37. Gerritsen MG, Durlofsky LJ (2005) Modeling fluid flow in oil reservoirs. Annual Review of Fluid Mechanics 37:211–238
38. Hajibeygi H, Jenny P (2009) Multiscale finite-volume method for parabolic problems arising from compressible multiphase flow in porous media. Journal of Computational Physics 228(14):5129–5147
39. Hesse MA, Mallison BT, Tchelepi HA (2008) Compact multiscale finite volume method for heterogeneous anisotropic elliptic equations. Multiscale Modeling and Simulation 7(2): 934–962
40. Hewett TA, Yamada T (1997) Theory for the semi-analytical calculation of oil recovery and effective relative permeabilities using streamtubes. Advances in Water Resources 20:279–292
41. Holden L, Nielsen BF (2000) Global upscaling of permeability in heterogeneous reservoirs: the output least squares (OLS) method. Transport in Porous Media 40:115–143
42. Hou TY, Wu XH (1997) A multiscale finite element method for elliptic problems in composite materials and porous media. Journal of Computational Physics 134:169–189
43. Hou TY, Westhead A, Yang D (2006) A framework for modeling subgrid effects for two-phase flows in porous media. Multiscale Modeling and Simulation 5(4):1087–1127
44. Hui M, Durlofsky LJ (2005) Accurate coarse modeling of well-driven, high-mobility-ratio displacements in heterogeneous reservoirs. Journal of Petroleum Science and Engineering 49:37–56

45. Jenny P, Lunati I (2009) Modeling complex wells with the multi-scale finite-volume method. Journal of Computational Physics 228(3):687–702
46. Jenny P, Lee SH, Tchelepi HA (2003) Multi-scale finite-volume method for elliptic problems in subsurface flow simulation. Journal of Computational Physics 187:47–67
47. Jiang L, Efendiev Y, Mishev I (2010) Mixed multiscale finite element methods using approximate global information based on partial upscaling. Computational Geosciences 14(2):319–341
48. Juanes R, Dub FX (2008) A locally conservative variational multiscale method for the simulation of porous media flow with multiscale source terms. Computational Geosciences 12(3):273–295
49. Kolyukhin D, Espedal M (2010) Modified adaptive local-global upscaling method for discontinuous permeability distribution. Computational Geosciences 14(4):675–689
50. Krogstad S, Durlofsky LJ (2009) Multiscale mixed-finite-element modeling of coupled wellbore/near-well flow. SPE Journal 14(1):78–87
51. Kyte JR, Berry DW (1975) New pseudo functions to control numerical dispersion. SPE Journal 15:269–275
52. Lambers J, Gerritsen M, Mallison B (2008) Accurate local upscaling with compact multipoint transmissibility calculations. Computational Geosciences 12:399–416
53. Lee SH, Zhou H, Tchelepi HA (2009) Adaptive multiscale finite-volume method for non-linear multiphase transport in heterogeneous formations. Journal of Computational Physics 228(24):9036–9058
54. Muggeridge AH, Cuypers M, Bacquet C, Barker JW (2002) Scale-up of well performance for reservoir flow simulation. Petroleum Geoscience 8:133–139
55. Nakashima T, Durlofsky LJ (2010) Accurate representation of near-well effects in coarse-scale models of primary oil production. Transport in Porous Media 83:741–770
56. Nielsen BF, Tveito A (1998) An upscaling method for one-phase flow in heterogeneous reservoirs. A weighted output least squares (WOLS) approach. Computational Geosciences 2:93–123
57. Niessner J, Helmig R (2007) Multi-scale modeling of three-phase-three-component processes in heterogeneous porous media. Advances in Water Resources 30(11):2309–2325
58. Oren P, Bakke S (2003) Reconstruction of Berea sandstone and pore-scale modelling of wettability effects. Journal of Petroleum Science and Engineering 39(3-4):177–199
59. Peaceman DW (1983) Interpretation of well-block pressures in numerical reservoir simulation with nonsquare grid blocks and anisotropic permeability. SPE Journal 23:531–543
60. Pickup GE, Ringrose PS, Jensen JL, Sorbie KS (1994) Permeability tensors for sedimentary structures. Mathematical Geology 26:227–250
61. Potsepaev R, Farmer CL, Fitzpatrick AJ (2009) Multipoint flux approximations via upscaling. Paper SPE 118994 presented at the SPE Reservoir Simulation Symposium, The Woodlands, Texas, February 2–4
62. Renard P, de Marsily G (1997) Calculating effective permeability: a review. Advances in Water Resources 20:253–278
63. Romeu R, Noetinger B (1995) Calculation of internodal transmissivities in finite difference models of flow in heterogeneous porous media. Water Resources Research 31:943–959
64. Stone HL (1991) Rigorous black oil pseudo functions. Paper SPE 21207 presented at the SPE Symposium on Reservoir Simulation, Anaheim, California, February 17–20
65. Suzuki K, Hewett TA (2002) Sequential upscaling method. Transport in Porous Media 46:179–212
66. Wallstrom TC, Hou S, Christie MA, Durlofsky LJ, Sharp DH, Zou Q (2002) Application of effective flux boundary conditions to two-phase upscaling in porous media. Transport in Porous Media 46:155–178
67. Wen XH, Gómez-Hernández JJ (1996) Upscaling hydraulic conductivities in heterogeneous media: an overview. Journal of Hydrology 183:ix–xxxii
68. Wen XH, Durlofsky LJ, Edwards MG (2003) Use of border regions for improved permeability upscaling. Mathematical Geology 35:521–547

69. White CD, Horne RN (1987) Computing absolute transmissibility in the presence of fine-scale heterogeneity. Paper SPE 16011 presented at the SPE Symposium on Reservoir Simulation, San Antonio, Texas, February 1–4

70. Wolfsteiner C, Lee SH, Tchelepi HA (2006) Well modeling in the multiscale finite volume method for subsurface flow simulation. Multiscale Modeling and Simulation 5:900–917

71. Wu XH, Efendiev YR, Hou TY (2002) Analysis of upscaling absolute permeability. Discrete and Continuous Dynamical Systems, Series B 2:185–204

72. Wu XH, Parashkevov R, Stone M, Lyons S (2008) Global scale-up on reservoir models with piecewise constant permeability field. Journal of Algorithms & Computational Technology 2:223–247

73. Yue X, E W (2007) The local microscale problem in the multiscale modeling of strongly heterogeneous media: effects of boundary conditions and cell size. Journal of Computational Physics 222(2):556–572

74. Zhang P, Pickup GE, Christie MA (2008) A new practical method for upscaling in highly heterogeneous reservoir models. SPE Journal 13:68–76

75. Zijl W, Trykozko A (2001) Numerical homogenization of the absolute permeability using the conformal-nodal and mixed-hybrid finite element method. Transport in Porous Media 44: 33–62

Sparse Tensor Approximation of Parametric Eigenvalue Problems

Roman Andreev and Christoph Schwab

Abstract We design and analyze algorithms for the efficient sensitivity computation of eigenpairs of parametric elliptic self-adjoint eigenvalue problems on high-dimensional parameter spaces. We quantify the analytic dependence of eigenpairs on the parameters. For the efficient approximate evaluation of parameter sensitivities of isolated eigenpairs on the entire parameter space we propose and analyze a sparse tensor spectral collocation method on an anisotropic sparse grid in the parameter domain. The stable numerical implementation of these methods is discussed and their error analysis is given. Applications to parametric elliptic eigenvalue problems with infinitely many parameters arising from elliptic differential operators with random coefficients are presented.

1 Introduction

Multiparametric eigenvalue problems (EVPs) arise in numerous applications: we mention only engineering (parametric design optimization of the spectrum of structures in solids, fluids and electromagnetics), uncertainty quantification, stability analysis of engineering systems and the like. Other applications arise in the perturbation analysis in physics. Accordingly, there is a sizable body of references devoted to eigenvalue perturbation analysis. The mathematical theory of perturbation evolved in close connection to these applications; seminal works are by Rellich and Kato, see [33, 39] and references therein.

In recent years, much attention has been devoted to the computational analysis of so-called *complex systems*; in the context of the results in the present paper, such engineering systems could be, e.g. deterministic initial boundary value problems

R. Andreev · C. Schwab (✉)
Seminar for Applied Mathematics, ETH Zürich, Switzerland
e-mail: andreevr@math.ethz.ch; schwab@math.ethz.ch

I.G. Graham et al. (eds.), *Numerical Analysis of Multiscale Problems*, Lecture Notes in Computational Science and Engineering 83, DOI 10.1007/978-3-642-22061-6_7, © Springer-Verlag Berlin Heidelberg 2012

depending on possibly a large number of design parameters. Alternatively, one could consider spectral problems for partial differential equations with random field input such as, e.g. diffusion problems with random heat conductivity. Adopting parametric representations of input random fields (e.g. by Karhunen-Loève expansions) renders the EVP of interest deterministic but depending on possibly a *countable number of parameters*.

Prior to entering the technical subject matter of this paper, let us briefly comment on the notion of "sparse tensor discretization" in the present context. Increasingly, in engineering and in the sciences, there is a need for the numerical solution of partial differential equation (PDE) models for a number of input parameters; in particular, models of so-called "complex systems" are often characterized by PDE models that depend on possibly countably many parameters. The task of *numerical solution of the PDE model on the entire parameter space* thus has to deal with the issue of numerical approximation in high dimension. Standard techniques of discretization are not efficient in high dimensional parameter spaces, due to the curse of dimensionality. One readily available alternative is Monte Carlo sampling which yields a convergence rate $N^{-1/2}$ in terms of the number N of samples, *independent of the dimension of the parameter domain that is sampled*. However, the rate $1/2$ is insufficient for achieving a low computational complexity, since each "sample" amounts to solving a PDE problem numerically. In addition, the parameter dependence in these problems is often very smooth (in fact, we show here that isolated eigenpairs depend analytically on the parameters), which is not exploited by Monte Carlo sampling. In recent years, therefore, *deterministic, high order* discretization techniques in parameter spaces have been proposed. They exploit the idea of analytic parameter dependence for expanding the unknown, parametric solution into multivariate polynomials of these parameters. Coefficients in these expansions are determined either by spectral collocation or by Galerkin projections. Without asserting completeness, we mention only [3–5, 16, 17, 20, 34, 46, 49, 50] and the references there for technical aspects and applications to a host of problems in engineering and in the sciences.

As a rule, these spectral approximation techniques exploit analytic parameter dependence in that they yield *exponential convergence in terms of the spectral approximation order*. The spectral projection methods are, however, still prone to the curse of dimensionality, and exponential convergence is de-facto lost in the range of computationally accessible spectral orders when the dimension of the parameter space becomes even moderately large (practically, 10 and higher). Still, for parametric problems which depend only on a few parameters, the spectral approach has been repeatedly demonstrated in engineering applications to be superior to, for example, Monte Carlo sampling, see e.g. [20, 47, 48] and the references there.

For high dimensional parameter spaces, however, several essential modifications of this argument are required: on the one hand, it is necessary to track the size of domains of analyticity in dependence on the dimension, to exploit *sparsity* in polynomial expansions of the parametric solution. This has been proposed, for example, in [17, 35] and has shown to yield substantial improvements in computational efficiency and allowed, in particular, to treat numerically problems in two spatial dimensions depending on several hundred parameters. In these algorithms, however, the *level of discretization in space and time is equal for all polynomial coefficients resp. for all collocation points*, so that these anisotropic discretization approaches exploit sparsity in polynomial expansions, in some sense. They are *tensor product discretizations* in the sense that they can be viewed as a tensorization of a (sparse, anisotropic) discretization scheme in parameter space and a family of (space-time) discretizations in physical space. Still, when the numerical solution of a single instance of the physical problem already entails massive computational effort (we think, for example, of parametric sensitivity studies in global climate simulations), even the approaches which exploit sparsity in parametric expansions of the solution entail a prohibitive computational effort and more sophisticated approaches are required.

Numerical analysis, in particular a careful study of the smoothness of solutions in dependence on the parameters and their regularity in physical space and time, and the approximation properties of discretizations in parameter as well as in physical space, pointed the way to more efficient discretizations. In [43,44], it was shown for the first time that sparse discretizations of parametric PDE problems which depend on a countable number of parameters could achieve arbitrary high, algebraic rates of convergence in terms of N, the number of instances of the physical problem to be solved, with constants that are independent of the dimension, i.e., converging at a dimensionally robust rate and outperforming Monte Carlo Methods, in terms of the rate of convergence.

The analysis in [43, 44] was performed under rather restrictive assumptions on the parameter dependence, in particular, on *exponential growth of the size of domain of analyticity of the solutions' parameter domains*. This analysis was subsequently considerably refined in [12, 13] and extended to parabolic and wave equations with random coefficients in [29, 30] to algebraically growing analyticity domains, with an in a sense optimal sparsity relation between parametric inputs and outputs.

This analysis paved the way, on the one hand, for the *coarsening of discretizations in physical space and time for "high order" terms in the polynomial expansions* (which are, a-priori, known to be relatively small). On the other hand, the results in [12, 13] the design of *adaptive, optimal* algorithms in [21–24].

The *adaptation of the discretization level in physical space to the relative coefficient size in the solution's polynomial expansion* can therefore be viewed as *sparse tensor product discretization* and is, in fact, a direct generalization of the idea of *sparse grids* (see e.g. [11] and the references there) well-known from high-dimensional numerical integration, and going back to [41].

Importantly, also in the context of collocation schemes, such sparse tensor discretizations can be applied with any given forward solver for instances of problems of interest, while essentially preserving the theoretically best possible convergence rates, as was demonstrated for elliptic problems with parametric diffusion coefficients in [6, 7]. A concrete algorithmic strategy for the determination of quasioptimal sets of "active" polynomial coefficients based on *a-priori information on the parametric dependence* of the inputs in the PDE has been given in [8], and has been used on connection with a stochastic Galerkin discretization.

In the present paper, we extend, among others, the approach to a *sparse tensor stochastic collocation* technique: given a predicted sets of active polynomial degrees in the solution's spectral expansion with respect to the parameters, we propose a family of sparse interpolation operators generalizing, among others, the Smolyak construction from [41] used in [35].

We use this interpolant for the Finite Element approximation of parametric eigenvalue problems of elliptic, self-adjoint operators.

In the present paper, we develop the numerical analysis of parametric eigenvalue problems for self-adjoint operators on possibly high dimensional parameter spaces in an abstract setting: we consider parametric operator families $A(y) \in \mathcal{L}(V, V^*)$ depending on a vector of real parameters $y = (y_1, y_2, \ldots)$. This setup covers differential- and integral operators acting between separable Hilbert spaces V and V^*, and, therefore, also multiparametric matrix pencils. We assume in the present paper that the dependence on the spectral parameter is linear and that the dependence on the problem parameter is affine.

The outline of the paper is as follows: in Sect. 2, we present an abstract variational formulation of self-adjoint parametric eigenvalue problems, with affine parameter dependence. We establish in particular holomorphic parameter dependence of isolated, simple eigenpairs, and give explicit bounds on the size of the domains of analyticity, in dependence on the input data. These results draw upon classical results on analytic dependence of spectra due to Rellich and Kato; see e.g. [33, 39] and the surveys [37, 38].

In Sect. 3, we review the abstract theory of Galerkin approximation of variational eigenvalue problems, drawing upon the basic reference [2] and the references there.

Section 4 then present the first and most important instance of the abstract theory, namely a class of second order elliptic eigenvalue problems where the differential operator depends in an affine fashion on countably many parameters. This situation typically arises in the context of elliptic PDEs whose operator has random coefficients, if the randomness is parametrized by a Karhunen-Loève expansion (see e.g. [17, 40]).

The core technical material of the present paper is contained in Sect. 5, namely the extension of the sparse tensor collocation operator to a quite arbitrary class of sets of "active" polynomial modes which we term "monotone", and for which the sparse tensor interpolation operators introduced here are proved to be unisolvent and well-conditioned, generalizing the analysis in [6–8]. Importantly, the analysis of the sparse tensor collocation operator in Sect. 5 is also applicable to source problems. Section 6 illustrates the foregoing analysis with some numerical experiments on a model eigenvalue problem, corresponding to the source problem which was considered in [8]. Section 7 briefly summarizes the main conclusions of the present work and indicates several generalizations which will, however, be presented elsewhere.

2   Parametric Eigenvalue Problems

We present a class of abstract eigenvalue problems for parametric self-adjoint families of operators with real-analytic dependence on a vector of parameters and discuss the dependence of their eigenpairs on these parameters. This will be the foundation of the design and the analysis of sparse tensor approximation methods in subsequent sections.

We will specialize on operators of the particular form

$$A(y) = \bar{A} + y_1 B_1 + y_2 B_2 + \ldots$$

with a self-adjoint "principal part" $\bar{A}$. In first approximation, the affine dependence of $A(y)$ on the parameter vector $y = (y_1, y_2, \ldots)$ will also appear in the case of general smooth nonlinear dependence $y \mapsto A(y)$.

Specific examples are also provided in the next section. In particular, operators depending on a countable number of parameters arise in applications such as PDEs with spatially inhomogeneous random coefficients.

2.1   Variational Eigenvalue Problems

Let V and H be separable Hilbert spaces over $\mathbb{R}$ (or $\mathbb{C}$) with inner products $(\cdot, \cdot)_V$ and $(\cdot, \cdot)_H$ and norms $\| \cdot \|_V$ and $\| \cdot \|_H$, respectively. Unless stated otherwise, V and H are assumed to be infinite-dimensional. We assume V and H form the *Gel'fand triple* $V \subset H \cong H^* \subset V^*$ with dense and compact injections, where $H \cong H^*$ indicates identification of the "pivot space" H with its dual H^*. By $\langle \cdot, \cdot \rangle_{V \times V^*}$ we denote the duality pairing on V and V^*.

Let $b : V \times V \to \mathbb{C}$ be a bilinear (or sesquilinear) form for which there exist constants $\gamma > 0$ and $C_1 > 0$ such that

$$\forall u, v \in V : \quad |b(u, v)| \leq C_1 \|u\|_V \|v\|_V, \tag{1}$$

$$\inf_{0 \neq u \in V} \sup_{0 \neq v \in V} \frac{|b(u, v)|}{\|u\|_V \|v\|_V} \geq \gamma > 0 \tag{2}$$

and

$$\forall 0 \neq v \in V : \quad \sup_{u \in V} |b(u, v)| > 0. \tag{3}$$

We denote by $A \in \mathscr{L}(V, V^*)$ the operator corresponding to the form $b(\cdot, \cdot)$ via the identification $b(u, v) = \langle v, Au \rangle_{V \times V^*}$ for all $u, v \in V$. Then (1)–(3) imply that both, A and its adjoint A^* are isomorphisms between the space V and its dual V^*.

We call $\lambda \in \mathbb{C}$ *eigenvalue* of the form $b(\cdot, \cdot)$ if there exists an *eigenvector* $0 \neq w \in V$ associated to λ, such that

$$\forall v \in V : \quad b(w, v) = \lambda (w, v)_H, \tag{4}$$

in which case the pair $(\lambda, w) \in \mathbb{C} \times V$ is called *eigenpair* of $b(\cdot, \cdot)$.

By $\sigma(A)$ we denote the spectrum of an operator A [37, Ch. VI]. Conditions (1)–(3) and compactness of the embedding $V \subset H$ imply the existence of a unique compact linear operator $T \in \mathscr{L}(V, V)$ such that

$$\forall u, v \in V : \quad b(Tu, v) = (u, v)_H. \tag{5}$$

The pair $(\lambda, w) \in \mathbb{C} \times V$ satisfies (4) if and only if $\lambda Tw = w \neq 0$, i.e., if and only if the pair (λ^{-1}, w) is an eigenpair of the compact operator T. Note that by (2), the eigenvalue λ is non-zero.

Let $\mu \in \sigma(T)$, $\mu \neq 0$. The number $\lambda = \mu^{-1}$ is an eigenvalue of the form $b(\cdot, \cdot)$. The smallest integer α such that $\mathrm{Ker}((\mu - T)^{\alpha}) = \mathrm{Ker}((\mu - T)^{\alpha+1})$ is called *ascent of* $\mu - T$. The dimension $m = \dim \mathrm{Ker}((\mu - T)^{\alpha})$ is finite and is called *algebraic multiplicity of* λ. Vectors in $\mathrm{Ker}((\mu - T)^{\alpha})$, *the generalized eigenspace of T corresponding to* λ are called *generalized eigenvectors of T corresponding to* λ. The *geometric multiplicity of* μ is equal to $\dim \mathrm{Ker}(\mu - T)$. The b-adjoint of T, denoted by T_* is defined by $b(Tu, v) = b(u, T_*v)$ for all $u, v \in V$. A pair $(\lambda, v) \in \mathbb{C} \times V$ is called *adjoint eigenpair of the form* $b(\cdot, \cdot)$ if and only if (λ^{-1}, v) is an eigenpair of T_*, i.e., $v \neq 0$ and $b(u, v) = \lambda (u, v)_H$ for all $u \in V$. In this case v is called *adjoint eigenvector corresponding to* λ. *Generalized adjoint eigenvectors of T* are exactly the generalized eigenvectors of T_*. An eigenvalue $\lambda \in \sigma(A)$ is called *isolated* if $\mathrm{dist}(\lambda, \sigma(A) \backslash \{\lambda\}) > 0$. It is called *discrete* if it is isolated and if for self-adjoint A: it is of finite multiplicity, i.e., $\dim\{u \in V : Au = \lambda u\} < \infty$ see [37, Theorem VII.10]; for non-selfadjoint A: the spectral projection $P_\lambda = -\frac{1}{2\pi i} \oint_{|\mu-\lambda|=r} (A - \mu)^{-1} d\mu$, is finite dimensional [38, Ch. XII]. An eigenvalue $\lambda \in \sigma(A)$ is called *nondegenerate* if the respective dimension equals one.

2.2 Abstract Parametric Eigenvalue Problems

We consider a family of *real, parametric* eigenvalue problems, and *assume until further specification that V and H are Hilbert spaces over* $\mathbb{R}$ (rather than over $\mathbb{C}$). Assume we are given a family of bounded self-adjoint operators $A(y) \in \mathscr{L}(V, V^*)$ parameterized by a vector $y = (y_1, y_2, ...)$ of real numbers, which we assume to take values in bounded intervals, after rescaling $y_m \in [-1, 1]$. In many applications, we deal with a finite, but possibly large number M of parameters, whereas for applications from elliptic PDEs with random coefficients we allow countably many parameters. Accordingly, we assume $y \in U$, where

$$U = \begin{cases} [-1,1]^M, & M < \infty, \\ [-1,1]^{\mathbb{N}}, & M = \infty. \end{cases} \qquad (6)$$

For $M = \infty$ the summation "$\sum_{m=1}^M$" is understood an (unconditionally) convergent infinite sum. For U we consider a Hausdorff topology, i.e., the Euclidean topology if U is finite-dimensional; if U is infinite-dimensional we equip U with the topology and metric of $\ell^\infty(\mathbb{N})$. This setting fits the abstract framework of [28]. For all $y \in U$ we associate to $A(y)$ the bilinear form

$$b(y;\cdot,\cdot) : V \times V \to \mathbb{R}, \quad (u,v) \mapsto b(y;u,v) := \langle u, A(y)v \rangle_{V \times V^*} \qquad (7)$$

We assume that $y \mapsto A(y)$ is uniformly bounded on U:

$$\forall u,v \in V: \quad \langle u, A(y)v \rangle_{V \times V^*} \leq \sup_{y \in U} ||A(y)||_{\mathscr{L}(V,V^*)} ||u||_V ||v||_V, \qquad (8)$$

with $\sup_{y \in U} ||A(y)||_{\mathscr{L}(V,V^*)} < \infty$, and uniformly elliptic on U:

$$\exists \alpha > 0: \quad \forall y \in U: \quad \forall v \in V: \quad \langle v, A(y)v \rangle_{V \times V^*} \geq \alpha ||v||_V^2. \qquad (9)$$

This implies that for every $y \in U$ the operator $A(y)$ is boundedly invertible, i.e., for its inverse we have $||A^{-1}(y)||_{\mathscr{L}(V^*,V)} \leq \alpha^{-1}$. The compactness of the embedding $V \hookrightarrow H$ implies that the parametric EVP: given $y \in U$, find

$$\lambda(y) \in \mathbb{R} \quad \text{and} \quad 0 \neq w(y) \in V \quad \text{s.t.} \quad A(y)w(y) = \lambda(y)w(y) \qquad (10)$$

or, in variational form: given $y \in U$, find

$$\lambda(y) \in \mathbb{R}, \quad 0 \neq w(y) \in V: \quad \forall v \in V: \quad b(y;w(y),v) = \lambda(y)(w(y),v)_H \qquad (11)$$

admits, for every $y \in U$, countably many *real* eigenvalues $(\lambda_j(y))_{j \geq 1} \subset \mathbb{R}$ of finite multiplicity. *Here, and in the following, we always assume the eigenvalue sequences to be numbered in increasing magnitude, counting multiplicities, i.e., an eigenvalue of multiplicity k is listed k times.* The corresponding set of eigenfunctions $\{w_j(y)\}_{j \geq 1} \subset V$ forms a countable dense set in V, and therefore, by compact and dense embedding $V \hookrightarrow H$, we assume w.l.o.g. that for every $y \in U$ the sequence $(w_j(y))_{j \geq 1}$ forms a countable orthonormal basis of H.

2.3 Analyticity

We are particularly interested in the case where the dependence of $A(y)$ on the parameters y_m is *analytic* in a suitable sense and, more specifically, in the case when the dependence of $A(y)$ on each coordinate y_m is affine, possibly after linearization of $A(y)$ given smooth dependence on the parameter vector y: there exist $\bar{A}, B_m \in \mathcal{L}(V, V^*), m \geq 1$, such that

$$\forall y = (y_1, y_2, \ldots) \in U : \quad A(y) = \bar{A} + \sum_{m=1}^{M} y_m B_m \tag{12}$$

with convergence in $\mathcal{L}(V, V^*)$.

Remark 1. To ensure coercivity (9) of $A(y)$ in (12), it is sufficient that in (12) the "mean operator" $\bar{A}$ and the "fluctuations" B_m satisfy

$$\exists \bar{\alpha} > 0 \quad \text{and} \quad \exists \kappa < 1 : \quad \begin{cases} \langle v, \bar{A}v \rangle_{V \times V^*} \geq \bar{\alpha} \|v\|_V^2 \quad \forall v \in V, \\ \sum_{m=1}^{M} \|B_m\|_{\mathcal{L}(V,V^*)} \leq \kappa \bar{\alpha}. \end{cases} \tag{13}$$

Indeed, condition (13) implies (9) with $\alpha = \bar{\alpha}(1-\kappa) > 0$: for any $y \in U$ and $v \in V$ we have

$$b(y; v, v) \geq b(0; v, v) - \left(\sum_{m=1}^{M} \|B_m\|_{\mathcal{L}(V,V^*)} \right) \|v\|_V^2 \geq \bar{\alpha}(1 - \kappa)\|v\|_V^2.$$

As we will show, under certain conditions the eigenpairs $(\lambda_m(y), w_m(y))_{m \geq 1}$ depend *analytically* on the parameter vector y. To make this precise, we first recall definitions and facts on Hilbert space valued analytic functions. To this end, *from now on we assume that V and H are complex separable Hilbert spaces* and extend all inner products and duality pairings sesquilinearly for complex valued arguments. We emphasize that this extension is for purposes of analysis only; the parametric eigenvalue problems under consideration here are *real and self-adjoint* (for non-selfadjoint operators some of our results require essential modifications in their statement and proof).

We recall some definitions. Assume initially that $M < \infty$. Let X be a Banach space over $\mathbb{C}$ and let $E \subset \mathbb{C}^M$ be an open, bounded and connected domain, $M \in \mathbb{N}$. A function $x : E \to X$ is said to be:

- *(Strongly jointly) analytic in E* if for each $a \in E$ there is $\{c_k\}_{k \in \mathbb{N}_0^M} \subset X$ such that the Taylor series $\sum_{k \in \mathbb{N}_0^M} c_k \prod_{m=1}^{M} (z_m - a_m)^{k_m}$ is summable to $x(z)$ for $z \in E$ sufficiently close to a.
- *(Strongly) holomorphic in E* if for each $a \in E$ each first order partial derivative $\lim_{0 \neq h \to 0} (x(a + he_m) - x(a))/h$ exists in X where $e_m \in \mathbb{C}^M$ is the mth standard unit vector.
- *Weakly analytic in E* if for each $\ell \in X^*$ the function $\ell(x(\cdot))$ is a $\mathbb{C}$-valued analytic function in E (equivalently, holomorphic by Hartogs' theorem [31, Theorem 2.2.8]).

For X endowed with a locally convex sequentially complete topology (e.g. Banach space) these notions coincide, see [28, Theorem 2.1.3], also [37, Theorem VI.4].

In the case $M = \infty$ we call $x : E \to X$ *(jointly) analytic* on a set $E \subset \mathbb{C}^M$ which is open w.r.t. $\ell^\infty(\mathbb{N})$, if X is locally convex sequentially complete and if the series

$$f(z) = f(a) + \sum_{\substack{\nu \in \mathbb{N}_0^\mathbb{N} \\ \# \operatorname{supp} \nu < \infty}} \frac{1}{\nu!} (D_a^\nu f)(z-a)^\nu$$

is uniformly summable for $z-a$ in any compact subset of the largest *balanced* subset of the set $E - a$ (here, $\tilde{E} \subset E - a$ is called balanced if $z \in \tilde{E} \wedge |\zeta| \leq 1 \Rightarrow \zeta z \in \tilde{E}$).

2.3.1 Case $M = 1$

The case $M = 1$ is of independent interest, and also serves as a building block for the multiparameter case. Therefore, we recapitulate the pertinent results here. For single parameter, regular[1] analytic spectral perturbation theory we refer to [33, Chaps. I and II] and [38, Chap. XII]. For $M = 1$, the operator $A(y)$ in (12) takes the form

$$A(y) = \bar{A} + yB, \quad y \in U = [-1, 1]. \tag{14}$$

We can extend $A(\cdot)$ to an entire, operator-valued function by allowing $y \in \mathbb{C}$. In this case, the dependence of the eigenpairs $(\lambda_j(y), w_j(y))_{j \geq 1}$ on the parameter y is well understood. Although the dependence of $A(y)$ on y in (14) is analytic, the eigenpairs of $A(y)$ do not necessarily inherit this analytic dependence:

Example 1 ([33], Sect. II.1 or [38], Sect. XII.1). For $V = \mathbb{C}^2$ consider

$$A(z) = \begin{pmatrix} 1 & 0 \\ 0 & -1 \end{pmatrix} + z \begin{pmatrix} 0 & 1 \\ 1 & 0 \end{pmatrix} \in \mathscr{L}(V, V^*).$$

Then $A(z)$ has eigenvalues $\lambda_\pm(z) = \pm\sqrt{1+z^2}$ and is real and self-adjoint for real z. Evidently, *for $|z| < 1$, the eigenvalues are complex-analytic functions of z.* However, *even though the map $z \mapsto A(z)$ is entire analytic and $\mathbb{R} \ni z \mapsto \lambda_\pm(z)$ are real-analytic functions, the maps $\mathbb{C} \ni z \mapsto \lambda_\pm(z)$ exhibit singularities* as complex-analytic functions at $z = \pm i$. Note that $A(z)$ is symmetric for $\operatorname{Im} z = 0$, but is not Hermitean for any $\operatorname{Im} z \neq 0$, and even not diagonalizable for $z = \pm i$.

We now consider in (10) the parametric operator $A(y)$ as restriction to U of the analytic, $\mathscr{L}(V, V^*)$-valued function

---

[1]as opposed to "asymptotic", see [38], Sect. XII.2 and XII.3.

$$A : \mathbb{C} \to \mathscr{L}(V, V^*), \quad z \mapsto A(z) := \bar{A} + zB. \tag{15}$$

Note that the extension of the self adjoint $A(y)$, $y \in \mathbb{R}$, obtained in this fashion is *not* necessarily Hermitean (cf. Example 1).

We will assume that (13) holds, which in the case $M = 1$ becomes the following.

Assumption 1. *There exist $\bar{\alpha} > 0$ and $\kappa < 1$ s.t.*

$$\mathrm{Re}\langle v, \bar{A}v \rangle_{V \times V^*} \geq \bar{\alpha}||v||_V^2 \quad \forall v \in V \quad \text{and} \quad ||B||_{\mathscr{L}(V,V^*)} \leq \kappa\bar{\alpha}. \tag{16}$$

Under Assumption 1, the variational form of the eigenvalue problem for $A(z)$ in (15) satisfies the assumptions (1)–(3):

Proposition 1. *Let Assumption 1 hold. Then, for any $0 < \delta < 1$ and all $z \in \mathbb{C}$ with $|z| \leq \kappa^{-1}\delta$ the sesquilinear form $b(z; \cdot, \cdot) : V \times V \to \mathbb{C}$, $b(z; u, v) := \langle u, A(z)v \rangle_{V \times V^*}$ satisfies (1)–(3) with*

$$\gamma = \bar{\alpha}(1 - \delta) > 0, \quad C_1 = ||\bar{A}||_{\mathscr{L}(V,V^*)} + \kappa^{-1}\delta||B||_{\mathscr{L}(V,V^*)}, \tag{17}$$

in particular, for any $\kappa < \delta$. Moreover, for all $z \in \mathbb{C}$ with $|z| \leq \kappa^{-1}\delta$ the operator $A(z)$ is boundedly invertible with $\sup_{|z| \leq \kappa^{-1}\delta} ||A^{-1}(z)||_{\mathscr{L}(V^*,V)} \leq (\bar{\alpha}(1 - \delta))^{-1}$.

By compactness of the embedding $V \hookrightarrow H$, for every $y \in U$, the spectrum $\sigma(A(y))$ of the self-adjoint operator $A(y)$ is discrete and consists of at most countably many real eigenvalues $\lambda_j(y)$, $j = 1, 2, \ldots$ of finite multiplicity which accumulate only at infinity. As $A(y)$ is symmetric, the algebraic and geometric multiplicities of $\lambda_j(y)$ coincide.

We are interested in the relation of $\lambda(y) \in \sigma(A(y))$ to eigenvalues $\lambda(z) \in \mathbb{C}$ of the complex extension $A(z)$ in (15) of $A(y)$.

Theorem 1. *For $z \in \mathbb{C}$, consider the family (15) of linear operators where $\bar{A}$, B satisfy Assumption 1. Fix $y \in U = [-1, 1]$. Let $\lambda(y) \in \sigma(A(y)) \subset \mathbb{R}$ be a discrete eigenvalue of $A(y)$ of multiplicity $m \in \mathbb{N}$. Then the following holds:*

(a) *There exist m (not necessarily distinct) complex-valued functions of z which are single-valued and analytic near $z = y$, denoted by $\lambda^{(1)}, \lambda^{(2)}, \ldots, \lambda^{(m)}$, such that $\lambda^{(j)}(y) = \lambda(y)$, $j = 1, \ldots, m$ and $\lambda^{(j)}(z)$, $j = 1, \ldots, m$ are discrete eigenvalues of $A(z)$ near $z = y$,*

(b) *There are no other eigenvalues of $A(z)$ near $\lambda(z)$,*

(c) *There are m complex-analytic V-valued functions $w^{(1)}, \ldots, w^{(m)}$, such that near $z = y$, $w^{(1)}(z), \ldots, w^{(m)}(z)$ are corresponding eigenvectors of $A(z)$,*

(d) *The domains of analyticity contain discs $\{z \in \mathbb{C} : |z - y| < \epsilon\}$ where $\epsilon = \epsilon(\kappa) \sim \kappa^{-1}$ as $\kappa \to 0$.*

Proof. Parts *(a)* and *(b)* are in Theorem XII.13 in [38]. Part *(c)* is Problem 17 in Sect. XII of [38] applied to the projector $P(z - y)$ in Theorem XII.13 of [38]. Part *(d)* follows by a scaling argument. $\square$

The following quantitative bound on the parameter range ensuring ellipticity will be useful later for obtaining uniform bounds on convergence radii.

Theorem 2. *Let Assumption 1 hold. Fix $y \in U$. Let $\lambda(y) \in \sigma(A(y))$ be isolated and nondegenerate. Define the spectral gap $\gamma = \gamma(y, \lambda(y)) = \mathrm{dist}(\lambda(y), \sigma(A(y)) \setminus \{\lambda(y)\})$.*

Then there exists $z \mapsto \lambda_y(z) \in \sigma(A(z))$ with $\lambda_y(y) = \lambda(y)$ which is complex-analytic in the disc $E(y, \kappa, \gamma/\lambda(y)) = \{z \in \mathbb{C} : |z - y| < \frac{1}{2}\frac{\kappa^{-1} - |y|}{1 + \lambda(y)/\gamma}\}$. Moreover, $\lambda_y(z) \in \sigma(A(z))$ is isolated and nondegenerate for each $z \in E(y, \kappa, \gamma/\lambda(y))$.

Proof. We have $\|Bv\|_{V*} \leq \kappa\bar{\alpha}\|v\|_V \leq \frac{\kappa}{1-\kappa|y|}\|A(y)v\|_{V*}$ for all $v \in V$ by coercivity $\|A(y)v\|_{V*} \geq \bar{\alpha}(1 - |y|\kappa)\|v\|_V$. Theorem XII.8 of [38] (Kato-Rellich) and Theorem XII.11 of [38] show the claimed analyticity inside the circle of radius

$$r = [a + \varepsilon^{-1}[b + a(\lambda(y) + \varepsilon)]]^{-1} = \frac{1}{2}\frac{\kappa^{-1} - |y|}{1 + \lambda(y)/\gamma} \tag{18}$$

with $a = \frac{\kappa}{1-\kappa|y|}$, $b = 0$ and $\varepsilon = \frac{1}{2}\gamma$. $\qquad\square$

Remark 2. Note that

$$\frac{1}{1 + \delta^{-1}} = \begin{cases} \delta + \mathcal{O}(\delta^2), & \text{for} \quad \delta \to 0, \\ 1 - \delta^{-1} + \mathcal{O}(\delta^{-2}), & \text{for} \quad \delta \to \infty. \end{cases} \tag{19}$$

Thus, as the ratio $\delta = \gamma/\lambda(y)$ becomes small, the size of the domain analyticity is critically restricted. If γ is large compared to $\lambda(y)$, the size of the domain of analyticity is essentially given by κ^{-1}, cf. Theorem 1.

An analytic continuation argument yields the "y-uniform" version of Theorem 2.

Corollary 1. *Suppose that $U \ni y \mapsto \lambda(y) \in \sigma(A(y))$ is continuous and such that for each $y \in U$, $\lambda(y)$ is isolated and nondegenerate. Let $\delta > 0$ be such that*

$$\forall y \in U : \quad \mathrm{dist}(\lambda(y), \sigma(A(y)) \setminus \{\lambda(y)\}) \geq \delta\lambda(y).$$

Then we can extend $y \mapsto \lambda(y)$ to a functions $z \mapsto \lambda(z)$ which is analytic on $E(\kappa, \delta) = \bigcup_{y \in U} E(y, \kappa, \delta)$, s.t. $\lambda(z) \in \sigma(A(z))$ is isolated and nondegenerate for all $z \in E(\kappa, \delta)$.

Proof. We need the following special case of Lemma 1 below: for all $y \in U$ there exists $r > 0$ s.t. for any two continuous functions $f_1, f_2 : B_r(y) \to \sigma(A(z))$,

$$f_1(y) = \lambda(y) = f_2(y) \quad \forall y \in B_r(y) \cap U \implies f_2(z) = f_2(z) \quad \forall z \in B_r(y),$$

where $B_r(y) = \{z \in \mathbb{C} : |z - y| < r\}$.

We only sketch the rest of the proof, omitting the technical details. Let $U \subset K_1 \subset K_2 \subset \ldots \subset E(\kappa, \delta)$ be a monotonic sequence of compact sets such that $E(\kappa, \delta) = \bigcup_{n \in \mathbb{N}} K_n$. On every K_n an extension λ_n of λ can be constructed using Theorem 2 which is complex-analytic on any open $B \subset K_n$ and unique by Lemma 1. Moreover, all λ_n agree on their respective domains of definition, giving rise to a complex-analytic extension of λ to $E(\kappa, \delta)$, again unique by Lemma 1. $\square$

2.3.2   Case $M \geq 1$

In generalizing the above results to $M \geq 2$ parameters, care is necessary, as the following example due to Rellich shows, cf. [38, p. 60].

Example 2. Consider the two-parameter family of *symmetric* 2×2 matrices

$$A(y_1, y_2) = y_1 \begin{pmatrix} 1 & 0 \\ 0 & -1 \end{pmatrix} + y_2 \begin{pmatrix} 0 & 1 \\ 1 & 0 \end{pmatrix}.$$

Then A is linear, hence entire analytic in y_1 and y_2, and symmetric for real y; however, its eigenvalues $\lambda(y_1, y_2) = \pm\sqrt{y_1^2 + y_2^2}$ are *not real-analytic* with respect to y_1, y_2 in any vicinity of the origin. Note also that $A(0, 0)$ has a double eigenvalue 0.

In this section we derive sufficient conditions which preclude this kind of singularity. For technical reasons we focus on eigenvalues which are isolated and nondegenerate. The following theorem is a local result on holomorphic dependence of an isolated and nondegenerate eigenvalue of the complex extension of $U \ni y \mapsto A(y)$ given in (12) to $\mathbb{C}^M \ni z \mapsto A(z) := \bar{A} + \sum_{m=1}^{M} z_m B_m$ on the parameters.

Theorem 3. *Suppose $z \in \mathbb{C}^M$ is such that $A(z)$ has an isolated and nondegenerate eigenvalue $\lambda(z)$. Pick $m \in \mathbb{N}$, $m \leq M$, and write $e_m = (0, \ldots, 0, 1, 0, \ldots)$, nonzero in mth coordinate. Then $\exists \varepsilon_m(z) > 0$ s.t. for $E_m := \{\zeta_m \in \mathbb{C} : |\zeta_m| < \varepsilon_m(z)\}$:*

1. There exists a unique complex-analytic function

$$E_m \ni \zeta_m \mapsto \lambda(z + e_m \zeta_m) \in \sigma(A(z + e_m \zeta_m)),$$

2. $\lambda(z + e_m \zeta_m)$ is isolated and nondegenerate for all $\zeta_m \in E_m$,
3. There exists a complex-analytic V-valued function

$$E_m \ni \zeta_m \mapsto w(z + e_m \zeta_m) \in V$$

such that $w(z + e_m \zeta_m)$ is a corresponding eigenvector for all $\zeta_m \in E_m$.

Proof. Note first that for any $r > 0$ the complex-analytic operator-valued function $\zeta_m \mapsto A(z + e_m \zeta_m)$ is uniformly bounded on the closed disk $\{\zeta_m \in \mathbb{C} : |\zeta_m| \leq r\}$.

Thus, by [38, Exercise XII.8], the Kato-Rellich theorem [38, Theorem XII.8] applies.   □

We now prove the following lemma ensuring uniqueness of the extension.

Lemma 1. *Let $E \subset U$ be connected by polygonal paths in the following sense: for any $z_a, z_b \in E$ there exist $n \in \mathbb{N}$ and $z_0, \ldots, z_n \in E$ with $z_0 = z_a$, $z_n = z_b$ and*

$$[z_{k-1}, z_k] := \{t z_{k-1} + (1-t)z_k : t \in [0,1]\} \subset E \quad \text{for all} \quad k = 1, \ldots, n.$$

Take continuous functions $f, g : E \to \mathbb{C}$ with $f(z), g(z) \in \sigma(A(z))$ for all $z \in E$ such that $f(z)$ and $g(z)$ are isolated and nondegenerate for all $z \in E$. If $f(z) = g(z)$ for some $z \in E$, then $f(z) = g(z)$ for all $z \in E$.

Proof. Suppose to the contrary that for some $z_a, z_b \in E$ we have $f(z_a) = g(z_a)$ and $f(z_b) \neq g(z_b)$. Let $\phi : [0,1] \to E$ be a continuous function with $\phi(0) = z_a$ and $\phi(1) = z_b$. Without loss of generality we can assume that:

- $\phi(t) = z_a + t(z_b - z_a)$.
- $\forall \epsilon > 0$ there exists $t \in (0, \epsilon)$ with $f(\phi(t)) \neq g(\phi(t))$ (to this end note that $\{z \in E : f(z) = g(z)\}$ is closed in E).
- The family of operators $\zeta \mapsto \tilde{A}(\zeta) = A(z_a + \zeta(z_b - z_a))$, $|\zeta| < 2$ is well-defined and is complex-analytic.

By the Kato-Rellich theorem [38, Theorem XII.8], there exists exactly one point $\mu(\zeta) \in \sigma(\tilde{A}(\zeta))$ close to $\mu(0) = f(z_a) = g(z_a)$ for ζ sufficiently small, w.l.o.g. for $|\zeta| < 2$ by rescaling $(z_b - z_a)$. But this contradicts $f(z_b) \neq g(z_b)$ if we set $\zeta = 1$.   □

In the following (Theorem 4) we identify the range of $z \in \mathbb{C}^M$ close to the parameter set U such that the conditions of Theorem 3 apply. We start with a Lemma.

Lemma 2. *Fix $y \in U = [-1, 1]^M$ and $\zeta \in \mathbb{C}^M$. Assume that for some $p \in (0, 1]$*

$$\beta(p) := \sum_{m \geq 1} \|B_m\|_{\mathscr{L}(V, V^*)}^p \tag{20}$$

is finite. Assume $\kappa := \frac{1}{\alpha}\beta(p) \sup_{m \geq 1} |\zeta_m| \|B_m\|_{\mathscr{L}(V, V^*)}^{1-p} < \infty$, where $\alpha > 0$ is as in (9). Then

$$\left\| \sum_{m \geq 1} \zeta_m B_m \right\|_{\mathscr{L}(V, V^*)} \leq \beta(p) \sup_{m \geq 1} |\zeta_m| \|B_m\|_{\mathscr{L}(V, V^*)}^{1-p}, \tag{21}$$

and moreover, for $\bar{\alpha} := \alpha$, κ, $\bar{A} := A(y)$ and $B := \sum_{m \geq 1} \zeta_m B_m$ we have (16).

Proof. Equation (21) follows by the triangle and the Hölder inequalities. Thus, $\|B\|_{\mathscr{L}(V,V^*)} \leq \kappa\bar{\alpha}$ and (16) holds by the assumption that $\bar{\alpha} = \alpha$ and $\bar{A} = A(y)$. $\quad\square$

The results of this section are now combined in the following theorem.

Theorem 4. *For the family of operators $A(y)$, $y \in U = [-1,1]^M$, assume that i) $A(y)$ is of the form (12), ii) (8) and (9) hold, iii) for some $p \in (0,1]$, the perturbations are p-summable as in (20), iv) $\lambda(y) \in \sigma(A(y))$ is an isolated and nondegenerate eigenvalue for all $y \in U$ with corresponding eigenfunction $w(y) \in V$, $\|w(y)\|_H = 1$ and v) the function $U \ni y \mapsto (\lambda(y), w(y)) \in \mathbb{R} \times V$ is continuous. Finally, assume that vi) there exists $\delta > 0$ such that $\mathrm{dist}(\lambda(y), \sigma(A(y)) \setminus \{\lambda(y)\}) \geq \delta\lambda(y)$ for all $y \in U$. Let $\varepsilon \in (0,1)$. Define $\tau_m := (1-\varepsilon)\frac{\alpha\|B_m\|_{\mathscr{L}(V,V^*)}^{p-1}}{2\beta(p)(1+\delta^{-1})}$ for $m \geq 1$, and*

$$E(\tau) := \{z \in \mathbb{C}^M : \mathrm{dist}(z_m, [-1,1]) < \tau_m\}. \tag{22}$$

Then (λ, w) can be extended to a jointly complex-analytic function on $E(\tau)$.

Proof. For any $z \in E(\tau)$ we first identify a candidate $\lambda(z) \in \sigma(A(z))$ which is isolated and nondegenerate. Fix $z \in E(\tau)$. Take $y = y(z) \in U$ with $|z_m - y_m| < \tau_m$, $m \geq 1$. Define $\zeta = z - y$. Obviously, $|\zeta_m| < \tau_m$ and therefore

$$\beta(p)\sup_{m\geq 1}|\zeta_m|\|B_m\|_{\mathscr{L}(V,V^*)}^{1-p} \leq (1-\varepsilon)\frac{\alpha}{2(1+\delta^{-1})}.$$

Thus, for $\kappa := \frac{1}{\alpha}\beta(p)\sup_{m\geq 1}|\zeta_m|\|B_m\|_{\mathscr{L}(V,V^*)}^{1-p}$ we have $(1-\varepsilon)^{-1} \leq \frac{1}{2\kappa}\frac{1}{1+\delta^{-1}}$. Consider the complex-analytic operator-valued function $t \mapsto A(y+t\zeta) = A(y) + t\sum_{m\geq 1}\zeta_m B_m$. By Corollary 1 and Lemma 2 there exists a complex-valued function $t \mapsto \tilde{\lambda}(y+t\zeta) \in \sigma(A(y+t\zeta))$ which depends holomorphically on the parameter t in the disk

$$\{t \in \mathbb{C} : |t| < (1-\varepsilon)^{-1}\} \subset \{t \in \mathbb{C} : |t| < \frac{1}{2\kappa}\frac{1}{1+\delta^{-1}}\}$$

and which is such that $\tilde{\lambda}(y+t\zeta)$ is isolated and nondegenerate eigenvalue, whenever $|t| < \frac{1}{2\kappa}\frac{1}{1+\delta^{-1}}$, in particular for $t = 1$. Thus, $\tilde{\lambda}(y+\zeta)$ is a candidate for the holomorphic extension $\lambda(z)$ of the parametric eigenvalue. By Lemma 1, $\tilde{\lambda}(y+\zeta)$ is, in fact, the same eigenvalue for any choice of $y \in U$, $\zeta \in \mathbb{C}^M$ satisfying $y + \zeta = z$. Therefore, $\lambda(z) := \tilde{\lambda}(y+\zeta)$ is well-defined. Similar considerations apply to the eigenfunction $w(z)$. In the remainder of the proof, we distinguish two cases:

Case $M < \infty$. By Theorem 3 and Lemma 1, the function (λ, w) is separately complex-analytic in $E(\tau)$. The classical Hartogs' theorem [31, Theorem 2.2.8] implies joint complex-analyticity of (λ, w) on $E(\tau)$.

Case $M = \infty$. The function (λ, w) is analytic on $E(\tau)$ in the sense of Definition 2.3.1 of [28]. The claim therefore follows using [28, Prop. 3.1.2] and [28, Theorem 3.1.5].   □

Remark 3. Note that summability with $p < 1$ is required in (20) for the analyticity region $E(\tau)$ in (22) to increase with dimension m.

3 Galerkin Discretization of Abstract Variational Eigenvalue Problems

To prepare the convergence analysis of the Smolyak-Galerkin approximation of parametric eigenvalue problems we recapitulate abstract results for the Galerkin method for elliptic eigenvalue problems from [2].

Variational approximations of the abstract eigenvalue problem (4) are defined in terms of a one-parameter family $\{V_h\}_{h>0} \subset V$ of finite-dimensional subspaces of increasing dimension $N(h) \to \infty$ as $h \to 0$, which is dense in V, i.e., for all $u \in V$: $\lim_{h\to 0} \inf_{\chi \in V_h} \|u - \chi\|_V = 0$. We assume that

$$\inf_{0 \neq u \in V_h} \sup_{0 \neq v \in V_h} \frac{|a(u,v)|}{\|u\|_V \|v\|_V} \geq \gamma(h) > 0. \tag{23}$$

We further assume *stability* of $\{V_h\}_{h>0}$ in the sense that

$$\forall u \in V: \quad \lim_{h\to 0} \gamma(h)^{-1} \inf_{\chi \in V_h} \|u - \chi\|_V = 0. \tag{24}$$

We define $P_h : V \to V_h$ as the projector characterized by

$$\forall u \in V, v \in V_h: \quad a(P_h u, v) = a(u, v). \tag{25}$$

The projection P_h is quasi-optimal in V:

$$\|u - P_h u\|_V \leq \left(1 + \frac{C_1}{\gamma(h)}\right) \inf_{\chi \in V_h} \|u - \chi\|_V. \tag{26}$$

Galerkin approximations (λ_h, w_h) of eigenpairs are obtained by restricting the abstract eigenvalue problem (4) to V_h: find

$$\lambda_h \in \mathbb{C} \quad \text{and} \quad 0 \neq w_h \in V_h \quad \text{s.t.} \quad \forall v \in V_h: \quad a(w_h, v) = \lambda_h (w_h, v)_H. \tag{27}$$

Equation (27) is a matrix eigenvalue problem of dimension $N(h) = \dim V_h$, so that there exist $N(h)$, in general complex, eigenvalues $\lambda_j(h)$, $j = 1, \ldots, N(h)$. By (23), the eigenvalues are non-zero. The pair (λ_h, w_h) is an eigenpair of (27) if and only if (λ_h^{-1}, w_h) is an eigenpair of the compact operator $T_h : V \to V_h$ defined by

$$\forall u \in V, v \in V_h : \quad a(T_h u, v) = (u, v)_H. \tag{28}$$

The operator T_h can be written as $P_h T$.

Let now $\lambda \in \mathbb{C}$ be an eigenvalue of (4) with algebraic multiplicity m and ascent $\alpha(\lambda)$, i.e., $\lambda^{-1} \in \mathbb{C}$ is an eigenvalue of the operator T with algebraic multiplicity m and $\alpha(\lambda)$ is the ascent of $\lambda^{-1} - T$. Since T is compact, $T_h = P_h T \to T$ in norm in $\mathscr{L}(V, V)$ by (26) and, as $h \to 0$, there are m discrete eigenvalues $\lambda_1(h), \dots, \lambda_m(h)$ of (27) that converge to λ. Define

$$\mathscr{M}(\lambda) := \{w : w \text{ is a generalized unit eigenvector of (4) for } \lambda\},$$

$$\mathscr{M}^*(\lambda) := \{w : w \text{ is a generalized adjoint unit eigenvector of (4) for } \lambda\}$$

and their approximation bounds

$$\epsilon_h(\lambda) = \sup_{u \in \mathscr{M}(\lambda)} \inf_{\chi \in V_h} \|u - \chi\|_V \quad \text{and} \quad \epsilon_h^*(\lambda) = \sup_{u \in \mathscr{M}^*(\lambda)} \inf_{\chi \in V_h} \|u - \chi\|_V. \tag{29}$$

There holds the following general result on the asymptotic eigenvalue error analysis.

Theorem 5. *Consider the variational eigenvalue problem* (4) *and its variational approximation* (27). *For* $\lambda \in \sigma(A)$, *let* $\alpha(\lambda)$ *denote the ascent of* $\lambda^{-1} - T$. *Then:*

1. *There exists a constant* $C > 0$ *such that, as* $h \to 0$,

$$\left| \lambda - \hat{\lambda}(h) \right| \leq C\gamma(h)\epsilon_h(\lambda)\epsilon_h^*(\lambda) \quad \text{where} \quad \hat{\lambda}(h) = \frac{1}{m} \sum_{j=1}^{m} \lambda_j(h), \tag{30}$$

2. *And*

$$\left| \lambda - \check{\lambda}(h) \right| \leq C\gamma(h)^{-1}\epsilon_h(\lambda)\epsilon_h^*(\lambda) \quad \text{where} \quad \check{\lambda}(h) = \left(\frac{1}{m} \sum_{j=1}^{m} \frac{1}{\lambda_j(h)} \right)^{-1}, \tag{31}$$

3.

$$\left| \lambda - \lambda_j(h) \right| \leq C(\gamma(h)^{-1}\epsilon_h(\lambda)\epsilon_h^*(\lambda))^{1/\alpha(\lambda)}. \tag{32}$$

4. *Let* $\lambda(h)$ *be an eigenvalue of* (27) *such that* $\lim_{h \to 0} \lambda(h) = \lambda \in \sigma(A)$. *Suppose further that for each* $h > 0$ *the vector* $w_h \in V_h$ *is a unit vector which satisfies* $(\lambda(h)^{-1} - T_h)^k w_h = 0$ *for some positive integer* $k \leq \alpha(\lambda)$. *Then, for any* $\ell \in \mathbb{N}_0$ *with* $k \leq \ell \leq \alpha(\lambda)$ *there is a vector* $u_h \in V$ *such that*

$$(\lambda^{-1} - T)^\ell u_h = 0 \quad \text{and} \quad \|u_h - w_h\|_V \leq C(\gamma(h)^{-1}\epsilon_h(\lambda))^{(\ell-k+1)/\alpha(\lambda)}. \tag{33}$$

4 Parametric Elliptic EVPs

Let $D \subset \mathbb{R}^d$ be a bounded Lipschitz domain. We consider the parametric diffusion operator $A(y)$ given by

$$(A(y)\zeta)(x) = -\nabla_x \cdot (a(x, y)\nabla_x\zeta(x)), \quad x \in D, \quad y \in U \qquad (34)$$

for a *family of diffusion coefficients* $y \mapsto a(\cdot, y)$, which is parametrized by a vector $y = (y_1, y_2, \ldots)$ belonging to the set U of admissible parameters, defined in (6). The parameters $y_m, m = 1, 2, \ldots$ could denote design parameters of an engineering system, or states of an optimal controller; they could also denote random coefficients in a wavelet or Karhunen-Loève expansion of the random field $y \mapsto a(\cdot, y)$. We present examples for the latter scenario below.

4.1 Assumptions on the Data

We suppose that in addition to being bounded and Lipschitz the domain $D \subset \mathbb{R}^d$ is simply connected. We denote by $(\cdot, \cdot)$ the $L^2(D)$ scalar product which extends uniquely by continuity to the duality pairing on $H_0^1(D) \times H^{-1}(D)$, again denoted by $(\cdot, \cdot)$.

The parameter dependence of the eigenvalue problem enters through the diffusion coefficient a. We assume that $a : U \to L^\infty(D)$ is uniformly positive, i.e., that there exist constants $0 < a_{min} \leq a_{max} < \infty$ such that

$$\forall y \in U : \quad 0 < a_{min} \leq \operatorname*{ess\,inf}_{x \in D} a(x, y) \leq \|a(\cdot, y)\|_{L^\infty(D)} \leq a_{max} < \infty. \qquad (35)$$

In particular, (35) implies that (23) holds with $\gamma(h) = a_{min} > 0$ for any closed subspace $\{0\} \neq V_h \subseteq V$. For every $y \in U$ consider the parametric Dirichlet eigenvalue problem:

$$\text{find} \quad \lambda(y) \in \mathbb{R} \quad \text{and} \quad 0 \neq w(y) \in H_0^1(D) \quad \text{s.t.} \quad A(y)w(u) = \lambda(y)w(y). \qquad (36)$$

Of particular interest is the case when the diffusion coefficient depends affinely on the parameter vector $y \in U$ (for background on the following assumption see [8, 13]):

Assumption 7. *The parametric diffusion coefficient is of the form*

$$a(x, y) = \bar{a}(x) + \sum_{m \geq 1} y_m \psi_m(x). \qquad (37)$$

with $y_m \in [-1, 1]$, *and*

$$\sum_{m\geq 1} ||\psi_m||_{L^\infty(D)} \leq \frac{\kappa}{1+\kappa}\bar{a}_{\min} \tag{38}$$

with $\bar{a}_{\min} = \text{ess}\inf_D \bar{a}$, $\kappa > 0$. Further, we assume

$$\sum_{m\geq 1} ||\psi_m||^p_{W^{1,\infty}(D)} < \infty \tag{39}$$

with some $p \in (0, 1)$.

In particular, $U \ni y \mapsto a(\cdot, y) \in W^{1,\infty}(D)$ and for $\beta(p)$, given by (20), we have $\beta(p) = \sum_{m\geq 1} ||B_m||^p_{\mathscr{L}(V,V^*)} \leq \sum_{m\geq 1} ||\psi_m||^p_{L^\infty(D)} < \infty$, where we define B_m by $(B_m u, v) = (\psi_m u, v)$ for all $u, v \in H_0^1(D)$.

Remark 4. Equation (39) implies for (36) that $\Delta w(y) \in L^2(D)$ for all $y \in U$. If $D \subset \mathbb{R}^d$ is convex, this in turn implies that $w(y) \in W = H^2(D) \cap H_0^1(D)$ for all $y \in U$. For polygonal domains $D \subset \mathbb{R}^2$ with straight sides, the same statement holds for W being a weighted Sobolev space, see [1]. For simplicity we assume in the following that $D \subset \mathbb{R}^d$ is convex.

We obtain the following corollary to Theorem 4.

Corollary 2. *In the setting of Theorem 4 with $A(y)$ given by (34) let Assumption 7 hold. Let $D \subset \mathbb{R}^d$ be convex. Then, the eigenpair (λ, w) can be extended to a jointly complex-analytic function on $E(\tilde{\tau})$ with values in $\mathbb{C} \times (H^2(D) \cap H_0^1(D))$, where $\tilde{\tau}$ is given by $\tilde{\tau}_m = \min\{\tau_m, c||\psi_m||^{p-1}_{W^{1,\infty}(D)}\}$ with $c > 0$ being arbitrary, and τ and $E(\cdot)$ are as in (22).*

Proof. Assume $z \in \mathbb{C}^M$ is such that $\text{dist}(z, [-1, 1]) < \tau_m \in (0, \infty)$, $m \geq 1$. Then (36) implies

$$-a(\cdot, z)\Delta w(z) = \lambda(z)w(z) + \nabla w(z) \cdot \nabla a(\cdot, z). \tag{40}$$

The mappings $z \mapsto a(\cdot, z)$ and $z \mapsto \nabla a(\cdot, z)$ are jointly complex-analytic with values in $L^\infty(D)$ by (39). Given Assumption 7, Theorem 4 holds for $z \in E(\tilde{\tau})$ where $\tilde{\tau}_m = \min\{\tau_m, c||\psi_m||^{p-1}_{W^{1,\infty}(D)}\}$ with $c > 0$ arbitrary and τ and $E(\cdot)$ are as in (22), and thus the pair $(\lambda(z), w(z))$ is jointly complex-analytic in an open neighborhood $\mathcal{O}$ of z with values in $\mathbb{R} \times H_0^1(D)$. Therefore $z \mapsto w(z)$ is jointly complex-analytic on $\mathcal{O}$ with values in $L^2(D)$. Since $\sup_{z\in\mathcal{O}} ||a(\cdot, z)||_{W^{1,\infty}(D)} < \infty$ and ess $\inf_{(x,z)\in D\times\mathcal{O}} a(x, z) > 0$, from (40) we obtain $\Delta w(z) \in L^2(D)$ for all $z \in \mathcal{O}$. Moreover, $z \mapsto \Delta w(z)$ is jointly complex-analytic on $\mathcal{O}$ with values in $L^2(D)$, as can be easily checked using (40).   □

The weak formulation of the parametric EVP is obtained in the usual way by testing (36) in $H_0^1(D)$ and integrating by parts. This results in the parametric eigenvalue probem: for every $y \in U$, find $\lambda(y) \in \mathbb{R}$ and $0 \neq w(y) \in H_0^1(D)$ such that

$$(\bar{a}\nabla u(y), \nabla \eta) + \sum_{m \geq 1} y_m (\psi_m u(y), \eta) = \lambda(y)(u(y), \eta) \quad \forall \eta \in H_0^1(D). \qquad (41)$$

By assumption (35) and the Lax-Milgram lemma, for every $y \in U$ the operator $A(y) \in \mathscr{L}(H_0^1(D), H^{-1}(D))$ is boundedly invertible and by the compactness of the embedding $L^2(D) \subset H^{-1}(D)$ for every $y \in U$ the inverse $T(y) = A^{-1}(y)$: $L^2(D) \rightarrow H_0^1(D)$ is compact. Hence, for every $y \in U$, the elliptic operator $A(y)$ admits a countable set of eigenvectors $\{w_j(y)\}_{j \geq 1} \subset H_0^1(D)$ forming an orthonormal basis of $L^2(D)$ with corresponding eigenvalues $0 < \lambda_1(y) \leq \lambda_2(y) \leq \ldots$ and $\lambda_j \rightarrow \infty$ as $j \rightarrow \infty$ which we assume to be numbered in increasing order, counting multiplicity. Due to the self-adjointness of $A(y)$, the eigenvalues' dependence on $y \in U$ is Lipschitz [33, Sect. V.4.3, Theorem 4.10]. This Lipschitz dependence (which fails to hold for general, non self-adjoint operators) allows us to speak of parametric eigenvalue families.

Remark 5. Enumeration of the eigenvalues $\lambda_j(y)$ can be done in two ways: first, as stated above, for every $y \in U$ in increasing eigenvalue magnitude, and second, in increasing eigenvalue magnitude with respect to a reference value of y, e.g. $y = 0 \in U$. Due to possible crossings of eigenvalues, the two eigenvalue numberings may differ.

Remark 6. For every $y \in U$ the *fundamental* or *spectral gap* $\lambda_2(y) - \lambda_1(y)$ is strictly positive and $\lambda_1(y)$ is nondegenerate by the Krein-Rutman theorem see e.g. [27, Theorem 1.2.6]. Thus there exists $\delta > 0$ such that

$$\text{dist}(\lambda_1(y), \sigma(A(y)) \setminus \{\lambda_1(y)\}) \geq \delta\lambda_1(y) \quad \text{for all} \quad y \in U. \qquad (42)$$

We are in particular interested in computing for all $y \in U$ the "ground state", i.e., the smallest eigenvalue $\lambda_1(y)$ and a corresponding eigenfunction $w_1(y)$ of (41). *This shall be understood from now on.*

4.2 Multilevel Finite Element Spaces in D

We discretize the space $V = H_0^1(D)$ by means of a dense, nested sequence of subspaces

$$\{0\} = V_{-1} \subset V_0 \subset V_1 \subset \ldots \subset V = H_0^1(D),$$

of finite dimensions $n_\ell = \dim V_\ell < \infty$. For instance, V_ℓ, $\ell \geq 0$ could consist of continuous functions, which are piecewise linear w.r.t. a mesh $\mathcal{M}_\ell$, $\ell \geq 1$ resulting from a uniform refinement of $\mathcal{M}_{\ell-1}$, with $\mathcal{M}_0$ being a given finite, regular, simplicial mesh. Other subspace sequences, such as spectral, isoparametric or p-version finite element discretizations could be considered as well. With respect to a basis $\{\phi_i^{(\ell)}\}_{i=1}^{n_\ell}$ for V_ℓ, the abstract Galerkin discretization (27) becomes a family of *parametric matrix eigenvalue problems*, the Galerkin projections of (41) onto V_ℓ: for $y \in U$, find $\lambda_\ell(y) \in \mathbb{R}$ and $0 \neq u_\ell(y) \in V_\ell$ such that

$$(\bar{a}\nabla u(y), \nabla\eta) + \sum_{m\geq 1} y_m(\psi_m\nabla u(y), \nabla\eta) = \lambda_\ell(y)(u(y), \eta) \quad \forall \eta \in V_\ell, \qquad (43)$$

where, for each $y \in U$ and $\ell = 0, 1, \ldots$, the pair $(\lambda_\ell(y), u_\ell(y)) \in \mathbb{R} \times V_\ell$ denotes the Galerkin approximation of the *first eigenpair* in V_ℓ.

The algebraic structure of (43) is particularly convenient for computation. Indeed, denoting by $\mathbf{A}_0^{(\ell)}$, $\mathbf{A}_m^{(\ell)}$, $m \geq 1$ the stiffness matrices and by $\mathbf{M}^{(\ell)}$ the mass matrix w.r.t. the basis $\{\phi_i^{(\ell)}\}_{i=1}^{n_\ell}$, given by

$$(\mathbf{A}_0^{(\ell)})_{ij} = (\bar{a}\nabla\phi_j^{(\ell)}, \nabla\phi_i^{(\ell)}), \quad (\mathbf{A}_m^{(\ell)})_{ij} = (\psi_m\nabla\phi_j^{(\ell)}, \nabla\phi_i^{(\ell)}), \quad \mathbf{M}_{ij}^{(\ell)} = (\phi_j^{(\ell)}, \phi_i^{(\ell)}),$$

the problem (43) reads *for any given* $y \in U$: find

$$(\lambda_\ell(y), \mathbf{u}_\ell(y)) \in \mathbb{R} \times \mathbb{R}^{n_\ell} \quad \text{s.t.} \quad (\mathbf{A}_0^{(\ell)} + \sum_{m\geq 1} y_m\mathbf{A}_m^{(\ell)})\mathbf{u}_\ell(y) = \lambda_\ell(y)\mathbf{M}^{(\ell)}\mathbf{u}_\ell(y).$$
$$(44)$$

Thus, the matrices need only be assembled once if a hierarchic basis is used, and once on each level of interest $\ell \in \mathbb{N}_0$ if standard hat functions are used. Note that the method presented below uses (44) only with finitely supported sequences $(y_m)_{m\geq 1}$.

Remark 7. For the ground state we know that the eigenvalue is separated and that the corresponding eigenvector can be chosen to be positive, see Remark 6, uniformly in the choice of the parameter by Lipschitz dependence of the eigenpair on the parameter. Therefore, we normalize the approximate parametric eigenvector $\mathbf{u}_\ell(y)$ at any given $y \in U$ by imposing the normalization $\mathbf{u}_\ell(y)^\top\mathbf{M}^{(\ell)}\mathbf{u}_\ell(y) = 1$ and positivity, $\int_D \sum_{i=1}^{n_\ell} \phi_i^{(\ell)}(x)(\mathbf{u}_\ell(y))_i \, dx \geq 0$.

We make the following assumption on the ansatz spaces V_ℓ, which is satisfied by the usual finite element ansatz spaces (see, e.g., [10]).

Assumption 8. *There exists a* $t_0 > 0$ *and* $C > 0$ *such that for all* $t \in [0, t_0]$:

$$\inf_{v_h \in V_\ell} ||v_h - v||_{H_0^1(D)} \leq C2^{-t\ell}||v||_W \qquad (45)$$

for all $\ell \in \mathbb{N}_0$ *and* $v \in W$. *Here,* $W = H_0^1(D) \cap H^{1+t}(D)$, *cf. Remark 4. In particular,* (45) *implies* (24).

4.3 Numerical Solution of the Parametric EVP

In this section we discuss the numerical solution of the parametric matrix eigenvalue problems (44) for fixed $y \in U$. By Assumption (35), for each $y \in U$ and $\ell \in \mathbb{N}_0$, the matrix $\mathbf{A}_0^{(\ell)} + \sum_{m \geq 1} y_m \mathbf{A}_m^{(\ell)}$ is symmetric positive definite. Setting $y_m = 0$, for all $m > M$, we conclude that the truncated matrix $\mathbf{A}_0^{(\ell)} + \sum_{m=1}^{M} y_m \mathbf{A}_m^{(\ell)}$ has the same property uniformly w.r.t. $M \in \mathbb{N}$. Hence, for any $y \in U$, $\ell \in \mathbb{N}_0$ and $M \in \mathbb{N}$, the truncated problem

$$\mathbf{u}_\ell^{(M)}(y) \in \mathbb{R}^{n_\ell}: \quad (\mathbf{A}_0^{(\ell)} + \sum_{m=1}^{M} y_m \mathbf{A}_m^{(\ell)}) \mathbf{u}_\ell^{(M)}(y) = \lambda_\ell^{(M)}(y) \mathbf{M}^{(\ell)} \mathbf{u}_\ell^{(M)}(y) \quad (46)$$

is a symmetric positive definite generalized matrix eigenvalue problem. By the same reasoning, the first eigenvalue of the truncated system (46) is simple, i.e., of single multiplicity, uniformly in $y \in U$ for all $\ell \geq 0$ sufficiently large, see Corollary 3. We identify $\mathbf{u}_\ell(y_1, y_2, \ldots, y_M, 0, \ldots) = \mathbf{u}_\ell^{(M)}(y_1, \ldots, y_M)$ and denote by $u_\ell(y) \in V_\ell$ the function

$$u_\ell(y) = \sum_{i=1}^{n_\ell} \phi_i^{(\ell)} (\mathbf{u}_\ell(y))_i.$$

Any numerical method for symmetric generalized eigenvalue problems applies to (46), see, e.g. [14, 25, 26, 42]. For our computations we use the JDBSYM library [19], which implements a variant of the Jacobi-Davidson method.

4.4 Abstract Error Bounds

Based on the abstract eigenvalue approximation theory in Sect. 3, we obtain the following a-priori error bounds, cf. [2, Sect. 8].

Proposition 2. *Consider the eigenvalue problem (36) (or in variational form (11)) with the operator $A(y)$ being the parametric diffusion operator given by (34). Assume (35), and let Assumption 7 hold. For some $m \geq 1$ let $(\lambda_m(y), w_m(y))$ be an eigenpair of the EVP (36) with eigenspace of multiplicity one for all $y \in U$, i.e., eigenvalue crossings are excluded.*

Then there exist $C > 0$ and $h_0 > 0$ such that the Galerkin eigenvalue approximation from the finite element space V_ℓ is quasi-optimal uniformly in $y \in U$: for every $y \in U$ and every $0 < h < h_0$

$$||w_m(y) - w_m^h(y)||_{H_0^1(D)} \leq C \epsilon_h(\lambda_m(y)) = C \inf_{v_h \in V_\ell} ||w_m(y) - v_h||_{H_0^1(D)}. \quad (47)$$

For the Galerkin eigenvalue approximations we have

$$|\lambda_m(y) - \lambda_m^h(y)| \leq C(\epsilon_h(\lambda_m(y)))^2 \leq C \inf_{v_h \in V_\ell} ||w_m(y) - v_h||_{H_0^1(D)}^2. \quad (48)$$

Using Remark 6, Assumption 8 and $\sup_{y \in U} ||w_m(y)||_W < \infty$ we obtain the following corollary.

Corollary 3. *For $h > 0$ sufficiently small, the spectral gap of the discretized problem, $\lambda_2^h(y) - \lambda_1^h(y)$ is strictly positive uniformly in $y \in U$. Moreover, for $0 \le t \le t_0$ as in Assumption 8, we have*

$$\sup_{y \in U} \|w_1(y) - w_1^h(y)\|_{H_0^1(D)} \le C h^t \quad and \quad \sup_{y \in U} |\lambda_1(y) - \lambda_1^h(y)| \le C h^{2t} \quad (49)$$

where $C < \infty$ is independent of h for sufficiently small $h > 0$.

5   Sparse Composite Collocation Method

In this section we introduce a generalization of the sparse grid collocation operator introduced in [6, 35]. We develop a formalism and approximation results, see e.g. Lemma 4, which in particular enable comparison with (multilevel) Monte Carlo methods.

5.1   Sparse Collocation Operator: Basic Properties

In this section we define the *sparse collocation operator* based on general *monotone* index sets, and investigate some of its properties such as *unisolvency*. We start by collecting the necessary definitions. Let L_n denote the univariate Lagrange polynomial of degree $n \in \mathbb{N}_0$ scaled such that $\int_{-1}^1 |L_n(t)|^2 \frac{dt}{2} = 1$. For $n \in \mathbb{N}_0$ let i_n denote the Lagrange interpolation operator of degree n which maps a continuous function $v \in C^0([-1, 1])$ to a polynomial $i_n v$ of degree n with $(i_n v)(z) = v(z)$ in the Gauss-Legendre points $z \in \{-1 < z_0^n < \ldots < z_n^n < 1 : L_{n+1}(z_k^n) = 0\}$. Denote by $\{w_k^n\}_{k=0}^n \subset (0, \infty)$ the corresponding Gauss-Legendre quadrature weights [18]. Define $j_n = i_n - i_{n-1}$, $n \ge 0$ where, by convention, $i_{-1} := 0$. Let $\mathscr{F} \subset \mathbb{N}_0^{\mathbb{N}}$ denote the collection of all multiindices $v \in \mathbb{N}_0^{\mathbb{N}}$ such that for each $v \in \mathscr{F}$ the *support* $\operatorname{supp} v = \{m \in \mathbb{N} : v_m \ne 0\}$ is finite. For a multiindex $v \in \mathscr{F}$ we denote by L_v the tensorized Legendre polynomial $L_v = L_{v_1} \otimes L_{v_2} \otimes \cdots$, that is $L_v(y) = \prod_{m \ge 1} L_{v_m}(y_m)$ for all $y \in U$. For a multiindex set $\Lambda \subset \mathscr{F}$ we denote by $\mathbb{P}_\Lambda$ the span of $\{L_v : v \in \Lambda\}$. Note that $\mathbb{P}_{\mathscr{F}}$ is dense in $L_\pi^2(U)$, where π is the uniform probability measure. Below we frequently use the notation $1_A = 1$ if A is true and $1_A = 0$ otherwise, as well as $\tilde{v} \le v$ iff $\tilde{v}_m \le v_m$ for all $m \ge 1$. For any finite multiindex set $\Lambda \subset \mathscr{F}$ we define the *sparse collocation operator*

$$I_\Lambda = \sum_{v \in \Lambda} \bigotimes_{m \ge 1} j_{v_m} \quad (50)$$

with the convention $I_\emptyset := 0$. Due to Λ being finite, for each of the finitely many $v \in \Lambda$ the tensorized operator $\bigotimes_{m \ge 1} j_{v_m}$ has only finitely many non-trivial increment factors j_{v_m}, $m \in \operatorname{supp} v$, while $j_{v_m} = j_0 = i_0$ for $m \notin \operatorname{supp} v$. This implies that

I_Λ is well defined on bounded continuous functions on U. The operator I_Λ has appeared in various variants, i.e., for specific choices of Λ in the literature, see [3, 4, 6, 7, 35, 36] and references therein. In particular, the generic definition (50) accommodates all formulas of [5].

Definition 1. A multiindex set $\Lambda \subseteq \mathscr{F}$ is called *monotone* if the following implication holds:

$$\forall \nu \in \Lambda : \quad \tilde{\nu} \in \mathscr{F} \wedge \tilde{\nu} \le \nu \Rightarrow \tilde{\nu} \in \Lambda.$$

In particular, $\{\}$, $\{0\}$ and $\mathscr{F}$ are monotone. Monotone multiindex sets Λ for I_Λ defined by (50) are of particular interest, as the following lemma shows.

Lemma 3. *Let $\Lambda \subset \mathscr{F}$ be finite and monotone. Then $I_\Lambda p = p$ for all $p \in \mathbb{P}_\Lambda$. Moreover, I_Λ is unisolvent on $\mathbb{P}_\Lambda$, i.e., $\forall p \in \mathbb{P}_\Lambda$: $I_\Lambda p = 0$ implies $p = 0$. Further, defining $\mathbb{Q}_\Lambda$ as the span of the monomials $\{U \ni y \mapsto y^\nu : \nu \in \Lambda\}$, we have $\dim \mathbb{P}_\Lambda = \#\Lambda = \dim \mathbb{Q}_\Lambda < \infty$ and moreover $\mathbb{Q}_\Lambda = \mathbb{P}_\Lambda$.*

Proof. Observe that $I_{\{\nu'\}} L_\nu = \bigotimes_{m \ge 1} j_{\nu'_m} L_{\nu_m} = 0$ if $\nu'_m > \nu_m$ for at least one $m \in \mathbb{N}$. This allows to write for $\nu \in \Lambda$

$$I_\Lambda L_\nu = \sum_{\nu' \le \nu} \bigotimes_{m \ge 1} j_{\nu'_m} L_{\nu_m} = \bigotimes_{m \ge 1} \sum_{\nu'_m = 0}^{\nu_m} j_{\nu'_m} L_{\nu_m} = \bigotimes_{m \ge 1} L_{\nu_m} = L_\nu, \qquad (51)$$

which shows the first assertion. Let now $p \in \mathbb{P}_\Lambda = \mathrm{span}\{L_\nu : \nu \in \Lambda\}$ s.t. $I_\Lambda p = 0$. By (51), linearity of I_Λ and linear independence of the set $\{L_\nu\}_{\nu \in \Lambda}$ imply $p = 0$. The statement $\dim \mathbb{P}_\Lambda = \#\Lambda = \dim \mathbb{Q}_\Lambda < \infty$ is obvious. Since Λ is monotone, we have $L_\nu \in \mathbb{Q}_\Lambda$, for all $\nu \in \Lambda$, and hence $\mathbb{P}_\Lambda \subseteq \mathbb{Q}_\Lambda$. Now $\dim \mathbb{P}_\Lambda = \dim \mathbb{Q}_\Lambda < \infty$ implies $\mathbb{P}_\Lambda = \mathbb{Q}_\Lambda$. $\qquad \square$

Note that in general, however, I_Λ is not interpolatory, as the cardinality of the set of collocation nodes in (50) may be larger than $\dim \mathbb{P}_\Lambda$, as the following example shows.

Example 3. Consider $\Lambda = \{\nu_0 = 0, \nu_1 = (1, 0, 0, \ldots), \nu_2 = (0, 1, 0, 0, \ldots)\}$ consisting of *three* multiindices. The corresponding sparse collocation operator I_Λ is then based on the following *five* collocation nodes: $z_0^{\nu_0} = 0$, $z_-^{\nu_1} = (-\frac{1}{\sqrt{3}}, 0, \ldots)$, $z_+^{\nu_1} = (+\frac{1}{\sqrt{3}}, 0, \ldots)$, $z_-^{\nu_2} = (0, -\frac{1}{\sqrt{3}}, 0, \ldots)$ and $z_+^{\nu_2} = (0, +\frac{1}{\sqrt{3}}, 0, \ldots)$.

The following observation will be useful: for any monotone $\Lambda \subset \mathscr{F}$ we have

$$\forall \nu \in \Lambda : \quad \prod_{m \ge 1} (\nu_m + 1) = \#\{\tilde{\nu} \in \Lambda : \tilde{\nu} \le \nu\}. \qquad (52)$$

Lemma 4 (Number of collocation points). *Let $\Lambda \subset \mathscr{F}$ be monotone and finite. Then the number collocation points in I_Λ is at most $(\#\Lambda)^2$.*

Proof. We can write I_Λ as $I_\Lambda = \sum_{v \in \Lambda} c_v^\Lambda \bigotimes_{m \geq 1} i_{v_m}$, where each term requires $\prod_{m \geq 1}(v_m + 1)$ collocation points by definition of i_n, $n \geq 0$. The number of collocation points in I_Λ can therefore be estimated as

$$\sum_{v \in \Lambda} \prod_{m \geq 1}(v_m + 1) \overset{(52)}{=} \sum_{v \in \Lambda} \#\{\tilde{v} \in \Lambda : \tilde{v} \leq v\} \leq \sum_{v \in \Lambda} \#\Lambda = (\#\Lambda)^2.$$

$\square$

Remark 8. Note that the upper bound $(\#\Lambda)^2$ is independent of the "effective dimension" $\#\bigcup_{v \in \Lambda} \operatorname{supp} v$. Moreover, the exponent 2 in the upper bound $(\#\Lambda)^2$ on the number of collocation points in I_{Λ_ℓ} is sharp, as can be seen from the sequence

$$\Lambda_\ell = \{(v_1, \ldots, v_M, 0 \ldots) \in \mathcal{F} : \|v\|_{\ell^1(\mathbb{N})} \leq \ell\}, \quad \ell = 0, 1, \ldots$$

where $M \geq 1$ is fixed. Indeed, by [6, Sect. 6.2], we have $\#\Lambda_\ell = \binom{M + \ell}{M}$ while the number of collocation points in I_{Λ_ℓ} is given by $N_{\Lambda_\ell} = \binom{2M + \ell}{2M}$. This yields

$$N_{\Lambda_\ell} \geq \left(\frac{2M + \ell}{2M}\right)^{2M} \gtrsim (2M + \ell)^{2M} \geq (M + \ell)^{2M} \geq \left(\frac{(M + \ell)^M}{M!}\right)^2 \geq (\#\Lambda_\ell)^2$$

with the implied constant independent of $\ell \geq 0$. Thus, $N_{\Lambda_\ell} \sim (\#\Lambda_\ell)^2$ as $\ell \to \infty$.

Remark 9. For $\ell \geq 0$ let $n_\ell^\star := 1_{(\ell \geq 1)} 2^\ell$. For a monotone and finite $\Lambda \subset \mathcal{F}$ define $\Lambda^\star := \bigcup_{v \in \Lambda}\{\tilde{v} \in \mathcal{F} : \forall m \geq 1 : \tilde{v}_m \leq \min\{n_\ell^\star \geq v_m : \ell \geq 0\}\}$. For each $\ell \geq 0$, $n_\ell^\star + 1$ is the number of nodes in the Clenshaw-Curtis quadrature rule of order $n_\ell^\star + 1$. The nodes of these Clenshaw-Curtis quadrature rules are *nested*, cf. [35]. Thus, if we defined i_n to be the Lagrange interpolation operator based on the nodes of the Clenshaw-Curtis quadrature rule of order $\min\{n_\ell^\star \geq n : \ell \geq 0\} + 1$, for monotone and finite Λ the number of collocation points in $I_{\Lambda^\star}$ would be exactly $\#\Lambda^\star = \dim \mathbb{P}_{\Lambda^\star}$, as opposed to the possibly quadratic growth in Remark 8.

The following is a characterization of I_Λ for monotone multiindex sets $\Lambda \subset \mathcal{F}$ by an expansion into tensorized multivariate Legendre polynomials.

Lemma 5. *Let $\Lambda \subset \mathcal{F}$ be monotone and finite, and assume $c_v^\Lambda \in \mathbb{R}$, $v \in \Lambda$ are such that $I_\Lambda = \sum_{v \in \Lambda} c_v^\Lambda \bigotimes_{m \geq 1} i_{v_m}$. Then for $f \in C^0(\overline{U}; \mathbb{R})$ we have $I_\Lambda f = \sum_{v' \in \Lambda} d_{v'}^\Lambda(f) L_{v'}$, where $d_{v'}^\Lambda(f)$, $v' \in \Lambda$, is defined by*

$$d_{v'}^\Lambda(f) = \sum_{\substack{v \in \Lambda \\ v \geq v'}} c_v^\Lambda \sum_{\eta \leq v} w_\eta^v L_{v'}(z_\eta^v) f(z_\eta^v),$$

where $w_\eta^v = \prod_{m \geq 1} w_{\eta_m}^{v_m} \in \mathbb{R}$ and $z_\eta^v = (z_{\eta_1}^{v_1}, z_{\eta_2}^{v_2}, \ldots) \in U$, $\eta \leq v \in \Lambda$.

Proof. We have the univariate formula $i_n v = \sum_{n' \leq n} d_{n'}(v) L_{n'}$ with coefficients $d_{n'}(v) = \sum_{k \leq n'} w_k^n L_{n'}(z_k^n) v(z_k^n)$ as can be immediately checked using the fact that the Gauss-Legendre quadrature formula $p \mapsto \sum_{k \leq n} w_k^n p(z_k^n)$ integrates polynomials $p = L_{n'} L_{n''}$ of degree at most $2n+1$ exactly [18, (1.4.7) and (1.4.14)]. Thus we obtain

$$I_\Lambda f = \sum_{v \in \Lambda} c_v^\Lambda \left(\bigotimes_{m \geq 1} i_{v_m} \right) f = \sum_{v \in \Lambda} c_v^\Lambda \sum_{v' \leq v} L_{v'} \sum_{\eta \leq v} w_\eta^v L_{v'}(z_\eta^v) f(z_\eta^v),$$

which, after exchanging the summation, yields the claim. $\qquad\square$

5.2 Multiindex Sets $\Lambda(\mu, \varepsilon)$

In what follows, we specialize our considerations to multiindex sets Λ of a particular structure, motivated by a-priori estimates obtained in [12] for coefficients $a(x, y)$ satisfying Assumption 7. To describe the structure of these sets, let c_0 denote the collection of non-increasing sequences of reals less than one and tending to zero:

$$c_0 = \{\mu = (\mu_1, \mu_2, \ldots) \in [0, 1)^{\mathbb{N}} : 1 > \mu_1 \geq \mu_2 \geq \cdots \quad \text{and} \quad \lim_{m \to \infty} \mu_m = 0\} .$$

For $\mu \in c_0$, $v \in \mathcal{F}$ we write $\mu^v = \prod_{m \in \mathbb{N}} \mu_m^{v_m}$ (with $0^0 := 1$) and for $\varepsilon > 0$ we define

$$\Lambda(\mu, \varepsilon) = \{v \in \mathcal{F} : \mu^v \geq \varepsilon\}. \tag{53}$$

Clearly, for any $\mu \in c_0$ and any $\varepsilon > 0$ the multiindex set $\Lambda(\mu, \varepsilon)$ is finite and monotone. The multiindex sets $\Lambda(\mu, \varepsilon)$ were introduced in [9] and investigated in the context of the stochastic Galerkin method for elliptic stochastic PDEs in [8]. The next result give precise asymptotics for the cardinality of the sets $\Lambda(\mu, \varepsilon)$ defined in (53) for a sequence $\mu \in c_0$ which models algebraic decay of terms in the expansions given in (12) and (37).

Lemma 6. *For $\mu \in c_0$ given by $\mu_m = (1 + m)^{-\sigma}$ with $\sigma > 1$,*

1. As $\varepsilon \to 0$, the cardinality of the set $\Lambda(\mu, \varepsilon)$ in (53) equals

$$\#\Lambda(\mu, \varepsilon) = F(\varepsilon^{-1/\sigma}) \quad \text{where} \quad F(x) = x \frac{e^{2\sqrt{\log x}}}{2\sqrt{\pi}(\log x)^{3/4}}(1 + \mathcal{O}(1/\log x))$$

2. We have $\sup_{\varepsilon > 0} \sum_{v \in \Lambda(\mu, \varepsilon)} (\mu^v)^p < \infty$ if and only if $p > 1/\sigma$.

Here and throughout, the function log *always denotes the natural logarithm.*

Proof. See [8, Prop. 4.5] for the proof of the first claim. The second is a special case of [13, Lemma 7.1]. We provide a direct argument for completeness here. Let $0 < \delta < p\sigma - 1$, $c(\delta) > 0$ be such that $F(x) \leq c(\delta)x^{1+\delta}$ for $x > 1$. Take $\gamma = \sigma/(p\sigma - 1 - \delta)$ and let I_n be the interval $I_n = [2^{-\gamma n}, 2^{-\gamma(n-1)})$. Note that

$$\#\{\nu \in \mathscr{F} : \mu^\nu \in I_n\} \leq \#\Lambda(\mu, 2^{-\gamma n}) = F(2^{\gamma n/\sigma}) \leq c(\delta)2^{\gamma n/\sigma(1+\delta)}.$$

We compute

$$\sup_{\varepsilon>0} \sum_{\nu\in\Lambda(\mu,\varepsilon)} (\mu^\nu)^p = \lim_{\varepsilon\to 0} \sum_{\nu\in\Lambda(\mu,\varepsilon)} (\mu^\nu)^p = \sum_{n\geq 1} \sum_{\substack{\nu\in\mathscr{F}\\ \mu^\nu\in I_n}} (\mu^\nu)^p$$

$$\leq \sum_{n\geq 1} c(\delta)2^{\gamma n/\sigma(1+\delta)}2^{-p\gamma(n-1)} = c(\delta)2^{p\gamma} \sum_{n\geq 1} 2^{-\gamma n(p-(1+\delta)/\sigma)},$$

which equals $c(\delta)2^{p\gamma} < \infty$. For $0 < p \leq 1/\sigma$, the sum $\sum_{m\geq 1} \mu_m^p$ obviously diverges. □

In Example 3 we showed that, in general, $\#\Lambda > \dim \mathbb{P}_\Lambda$, i.e., the number of deterministic problems to be solved for the determination of I_Λ is larger than the number of monomials in $\mathbb{P}_\Lambda$ which determine the precision of I_Λ, cf. (56). To facilitate comparison with Monte-Carlo methods, we quantify the convergence of I_Λ in terms of the number of deterministic problems to be solved. To this end, we bound $\dim \mathbb{P}_\Lambda = \#\Lambda$ for several classes of monotone index sets Λ.

Definition 2. For $\mu \in c_0$ and $0 < \varepsilon \leq 1$, define

$$\mathscr{B}(\mu,\varepsilon) := \max\{m \geq 1 : \mu_m \geq \varepsilon\} \sum_{\nu\in\Lambda(\mu,\varepsilon)} 4^{\#\operatorname{supp}\nu} \prod_{m\in\operatorname{supp}\nu} \frac{1+\mu_m}{1-\mu_m} \qquad (54)$$

where $\Lambda(\mu, \varepsilon)$ is as in (53) and

$$\varkappa^\star(\mu) := \inf\{\varkappa > 0 : \sup_{0<\varepsilon\leq 1} \varepsilon^\varkappa \mathscr{B}(\mu,\varepsilon) < \infty\},$$

which may be infinite. We refer to $\varkappa^\star(\mu)$ as *asymptotic overhead order* of $\mu \in c_0$.

The class of sequences $\mu \in c_0$ which have finite asymptotic overhead order $\varkappa^\star(\mu) < \infty$ includes some important families, as we show in the following.

Lemma 7 (Asymptotic overhead order for algebraic decay). *For the model sequence $\mu_m = (1 + m)^{-\sigma}$ with algebraic decay with fixed order $\sigma > 1$ the asymptotic overhead order $\varkappa^\star(\mu)$ is bounded by $\varkappa^\star(\mu) \leq 2(1 + \log 4)/\sigma$.*

Proof. Let $0 < \varepsilon \leq \mu_1$. Clearly $\max\{m \geq 1 : \mu_m \geq \varepsilon\} \leq \varepsilon^{-1/\sigma}$. Using [8, Lemma 4.8] we therefore have $\#\operatorname{supp}\nu \leq 2\log\left(\varepsilon^{-1/\sigma}\right)$ if μ^ν is sufficiently small for $\nu \in \Lambda(\mu, \varepsilon)$ to hold. Thus, for $\nu \in \Lambda(\mu, \varepsilon)$ and any fixed $\delta > 0$ we obtain

$$4^{\# \, \text{supp} \, \nu} \prod_{m \in \text{supp} \, \nu} \frac{1 + \mu_m}{1 - \mu_m} \lesssim (4 + \delta)^{-2/\sigma \log \varepsilon} = \varepsilon^{-2 \log(4+\delta)/\sigma}$$

μ^{ν} is sufficiently small and trivially the same bound otherwise. From Lemma 6 we have $\# \Lambda(\mu, \varepsilon) \lesssim \varepsilon^{-(1+\delta)/\sigma}$, and thus, for $\varepsilon \to 0$

$$\mathcal{B}(\mu, \varepsilon) \leq \max\{m \geq 1 : \mu_m \geq \varepsilon\} \# \Lambda(\mu, \varepsilon) \max_{\nu \in \Lambda(\mu, \varepsilon)} 4^{\# \, \text{supp} \, \nu} \prod_{m \in \text{supp} \, \nu} \frac{1 + \mu_m}{1 - \mu_m} \lesssim \varepsilon^{-\varkappa_\delta}$$

with $\varkappa_\delta = 1/\sigma + (1 + \delta)/\sigma + 2 \log(4 + \delta)/\sigma$. Thus $\varkappa^{\star}(\mu) \leq \varkappa_\delta$. Since $\delta > 0$ was arbitrary, the claim follows. $\qquad\qquad\square$

Lemma 8 (Isotropic sparse tensor product asymptotic overhead order). *Assume $\mu \in c_0$ is of the form $\mu_1 = \mu_2 = \cdots = \mu_M = \mu_0 > 0$, $\mu_m = 0$ for $m > M$, where $M \in \mathbb{N}$ is fixed. Then the asymptotic overhead order of μ is $\varkappa^{\star}(\mu) = 0$.*

Proof. Take any $\varkappa > 0$. With $L = \log \varepsilon / \log \mu_0$ (for simplicity an integer) we have [45]

$$\# \Lambda(\mu, \varepsilon) = \binom{M + L}{M} \leq \frac{(M + L)^M}{M!} \lesssim (-\log \varepsilon)^M \lesssim \varepsilon^{-\varkappa} \quad \text{for} \quad \varepsilon \to 0$$

and the other terms in (54) are bounded, independently of $\varepsilon > 0$. $\qquad\qquad\square$

Lemma 7 and Lemma 8 give rise to an even larger family of sequences μ for which the asymptotic overhead order of μ can be estimated.

Proposition 3. *Let $\mu^{(1)} \in c_0$ for which the asymptotic overhead order of $\mu^{(1)}$ is finite, i.e., $\varkappa := \varkappa^{\star}(\mu^{(1)}) < \infty$. Let $M \geq 1$ and $\mu \in c_0$ be a sequence with $\mu_m^{(1)} \geq \mu_m$ for $m > M$. Then we have $\varkappa^{\star}(\mu) \leq \varkappa^{\star}(\mu^{(1)})$.*

Proof. We assume w.l.o.g. that $0 < \varepsilon \leq \min\{\mu_1, \mu_1^{(1)}\}$. Define $\mu^{(0)} \in c_0$ by $\mu_m^{(0)} = 1_{(m \leq M)} \mu_1$, $m \geq 1$. Let H_ε be the map $H_\varepsilon : \Lambda(\mu^{(0)}, \varepsilon) \times \Lambda(\mu^{(1)}, \varepsilon) \to \Lambda(\mu, \varepsilon)$ given by

$$h = (\nu^{(0)}, \nu^{(1)}) \mapsto H_\varepsilon(h) = 1_{(\mu^\nu \geq \varepsilon)} \nu \quad \text{with} \quad \nu_m = \begin{cases} \nu_m^{(0)} & m \leq M \\ \nu_m^{(1)} & m > M. \end{cases}$$

Observe that H_ε is well-defined and surjective, and thus

$$\# \Lambda(\mu, \varepsilon) \leq \# \Lambda(\mu^{(0)}, \varepsilon) \# \Lambda(\mu^{(1)}, \varepsilon)$$

due to $\varepsilon < 1$. Noticing further that

$$\max\{m \geq 1 : \mu_m \geq \varepsilon\} \leq \max\{m \geq 1 : \mu_m^{(0)} \geq \varepsilon\} \max\{m \geq 1 : \mu_m^{(1)} \geq \varepsilon\}$$

and for all $h = (v^{(0)}, v^{(1)}) \in H_\varepsilon^{-1}(\Lambda(\mu, \varepsilon))$ we have

$$\text{supp } H_\varepsilon(h) \subseteq \text{supp } v^{(0)} \cup (\text{supp } v^{(1)} \cap \{m \in \mathbb{N} : m > M\})$$

and thus also

$$\prod_{m \in \text{supp } H_\varepsilon(h)} \frac{1 + \mu_m}{1 - \mu_m} \leq \left(\prod_{m \in \text{supp } v^{(0)}} \frac{1 + \mu_m}{1 - \mu_m} \right) \left(\prod_{\substack{m \in \text{supp } v^{(1)} \\ m > M}} \frac{1 + \mu_m}{1 - \mu_m} \right)$$

$$\leq \left(\prod_{m \in \text{supp } v^{(0)}} \frac{1 + \mu_m^{(0)}}{1 - \mu_m^{(0)}} \right) \left(\prod_{m \in \text{supp } v^{(1)}} \frac{1 + \mu_m^{(1)}}{1 - \mu_m^{(1)}} \right),$$

we conclude from Lemma 8 that $\mathcal{B}(\mu) \leq \mathcal{B}(\mu^{(0)}) \mathcal{B}(\mu^{(1)}) \lesssim \varepsilon^{-\delta} \varepsilon^{-\varkappa}$ as $\varepsilon \to 0$ for any $\delta > 0$. This shows the claim. $\qquad \square$

5.3   Sparse Collocation Operator: Approximation Properties

For complex-analytic functions on product domains, e.g. as described in Theorem 4, we obtain the following theorem on the approximation property of I_Λ for multiindex sets Λ of type (53). We will need the following elementary inequality [12, (3.13)]: for all $0 < p \leq q \leq \infty$, $M \geq 1$ and $\mu \in c_0 \cap \ell^p(\mathbb{N})$,

$$\sum_{m > M} \mu_m^q \leq M^{1 - \frac{q}{p}} ||\mu||_{\ell^p(\mathbb{N})}^q. \tag{55}$$

Theorem 6. *Let $\{\rho_m\}_{m \geq 1}$ be a sequence with $\rho_m > 1$ and $\rho_m \to \infty$. For each $m \geq 1$ let $\mathcal{E}_{\rho_m}$ denote the ellipse in $\mathbb{C}$ with sum of semiaxes ρ_m (as in [15, p. 312]). Define $\mathcal{E} := \prod_{m \geq 1} \mathcal{E}_{\rho_m}$. Let $v : \mathcal{E} \to \mathbb{C}$ be a jointly complex-analytic function. Define $\mu \in c_0$ by $\mu_m = \sup_{m' \geq m} \frac{1}{\rho_{m'}}$, $m \in \mathbb{N}$. Assume $\sup_{m \geq 1} \mu_m m^\sigma < \infty$ for some $\sigma > 0$ and that $\varkappa^\star(\mu) < \infty$. Take $\varkappa > \varkappa^\star(\mu)$. Then there exists $C > 0$ such that*

$$||I_{\Lambda(\mu, \varepsilon)} v - v||_{L_\pi^2(U; \mathbb{C})} \leq C \left(\varepsilon^{1 - \varkappa} + \varepsilon^{1 - 1/\sigma} \right) ||v||_{C^0(\overline{\mathcal{E}}; \mathbb{C})} \tag{56}$$

for all $0 < \varepsilon \leq \mu_1$, where π is the uniform probability measure on U.

Proof. As in [6, 36] we use the following approximation property of the univariate interpolation operator i_n for complex-analytic functions f on $\mathcal{E}_{\rho_m}$:

$$||f - i_n f||_{L^2((-1,1); \frac{dt}{2})} \leq C(\rho_m) \mu_m^n ||f||_{C^0(\overline{\mathcal{E}_{\rho_m}})}, \quad n = 0, 1, \ldots$$

where $C(\rho_m) = \frac{4}{\rho_m - 1} \lesssim \mu_m \to 0$ as $m \to \infty$. In particular, for $n = 1, 2, \ldots$,

$$\|j_n f\|_{L^2((-1,1);\frac{dt}{2})} \leq C(\rho_m)\mu_m^n \left(1 + \frac{1}{\mu_m}\right)\|f\|_{C^0(\overline{\mathscr{E}_{\rho_m}})} \tag{57}$$

and there exists $C < \infty$ with $\prod_{m \in \mathrm{supp}\, \nu} C(\rho_m) \leq C$ for any $\nu \in \mathscr{F}$. Clearly,

$$\|j_0 f\|_{L^2((-1,1);\frac{dt}{2})} \leq \|f\|_{C^0([-1,1])}. \tag{58}$$

Another observation used in the following is this: for $0 < \varepsilon^* \leq \varepsilon \leq 1$ setting $n_0 = \max\{n \geq 0 : \mu_m^n \geq \varepsilon^*\}$ yields $\mu_m^{n_0+1} < \varepsilon^*$, and thus for $m \geq 1$

$$\left\| f - \sum_{n \geq 0} \mathbf{1}_{(\mu_m^n \geq \varepsilon^*)} j_n f \right\|_{L^2((-1,1);\frac{dt}{2})} \leq C(\rho_m)\varepsilon^* \mu_m^{-1}\|f\|_{C^0(\overline{\mathscr{E}_{\rho_m}})}. \tag{59}$$

In the remaining part of the proof we assume w.l.o.g. that $\|v\|_{C^0(\overline{\mathscr{E}};\mathbb{C})} \leq 1$. Observe that (55) with $q = 1$ and $p = 1/\sigma$ yields

$$\|\mathrm{Id}^M \otimes (I_{\{0\}} - \mathrm{Id}^{M*})v\|_{L^2_\pi(U;\mathbb{C})} \leq \sum_{m > M} C(\rho_m) \leq C \sum_{m > M} \mu_m$$

$$\leq C\|\mu\|_{\ell^{1/\sigma}(\mathbb{N})} M^{1-\sigma}$$

where Id^m is the identity on m-variate functions and $\mathrm{Id} =: \mathrm{Id}^m \otimes \mathrm{Id}^{m*}$, for all $M \geq 1$, in particular for $M := \max\{m \geq 0 : \mu_m \geq \varepsilon\} \lesssim \varepsilon^{-1/\sigma}$ as $\varepsilon \to 0$. Moreover, for any $m \in \mathbb{N}$, fixing $z_j \in \mathscr{E}_{\rho_j}$, $j \neq m$, the bound $\|v\|_{C^0(\overline{\mathscr{E}};\mathbb{C})} \leq 1$ implies $\|z_m \mapsto v(z_1, z_2, \ldots)\|_{C^0(\overline{\mathscr{E}_{\rho_m}};\mathbb{C})} \leq 1$. Thus, by stability (58), the claim (56) follows from

$$I_{\Lambda(\mu,\varepsilon)} - \mathrm{Id} = J_{\Lambda(\mu,\varepsilon)} \otimes I_{\{0\}} + \mathrm{Id}^M \otimes (I_{\{0\}} - \mathrm{Id}^{M*}),$$

where $J_{\Lambda(\mu,\varepsilon)} = \sum_{\nu \in \Lambda(\mu,\varepsilon)} \bigotimes_{m=1}^M j_{\nu_m}$, once $\|J_{\Lambda(\mu,\varepsilon)}\|_{L^2_\pi(U;\mathbb{C})} \lesssim \varepsilon^{1-\varkappa}$ has been established. To do so, we now write, similarly to [6,45], using $\mu^* \in c_0$, $\mu_m^* = \mu_{m+1}$, $m \geq 1$ for any $0 < \varepsilon \leq 1$

$$J_{\Lambda(\mu,\varepsilon)} = \sum_{\nu^* \in \Lambda(\mu^*,\varepsilon)} \left(\sum_{n \geq 0} \mathbf{1}_{(\mu_1^n \geq \varepsilon/(\mu^*)^{\nu^*})} j_n\right) \otimes j_{\mu_1^*} \otimes \cdots \otimes j_{\mu_{M-1}^*}$$

and use the decomposition

$$J_{\Lambda(\mu,\varepsilon)} - \mathrm{Id}^M = \mathrm{Id}^1 \otimes (J_{\Lambda(\mu^*,\varepsilon)} - \mathrm{Id}^{M-1}) - J^*_{\Lambda(\mu,\varepsilon)}, \tag{60}$$

where

$$
J_{\Lambda(\mu,\varepsilon)}^{\star} = \sum_{v^{\star}\in\Lambda(\mu^{\star},\varepsilon)} \left(\mathrm{Id}^1 - \sum_{n\geq 0} 1_{(\mu_1^n \geq \varepsilon/(\mu^{\star})^{v^{\star}})} j_n \right) \otimes j_{\mu_1^{\star}} \otimes \cdots \otimes j_{\mu_{M-1}^{\star}}. \tag{61}
$$

For v as in the statement of the claim and $\varepsilon > 0$ we now estimate using (59) and (57)

$$
\begin{aligned}
||J_{\Lambda(\mu,\varepsilon)}^{\star}v||_{L_{\pi}^2(U;\mathbb{C})} &\leq \sum_{v^{\star}\in\Lambda(\mu^{\star},\varepsilon)} C(\rho_1)\frac{\varepsilon\mu_1^{-1}}{(\mu^{\star})^{v^{\star}}} \prod_{m\in\mathrm{supp}\,v^{\star}} C(\rho_{m+1})(\mu_m^{\star})^{v_m^{\star}}\left(1+\frac{1}{\mu_m^{\star}}\right) \\
&\leq \varepsilon C(\rho_1)\mu_1^{-1} \sum_{v^{\star}\in\Lambda(\mu^{\star},\varepsilon)} \prod_{m\in\mathrm{supp}\,v^{\star}} C(\rho_{m+1})\left(1+\frac{1}{\mu_{m+1}}\right) \\
&\leq \varepsilon \sum_{v\in\Lambda(\mu,\varepsilon)} \prod_{m\in\mathrm{supp}\,v} 4\frac{1+\mu_m}{1-\mu_m}.
\end{aligned}
$$

Repeating this argument using the decomposition (60) we obtain (56). $\qquad\square$

Remark 10. The preceding result remains true for Banach space valued complex-analytic functions $v : \mathcal{E} \to X$.

The bound (56) implies convergence $I_{\Lambda(\mu,\varepsilon)}v \to v$ in $L_{\pi}^2(U)$ as $\varepsilon \to 0$ with a rate $r = \min\{1-\varkappa, 1-1/\sigma\}$ if $\mu \in c_0$ has asymptotic overhead order $\varkappa^{\star}(\mu) < \varkappa < 1$. Examples include sequences $\mu \in c_0$ with $\mu_m \sim m^{-\sigma}$ for a sufficiently large $\sigma > 0$, in which case, using Lemma 6 we obtain $||I_{\Lambda(\mu,\varepsilon)}v - v||_{L_{\pi}^2(U;\mathbb{C})} \lesssim \varepsilon^r \lesssim (\#\Lambda(\mu,\varepsilon))^{-r\sigma}$ as $\varepsilon \to 0$. *In particular, in this case the so-called curse of dimension does not appear.* The "effective dimension" $\#\bigcup_{v\in\Lambda(\mu,\varepsilon)} \mathrm{supp}\,v$, however, appears in the computation cost of each sample, and thus the problem (of, e.g. computing the average $\int_U v(y)d\pi(y)$) is *tractable* in the sense of [45].

Remark 11. For a particular choice of $\mu \in c_0$ (as described below) and $\varepsilon > 0$ the multiindex set $\Lambda(\mu,\varepsilon)$ can be identified with the index set suggested in [35],

$$
X_\alpha(w, N) = \{\iota \in \mathbb{N}_+^N : \sum_{n=1}^N (\iota_n - 1)\alpha_n \leq w \min_{1\leq n\leq N}\alpha_n\}
$$

where $w \in \mathbb{R}$, $N \in \mathbb{N}$, $\alpha \in \mathbb{R}_+^N$. Indeed, assuming without loss of generality that α is increasing and setting $\mu_m = e^{-c\alpha_m}$, $\varepsilon = \mu_M$ for $m = 1,\ldots,M$, $\mu_m = 0$ for $m > M$ for a suitable $c > 0$ and $w = \log\varepsilon/\log\mu_1$ it is easy to check that $v \in \Lambda(\mu,\varepsilon)$ iff $i \in X_\alpha(w, M)$, where we identify $v_m = \iota_m - 1$, $m = 1,\ldots,M$.

5.4   Sparse Composite Collocation Operator

We now compose the multilevel finite element discretization of the eigenvalue problem from Sect. 4.2 with the sparse collocation operator (50) based on hierarchic multiindex sets of the form (53) in a sparse way. For a given $\mu \in c_0$ let $(\varepsilon_j)_{j\geq 0} \in c_0$ be a sequence of *thresholds*. With $(\varepsilon_j)_{j\geq 0}$ we associate a sequence of *nested* multiindex sets $\Lambda_j = \Lambda(\mu, \varepsilon_j)$, $\Lambda_{-1} := \emptyset$. For $v \in \mathscr{F}$ we define $k(v) := \inf\{k \geq 0 : v \in \Lambda_k\}$, which may be infinite. We consider the following *sparse (tensor) composite collocation operator*, proposed for isotropic collocation in [6] (cf. also [6, Remark 6.2.5]):

$$\lambda \mapsto \hat{\lambda}_L := \sum_{0\leq k+\ell\leq L} (I_{\Lambda_k} - I_{\Lambda_{k-1}})(\lambda_\ell - \lambda_{\ell-1}) \tag{62}$$

and

$$u \mapsto \hat{u}_L := \sum_{0\leq k+\ell\leq L} (I_{\Lambda_k} - I_{\Lambda_{k-1}})(u_\ell - u_{\ell-1}). \tag{63}$$

Recall from Sect. 4.3 that (λ, u) is the "ground state" and (λ_ℓ, u_ℓ) its Galerkin approximation in the finite element space V_ℓ.

We obtain the following result on the approximation property of the sparse composite collocation operator $u \mapsto \hat{u}_L$.

Theorem 7. *Let $W \subset H_0^1(D)$ and $t \geq 0$ be as in Assumption 8. Let $D \subset \mathbb{R}^d$ be convex according to Remark 4. Let $\{\rho_m\}_{m\geq 1}$ be a sequence with $\rho_m > 1$ and $\rho_m \to \infty$ as $m \to \infty$. Assume that $u : \mathscr{E} = \prod_{m\geq 1} \mathscr{E}_{\rho_m} \to W$ is jointly complex-analytic. Define μ_m, $m \geq 1$ as in Theorem 6. Assume that $\mu_m m^\sigma \to 0$ as $m \to \infty$ for some $\sigma > 0$. If $\sigma > 0$ is sufficiently large there exist $\varkappa < 1$ and $C > 0$ such that for $L \geq 0$*

$$\|u - \hat{u}_L\|_{L_\pi^2(U;H_0^1(D))} \leq C \left(\sum_{k=0}^{L} \varepsilon_k^{1-\varkappa} 2^{-t(L-k)} + \varepsilon_L^{1-\varkappa} \right) \|u\|_{C^0(\overline{\mathscr{E}};W)}. \tag{64}$$

Proof. If $\sigma > 0$ is sufficiently large, by Prop. 3 and Lemma 7 the asymptotic overhead order of μ satisfies $\varkappa^\star(\mu) < 1$. The rest follows as in [7, Sect. 6.3] using Theorem 6 and Corollary 3. $\qquad\square$

Corollary 4. *Assume $\varkappa^\star(\mu) < \varkappa < 1$. Set $\varepsilon_k := 2^{-tk/(1-\varkappa)}$, $k \in \mathbb{N}$. Then*

$$\|u - \hat{u}_L\|_{L_\pi^2(U;H_0^1(D))} \lesssim L 2^{-tL} \|u\|_{C^0(\overline{\mathscr{E}};W)}. \tag{65}$$

Next, we estimate the computational effort of the approximation $u \mapsto \hat{u}_L$. As in [6, 35, 45] for computational purposes we rewrite (62) in terms of the interpolation operators $\bigotimes_{m\geq 1} i_{v_m}$. We will need the following lemma.

Lemma 9. *We have for $L \geq 0$*

$$\hat{\lambda}_L = \sum_{v \in \Lambda_L} I_{\{v\}} \lambda_{L-k(v)} \quad and \quad \hat{u}_L = \sum_{v \in \Lambda_L} I_{\{v\}} u_{L-k(v)}.$$

Proof. For $\hat{\lambda}_L$, $L \geq 0$ we compute

$$\hat{\lambda}_L = \sum_{\ell=0}^{L} I_{\Lambda_{L-\ell}}(\lambda_\ell - \lambda_{\ell-1}) = \sum_{k=0}^{L} I_{\Lambda_k}(\lambda_{L-k} - \lambda_{L-(k+1)})$$

$$= \sum_{k=0}^{L} \sum_{v \in \Lambda_k} I_{\{v\}}(\lambda_{L-k} - \lambda_{L-(k+1)}) = \sum_{k=0}^{L} \sum_{v \in \Lambda_L} 1_{(k(v) \leq k)} I_{\{v\}}(\lambda_{L-k} - \lambda_{L-(k+1)}),$$

and thus

$$\hat{\lambda}_L = \sum_{v \in \Lambda_L} I_{\{v\}} \sum_{k=0}^{L} 1_{k(v) \leq k}(\lambda_{L-k} - \lambda_{L-(k+1)})$$

$$= \sum_{v \in \Lambda_L} I_{\{v\}} \sum_{k=k(v)}^{L} (\lambda_{L-k} - \lambda_{L-(k+1)}) = \sum_{v \in \Lambda_L} I_{\{v\}} \lambda_{L-k(v)},$$

and similarly for $\hat{u}_L$. $\qquad\qquad\qquad\qquad\qquad\qquad\qquad\qquad\qquad\qquad\qquad\quad\square$

Note that the non-composite case $I_{\Lambda_L} \lambda_L$ corresponds to setting $k(\cdot) = 0$ in the above. Using Lemma 9, the definition of $I_{\{v\}}$, and factoring out the tensor product $(i_{v_1} - i_{v_1-1}) \otimes (i_{v_2} - i_{v_2-1}) \otimes \cdots$ we obtain:

$$\hat{\lambda}_L = \sum_{v \in \Lambda_L} \bigotimes_{m \geq 1} (i_{v_m} - i_{v_m-1}) \lambda_{L-k(v)}$$

$$= \sum_{v \in \Lambda_L} \sum_{\substack{\eta \in \{0,1\}^{\mathbb{N}} \\ \text{supp}\,\eta \subseteq \text{supp}\,v}} (-1)^{||\eta||_{\ell^1(\mathbb{N})}} \bigotimes_{m \geq 1} i_{v_m - \eta_m} \lambda_{L-k(v)}$$

$$= \sum_{v \in \Lambda_L} \sum_{\tilde{v} \in \Lambda_L} 1_{(\tilde{v} \leq v)} 1_{(||\tilde{v}-v||_{\ell^\infty(\mathbb{N})} \leq 1)} (-1)^{||\tilde{v}-v||_{\ell^1(\mathbb{N})}} \bigotimes_{m \geq 1} i_{\tilde{v}_m} \lambda_{L-k(v)}.$$

Finally, exchanging the sums and renaming the multiindices we obtain

$$\hat{\lambda}_L = \sum_{v \in \Lambda_L} \sum_{\tilde{v} \in \Lambda_L} 1_{(v \leq \tilde{v})} 1_{(||\tilde{v}-v||_{\ell^\infty(\mathbb{N})} \leq 1)} (-1)^{||\tilde{v}-v||_{\ell^1(\mathbb{N})}} \bigotimes_{m \geq 1} i_{v_m} \lambda_{L-k(\tilde{v})} \qquad (66)$$

and similarly

$$\hat{u}_L = \sum_{v \in \Lambda_L} \sum_{\tilde{v} \in \Lambda_L} 1_{(v \leq \tilde{v})} 1_{(||\tilde{v}-v||_{\ell^\infty (\mathbb{N})} \leq 1)} (-1)^{||\tilde{v}-v||_{\ell^1 (\mathbb{N})}} \bigotimes_{m \geq 1} i_{v_m} u_{L-k(\tilde{v})}. \qquad (67)$$

Remark 12. Standard hat function discretizations can be used to compute $\hat{u}_L$ in (67). Indeed, let $\hat{u}_L^{(\ell)}$, $\ell = 0, \ldots, L$ be the contribution of u_ℓ in (67) and let P_ℓ denote the prolongation operator $V_{\ell-1} \to V_\ell$. We then have

$$\hat{u}_L = \hat{u}_L^{(L)} + P_L(\hat{u}_L^{(L-1)} + P_{L-1}(\ldots P_2(\hat{u}_L^{(1)} + P_1 \hat{u}_L^{(0)}) \ldots)),$$

with total cost being proportional to the dimension of the ansatz space V_L.

An efficient algorithm for computing Λ_L has been given in [8]. Observe that for $v \in \Lambda_L = \Lambda(\mu, \varepsilon_L)$ we have $\tilde{v} \in \Lambda_L$ and $\tilde{v} \geq v$ iff $\eta = \tilde{v} - v$ satisfies $\eta \in \Lambda(\mu, \varepsilon_L/\mu^v)$. Thus, the same algorithm (with a straightforward modification to take the constraint $||\tilde{v} - v||_{\ell^\infty (\mathbb{N})} \leq 1$ and the variable level indicator $k(\tilde{v})$ into account) can be used to compute the coefficients of the terms $\bigotimes_{m \geq 1} i_{v_m} \lambda_{L-k(\tilde{v})}$ in (66) and $\bigotimes_{m \geq 1} i_{v_m} u_{L-k(\tilde{v})}$ in (67) efficiently.

Finally, the total computational effort for the application of the sparse composite collocation operator to the eigenpair of the ground state (66) and (67) can be estimated using the following lemma.

Lemma 10. *For $k \geq 0$ assume $\#\Lambda_k \lesssim 2^{d_1 k/2}$, $d_1 > 0$ and that the work for determination of the numerical solution u_k on one collocation node is bounded by a constant multiple of $2^{d_2 k}$, $d_2 > 0$. Then the computational effort for the numerical realization of the sparse composite operator applied to the eigenpair of the ground state is bounded by an absolute multiple of $L 2^{\max\{d_1, d_2\}L}$ as $L \to \infty$.*

Proof. We compute from (67) using the fact $v \leq \tilde{v} \Rightarrow k(v) \leq k(\tilde{v})$ and (52)

$$\sum_{v \in \Lambda_L} \left(\prod_{m \geq 1} (v_m + 1) \right) L 2^{d_2(L-k(v))} \lesssim L \sum_{k=0}^{L} \sum_{v \in \Lambda_k \backslash \Lambda_{k-1}} \left(\prod_{m \geq 1} (v_m + 1) \right) 2^{d_2(L-k)}$$

$$\leq L \sum_{k=0}^{L} (\#\Lambda_k - \#\Lambda_{k-1}) \#\Lambda_k 2^{d_2(L-k)}$$

$$\leq L \sum_{k=0}^{L} 2^{d_1 k} 2^{d_2(L-k)}$$

which shows the claim. $\square$

We collect the foregoing in the following theorem.

Theorem 8. *For the parametric eigenvalue problem* (10) *assume the particular form* (34) *with* (35) *and let Assumption 7 hold. Assume further that $D \subset \mathbb{R}^d$ is open, bounded and convex. For $y \in U$ let $(\lambda(y), u(y)) = (\lambda_1(y), w_1(y))$ denote the eigenpair with smallest eigenvalue. Let $\varepsilon \in (0, 1)$, $c > 0$, and $\delta > 0$ is as in* (42). *Define* $\tau_m := \min\{(1 - \varepsilon)\frac{a_{\min}||\psi_m||_{L^\infty(D)}^{p-1}}{2\beta(p)(1+\delta^{-1})}, c||\psi_m||_{W^{1,\infty}(D)}^{p-1}\}$, $m \geq 1$, *and* $E(\tau)$ *as in* (22). *Let* $\rho_m := \tau_m + \sqrt{1 + \tau_m^2}$, $m \geq 1$ *and* $\mu_m := \sup_{m' \geq m} \frac{1}{\rho_{m'}}$, $m \geq 1$. *Let* $0 \leq t \leq 1$ *and W be as in Assumption 8. If $p \in (0, 1)$ is sufficiently small then for all $0 < \sigma < \frac{1}{p} - 1$ there exist $0 < \varkappa < 1$ and $C_1 > 0$ such that:*

1. *$\sup_{m \geq 1} \mu_m m^\sigma < \infty$ and μ has asymptotic overhead order $\varkappa^\star(\mu) \leq \varkappa$,*
2. *Defining $\varepsilon_k = 2^{-tk/(1-\varkappa)}$ and $\Lambda_k = \Lambda(\mu, \varepsilon_k)$ we have $\#\Lambda_k \lesssim \varepsilon_k^{-1/\sigma}$,*
3. *The sparse composite approximation* (63) *satisfies for $L \geq 0$*

$$||u - \hat{u}_L||_{L^2_\pi(U;H^1_0(D))} \leq C(u)L2^{-tL} = C(u)L\varepsilon_L^{1-\varkappa}$$

and

$$||u - \hat{u}_L||_{L^2_\pi(U;H^1_0(D))} \leq C(u)(\#\Lambda_L)^{-\sigma(1-\varkappa)} \leq C(u)N_{\Lambda_L}^{-\sigma(1-\varkappa)/2}$$

where $C(u) = C_1||u||_{C^0(\overline{E(\tau)};W)} < \infty$ with $C_1 > 0$ independent of $L \geq 0$, and N_{Λ_L} denotes the number of collocation points in I_{Λ_L}.

6 Numerical Examples

In the numerical examples we approximate the parametric eigenpairs by tensorized polynomials using the sparse collocation method as described in Sect. 5.

We take an elliptic stochastic operator expanded in its Karhunen-Loève series as a model example [8]. We set $D = (-1, 1) \subset \mathbb{R}$ and $U = [-1, 1]^\infty$ and let the diffusion coefficient in (34) be $a(x, y) = \bar{a}(x) + \sum_{m \geq 1} y_m \psi_m(x)$, $(x, y) \in D \times U$, where $\bar{a}$ and $\{\psi_m\}_{m \geq 1} \subset L^\infty(D)$. Specifically, we set $\bar{a} \equiv 1$ and $\psi_m(x) = \frac{\cos(\pi m x)}{(m+1)^3}$, $x \in D$, $m \geq 1$. This implies $||\psi_m||_{L^\infty(D)} = \frac{1}{(m+1)^3}$, $||\nabla\psi_m||_{L^\infty(D)} = \frac{\pi m}{(m+1)^3}$ such that $||\psi_m||_{W^{1,\infty}(D)} = \frac{\pi}{(m+1)^2} + O(m^{-3})$ as $m \to \infty$. Hence, for all $p > \frac{1}{2}$ we have $(||\psi_m||_{W^{1,\infty}(D)})_{m \geq 1} \in \ell^p(\mathbb{N})$, which implies (39). In the computation we set $p := 0.6$, $\delta := 2.8$ (empirical estimate from a few samples), $a_{\min} := \inf_{x \in D} \bar{a} - \beta(1)$, $\varepsilon = 0$ and $c = 10$ in the definition of μ, see Theorem 8; for $\beta(p)$ we employ the approximation $\sum_{m=1}^{M_0} ||\psi_m||_{L^\infty(D)}^p$ with $M_0 = 10^5$ terms. This now completely defines the multiindex sets $\Lambda(\mu, \varepsilon)$ for all $\varepsilon > 0$.

Approximate mean $\mathbb{E}[u] = \int_U u(y)d\pi(y)$ and variance $\mathbb{E}[u^2] - \mathbb{E}[u]^2$ of the first eigenfunction u are shown in Fig. 1.

In Fig. 2 convergence of the mean $\mathbb{E}[I_{\Lambda(\mu,\varepsilon)}\lambda] \to \mathbb{E}[\lambda]$ and the number of collocation points in $I_{\Lambda(\mu,\varepsilon)}$ as functions of ε are investigated. We observe that the error $|\mathbb{E}[I_{\Lambda(\mu,\varepsilon)}\lambda] - \mathbb{E}[\lambda]|$ decays like ε^4 as $\varepsilon \to 0$, see Fig. 2 (left). Fig. 2 (right)

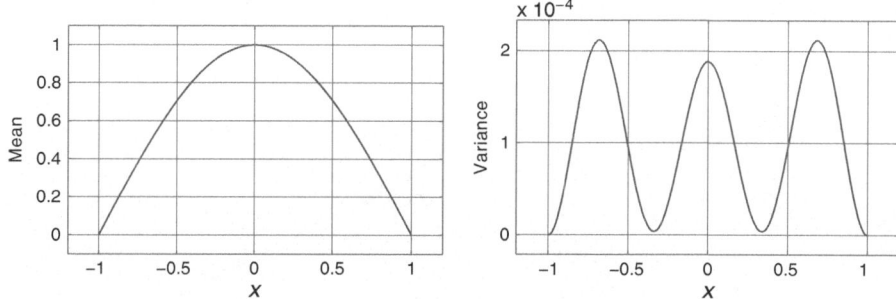

Fig. 1 Mean and variance of the ground state of the parameteric diffusion equation (36) as described in Sect. 6

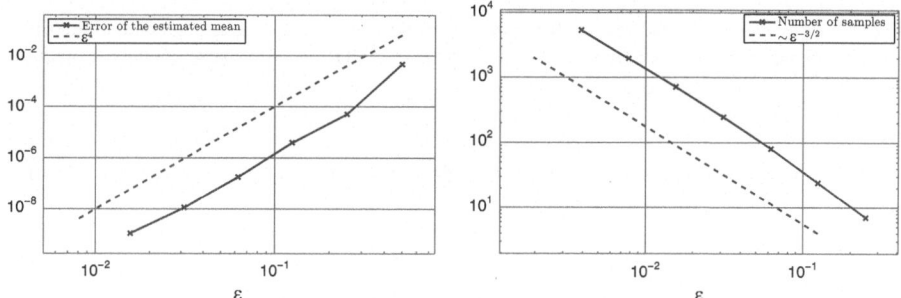

Fig. 2 Convergence of the mean of the computed parametric eigenvalue $\mathbb{E}\left[I_{\Lambda(\mu,\varepsilon)}\lambda\right] \to \mathbb{E}\left[\lambda\right]$ and the corresponding number of collocation points in $I_{\Lambda(\mu,\varepsilon)}$ as $\varepsilon \to 0$. See Sect. 6 for details

suggests that $N_{\Lambda(\mu,\varepsilon)}$, the number of collocation points in $I_{\Lambda(\mu,\varepsilon)}$ behaves like $\varepsilon^{-3/2}$ as $\varepsilon \to 0$. Combining these yields $\left|\mathbb{E}\left[I_{\Lambda(\mu,\varepsilon)}\lambda\right] - \mathbb{E}\left[\lambda\right]\right| \sim N_{\Lambda(\mu,\varepsilon)}^{-8/3}$.

In order to relate to the above theory, and verify convergence of the parametric eigenvalue in $L_\pi^2(U)$, we employ the parameterization via the Legendre polynomials, see Lemma 5, which allows the computation of the $L_\pi^2(U)$ norm to machine precision. We consider finite element spaces based either on first or second order splines on an equidistant mesh, and compare the "full tensor product" collocation approximation $I_{\Lambda_\ell}\lambda_\ell$ and the sparse composite collocation approximation $\hat{\lambda}_\ell$ for $\ell = 0, 1, 2$ against an overkill reference solution. In the computation we set $\varepsilon_\ell = 2^{-\ell}$ for simplicity. The first order spaces have 15, 31, 63, the second order space have 16, 32, 64 degrees of freedom on levels $\ell = 0, 1, 2$ respectively. The results are shown in Fig. 3, showing the error of the approximate parametric eigenvalue versus the total number of degrees of freedom in space, that is the sum of degrees of freedom of all EVPs solved; we observe algebraic convergence of the sparse composite collocation method in the total number of degrees of freedom, as can be expected from Theorem 8. As can be seen from Fig. 3, sparse composite collocation approximation $\hat{\lambda}_\ell$ enjoys an improved convergence rate (as $\ell \to \infty$)

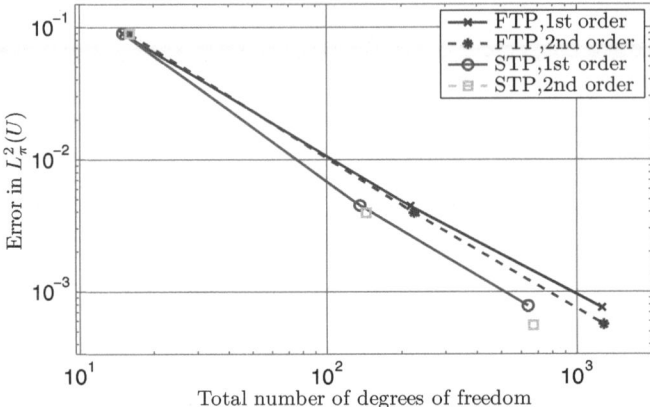

Fig. 3 Convergence of the eigenvalue in mean square sense for first and second order finite element, using the collocation operator $I_{\Lambda_\ell} \lambda_\ell$, $\ell = 0, 1, 2$ (FTP) and the sparse composite operator $\hat{\lambda}_\ell$, $\ell = 0, 1, 2$ (STP), see Sect. 6

over the "full tensor product" collocation approximation $I_{\Lambda_\ell} \lambda_\ell$ in terms of the total number of degrees of freedom.

7 Summary

We have quantified the analytic dependence of an isolated eigenpair of a linear operator depending affinely on a vector of parameters in an abstract setting. We have then specialized the discussion on stochastic differential operators expanded in its Karhunen-Loève series. Analyticity has been used to prove convergence of the sparse composite collocation operator applied to the discretization of an isolated, simple eigenpair. In particular, the approximation of the parametric eigenvector was shown to allow for the same gains in complexity as in the case of source problems. Our numerical example of an infinite dimensional paramateric eigenvalue problem confirms algebraic convergence of the sparse composite collocation method in the total number of degrees of freedom, and shows that sparse composite tensorization can be an effective tool to reduce the complexity of the problem.

In the sequel we will address the case of non-selfadjoint operators, as well as eigenpair computation with eigenvalue "crossings".

We finally remark that the extension of most results presented here to parametric spectral problems of polynomial and, more generally, holomorphic operator pencils with general, analytic parameter dependence can be achieved with similar tools, based on a suitable version of the implicit function theorem for holomorphic functions (see, e.g. [32] for a statement and proof). This topic will be addressed in forthcoming work.

Acknowledgements Supported by SNF grant PDFMP2-127034/1 and by ERC AdG grant STAHDPDE 247277.

References

1. I. BABUŠKA AND B. Q. GUO, *Regularity of the solution of elliptic problems with piecewise analytic data. I. Boundary value problems for linear elliptic equation of second order*, SIAM J. Math. Anal., 19 (1988), pp. 172–203.
2. I. BABUŠKA AND J. OSBORN, *Eigenvalue problems*, in Handb. Numer. Anal., Vol. II, Handb. Numer. Anal., II, North-Holland, Amsterdam, 1991, pp. 641–787.
3. I. BABUŠKA, R. TEMPONE, AND G. E. ZOURARIS, *Galerkin finite element approximations of stochastic elliptic partial differential equations*, SIAM J. Num. Anal., 42 (2002), pp. 800–825.
4. I. BABUŠKA, F. NOBILE, AND R. TEMPONE, *A stochastic collocation method for elliptic partial differential equations with random input data*, SIAM J. Num. Anal., 45 (2007), pp. 1005–1034.
5. J. BÄCK, F. NOBILE, L. TAMELLINI, AND R. TEMPONE, *Stochastic Galerkin and collocation methods for PDEs with random coefficients: a numerical comparison*, Tech. Report 09-33, ICES, 2009.
6. M. BIERI, *A sparse composite collocation finite element method for elliptic sPDEs*, Tech. Report 2009-08, Seminar for Applied Mathematics, ETH Zürich, 2009. SIAM Journ. Numer. Anal. (to appear 2011). Available via http://www.sam.math.ethz.ch/reports/2009/08.
7. M. BIERI, *Sparse tensor discretizations of elliptic PDEs with random input data*, PhD thesis, ETH Zürich, 2009. Diss ETH No. 18598 Available via http://e-collection.ethbib.ethz.ch/.
8. M. BIERI, R. ANDREEV, AND CH. SCHWAB, *Sparse tensor discretization of elliptic spdes*, SIAM J. Sci. Comput., 31 (2009), pp. 4281–4304.
9. M. BIERI AND CH. SCHWAB, *Sparse high order FEM for elliptic sPDEs*, Comp. Meth. Appl. Mech. Engrg., 198 (2009), pp. 1149–1170.
10. D. BRAESS, *Finite Elemente*, Springer, Berlin, 3rd ed., 2002.
11. HANS-JOACHIM BUNGARTZ AND MICHAEL GRIEBEL, *Sparse grids*, Acta Numer., 13 (2004), pp. 147–269.
12. A. COHEN, R. DEVORE, AND CH. SCHWAB, *Analytic regularity and polynomial approximation of parametric and stochastic elliptic PDEs*, Tech. Report 2010-03, Seminar for Applied Mathematics, ETH Zürich, 2010.
13. ALBERT COHEN, RONALD DEVORE, AND CHRISTOPH SCHWAB, *Convergence rates of best n-term galerkin approximations for a class of elliptic spdes*, Found. Comput. Math., 10 (2010), pp. 615–646. 10.1007/s10208-010-9072-2.
14. W. DAHMEN, T. ROHWEDDER, R. SCHNEIDER, AND A. ZEISER, *Adaptive eigenvalue computation: complexity estimates*, Numer. Math., 110 (2008), pp. 277–312.
15. P. J. DAVIS, *Interpolation and approximation*, Introductions to higher mathematics, Blaisdell Publishing Company, 1963.
16. J. FOO, X. WAN, AND G.E. KARNIADAKIS, *The multi-element probabilistic collocation method: analysis and simulation*, Journal of Computational Physics, (2008), pp. 9572–9595.
17. P. FRAUENFELDER, CH. SCHWAB, AND R.-A. TODOR, *Finite elements for elliptic problems with stochastic coefficients*, Comp. Meth. Appl. Mech. Engrg., 194 (2005), pp. 205–228.
18. W. GAUTSCHI, *Orthogonal polynomials. Computation and Approximation*, Numer. Math. Sci. Comput., Oxford University Press Inc., 2004.
19. R. GEUS, *The Jacobi-Davidson algorithm for solving large sparse symmetric eigenvalue problems with application to the design of accelerator cavities*, PhD thesis, ETH Zürich, 2002. Diss. Nr. 14734.
20. ROGER G. GHANEM AND POL D. SPANOS, *Stochastic Finite Elements, a Spectral Approach*, Dover Publications Inc., New York, revised ed., 2003.

21. CLAUDE JEFFREY GITTELSON, *Adaptive Galerkin Methods for Parametric and Stochastic Operator Equations*, PhD thesis, ETH Zürich, 2011. ETH Dissertation No. 19533.
22. CLAUDE JEFFREY GITTELSON, *An adaptive stochastic Galerkin method*, Tech. Report 2011-11, Seminar for Applied Mathematics, ETH Zürich, 2011.
23. CLAUDE JEFFREY GITTELSON, *Adaptive stochastic Galerkin methods: Beyond the elliptic case*, Tech. Report 2011-12, Seminar for Applied Mathematics, ETH Zürich, 2011.
24. CLAUDE JEFFREY GITTELSON, *Stochastic Galerkin approximation of operator equations with infinite dimensional noise*, Tech. Report 2011-10, Seminar for Applied Mathematics, ETH Zürich, 2011.
25. G. GOLUB AND C. VAN LOAN, *Matrix computations*, The Johns Hopkins University Press, London, 1996.
26. W. HACKBUSCH, *On the computation of approximate eigenvalues and eigenfunctions of elliptic operators by means of a multi-grid method*, SIAM J. Numer. Anal., 16 (1979), pp. 201–215.
27. A. HENROT, *Extremum Problems for Eigenvalues of Elliptic Operators*, vol. 8 of Frontiers in Mathematics, Birkhäuser Basel, 2006.
28. M. HERVÉ, *Analyticity in Infinite Dimensional Spaces*, vol. 10 of De Gruyter studies in mathematics, Walter de Gruyter, 1989.
29. VIET-HA HOANG AND CHRISTOPH SCHWAB, *Analytic regularity and gpc approximation for parametric and random 2nd order hyperbolic PDEs*, Tech. Report 2010-19, Seminar for Applied Mathematics, ETH Zürich, 2010.
30. VIET-HA HOANG AND CHRISTOPH SCHWAB, *Sparse tensor Galerkin discretization for parametric and random parabolic PDEs I: Analytic regularity and gpc-approximation*, Tech. Report 2010-11, Seminar for Applied Mathematics, ETH Zürich, 2010.
31. L. HÖRMANDER, *An Introduction to Complex Analysis in Several Variables*, The University Series in Higher Mathematics, D. van Nostrand Company, 1st ed., 1966.
32. L. HÖRMANDER, *An Introduction to Complex Analysis in Several Variables*, North Holland Mathematical Library, North Holland, 3rd ed., 1990.
33. T. KATO, *Perturbation theory for linear operators*, vol. 132 of Grundlehren Math. Wiss., Springer Berlin, Heidelberg, New-York, 2 ed., 1976.
34. XIANG MA AND NICHOLAS ZABARAS, *An adaptive hierarchical sparse grid collocation algorithm for the solution of stochastic differential equations*, J. Comput. Phys., 228 (2009), pp. 3084–3113.
35. F. NOBILE, R. TEMPONE, AND C.G. WEBSTER, *An anisotropic sparse grid stochastic collocation method for elliptic partial differential equations with random input data*, SIAM J. Num. Anal., 46 (2008), pp. 2411–2442.
36. F. NOBILE, R. TEMPONE, AND C.G. WEBSTER, *A sparse grid stochastic collocation method for elliptic partial differential equations with random input data*, SIAM J. Num. Anal., 46 (2008), pp. 2309–2345.
37. M. REED AND B. SIMON, *Methods of modern mathematical physics. I. Functional analysis*, Academic Press, New York, 1972.
38. M. REED AND B. SIMON, *Methods of modern mathematical physics. IV. Analysis of operators*, Academic Press [Harcourt Brace Jovanovich Publishers], New York, 1978.
39. F. RELLICH, *Perturbation theory of eigenvalue problems*, Notes on mathematics and its applications, Gordon and Breach, New York, London, Paris, 1969.
40. CH. SCHWAB AND R.-A. TODOR, *Karhunen-Loève approximation of random fields by generalized fast multipole methods*, Journal of Computational Physics, 217 (2006), pp. 100–122.
41. S.A. SMOLYAK, *Quadrature and interpolation formulas for tensor products of certain classes of functions*, Sov. Math. Dokl., 4 (1963), pp. 240–243.
42. D. C. SORENSEN, *Numerical methods for large eigenvalue problems*, Acta Numer., 11 (2002), pp. 519–584.
43. R.A. TODOR, *Sparse Perturbation algorithms for elliptic PDE's with stochastic data*, PhD thesis, ETH Zürich, 2005. Diss. Nr. 16192.

44. R.-A. TODOR AND CH. SCHWAB, *Convergence rates for sparse chaos approximations of elliptic problems with stochastic coefficients*, IMA J. Num. Anal., 27 (2007), pp. 232–261.
45. G. W. WASILKOWSKI AND H. WOŹNIAKOWSKI, *Explicit cost bounds of algorithms for multivariate tensor product problems*, J. Complexity, 11 (1995), pp. 1–56.
46. DONGBIN XIU, *Fast numerical methods for stochastic computations: a review*, Commun. Comput. Phys., 5 (2009), pp. 242–272.
47. DONGBIN XIU AND JAN S. HESTHAVEN, *High-order collocation methods for differential equations with random inputs*, SIAM J. Sci. Comput., 27 (2005), pp. 1118–1139 (electronic).
48. DONGBIN XIU AND GEORGE EM KARNIADAKIS, *Modeling uncertainty in steady state diffusion problems via generalized polynomial chaos*, Comput. Methods Appl. Mech. Engrg., 191 (2002), pp. 4927–4948.
49. DONGBIN XIU AND GEORGE EM KARNIADAKIS, *The Wiener-Askey polynomial chaos for stochastic differential equations*, SIAM J. Sci. Comput., 24 (2002), pp. 619–644 (electronic).
50. DONGBIN XIU AND DANIEL M. TARTAKOVSKY, *Numerical methods for differential equations in random domains*, SIAM J. Sci. Comput., 28 (2006), pp. 1167–1185 (electronic).

Mixed Multiscale Methods for Heterogeneous Elliptic Problems

Todd Arbogast

Abstract We consider a second order elliptic problem written in mixed form, i.e., as a system of two first order equations. Such problems arise in many contexts, including flow in porous media. The coefficient in the elliptic problem (the permeability of the porous medium) is assumed to be spatially heterogeneous. The emphasis here is on accurate approximation of the solution with respect to the scale of variation in this coefficient. Homogenization and upscaling techniques alone are generally inadequate for this problem. As an alternative, multiscale numerical methods have been developed. They can be viewed in one of three equivalent frameworks: as a Galerkin or finite element method with nonpolynomial basis functions, as a variational multiscale method with standard finite elements, or as a domain decomposition method with restricted degrees of freedom on the interfaces. We treat each case, and discuss the advantages of the approach for devising effective local multiscale methods. Included is recent work on methods that incorporate information from homogenization theory and effective domain decomposition methods.

1 Elliptic Systems with a Heterogeneous Coefficient

We consider a second order elliptic problem, which we write in *mixed form,* i.e., as the following system of two first order equations:

T. Arbogast (✉)
The University of Texas at Austin, Institute for Computational Engineering and Sciences,
1 University Station C0200, Austin, TX 78712, USA

The University of Texas at Austin, Mathematics Department, 1 University Station C1200, Austin, TX 78712–0257, USA
e-mail: arbogast@ices.utexas.edu

I.G. Graham et al. (eds.), *Numerical Analysis of Multiscale Problems*, Lecture Notes in Computational Science and Engineering 83, DOI 10.1007/978-3-642-22061-6_8, © Springer-Verlag Berlin Heidelberg 2012

$$\mathbf{u} = -a_\epsilon \nabla p \quad \text{in } \Omega \subset \mathbb{R}^d \quad \text{(Darcy's law)}, \tag{1}$$

$$\nabla \cdot \mathbf{u} = f \qquad \text{in } \Omega \qquad \text{(Conservation)}, \tag{2}$$

$$\mathbf{u} \cdot \nu = 0 \qquad \text{on } \partial \Omega. \tag{3}$$

The system models, e.g., incompressible, single phase flow in a porous medium in which case the unknowns are p, the fluid pressure, and $\mathbf{u}$, the Darcy velocity of the fluid, while the known parameters are a_ϵ, the medium permeability, and f, the source or sink term (i.e., the wells) [17, 18, 29]. The first equation is the empirical *Darcy Law* relating the velocity to the pressure gradient, the second equation expresses the mass conservation principle, and we take a boundary condition representing no normal flow for simplicity of exposition. In this discussion, we assume that a_ϵ is *heterogeneous* on a scale ϵ.

Our objective is to find an *accurate* approximation of $\mathbf{u}$ and p while we respect the principle of mass *conservation,* which is a critical property in many applications. In fact, we are most interested in an accurate, conservative approximation to $\mathbf{u}$, since the velocity controls the transport of mass, such as a contaminant in the groundwater or the macroscopic mixing of multiple phases.

We can rewrite the system (1)–(3) in *mixed variational form* as follows. Let $(\cdot, \cdot)_\omega$ denote the $L^2(\omega)$ or $(L^2(\omega))^d$ inner product, wherein we omit ω when it is Ω, and define the Hilbert spaces $H(\mathrm{div}; \Omega) := \{\mathbf{v} \in (L^2(\Omega))^d : \nabla \cdot \mathbf{v} \in L^2(\Omega)\}$ and

$$\mathbf{V} := H_0(\mathrm{div}; \Omega) := \{\mathbf{v} \in H(\mathrm{div}; \Omega) : \mathbf{v} \cdot \nu = 0 \text{ on } \partial \Omega\},$$

where $\|\mathbf{v}\|_{\mathbf{V}}^2 = \|\mathbf{v}\|_0^2 + \|\nabla \cdot \mathbf{v}\|_0^2$, $\|\psi\|_0^2 = (\psi, \psi)$. Using integration by parts to rewrite

$$-(\nabla p, \mathbf{v}) = (p, \nabla \cdot \mathbf{v}),$$

the problem is equivalent to
    Find $p \in W = L^2(\Omega)/\mathbb{R}$ and $\mathbf{u} \in \mathbf{V}$ such that

$$(a_\epsilon^{-1} \mathbf{u}, \mathbf{v}) - (p, \nabla \cdot \mathbf{v}) = 0 \qquad \forall \, \mathbf{v} \in \mathbf{V} \quad \text{(Darcy's law)}, \tag{4}$$

$$(\nabla \cdot \mathbf{u}, w) \qquad\qquad = (f, w) \quad \forall \, w \in W \quad \text{(Conservation)}. \tag{5}$$

We remark that the mixed form preserves the conservation equation, and so allows locally conservative approximations.

We have a saddle-point problem, since it has both positive and negative eigenvalues. There is a well-developed abstract theory for the well-posedness of such mixed variational forms [13, 21, 22, 26].

Theorem 1. [Babuška, 1973; Brezzi, 1974] *For the abstract saddle-point problem*
    Find $p \in W$ and $\mathbf{u} \in \mathbf{V}$ such that

$$A(\mathbf{u}, \mathbf{v}) - (p, \nabla \cdot \mathbf{v}) = G(\mathbf{v}) \quad \forall \, \mathbf{v} \in \mathbf{V},$$

$$(w, \nabla \cdot \mathbf{u}) \qquad\qquad = F(w) \quad \forall \, w \in W,$$

suppose A is a continuous, symmetric bilinear form, coercive on $\mathbf{V} \cap \ker(\nabla \cdot)$, *and that there exists* $\gamma > 0$ *such that*

$$\inf_{w \in W} \sup_{\mathbf{v} \in \mathbf{V}} \frac{(w, \nabla \cdot \mathbf{v})}{\|w\|_W \|\mathbf{v}\|_{\mathbf{V}}} \geq \gamma. \tag{6}$$

Then there exists a unique solution $(p, \mathbf{u}) \in W \times \mathbf{V}$, *and*

$$\|p\|_W + \|\mathbf{u}\|_{\mathbf{V}} \leq C\{\|F\|_{W^*} + \|G\|_{\mathbf{V}^*}\}.$$

In our case, we satisfy the inf-sup condition (6), and

$$A(\mathbf{u}, \mathbf{v}) = (a_\epsilon^{-1} \mathbf{u}, \mathbf{v})$$

is continuous and coercive provided that the tensor $a_\epsilon \in (L^\infty(\Omega))^{d \times d}$ is uniformly positive definite: there are $\alpha_* > 0$ and $\alpha^* < \infty$ such that

$$\alpha_* |\xi|^2 \leq \xi^T a_\epsilon(x) \xi \leq \alpha^* |\xi|^2 \quad \forall \, \xi \in \mathbb{R}^d, \, a.e. \, x \in \Omega. \tag{7}$$

Thus we have a well-posed problem for, say, $f \in L^2(\Omega)$.

The complication comes from the problem of scale. Because a_ϵ varies on the spatial scale ϵ,

$$|\mathbf{u}| = \mathcal{O}(1) \quad \text{but} \quad |D^k \mathbf{u}| = \mathcal{O}(\epsilon^{-k}).$$

Therefore, to approximate the solution accurately, we need to *resolve* the spatial scale ϵ, using a fine computational mesh of spacing $h_f < \epsilon$. This is not always computationally feasible, since it would require a mesh with many orders of magnitude more elements than can be handled on the world's largest supercomputers. Instead we consider four multiscale numerical techniques, as follows.

1. Homogenization and upscaling [Bensoussan, Lions, and Papanicolaou, 1978; Sanchez-Palencia, 1980]. We replace the coefficient a_ϵ in the differential equation by one that is easier to resolve.
2. Multiscale finite elements [Babuška and Osborn, 1983; Babuška, Caloz, and Osborn 1994; Hou and Wu 1997; Chen and Hou 2003]. We define the finite element space to better capture the fine scales.
3. Variational multiscale method [Hughes, 1995; Arbogast, Minkoff, and Keenan, 1998; Arbogast, 2000; Arbogast and Boyd, 2006]. We modify the variational form to better capture the fine scales.
4. Domain decomposition and mortar methods [Schwarz, 1870; Arbogast, Pencheva, Wheeler, and Yotov 2007]. We divide the problem into weakly coupled small subdomains that can be resolved.

2 Homogenization and Upscaling

We want to solve the problem on a coarse grid. *Upscaling* is the process of representing the system on a coarser scale by defining average or *effective macroscopic* parameters in place of the true parameters (in our case, a_ϵ). Perhaps the most well-developed mathematical theory of upscaling is *homogenization* [19, 45, 51, 62]. We begin with an overview of the philosophy of homogenization.

The solution $\mathbf{u}$ has high frequency "wiggles" due to the heterogeneity of a_ϵ, as illustrated in Fig. 1 Can we find the smooth "local average" $\bar{\mathbf{u}}(x)$ without knowing $\mathbf{u}(x)$? The wiggles are irregular, so they are hard to deal with.

The key assumption in homogenization theory is that the heterogeneity has *scale separation*, meaning the system separates into fine and coarse scales with some gap in scales; that is, Fig. 1 is somehow an accurate picture of the scales separating into fine wiggles and some coarse average. More precisely, we assume the heterogeneity is *periodic* of period ϵ, so that the wiggles are regular, and thus easily identified and removed. Later we will see that this is basically the *closure assumption* that allows the fine-scale details to be removed uniquely from the problem. We will let $\epsilon \to 0$, which should remove the wiggles (at least in some weak sense) and give us our macro-scale model for the average flow.

Let Y be a unit-sized reference parallelepiped cell domain, which we scale by ϵ. Then Ω is composed of translated copies of ϵY (see Fig. 2). The permeability a_ϵ is assumed to be, more generally, *locally periodic*, meaning that

$$a_\epsilon(x) = a(x, y), \tag{8}$$

where $a(x, y)$ is periodic in $y \in Y$ but varies slowly and smoothly in $x \in \Omega$.

Homogenization is very mathematical, and involves deep analysis of partial differential equations [51]. Fortunately there is a simpler, more physical view of homogenization, called *formal homogenization* [19, 62].

Fig. 1 The solution $\mathbf{u}$ with high frequency wiggles compared to its "local average" $\bar{\mathbf{u}}$. Can we find $\bar{\mathbf{u}}$ without knowing $\mathbf{u}$?

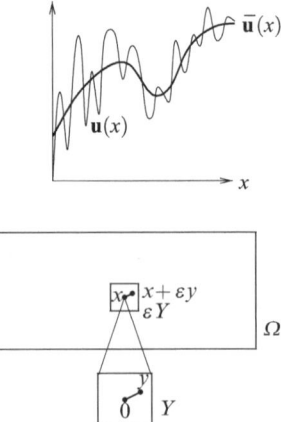

Fig. 2 The domain Ω composed of ϵ-scaled reference parallelepipeds Y. General location $x \in \Omega$ is modified by $y \in Y$ to give position $x + \epsilon y$. Scaling by $1/\epsilon$ focuses in on the details

As depicted in Fig. 2, we represent the idea of scale separation by assuming that the space variable $x_{\text{absolute}} \in \Omega$ has both a slow (denoted by $x \in \Omega$) and a fast (denoted by $y \in Y$) component, where these scale as

$$x_{\text{absolute}} \sim x + \epsilon y.$$

At any point x_{absolute}, x is the macroscopic position that ignores fine scales and y allows us to "see" the local details under dilation by $1/\epsilon$. In this way, we can quantify how these local details affect larger scales. The local details disappear as $\epsilon \to 0$, but not necessarily their coarse-scale effects.

Formal Assumption. *Assume without proof that the true solution $p(x)$ can be expanded into a power series involving ϵ as*

$$p(x) \sim p_0(x, y) + \epsilon \, p_1(x, y) + \epsilon^2 p_2(x, y) + \cdots ,$$

where in $y = x/\epsilon$ and each p_k is periodic in $y \in Y$.

Note that the gradient operator scales as

$$\nabla \sim \nabla_x + \epsilon^{-1} \nabla_y.$$

We expect that

$$p = p_\epsilon \to p_0 \quad \text{as } \epsilon \to 0.$$

Substitute the formal expansion into the equations (1)–(2) to obtain

$$-(\epsilon^{-1}\nabla_y + \nabla_x) \cdot a(x, y)(\epsilon^{-1}\nabla_y + \nabla_x) \sum_{k=0}^{\infty} \epsilon^k \, p_k(x, y) = f.$$

Equating terms with like powers of ϵ leads to the following conclusions (for more details, see, e.g., [19, 45, 62]).

1. The ϵ^{-2} terms and periodicity in y imply that $p_0(x, y) = p_0(x)$ only (i.e., homogenization removes y, as we had hoped).
2. The ϵ^{-1} terms imply the existence and form of a *closure operator,* which is

$$p_1(x, y) := \sum_{j=1}^{d} \omega_j(x, y) \, \frac{\partial p_0(x)}{\partial x_j},$$

where the ω_j solve the local *cell problems,* one for each coordinate direction $j = 1, ..., d$,

$$-\nabla_y \cdot \left[a(x, y)\nabla_y\omega_j(x, y) \right] = \nabla_y \cdot \left[a(x, y)\mathbf{e}_j \right] \quad \text{in } \Omega \times Y, \qquad (9)$$

$$\omega_j(x, y) \text{ is periodic in } y, \qquad (10)$$

where $\mathbf{e}_j$ is the standard unit vector in the jth direction.

3. By averaging over the cell Y, the ϵ^0 terms give the *homogenized equations*

$$\mathbf{u}_0 = -a_0 \nabla p_0 \quad \text{in } \Omega \qquad \text{(Homogenized Darcy's law)}, \tag{11}$$

$$\nabla \cdot \mathbf{u}_0 = f \qquad \text{in } \Omega \qquad \text{(Conservation)}, \tag{12}$$

$$\mathbf{u}_0 \cdot \nu = 0 \qquad \text{on } \partial\Omega, \tag{13}$$

wherein $a_0(x)$ can be computed as the tensor

$$a_{0,ij}(x) := \frac{1}{|Y|} \int_Y a(x,y) \left(\frac{\partial \omega_j}{\partial y_i}(x,y) + \delta_{ij} \right) dy. \tag{14}$$

Lemma 1. *The homogenized permeability a_0 is symmetric and positive definite.*

Therefore a_0 has d principle eigenvectors and only positive eigenvalues, which is required of a permeability tensor in Darcy's Law (11) on physical grounds.

Lemma 2. [Voigt-Reiss Inequality] *The homogenized permeability a_0 lies between the harmonic and arithmetic averages. More precisely, if*

$$\hat{a} := \left(\frac{1}{|Y|} \int_Y (a(x,y))^{-1} \, dy \right)^{-1} \quad \text{and} \quad \bar{a} := \frac{1}{|Y|} \int_Y a(x,y) \, dy,$$

then

$$\xi^T \hat{a}(x)\xi \leq \xi^T a_0(x)\xi \leq \xi^T \bar{a}(x)\xi \quad \forall \, \xi \in \mathbb{R}^d, \text{ a.e. } x \in \Omega.$$

Thus we have the homogenized permeability tensor $a_0(x)$ from (14), and we can compute $p_0(x)$ from (11)–(13), which is well-posed by Theorem 1 and the remarks following. In fact, one can prove the following theorem on convergence [7, 8, 51], essentially justifying the first two terms in the formal asymptotic expansion.

Theorem 2. *Let p_ϵ and $\mathbf{u}_\epsilon$ solve (4)–(5), with a_ϵ satisfying the local periodicity condition (8), and let p_0 and $\mathbf{u}_0$ solve (11)–(13). If $p_0 \in H^2(\Omega) \cap W^{1,\infty}(\Omega)$ and the first order corrector is defined as*

$$p_\epsilon^1 := p_0 + \epsilon \sum_{j=1}^d \omega_j(x, x/\epsilon) \frac{\partial p_0(x)}{\partial x_j} = p_0(x) + \epsilon \, p_1(x, x/\epsilon),$$

then there is $C > 0$, independent of ϵ, such that

$$\| p_\epsilon - p_0 \|_0 \leq C\epsilon \quad \text{and} \quad \| \nabla(p_\epsilon - p_\epsilon^1) \|_0 \leq C \sqrt{\epsilon}.$$

Moreover, let $\alpha_0 = a_0^{-1}$ and define the fixed tensor $\mathscr{A} := a \, (I + D\omega)\alpha_0$, i.e.,

$$\mathscr{A}_{ij}(x,y) := \sum_{k,\ell} a_{ik}(x,y) \left(\delta_{k\ell} + \frac{\partial \omega_\ell(x,y)}{\partial y_k} \right) \alpha_{0,\ell j},$$

which is independent of ϵ and the domain Ω. If $\mathscr{A}_\epsilon(x) = \mathscr{A}(x, x/\epsilon)$, then

$$\mathbf{u}_\epsilon(x) = \mathscr{A}_\epsilon(x)\,\mathbf{u}_0(x) + \theta_\epsilon^\Omega(x),$$

where

$$\|\theta_\epsilon^\Omega\|_0 \leq C\{\epsilon\|\mathbf{u}_0\|_1 + \sqrt{\epsilon|\partial\Omega|}\|\mathbf{u}_0\|_{0,\infty}\} = \mathscr{O}(\sqrt{\epsilon}).$$

Herein, $\|\cdot\|_{k,p,\omega}$ denotes the norm in the Sobolev space $W^{k,p}(\omega)$. We will omit p when it is 2 and ω when it is Ω. The theory of homogenization has seemingly solved our problem with heterogeneity, since we are dealing with the case where ϵ is small and so the error in Theorem 2 is negligible. However, there are serious limitations to this approach, especially in the mixed context where we are more concerned with accurate approximation of $\mathbf{u} = \mathbf{u}_\epsilon$.

First, p_0 is approximated coarsely, and so has no microstructure. Thus

$$\mathbf{u}_0 = -a_0\nabla p_0 \not\approx \mathbf{u}_\epsilon.$$

We therefore need to use $p_\epsilon^1 \approx p_\epsilon$, which does contain the microstructure. However, even though

$$\mathbf{u}_\epsilon^1 = -a_\epsilon\nabla p_\epsilon^1 \approx \mathbf{u}_\epsilon \quad \text{and} \quad \mathscr{A}_\epsilon(x)\,\mathbf{u}_0(x) \approx \mathbf{u}_\epsilon,$$

we lose the divergence property, since

$$\nabla\cdot\mathbf{u}_\epsilon^1 \neq \nabla\cdot\mathbf{u}_\epsilon = f \quad \text{and} \quad \nabla\cdot\mathscr{A}_\epsilon\mathbf{u}_0 \neq \nabla\cdot\mathbf{u}_\epsilon = f.$$

This means that the local conservation principle is not satisfied. In fact, the error is $\mathscr{O}(1)$, so we are not even approximately mass conservative.

Second, a subtle question in the two-scale separation case arises: given $a_\epsilon(x)$, what is $a(x, y)$? In practice, one works on a coarse computational grid, and, given $x \in \Omega$, one treats Y as a portion of the mesh (one or more coarse elements) around x, and sets $a(x, y) = a_\epsilon(y)$ there. But it is not completely clear that this is appropriate.

Finally, and most importantly, we really want to develop techniques that apply to the non-two-scale separation cases. We thus turn to multiscale numerical techniques. However, we will use homogenization theory as a guide for the general case, since the two-scale separation case is the only one we completely understand.

3 Multiscale Numerics

Within the multiscale numerical approach, the objective is to solve the problem in a way that:

1. Does not fully incorporate the problem dynamics (i.e., solves some global coarse scale problem to resolution $h > \epsilon$).

2. Yet captures significant features of the solution, by taking into account the micro-structure (to resolution $h_f < \epsilon$).

Many techniques fall into this general class of methods. We note the following techniques and give some references, including what we believe are the first works in the area, although our list is very incomplete, since there is a vast amount of work in the area of multiscale numerics.

1. **Multiscale finite elements** began with the work of Babuška and Osborn in 1983 and experienced major advancements in the work of Hou and Wu starting in 1997 [14, 35, 46, 47, 64]. These methods were extended to mixed systems explicitly by Chen and Hou in 2003 [3, 4, 28], but they were actually defined implicitly for mixed systems much earlier as variational multiscale methods by Arbogast, Minkoff, and Keenan in 1998, as noted by Arbogast and Boyd in 2006 [8, 10].
2. **Variational multiscale analysis** began with the work of Hughes in 1995 [23, 48, 49], and was defined for the mixed case by Arbogast, Minkoff, and Keenan in 1998 [5, 6, 8, 10, 52, 56, 57].
3. **Multiscale multilevel and mortar methods,** in the context of homogenization or multiscale problems, can be considered to be implicit in the work of Moulton, Dendy, and Hyman in 1998, and were further developed by Xu and Zikatanov in 2004 [41, 53–55, 65]. These were extended to the mixed case in the sense of multiscale mortar methods by Arbogast, Pencheva, Wheeler, and Yotov in 2007 [11]. A multiscale basis optimization technique was defined by Rath in 2006 [59, 60].
4. **Multiscale finite volumes and discontinuous Galerkin methods** were also developed. Multiscale finite volumes were first described by Jenny, Lee, and Tchelepi in 2003 [39, 43, 50], and multiscale discontinuous Galerkin methods were defined by Aarnes and Heimsund [1].
5. **Heterogeneous multiscale methods** were defined by E and Engquist in 2003 [32].

We discuss three of these techniques in detail herein: multiscale finite elements, the variational multiscale method, and multiscale mortar methods. Each of these take an overall multiscale strategy with four main components, as follows.

1. **Localization.** The full partial differential problem is decomposed into many small, local, coarse element subproblems (of scale $h > \epsilon$).
2. **Fine-scale effects.** The local subproblems are given appropriate boundary conditions and solved on the fine scale $h_f < \epsilon$ (to resolve variations in a_ϵ) to define a coarse scale multiscale finite element or finite volume basis.
3. **Global coarse-grid problem.** This h-scale coarse basis is used to approximate the solution globally.
4. **Fine-grid reconstruction.** The finite element basis encapsulates an h_f-scale fine representation of the solution.

Note that in these methods, the problem is fully resolved on the fine scale, but the problem is *not* fully coupled. The global problem is a reduced degree-of-freedom system. Computational efficiency comes from divide-and-conquer: small, localized subproblems are easily solved; and the coupled global problem has only a few degrees of freedom per coarse element, and so is relatively easily solved.

3.1 Nonmixed Multiscale Finite Elements

For simplicity, we introduce multiscale finite elements in the nonmixed case. Recall that the objective is to define finite elements tailored to the problem at hand to better capture the fine scales.

To make everything concrete, we begin with an example in one dimension. Consider the problem

$$-(ap')' = 0, \quad 0 < x < 1, \tag{15}$$

$$p(0) = 0 \text{ and } p(1) = 1, \tag{16}$$

where $a > 0$ is highly oscillatory, leading to an oscillatory true solution, as indicated in Fig. 3. Let $X = H_0^1(0, 1) = \{w \in H^1 : w(0) = w(1) = 0\}$. Then our problem has the variational form

Find $p \in X + x$ such that

$$(ap', w') = 0 \quad \forall \, w \in X.$$

We choose a uniform grid of five points $x_i = i/4, i = 0, 1, 2, 3, 4$. We illustrate the definition of finite elements beginning with the standard piecewise linear basis before defining the multiscale variant.

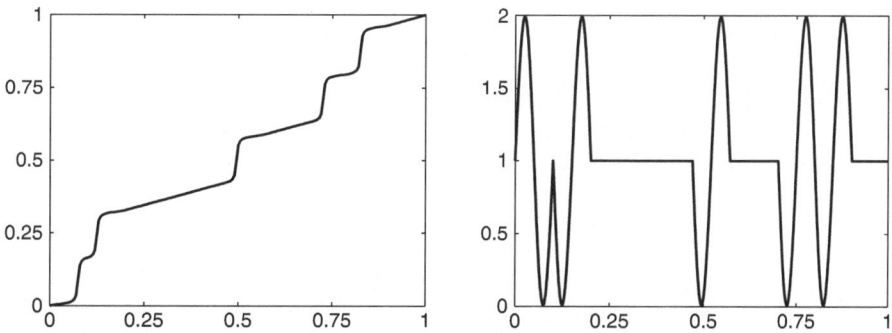

Fig. 3 The true solution p is shown on the left for the coefficient a on the right (although a becomes very small, it remains uniformly positive)

Standard finite element space $\bar{X}_h$. At x_i, $i = 1, 2, 3, 4$, we define $\bar{q}_i$ to be piecewise linear over the mesh such that $\bar{q}_i(x_j) = \delta_{ij}$. To state this constructively, let $x_5 = 1$. As illustrated in Fig. 4, we define $\bar{q}_i$, supported in (x_{i-1}, x_{i+1}), by:

1. Setting $\bar{q}_i$ on the boundary of each element separately, so $\bar{q}_i(x_j) = \delta_{ij}$;
2. Linearly interpolating over each element separately;
3. Joining the two pieces together continuously to form $\bar{q}_i$ and setting $\bar{X}_h = \text{span}\{\bar{q}_i\}$.

Multiscale finite element space X_h. Localize X to the element $E = (x_{i-1}, x_i)$ as $X(E) = H_0^1(E)$. As illustrated in Fig. 5, at x_i, $i = 1, 2, 3, 4$, we define q_i, supported in (x_{i-1}, x_{i+1}), by:

1. Setting q_i on the boundary of each element separately, so $q_i(x_j) = \delta_{ij}$;
2. Solving the homogeneous problem on each element E
   Find $q_i \in X(E) + \bar{q}_i(x)$ such that

$$(aq_i', w')_E = 0 \quad \forall\, w \in X(E),$$

   where E is (x_{i-1}, x_i) or (x_i, x_{i+1}), using the appropriate linear function $\bar{q}_i(x)$ for the boundary conditions;
3. Joining the two pieces together continuously to form q_i and setting $X_h = \text{span}\{q_i\}$.

We illustrate the finite element solutions that result from using standard and multiscale finite elements in Fig. 6. This is merely an illustration, since the multiscale finite elements reproduce the exact solution in one dimension, but *not* in

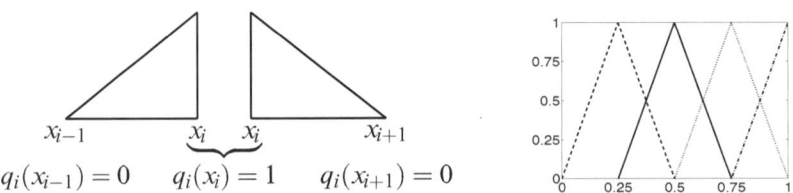

Fig. 4 Construction of standard piecewise linear basis functions in one dimension

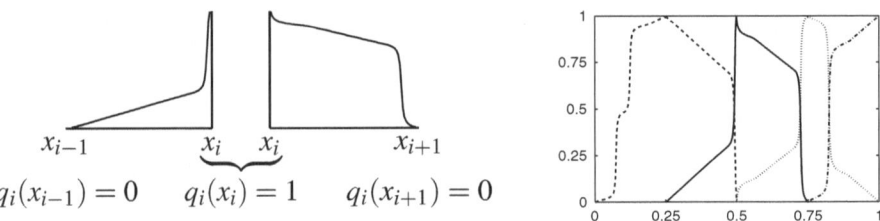

Fig. 5 Construction of multiscale basis functions in one dimension

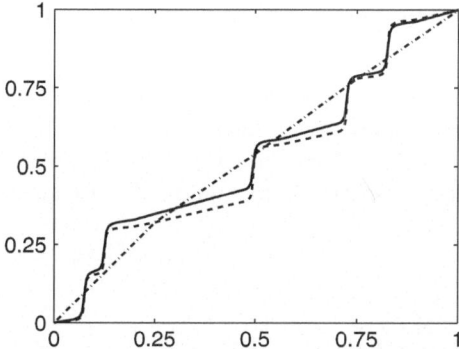

Fig. 6 The *solid line* is the true solution, the *dashed line* illustrates the multiscale finite element solution (actually, it is exact), and the *dashed-dotted line* is the standard finite element solution

higher dimensions. To avoid misleading the reader, an error has been displayed. The point is that standard elements simply cannot represent the microstructure, whereas multiscale elements have this ability.

For the general multi-dimensional problem

$$-\nabla \cdot a_\epsilon \nabla p = f \quad \text{in } \Omega, \tag{17}$$

$$-a_\epsilon \nabla p \cdot v = 0 \quad \text{on } \partial\Omega, \tag{18}$$

the standard variational problem is
Find $p \in X = H^1(\Omega)/\mathbb{R}$ such that

$$A_\epsilon(p, w) := (a_\epsilon \nabla p, \nabla w) = (f, w) \quad \forall\, w \in X. \tag{19}$$

General finite element construction. Let $\mathcal{T}_h$ be a finite element partition of Ω. Define standard $\bar{q}_i$ and multiscale q_i finite elements on an element $E \in \mathcal{T}_h$ as follows.

1. Set $\bar{q}_i$ and q_i on ∂E to be some simple polynomial. More generally, in the multiscale case, we can set $q_i = \ell_i(x)$ on ∂E, where ℓ_i is any appropriate function.
2. Use some polynomial interpolation over E to define $\bar{q}_i$. However, for multiscale elements, we solve the homogeneous problem on each element E
   Find $q_i \in X(E) + \ell_i(x)$ such that

$$A_\epsilon(q_i, w)_E = 0 \quad \forall\, w \in X(E),$$

i.e., we solve the Dirichlet problems (on a fine grid)

$$-\nabla \cdot a_\epsilon \nabla q_i = 0 \quad \text{in } E, \tag{20}$$

$$q_i = \ell_i \quad \text{on } \partial E. \tag{21}$$

3. Join the pieces together continuously to form $\bar{X}_h = \text{span}\{\bar{q}_i\}$ and $X_h = \text{span}\{q_i\}$.

We remark that the multiscale approach has a lot of flexibility in Steps 1 and 2, and there exist many variants of the above procedure.

Multiscale structure of X_h. When we set q_i to be $\bar{q}_i$ on each element boundary (i.e., we take $\ell_i = \bar{q}_i$), we can exhibit the multiscale structure of X_h by noting that

$$q_i = \bar{q}_i + (q_i - \bar{q}_i) =: \bar{q}_i + q_i'.$$

In this form, we define q_i' by

Find $q_i' \in X(E)$ such that

$$A_\epsilon(q_i', w')_E = -A_\epsilon(\bar{q}_i, w')_E \quad \forall\, w' \in X(E).$$

The $q_i' \in X(E) = H_0^1(E)$ are "bubble functions," localized to a coarse element by imposing homogeneous Dirichlet boundary conditions. They are fine-scale and contain the microstructure information. The $\bar{q}_i$ are coarse-scale.

Theorem 3. *Let $X_h' = \text{span}\{q_i'\}$. Then*

$$X_h = \text{span}\{\bar{q}_i + q_i'\} \subsetneq \bar{X}_h \oplus X_h'$$

is a Hilbert space direct sum decomposition into coarse and fine scales.

3.2 Nonmixed Variational Multiscale Method

Again for simplicity, we introduce the variational multiscale method in the non-mixed case, treating the system (17)–(18). Recall that the objective is to modify the variational form of the differential system (19) to better capture the fine scales.

We begin by separating the solution space $X = H^1(\Omega)/\mathbb{R}$ into coarse and fine scales using a Hilbert space direct sum decomposition. Let

$$X = \bar{X} \oplus X', \tag{22}$$

and separate the standard variational form into coarse and fine scales through the test functions as

Find $p = \bar{p} + p' \in \bar{X} \oplus X'$ such that

$$A_\epsilon(\bar{p} + p', \bar{w}) = (f, \bar{w}) \quad \forall\, \bar{w} \in \bar{X} \quad \text{(Coarse scales)}, \tag{23}$$

$$A_\epsilon(\bar{p} + p', w') = (f, w') \quad \forall\, w' \in X' \quad \text{(Fine scales)}. \tag{24}$$

Rewrite the fine scale equation as

$$A_\epsilon(p', w') = (f, w') - A_\epsilon(\bar{p}, w') \quad \forall\, w' \in X',$$

and note that this is a well defined problem for p'. It implicitly defines an *affine* upscaling operator taking $\bar{X}$ to X'. The linear part of the operator is $\hat{p}' : \bar{X} \to X'$, and it satisfies

$$A_\epsilon(\hat{p}'(\bar{q}), w') = -A_\epsilon(\bar{q}, w') \quad \forall\, w' \in X'. \tag{25}$$

The constant part of the upscaling operator is $\tilde{p}' \in X'$, which satisfies

$$A_\epsilon(\tilde{q}', w') = (f, w') \quad \forall\, w' \in X'. \tag{26}$$

The full upscaling operator is $\hat{p}'(\cdot) + \tilde{p}' : \bar{X} \to X'$, and given coarse scales, we can obtain fine scales as

$$p' = \hat{p}'(\bar{p}) + \tilde{p}'. \tag{27}$$

Now the coarse scale equation is simply

$$A_\epsilon\big(\bar{p} + \hat{p}'(\bar{p}), \bar{w}\big) = (f, \bar{w}) - A_\epsilon(\tilde{p}', \bar{w}) \quad \forall\, \bar{w} \in \bar{X},$$

and the effect of the fine scales is manifest within this coarse-scale variational problem. Taking $w' = \hat{p}'(\bar{w})$ in (25) enables us to symmetrize the form to

$$A_\epsilon\big(\bar{p} + \hat{p}'(\bar{p}), \bar{w} + \hat{p}'(\bar{w})\big) = (f, \bar{w}) - A_\epsilon(\tilde{p}', \bar{w}) \quad \forall\, \bar{w} \in \bar{X}. \tag{28}$$

If we define the bilinear and linear forms to be

$$B_\epsilon(\bar{p}, \bar{w}) = A_\epsilon\big(\bar{p} + \hat{p}'(\bar{p}), \bar{w} + \hat{p}'(\bar{w})\big) \quad \text{and} \quad \mathscr{F}(\bar{w}) = (f, \bar{w}) - A_\epsilon(\tilde{p}', \bar{w}),$$

then we have the modified variational form

$$B_\epsilon(\bar{p}, \bar{w}) = \mathscr{F}(\bar{w}) \quad \forall\, \bar{w} \in \bar{X}. \tag{29}$$

In the variational multiscale method, both the bilinear and linear forms are modified.
Choice of Hilbert space decomposition. To be useful for finite element approximation, we need to localize the fine scales. For $\mathscr{T}_h$ a (coarse) finite element partition of Ω, let

$$X' := \bigoplus_E X(E) = \bigoplus_E H_0^1(E),$$

and then $X = \bar{X} \oplus X'$, where

$$\bar{X} = X/X' \simeq \{q|_e : e \text{ is a coarse edge of } \mathscr{T}_h\}.$$

Thus $\bar{X}$ is determined by its values on $\partial E \; \forall \, E \in \mathscr{T}_h$.

Finite element approximation. We use a standard Galerkin finite element space $\bar{X}_h = \{\bar{q}_h\}$ and the multiscale fine space X'. That is, X' is localized and

$$\bar{X}_h \oplus X' \subsetneq \bar{X} \oplus X' = H^1/\mathbb{R}.$$

In practice, we must further approximate $X'|_E = X(E)$ on each element E by a finite element space on a very fine mesh $X_{h_f}(E)$. However, since E is small, we can make this approximation as accurate as we need, and so for simplicity we will assume that it is handled exactly.

We have three equivalent ways to describe the finite element approximation. The primary method is given by direct approximation of (29); however, it is instructive to instead start from the original two-scale decomposition (23)–(24). This leads to

Version 1. Find $p_h = \bar{p}_h + p'_h \in \bar{X}_h \oplus X'$ such that

$$A_\epsilon(p_h, w) = (f, w) \quad \forall w \in \bar{X}_h \oplus X'.$$

But $\bar{X}_h \oplus X'$ is a very large space. In fact, $\bar{p}_h$ and p'_h are related, and the solution is in a much smaller space.

Since Galerkin methods minimize energy, the multiscale solution minimizes energy in the large space $\bar{X}_h \oplus X'$. For these methods, if one specifies the value of the finite elements on ∂E, then the best approximation comes from using the finite element that minimizes energy within E.

Theorem 4. *If the multiscale finite elements are specified on ∂E for each element $E \in \mathcal{T}_h$, then the best approximation comes from using the multiscale finite element that minimizes energy within E.*

By solving for the upscaling operator as above, we obtain the equivalent form

Version 2. Find $\bar{p}_h \in \bar{X}_h$ such that

$$B_\epsilon(\bar{p}_h, \bar{w}) = \mathscr{F}(\bar{w}) \quad \forall \bar{w} \in \bar{X}_h.$$

Now $\bar{X}_h$ is very small, but we must find the upscaling operator to relate $\bar{q}_h$ and $p'_h(\bar{q}_h)$. Given a basis $\bar{X}_h = \text{span}\{\bar{q}_i\}$, we solve a local Dirichlet problem for each $\bar{q}_i$ on element E

$$A_\epsilon(\bar{q}_i + \hat{p}'(\bar{q}_i), w')_E = 0 \quad \forall w' \in X(E).$$

These are the same problems as in the multiscale finite element case, so $X_h = \text{span}\{\bar{q}_i + \hat{p}'(\bar{q}_i)\}$ are the same elements from Theorem 3, and we can reformulate the variational multiscale method as a multiscale finite element method

Version 3. Find $p_h \in X_h$ such that

$$A_\epsilon(p_h, w) = (f, w) - A_\epsilon(\tilde{p}', w) \quad \forall w \in X_h.$$

Theorem 5. *Up to treatment of f (i.e., $\tilde{p}'$), the variational multiscale and multiscale finite element methods are the same in this basic setting.*

Unlike multiscale finite elements, the variational multiscale method naturally handles nonzero f. Henceforth we will use this correction in the multiscale finite element method as well.

4 Mixed Variational Multiscale Method

The mixed case (4)–(5) is complicated by the fact that we treat directly both the scalar unknown $p \in W = L^2(\Omega)/\mathbb{R}$ and the vector unknown $\mathbf{u} \in \mathbf{V} = H_0(\text{div}; \Omega)$. We base our two-scale expansion of the solution space $W \times \mathbf{V}$ on the local mass conservation principle. Given a coarse computational mesh of elements $\mathcal{T}_h$ on Ω with element edges (or faces) $\mathcal{E}_h$, let the *pressure space* be $W = \bar{W} \oplus W'$, where

$$\bar{W} \subset \{\bar{w} \in W : \bar{w} \text{ is constant on each coarse element } E \in \mathcal{T}_h\},$$

$$W' := \bar{W}^\perp.$$

The *velocity space* is then $\mathbf{V} = \bar{\mathbf{V}} \oplus \mathbf{V}'$, where

$$\mathbf{V}' := \{\mathbf{v}' \in \mathbf{V} : \nabla \cdot \mathbf{v}' \in W', \ \mathbf{v}' \cdot \nu = 0 \text{ on } \partial E \ \forall \ E \in \mathcal{T}_h\},$$

$$\bar{\mathbf{V}} := \mathbf{V}/\mathbf{V}' \simeq \{\mathbf{v} \cdot \nu \text{ on } \partial E : E \in \mathcal{T}_h\}.$$

Note that $\mathbf{V}'$ is localized by imposing homogeneous Neumann boundary conditions, leaving $\bar{\mathbf{V}}$ with full normal velocity coupling on the coarse edges $e \in \mathcal{E}_h$. Note also that we have decomposed $\mathbf{V}$ according to coarse and fine scales related to mass conservation, since we can define $\bar{\mathbf{V}}$ explicitly in such a way that

$$\nabla \cdot \bar{\mathbf{V}} = \bar{W} \quad \text{(coarse conservation)},$$

$$\nabla \cdot \mathbf{V}' = W' \quad \text{(fine subgrid conservation)}.$$

We can now separate scales uniquely via the direct sum as

$$p = \bar{p} + p' \in \bar{W} \oplus W' \quad \text{and} \quad \mathbf{u} = \bar{\mathbf{u}} + \mathbf{u}' \in \bar{\mathbf{V}} \oplus \mathbf{V}'.$$

Moreover, we can separate the variational form (4)–(5) into coarse scales

$$(a_\epsilon^{-1}(\bar{\mathbf{u}} + \mathbf{u}'), \bar{\mathbf{v}}) - (\bar{p}, \nabla \cdot \bar{\mathbf{v}}) = 0 \quad \forall \ \bar{\mathbf{v}} \in \bar{\mathbf{V}}, \tag{30}$$

$$(\nabla \cdot \bar{\mathbf{u}}, \bar{w}) = (f, \bar{w}) \qquad \qquad \forall \ \bar{w} \in \bar{W}, \tag{31}$$

and fine scales

$$(a_\epsilon^{-1}(\bar{\mathbf{u}} + \mathbf{u}'), \mathbf{v}') - (p', \nabla \cdot \mathbf{v}') = 0 \quad \forall \, \mathbf{v}' \in \mathbf{V}', \tag{32}$$

$$(\nabla \cdot \mathbf{u}', w') = (f, w') \qquad\qquad \forall \, w' \in W'. \tag{33}$$

As noted in [6], we have the following well-posedness result.

Lemma 3. *The inf-sup condition holds over both $\bar{W} \times \bar{\mathbf{V}}$ and $W' \times \mathbf{V}'$, with constants independent of the coarse mesh and ϵ. Moreover, given $\bar{\mathbf{u}} \in \bar{\mathbf{V}}$, there exists a unique solution $(p', \mathbf{u}') \in W' \times \mathbf{V}'$ and*

$$\|p'\|_0 + \|\mathbf{u}'\|_{\mathbf{V}} \le C\{\|f\|_0 + \|\bar{\mathbf{u}}\|_0\}.$$

Thus we can define a *closure operator* relating fine scales to coarse scales from (32)–(33). This is an affine operator, with linear part defined for $\bar{\mathbf{v}} \in \bar{\mathbf{V}}$ as $(\hat{p}', \hat{\mathbf{u}}') \in W' \times \mathbf{V}'$, where

$$(a_\epsilon^{-1}(\bar{\mathbf{v}} + \hat{\mathbf{u}}'), \mathbf{v}') - (\hat{p}', \nabla \cdot \mathbf{v}') = 0 \quad \forall \, \mathbf{v}' \in \mathbf{V}', \tag{34}$$

$$(\nabla \cdot \hat{\mathbf{u}}', w') = 0 \qquad\qquad \forall \, w' \in W', \tag{35}$$

and constant part defined as $(\tilde{p}', \tilde{\mathbf{u}}') \in W' \times \mathbf{V}'$, where

$$(a_\epsilon^{-1}\tilde{\mathbf{u}}', \mathbf{v}') - (\tilde{p}', \nabla \cdot \mathbf{v}') = 0 \quad \forall \, \mathbf{v}' \in \mathbf{V}', \tag{36}$$

$$(\nabla \cdot \tilde{\mathbf{u}}', w') = (f, w') \qquad\qquad \forall \, w' \in W'. \tag{37}$$

That is,

$$p' = \hat{p}'(\bar{\mathbf{u}}) + \tilde{p}' \quad \text{and} \quad \mathbf{u}' = \hat{\mathbf{u}}'(\bar{\mathbf{u}}) + \tilde{\mathbf{u}}'.$$

Lemma 4. *The operator $\hat{\mathbf{u}}' : \bar{\mathbf{V}} \to \mathbf{V}' \cap \ker(\nabla \cdot)$ is bounded and linear.*

Using the upscaling operator to replace fine-scale quantities in (32)–(33), we obtain the upscaled variational problem (written in symmetric form)
   Find $(\bar{p}, \bar{\mathbf{u}}) \in \bar{W} \times \bar{\mathbf{V}}$ such that

$$(a_\epsilon^{-1}(\bar{\mathbf{u}} + \hat{\mathbf{u}}'(\bar{\mathbf{u}})), (\bar{\mathbf{v}} + \hat{\mathbf{u}}'(\bar{\mathbf{v}}))) - (\bar{p}, \nabla \cdot \bar{\mathbf{v}}) = -(a_\epsilon^{-1}\tilde{\mathbf{u}}', \bar{\mathbf{v}}) \quad \forall \, \bar{\mathbf{v}} \in \bar{\mathbf{V}}, \tag{38}$$

$$(\nabla \cdot \bar{\mathbf{u}}, \bar{w}) = (f, \bar{w}) \qquad\qquad \forall \, \bar{w} \in \bar{W}, \tag{39}$$

with the full solution given by

$$p = \bar{p} + \hat{p}'(\bar{\mathbf{u}}) + \tilde{p}' \quad \text{and} \quad \mathbf{u} = \bar{\mathbf{u}} + \hat{\mathbf{u}}'(\bar{\mathbf{u}}) + \tilde{\mathbf{u}}'.$$

Note that the equations maintain strict local conservation on both scales.
   We can also rewrite the problem as
      Find $(\bar{p}, \bar{\mathbf{u}}) \in \bar{W} \times \bar{\mathbf{V}}$ such that

$$(a_\epsilon^{-1}\bar{\mathbf{u}}, \bar{\mathbf{v}}) - (a_\epsilon^{-1}\hat{\mathbf{u}}'(\bar{\mathbf{u}}), \hat{\mathbf{u}}'(\bar{\mathbf{v}})) - (\bar{p}, \nabla \cdot \bar{\mathbf{v}}) = -(a_\epsilon^{-1}\tilde{\mathbf{u}}', \bar{\mathbf{v}}) \quad \forall \bar{\mathbf{v}} \in \bar{\mathbf{V}},$$

$$(\nabla \cdot \bar{\mathbf{u}}, \bar{w}) = (f, \bar{w}) \qquad \qquad \qquad \forall \bar{w} \in \bar{W}.$$

The positive diffusion term $(a_\epsilon^{-1}\bar{\mathbf{u}}, \bar{\mathbf{v}})$ has a negative subscale correction, which is therefore *antidiffusive* on the coarse scale. This is the main reason that effective parameters (merely replacing a_ϵ on E by some average quantity in the original equations) cannot work well. Rather, some multiscale ideas are needed.

Numerical Approximation. Choose any inf-sup stable mixed space $\bar{W}_h \times \bar{\mathbf{V}}_h$ on the coarse mesh [21, 26]. Then we approximate $p \approx p_h$ and $\mathbf{u} \approx \mathbf{u}_h$ as

Version 1. Find $(\bar{p}_h, \bar{\mathbf{u}}_h) \in \bar{W}_h \times \bar{\mathbf{V}}_h$ such that

$$\left(a_\epsilon^{-1}(\bar{\mathbf{u}}_h + \hat{\mathbf{u}}'(\bar{\mathbf{u}}_h)), \bar{\mathbf{v}}_h + \hat{\mathbf{u}}'(\bar{\mathbf{v}}_h)\right) - (\bar{p}_h, \nabla \cdot \bar{\mathbf{v}}_h) = -(a_\epsilon^{-1}\tilde{\mathbf{u}}', \bar{\mathbf{v}}_h) \quad \forall \bar{\mathbf{v}}_h \in \bar{\mathbf{V}}_h, \tag{40}$$

$$(\nabla \cdot \bar{\mathbf{u}}_h, \bar{w}_h) = (f, \bar{w}_h) \qquad \qquad \forall \bar{w}_h \in \bar{W}_h, \tag{41}$$

and then set

$$p_h = \bar{p}_h + \hat{p}'(\bar{\mathbf{u}}_h) + \tilde{p}' \quad \text{and} \quad \mathbf{u}_h = \bar{\mathbf{u}}_h + \hat{\mathbf{u}}'(\bar{\mathbf{u}}_h) + \tilde{\mathbf{u}}'. \tag{42}$$

By defining the space

$$\mathbf{V}_h := \{\bar{\mathbf{v}}_h + \hat{\mathbf{u}}'(\bar{\mathbf{v}}_h) : \bar{\mathbf{v}}_h \in \bar{\mathbf{V}}_h\} \subsetneq \bar{\mathbf{V}}_h + \mathbf{V}', \tag{43}$$

we can express the method as the multiscale finite element method

Version 2. Find $\bar{p}_h \in \bar{W}_h$ and $\mathbf{u}_h \in \mathbf{V}_h + \tilde{\mathbf{u}}'$ such that

$$(a_\epsilon^{-1}\mathbf{u}_h, \mathbf{v}_h) - (\bar{p}_h, \nabla \cdot \mathbf{v}_h) = 0 \quad \forall \mathbf{v}_h \in \mathbf{V}_h, \tag{44}$$

$$(\nabla \cdot \mathbf{u}_h, \bar{w}_h) = (f, \bar{w}_h) \qquad \forall \bar{w}_h \in \bar{W}_h, \tag{45}$$

with the reconstruction $p_h = \bar{p}_h + \hat{p}'(\bar{\mathbf{u}}_h) + \tilde{p}'$. Furthermore, after some manipulation of the equations, we can express the method in the non-computable form

Version 3. Find $p_h \in W$ and $\mathbf{u}_h \in \bar{V}_h + \mathbf{V}'$ such that

$$(a_\epsilon^{-1}\mathbf{u}_h, \mathbf{v}_h) - (p_h, \nabla \cdot \mathbf{v}_h) = 0 \quad \forall \mathbf{v}_h \in \bar{\mathbf{V}}_h + \mathbf{V}', \tag{46}$$

$$(\nabla \cdot \mathbf{u}_h, w) = (f, w) \qquad \forall w \in W. \tag{47}$$

5 Mixed Multiscale Finite Elements

The mixed multiscale finite element method is defined above in (44)–(45). Our task is to define the discrete spaces. We will define some of the simplest mixed multiscale finite elements that are commonly used. In all cases, we will take the *pressure space*

$$\bar{W}_h := \{\bar{w} \in L^2(\Omega) : \bar{w} \text{ is constant on each coarse element } E \in \mathcal{T}_h\}.$$

We deal with the fact that $W_h \not\subset W = L^2(\Omega)/\mathbb{R}$ in the usual way. Since

$$\bar{\mathbf{V}} \simeq \{\mathbf{v} \cdot \nu \text{ on } \partial E : E \in \mathcal{T}_h\},$$

we need only specify $\bar{\mathbf{v}} \in \bar{\mathbf{V}}$ on coarse element edges $e \in \mathcal{E}_h$. We obtain the corresponding multiscale finite element $\mathbf{v}_h$ by solving the local Neumann problem. For simplicity, we work in two dimensions on rectangular elements.

Raviart-Thomas mixed elements (RT0). The standard lowest order Raviart-Thomas vector variable space $\mathbf{V}_h^{\text{RTO}}$ [61] has one basis function $\mathbf{v}_e^{\text{RTO}}$ associated with each coarse element edge $e \in \mathcal{E}_h$. The degrees of freedom are the constants $\mathbf{v} \cdot \nu|_e$ for each edge $e \in \mathcal{E}_h$. For example, if $E = [0, 1]^2$ then we have the four pieces

$$\mathbf{v}_{\text{L}} = \begin{pmatrix} 1-x \\ 0 \end{pmatrix}, \quad \mathbf{v}_{\text{R}} = \begin{pmatrix} x \\ 0 \end{pmatrix}, \quad \mathbf{v}_{\text{B}} = \begin{pmatrix} 0 \\ 1-y \end{pmatrix}, \quad \mathbf{v}_{\text{T}} = \begin{pmatrix} 0 \\ y \end{pmatrix},$$

which are joined to neighbors across the edge for which $\mathbf{v} \cdot \nu = 1$. This is the standard polynomial definition.

We can also define these finite elements as the solution to two types of differential equations. The first we will call the *element definition*. For each edge $e \in \mathcal{E}_h$, let $E_{e,i}, i = 1, 2$, be the two elements that contain e. We solve on $E = E_{e,i}, i = 1, 2$,

$$\begin{cases} \mathbf{v}_e^{\text{RTO}} = -\nabla \phi_e^{\text{RTO}} & \text{in } E, \\ \nabla \cdot \mathbf{v}_e^{\text{RTO}} = \pm|e|/|E| & \text{in } E, \\ \mathbf{v}_e^{\text{RTO}} \cdot \nu = \begin{cases} 0 \text{ on } \partial E \setminus e, \\ 1 \text{ on } e. \end{cases} \end{cases} \tag{48}$$

However, it is equally valid to define $\mathbf{v}_e^{\text{RTO}}$ on the *dual-support* element $E_e = E_{e,1} \cup E_{e,2}$ by solving

$$\begin{cases} \mathbf{v}_e^{\text{RTO}} = -\nabla \phi_e^{\text{RTO}} & \text{in } E_e, \\ \nabla \cdot \mathbf{v}_e^{\text{RTO}} = \pm|e|/|E_{e,i}| & \text{in } E_{e,i}, \ i = 1, 2, \\ \mathbf{v}_e^{\text{RTO}} \cdot \nu = 0 & \text{on } \partial E_e. \end{cases} \tag{49}$$

We have the following convergence result [61]. Since these elements have no dependence on the scale ϵ, they are accurate only when $h < \epsilon$, i.e., h resolves the fine-scale heterogeneity.

Theorem 6. $\|\mathbf{u} - \mathbf{u}_h^{RT0}\|_0 \leq C \|\mathbf{u}\|_1 h = \mathcal{O}(h/\epsilon)$.

The main idea of multiscale finite elements is to use a_ϵ in their definition. In the boundary value problems above, we simply insert the coefficient a_ϵ.

Variational multiscale element (ME0) based on RT0. The simplest multiscale element (ME0) is due implicitly to Arbogast, Minkoff, and Keenan in 1998 [10] (cf. [8,28]). It is based on using RT0 as the coarse space in the variational multiscale method, or, equivalently, using the element definition of the RT0 element above. Define $\mathbf{v}_e^{ME0} \in \mathbf{V}_h^{ME0}$ for each coarse element edge $e \in \mathcal{E}_h$ by solving

$$
\begin{cases}
\mathbf{v}_e^{ME0} = -a_\epsilon \nabla \phi_e^{ME0} & \text{in } E, \\
\nabla \cdot \mathbf{v}_e^{ME0} = \pm |e|/|E| & \text{in } E, \\
\mathbf{v}_e^{ME0} \cdot \nu = \begin{cases} 0 \text{ on } \partial E \setminus e, \\ 1 \text{ on } e. \end{cases}
\end{cases}
\tag{50}
$$

We have the following convergence result [6, 8, 28].

Theorem 7. *In general,* $\|\mathbf{u} - \mathbf{u}_h^{ME0}\|_0 \leq C \|\mathbf{u}\|_1 h = \mathcal{O}(h/\epsilon)$. *In the two-scale separation case,* $\mathbf{u}_0$ *is the solution of the homogenized problem and*

$$
\|\mathbf{u} - \mathbf{u}_h^{ME0}\|_0 \leq C \left\{ \|\mathbf{u}_0\|_1 h + \|\mathbf{u}_0\|_0 \epsilon + \|\mathbf{u}_0\|_{0,\infty} \sqrt{\epsilon/h} \right\}.
$$

Since $\mathbf{u}_0$ is independent of ϵ,

$$
\|\mathbf{u} - \mathbf{u}_h^{ME0}\|_0 = \mathcal{O}\left(\min \left\{ h/\epsilon, \, h + \epsilon + \sqrt{\epsilon/h} \right\} \right).
$$

Multiscale dual-support (MD) elements. Elements based on RT0 can also be defined using the dual-support definition. This was done first by Aarnes et al. [3,4]. Define $\mathbf{v}_e^{MD} \in \mathbf{V}_h^{MD}$ for each coarse element edge $e \in \mathcal{E}_h$ by solving

$$
\begin{cases}
\mathbf{v}_e^{MD} = -a_\epsilon \nabla \phi_e^{MD} & \text{in } E_e, \\
\nabla \cdot \mathbf{v}_e^{MD} = \pm |e|/|E_{e,i}| & \text{in } E_{e,i}, \ i = 1, 2, \\
\mathbf{v}_e^{MD} \cdot \nu = 0 & \text{on } \partial E_e.
\end{cases}
\tag{51}
$$

The shape on $E_{e,1}$ depends on $E_{e,2}$, and vice-versa. Thus, as defined by Ciarlet [31], this is not a finite element. Nevertheless, we will consider it to be a finite element. It has a problem with convergence in the classical sense (i.e., the error should vanish as $h \to 0$) [7]. As noted in [7], the problem is related to anisotropy. For example, if one takes a constant

$$a_\epsilon(x) = a = Q \Lambda Q^T \quad \text{with } \Lambda = \begin{pmatrix} \lambda_1 & 0 \\ 0 & \lambda_2 \end{pmatrix} \text{ and } Q \text{ a rotation,}$$

we have a genuine anisotropy when the rotation is not a multiple of $\pi/2$ and $\lambda_1 \neq \lambda_2$, but no microstructure. On $e \in \mathscr{E}_h$, the MD element $\mathbf{v}_e^{\mathrm{MD}}$ has a nonconstant normal trace $\mathbf{v}_e^{\mathrm{MD}} \cdot \nu$. Therefore the space $\mathbf{V}_h^{\mathrm{MD}}$ cannot reproduce constants, so the method cannot converge in any reasonable sense as $h \to 0$.

It should be noted that these elements are designed to be effective when $h > \epsilon$. If for some reason one would take $h \to 0$, one should also change elements when h becomes smaller than $\mathscr{O}(\epsilon)$. Moreover, there are many variants of this basic element as defined herein which improve the convergence properties when $h > \epsilon$.

Second order accurate elements (BDM1 and ME1). The standard elements of Brezzi, Douglas, and Marini, defined in 1985, also use a piecewise constant scalar space $\bar{W}_h$, but they are formally second order accurate for the velocity space $\mathbf{V}_h^{\mathrm{BDM1}}$ [24, 25]. The BDM1 elements have two degrees of freedom per element edge. That is,

$$\mathbf{v}^{\mathrm{BDM1}} \cdot \nu|_e \text{ is a linear function for each edge } e \in \mathscr{E}_h.$$

However, we maintain the conservation property that $\nabla \cdot \mathbf{v}|_E$ is a constant on each element $E \in \mathscr{T}_h$ for all $\mathbf{v}^{\mathrm{BDM1}} \in \mathbf{V}_h^{\mathrm{BDM1}}$, so $\nabla \cdot \mathbf{V}_h^{\mathrm{BDM1}} = \bar{W}_h$. More precisely, on a rectangular element $E \in \mathscr{T}_h$, a finite element in $\mathbf{V}_h^{\mathrm{BDM1}}$ has eight degrees of freedom $a, b, ..., h$ as

$$\mathbf{v}^{\mathrm{BDM1}} = \begin{pmatrix} a + bx + cy + 2gxy - hx^2 \\ d + ex + fy - gy^2 + 2hxy \end{pmatrix}.$$

The multiscale variant, due to Arbogast in 2000 [5], is defined using BDM1 as the coarse space in the variational multiscale method, giving $\mathbf{V}_h^{\mathrm{ME1}}$ with two degrees of freedom per edge $e \in \mathscr{E}_h$. For $i = 1, 2$ and L_i a basis for linear polynomials on e, we construct $\mathbf{v}_{e,i}^{\mathrm{ME1}}$ by solving

$$\begin{cases} \mathbf{v}_{e,i}^{\mathrm{ME1}} = -a_\epsilon \nabla \phi_{e,i}^{\mathrm{ME1}} & \text{in } E, \\ \nabla \cdot \mathbf{v}_{e,i}^{\mathrm{ME1}} = \dfrac{1}{|E|} \displaystyle\int_e L_i & \text{in } E, \\ \mathbf{v}_{e,i}^{\mathrm{ME1}} \cdot \nu = \begin{cases} 0 & \text{on } \partial E \setminus e, \\ L_i & \text{on } e, \end{cases} \end{cases} \tag{52}$$

$E(i{=}1)$ $E(i{=}2)$

and joining the two pieces from each side of e.

We have the following convergence result [6, 8].

Theorem 8. *In general,* $\|\mathbf{u} - \mathbf{u}_h^{ME1}\|_0 \leq C\|\mathbf{u}\|_2 h^2 = \mathcal{O}(h^2/\epsilon^2)$. *In the two-scale separation case,* $\mathbf{u}_0$ *is the solution of the homogenized problem and*

$$\|\mathbf{u} - \mathbf{u}_h^{ME1}\|_0 \leq C\left\{\|\mathbf{u}_0\|_2 h^2 + \|\mathbf{u}_0\|_0 \epsilon + \|\mathbf{u}_0\|_{0,\infty}\sqrt{\epsilon/h}\right\} = \mathcal{O}\left(h^2 + \epsilon + \sqrt{\epsilon/h}\right).$$

Some additional elements. As noted earlier, the multiscale finite element approach allows great flexibility in their definition. We note here four main variants.

1. **Oversampled elements (OS).** Hou et al. [28, 46] proposed an *oversampling* technique for the local partial differential systems used to define multiscale finite elements. Instead of solving on $E \in \mathcal{T}_h$, one solves on a larger domain, and then restricts the solution back to E. This gives a nonconforming method, since the pieces do not join continuously across edges. The advantage of oversampling is that it increases the variability on ∂E and allows for better multiscale approximation.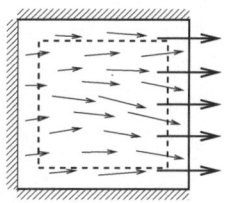

2. **Generalized finite elements and partition of unity methods.** Babuška et al. [14, 16, 64] advocate creating a multiscale finite element basis from local multiscale functions, perhaps defined by oversampling as above. However, instead of simply restricting back to the element $E \in \mathcal{T}_h$ (or E_e for $e \in \mathcal{E}_h$), one uses a partition of unity method so that the resulting elements are conforming.

3. **Reduced dimension-based elements.** Hou and Wu [46] proposed defining the multiscale boundary condition on $e \in \mathcal{E}_h$ by solving a reduced dimension problem. In the case of the nonmixed system, one would first solve the lower dimensional problem $-\nabla_e \cdot a_e \nabla_e p_e = 0$ on e for p_e. Then one sets the boundary condition in (20)–(21) to be $\ell_i = p_e$. Again this improves the variability on ∂E. It is not so clear that this technique applies to the mixed problem, since we need information normal to e (i.e., $\mathbf{v} \cdot \nu$).

4. **Local eigenfunction-based elements.** Efendiev, Galvis, and Wu, as well as Hetmaniuk and Lehoucq [33, 44] propose that, in the differential problems defining the multiscale finite elements, the boundary condition $\mathbf{v} \cdot \nu$ on $e \in \mathcal{E}_h$ be based on the solution of a local eigenfunction problem (solved in, e.g., E_e and restricted to e). The eigenbasis is the most efficient basis, and so should give a good definition of the local boundary condition on e. The energy minimizing extension into $E_{e,i}$, $i = 1, 2$, is then the best choice for finite element, as noted above in Theorem 4. Techniques for efficient definition of the multiscale basis, and for reducing its dimension, are given in [33].

5. **Homogenization-based elements (HE).** Arbogast [7] proposes using homogenization theory to define $\mathbf{v} \cdot \nu$ on $e \in \mathcal{E}_h$ and energy minimizing extension into $E_{e,i}$, $i = 1, 2$. The idea is easily seen from the homogenization theorem, Theorem 2. The microstructure is

$$\mathbf{u}_\epsilon \approx \mathscr{A}_\epsilon \mathbf{u}_0,$$

that is, $\mathbf{u}_\epsilon$ is a fixed, ϵ-scale multiple of a smooth function. Since we know how to approximate a smooth function (e.g., by a polynomial), we should define

$$\mathbf{V}_h \approx \{\mathscr{A}_\epsilon \mathbf{v} : \mathbf{v} \text{ is some nice smooth function}\}.$$

However, these finite elements lie outside $H(\mathrm{div}; \Omega)$, so we use the idea merely to define the normal velocity. For $e \in \mathscr{E}_h$, we approximate the smooth part, corresponding to $\mathbf{u}_0 \cdot v|_e$, by a constant vector. Thus we have two degrees of freedom per edge, and a basis for the *homogenization-based finite element space* (HE) $\mathbf{V}_h^{\mathrm{HE}}$ can be constructed by energy minimizing extension of the normal traces

$$\mathscr{A}_\epsilon \mathbf{e}_i \cdot v|_e \quad \text{to} \quad \mathbf{v}_{e,i}^{\mathrm{HE}}, \quad i = 1, 2.$$

6 A Multiscale Analysis of the Approximation Errors

In this section, we analyze the variational multiscale method (40)–(41), or, equivalently, the multiscale finite element method (44)–(45) with the $\tilde{\mathbf{u}}'$ correction term. Throughout this section, we will tacitly assume that a_ϵ is uniformly positive definite (7) and that $\nabla \cdot \bar{\mathbf{V}}_h \subset \bar{W}_h$. We noted already that the MD finite elements do not converge as $h \to 0$, but the ME0 and ME1 elements are better behaved. We assume that the upscaling operator (i.e., the local problems on E) is solved exactly, since it can be well resolved on a fine grid. Because our problems are well-posed (recall Lemma 3), small discretization errors on the subgrid scale will propagate boundedly in the error estimates that we present later. However, for completeness, we note in passing the following theorem, which accounts for subgrid approximation for standard multiscale elements [6].

Theorem 9. *In the variational multiscale method (40)–(41), suppose that inf-sup stable finite elements $\bar{W}_H \times \bar{\mathbf{V}}_H$ are used on the coarse scale H which approximate velocity to order L. Suppose also that the upscaling operator is approximated on a subgrid of element size $h < H$ by a mixed method $W_h \times \mathbf{V}_h$ approximating the velocity to order ℓ. If $\mathscr{P}_{W_h}$ denotes fine grid L^2-projection onto $W_h = \bar{W}_h + W_h'$, then*

$$\|a_\epsilon^{-1/2}(\mathbf{u} - \mathbf{u}_h)\|_0 \leq \inf_{\substack{v_h \in \bar{\mathbf{V}}_H + v_h' \\ \nabla \cdot v_h = \mathscr{P}_{W_h} f}} \|a_\epsilon^{-1/2}(\mathbf{u} - \mathbf{v}_h)\|_0 \leq C_\epsilon H^L,$$

$$\nabla \cdot \mathbf{u}_h = \mathscr{P}_{W_h} f.$$

Moreover, if pressure is approximated to coarse order M and fine order $m \leq M$, then

$$\|\mathscr{P}_{W_h} p - p_h\|_0 \leq C_\epsilon (H^{M-m} h^{m+1} + H^{L+1}),$$

$$\|p - p_h\|_0 \leq C_\epsilon (H^{M-m} h^m + H^{L+1}).$$

Of course, the constants suffer from the problem of scale. However, in the resolved case, the combination of coarse BDM1 spaces (i.e., ME1) with fine RT0 gives the nice estimates ($L = 2, l = M = m = 1$)

$$\|\mathbf{u} - \mathbf{u}_h\|_0 \leq C_\epsilon H^2, \quad \|\nabla \cdot (\mathbf{u} - \mathbf{u}_h)\|_0 \leq Ch, \quad \text{and} \quad \|p - p_h\|_0 \leq C_\epsilon(h + H^3),$$

which suggests the scaling $H = \sqrt{h}$, giving $\mathcal{O}(h)$ convergence.

We turn now to multiscale error analysis that quantifies the error in terms of h and ϵ. The proof is based on comparison to the homogenized solution, and so applies technically only to the two-scale separation case, i.e., when (8) holds, which we tacitly assume throughout this section. The style of proof is due to Hou et al. [35, 47] in groundbreaking work on the multiscale analysis of finite element methods. For the mixed case, see also [8, 28].

We begin with a general quasioptimality result [6].

Theorem 10. *For any* $\mathbf{v} \in \bar{\mathbf{V}}_h + \mathbf{V}'$ *such that* $\nabla \cdot \mathbf{v} = \nabla \cdot \mathbf{u}_\epsilon = f$,

$$\|a_\epsilon^{-1/2}(\mathbf{u}_\epsilon - \mathbf{u}_h)\|_0 \leq \|a_\epsilon^{-1/2}(\mathbf{u}_\epsilon - \mathbf{v})\|_0,$$

$$\nabla \cdot \mathbf{u}_h = f.$$

Proof. Since we assumed the upscaling operator is solved exactly, we obtain precisely that $\nabla \cdot \mathbf{u}_h = f$ from (47). This means that the multiscale method is locally conservative on the fully resolved fine scale.

The velocity error is bounded by subtracting (4) and (46) and taking a test function $\mathbf{v} - \mathbf{u}_h$, where $\mathbf{v} \in \bar{\mathbf{V}}_h + \mathbf{V}'$ has the required divergence. That we optimize over the larger space $\bar{\mathbf{V}}_h + \mathbf{V}'$ instead of $\mathbf{V}_h$ is a consequence of the fact that the upscaling operator is defined as energy minimizing extension. $\square$

We sketch now a simplified multiscale convergence proof [7] involving certain projection operators and four key results. We analyze here only ME0 and avoid complexities like oversampling, thus proving the multiscale part of Theorem 7. The first key result is quasioptimality, Theorem 10. The second key result is homogenization, Theorem 2, which says that $\mathbf{u}_\epsilon \approx \mathscr{A}_\epsilon \mathbf{u}_0$. Thus our goal is to find any $\mathbf{v}_\epsilon \approx \mathscr{A}_\epsilon \mathbf{u}_0$ in $\mathbf{V}_h^{\text{ME0}} + \tilde{\mathbf{u}}'$ with $\nabla \cdot \mathbf{v}_\epsilon = \nabla \cdot \mathbf{u}_\epsilon = f$.

The third key result involves dealing with the ϵ scale of our finite elements, so we define corresponding *homogenized finite elements*. We replace the true coefficient a_ϵ in the definition of the finite elements (50) with the corresponding homogenized one a_0, giving a finite element space

$$\mathbf{V}_{0,h}^{\text{ME0}} = \operatorname{span}_{e \in \mathscr{E}_h} \{\mathbf{v}_{0,e}^{\text{ME0}}\}.$$

Since our finite elements are defined by boundary value problems, the homogenization theorem applies, although we will see numerical resonance (i.e., factors of ϵ/h) in the estimate, which come from localizing to the element E_e of scale h.

Lemma 5. *For each $e \in \mathscr{E}_h$,*

$$\mathbf{v}_e^{MEO} = \mathscr{A}_\epsilon \mathbf{v}_{0,e}^{MEO} + \theta_\epsilon^{E_e, MEO},$$

$$\|\theta_\epsilon^{E_e, MEO}\|_{0, E_e} \leq C \left\{ \epsilon \|\mathbf{v}_{0,e}^{MEO}\|_{1, E_e} + \sqrt{\epsilon |\partial E_e|} \|\mathbf{v}_{0,e}^{MEO}\|_{0, \infty, E_e} \right\} = \mathscr{O}\left(\frac{\epsilon}{h} + \sqrt{\frac{\epsilon}{h}} \right) h^{d/2}.$$

We next define flux-based projection operators for both $\mathbf{V}_h^{MEO}$ and $\mathbf{V}_{0,h}^{MEO}$, which are related through the lemma above. The average normal flux across $e \in \mathscr{E}_h$ is

$$\gamma_e = \frac{1}{|e|} \int_e \mathbf{v} \cdot \nu_e \, ds.$$

The usual Raviart-Thomas projection is defined as

$$\pi^{RTO} \mathbf{v} := \sum_{e \in \mathscr{E}_h} \gamma_e \mathbf{v}_e^{RTO} \in \mathbf{V}_h^{RTO}.$$

Similarly, we define

$$\pi_\epsilon^{MEO} \mathbf{v} := \sum_{e \in \mathscr{E}_h} \gamma_e \mathbf{v}_e^{MEO} \in \mathbf{V}_h^{MEO} \quad \text{and} \quad \pi_0^{MEO} \mathbf{v} := \sum_{e \in \mathscr{E}_h} \gamma_e \mathbf{v}_{0,e}^{MEO} \in \mathbf{V}_{0,h}^{MEO},$$

leading us to the third key result.

Lemma 6. *Let $\mathscr{P}_{\bar{W}_h}$ be L^2-projection onto $\bar{W}_h$, the space of piecewise discontinuous constants. Then*

$$\nabla \cdot \pi_\epsilon^{MEO} \mathbf{v} = \nabla \cdot \pi_0^{MEO} \mathbf{v} = \nabla \cdot \pi^{RTO} \mathbf{v} = \mathscr{P}_{\bar{W}_h} \nabla \cdot \mathbf{v},$$

$$\|\pi_\epsilon^{MEO} \mathbf{v} - \mathscr{A}_\epsilon \pi_0^{MEO} \mathbf{v}\|_0 \leq C \|\mathbf{v}\|_1 \left(\epsilon/h + \sqrt{\epsilon/h} \right).$$

Proof. The divergence condition follows from the Divergence Theorem. For the estimate, note that

$$\pi_\epsilon^{MEO} \mathbf{v} - \mathscr{A}_\epsilon \pi_0^{MEO} \mathbf{v} = \sum_{e \in \mathscr{E}_h} \gamma_e (\mathbf{v}_e^{MEO} - \mathscr{A}_\epsilon \mathbf{v}_{0,e}^{MEO}) = \sum_{e \in \mathscr{E}_h} \gamma_e \theta_e^{E_e, MEO}.$$

Theorem 2 on homogenization gives us that

$$\|\pi_\epsilon^{MEO} \mathbf{v} - \mathscr{A}_\epsilon \pi_0^{MEO} \mathbf{v}\|_{0, E} \leq \sum_{e \subset \partial E} |\gamma_e| \|\theta_e^{E_e, MEO}\|_{0, E}$$

$$\leq C \sum_{e \subset \partial E} \left(h^{-d/2} \|\mathbf{v}\|_{1, E_e} \right) \left(\frac{\epsilon}{h} + \sqrt{\frac{\epsilon}{h}} \right) h^{d/2}$$

$$= C \sum_{e \subset \partial E} \|\mathbf{v}\|_{1,E_e} \left(\frac{\epsilon}{h} + \sqrt{\frac{\epsilon}{h}} \right),$$

and the proof is completed by squaring, summing over all $E \in \mathscr{T}_h$, and noting that there is finite overlap of the E_e. □

The fourth and final key result concerns smooth approximation by π_0^{ME0}. In general, this is not a nice operator. However, for the types of vector fields we consider, it approximates well.

Lemma 7. *If* $\mathbf{v}_0 = -a_0 \nabla \phi_0$, *then*

$$\|\mathbf{v}_0 - \pi_0^{\text{ME0}} \mathbf{v}_0\|_0 \le C \|\mathbf{v}_0\|_1 h.$$

Proof. Let

$$\psi = \mathbf{v} - \pi_0^{\text{ME0}} \mathbf{v} = -a_0 \nabla \left(\phi_0 - \sum_{e \subset \partial E} \gamma_e \phi_{0,e}^{\text{ME0}} \right) \quad \text{in } E,$$

which is a potential field satisfying the Neumann problem

$$\nabla \cdot \psi = \nabla \cdot \mathbf{v}_0 - \mathscr{P}_{\bar{W}_h} \nabla \cdot \mathbf{v}_0 \quad \text{in } E,$$

$$\psi \cdot \nu_e = \mathbf{v}_0 \cdot \nu_e - \gamma_e \qquad \text{on } e \subset \partial E.$$

The standard energy estimate (see, e.g., [36, 38]) gives the result. □

We are now ready to state and prove the discrete inf-sup condition. We need to use the elliptic regularity theorem (see, e.g., [21, 36, 38]), which requires that Ω have, e.g., a $C^{1,1}$ boundary or that Ω be convex [42].

Lemma 8. *If* Ω *supports elliptic regularity, then there is some* $\beta > 0$, *independent of* ϵ, *such that*

$$\sup_{\mathbf{v}_h \in \mathbf{V}_h^{\text{ME0}}} \frac{(\bar{w}_h, \nabla \cdot \mathbf{v}_h)}{\|\mathbf{v}_h\|_0 + \|\nabla \cdot \mathbf{v}_h\|_0} \ge \beta \|\bar{w}_h\|_0 \quad \forall \, \bar{w}_h \in \bar{W}_h.$$

Proof. Recall that $\bar{w}_h$ is orthogonal to constants. Solve

$$\nabla \cdot \mathbf{v}_0 = \bar{w}_h \qquad \text{in } \Omega, \tag{53}$$

$$\mathbf{v}_0 = -a_0 \nabla \phi_0 \quad \text{in } \Omega, \tag{54}$$

$$\mathbf{v}_0 \cdot \nu = 0 \qquad \text{on } \partial \Omega, \tag{55}$$

and note that elliptic regularity implies that

$$\|\mathbf{v}_0\|_1 \le C\|\bar{w}_h\|_0.$$

Take

$$\mathbf{v}_{h.} = \pi_\epsilon^{\mathrm{ME0}}\mathbf{v}_0 \in \mathbf{V}_h^{\mathrm{ME0}},$$

for which

$$\nabla \cdot \mathbf{v}_h = \mathscr{P}_{\bar{W}_h}\nabla \cdot \mathbf{v}_0 = \bar{w}_h.$$

Then, by the third and fourth key results,

$$\|\mathbf{v}_h\|_0 \le \|\pi_\epsilon^{\mathrm{ME0}}\mathbf{v}_0 - \mathscr{A}_\epsilon\pi_0^{\mathrm{ME0}}\mathbf{v}_0\|_0 + \|\mathscr{A}_\epsilon(\pi_0^{\mathrm{ME0}}\mathbf{v}_0 - \mathbf{v}_0)\|_0 + \|\mathscr{A}_\epsilon\mathbf{v}_0\|_0$$
$$\le C\|\mathbf{v}_0\|_1 \le C\|\bar{w}_h\|_0,$$

and the result follows. □

Theorem 11. *If Ω supports elliptic regularity and $p_0 \in H^2(\Omega) \cap W^{1,\infty}(\Omega)$, then*

$$\|\mathbf{u}_\epsilon - \mathbf{u}_h^{ME0}\|_0 + \|\mathscr{P}_{\bar{W}_h}p_\epsilon - \bar{p}_h\|_0$$
$$\le C\{(\epsilon + \epsilon/h + \sqrt{\epsilon/h} + h)\|\mathbf{u}_0\|_1 + \sqrt{\epsilon}\,\|\mathbf{u}_0\|_{0,\infty}\},$$
$$\nabla \cdot \mathbf{u}_h^{ME0} = f.$$

Proof. Since $\mathbf{u}_\epsilon \approx \pi_\epsilon^{\mathrm{ME0}}\mathbf{u}_0 + \tilde{\mathbf{u}}' \in \mathbf{V}_h^{\mathrm{ME0}} + \tilde{\mathbf{u}}'$ and $\nabla \cdot (\pi_\epsilon^{\mathrm{ME0}}\mathbf{u}_0 + \tilde{\mathbf{u}}') = \mathscr{P}_{\bar{W}_h}\nabla \cdot \mathbf{u}_0 + \nabla \cdot \tilde{\mathbf{u}}' = f$, by quasioptimality, key result one or Theorem 10, we have that

$$\|\mathbf{u}_\epsilon - \mathbf{u}_h^{\mathrm{ME0}}\|_0 \le C\|\mathbf{u}_\epsilon - \pi_\epsilon^{\mathrm{ME0}}\mathbf{u}_0 - \tilde{\mathbf{u}}'\|_0$$
$$\le C\{\|\mathbf{u}_\epsilon - \mathscr{A}_\epsilon\mathbf{u}_0\|_0 + \|\mathscr{A}_\epsilon(\mathbf{u}_0 - \pi_0^{\mathrm{ME0}}\mathbf{u}_0)\|_0$$
$$+ \|\mathscr{A}_\epsilon\pi_0^{\mathrm{ME0}}\mathbf{u}_0 - \pi_\epsilon^{\mathrm{ME0}}\mathbf{u}_0\|_0 + \|\tilde{\mathbf{u}}'\|_0\},$$

and the velocity estimate follows from the final three key results, Theorem 2, Lemmas 6 and 7, and the standard energy estimate of (36)–(37), which says that

$$\|\tilde{\mathbf{u}}'\|_0 \le C\|\mathscr{P}_{\bar{W}_h}^\perp f\|_{-1} \le C\|f\|_0 h \le C\|\mathbf{u}_0\|_1 h.$$

The divergence result follows in general, and the pressure result follows from the inf-sup condition, Lemma 8 and the difference of (4) and (46). □

We remark that a similar proof holds for ME1 [7]. We also note that recent work by Babuška and Lipton [15] and Efendiev, Galvis, and Wu [33] shows multiscale convergence for certain multiscale methods independently of the two-scale separation hypothesis (8).

7   Domain Decomposition and Mortar Methods

In this section, we discuss a restricted class of domain decomposition and mortar methods related to the mixed finite element methods considered earlier for our heterogeneous elliptic problem. In 1988, Glowinski and Wheeler [40] defined nonoverlapping domain decomposition for mixed methods by iterating on the Dirichlet-to-Neumann map. As depicted in Fig. 7, given the pressure λ on the subdomain interfaces Γ, one computes the flow locally. Based on the flux mismatch on Γ (i.e., the jump in $\mathbf{u} \cdot \nu$), one updates λ using, e.g., conjugate gradients. Once converged, the full fine-scale problem is solved. The technique allows great flexibility in handling interdomain multiphysics (different physical models in different subdomains) and is well suited to parallel computation. It allows us to handle interdomain multiscale aspects as well.

In 1994, Bernardi, Maday, and Patera [20] defined mortar methods to glue the subdomains together weakly when the subdomain meshes do not match. As illustrated in Fig. 8, Arbogast, Cowsar, Wheeler, and Yotov in 2000 [9] extended the mortar idea to mixed methods, using a continuous or discontinuous linear mortar λ. The idea was to use grid spacings of $\mathcal{O}(h)$ for all grids.

The mixed mortar method is similar to our previous multiscale techniques. It has the same four basic components noted in Sec. 3 above: localization to subdomains, fine-scale effects resolved on the subdomains, a global interface problem for the mortar unknowns, and fine-grid reconstruction over Ω. If the mortar resolves the computational meshes, i.e., the subdomain and mortar mesh spacings are $\mathcal{O}(h)$, $h < \epsilon$, we obtain a fully resolved and weakly but fully coupled approximation.

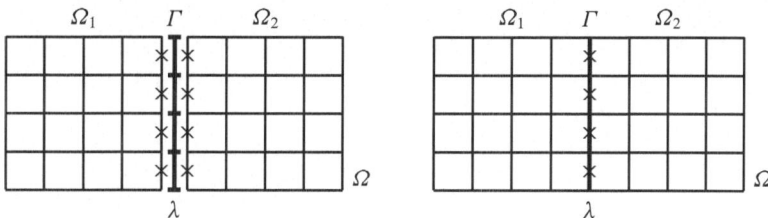

Fig. 7 Illustration of domain decomposition. The domain Ω on the right is shown separated on the left for clarity. On the interface Γ, λ resolves the computational mesh on both $\partial\Omega_1$ and $\partial\Omega_2$

Fig. 8 Illustration of mortar mixed methods. On the interface Γ, λ does not match the computational mesh on $\partial\Omega_1$ and/or $\partial\Omega_2$. The idea here is to consider that h_1, h_2, and h_3 are of the same order, so the problem is fully coupled

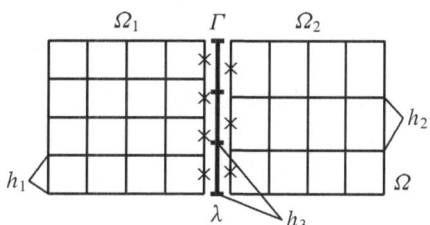

Fig. 9 Illustration of
multiscale mortar mixed
methods. Here, H is much
coarser than h_1 and h_2. We
use a higher order mortar
approximation to compensate
for the coarseness of the grid
and maintain good overall
accuracy

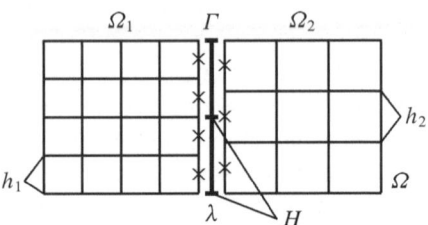

The idea behind the multiscale mortar mixed method is to relax the coupling dynamics as in multiscale methods. As illustrated in Fig. 9, we use the following four main components.

1. **Localization.** Divide Ω into many small subdomains (or coarse elements of scale H), over which the original partial differential system is imposed.
2. **Fine-scale effects.** The subdomains are given Dirichlet boundary conditions $p = \lambda$ on Γ and solved on the fine scale h to define the local solution.
3. **Global coarse-grid interface problem.** The weakly defined flux mismatch (jump in $\mathbf{u} \cdot v$) on Γ is used to define a better λ on scale $H > h$, and we iterate this and the previous step until convergence is attained.
4. **Fine-grid representation of the solution.** We obtain a fully resolved and well coupled approximate solution if λ is approximated in a higher order space.

To be more precise, let Ω be decomposed into nonoverlapping subdomains Ω_i, which correspond to coarse elements in our previous methods. Define the interfaces

$$\Gamma_{ij} := \partial\Omega_i \cap \partial\Omega_j, \quad \Gamma := \bigcup_{i<j} \Gamma_{ij}, \quad \text{and} \quad \Gamma_i := \partial\Omega_i \cap \Gamma.$$

With v_i denoting the outer unit normal to $\partial\Omega_i$, the differential problem (1)–(3) is equivalent to the decomposed system

$$a^{-1}\mathbf{u} = -\nabla p \quad \text{in } \Omega_i \quad \text{(subdomain Darcy's law)}, \tag{56}$$

$$\nabla \cdot \mathbf{u} = f \quad \text{in } \Omega_i \quad \text{(subdomain conservation)}, \tag{57}$$

$$\mathbf{u}|_{\Omega_i} \cdot v_i + \mathbf{u}|_{\Omega_j} \cdot v_j = 0 \quad \text{on } \Gamma_{ij} \quad \text{(conservation on interface } \Gamma\text{)}, \tag{58}$$

$$p|_{\Omega_i} = p|_{\Omega_j} \quad \text{on } \Gamma_{ij} \quad \text{(continuity of } p \text{ on } \Gamma\text{)}, \tag{59}$$

$$\mathbf{u} \cdot v = 0 \quad \text{on } \partial\Omega. \tag{60}$$

A variational form is
Find $p \in L^2(\Omega_i)$, $\mathbf{u} \in H_0(\text{div}; \Omega_i)$, and $\lambda = p \in H^{1/2}(\Gamma)$ such that

$$(a^{-1}\mathbf{u}, \mathbf{v})_{\Omega_i} - (p, \nabla \cdot \mathbf{v})_{\Omega_i} + \langle \lambda, \mathbf{v} \cdot v_i \rangle_{\Gamma_i} = 0 \quad \forall \mathbf{v} \in H_0(\text{div}; \Omega_i), \tag{61}$$

$$(\nabla \cdot \mathbf{u}, w)_{\Omega_i} = (f, w)_{\Omega_i} \qquad\qquad \forall\, w \in L^2(\Omega_i), \qquad (62)$$

$$\sum_i \langle \mathbf{u} \cdot v_i, \mu \rangle_{\Gamma_i} = 0 \qquad\qquad \forall\, \mu \in H^{1/2}(\Gamma), \qquad (63)$$

where we use $\langle \cdot, \cdot \rangle$ for interface inner products for emphasis and $H_0(\mathrm{div};\Omega_i) = \{v \in H(\mathrm{div};\Omega_i) : v \cdot v = 0 \text{ on } \partial\Omega\}$. The last equation enforces continuity of flux on Γ.

The multiscale mortar mixed method. For finite element approximation, define on each Ω_i a fine scale finite element partition $\mathcal{T}_h^{\Omega_i}$ of maximal element diameter $h < \epsilon$, and let $W_{h,i} \times \mathbf{V}_{h,i} \subset L^2(\Omega_i) \times H_0(\mathrm{div};\Omega_i)$ be any of the usual mixed finite element spaces. For the mortar, define a coarse scale finite element partition $\mathcal{T}_H^{\Gamma_{ij}}$ on each Γ_{ij} of maximal element diameter H, and let $M_{H,ij}$ be a space of continuous or discontinuous finite elements. For $W_h = \cup_i W_{h,i}$, $\mathbf{V}_h = \cup_i \mathbf{V}_{h,i}$, and $M_H = \cup_{ij} M_{H,ij}$, the multiscale mortar method is

Find $p_h \in W_h$, $\mathbf{u}_h \in \mathbf{V}_h$, and $\lambda_H \in M_H$ such that

$$(a^{-1}\mathbf{u}_h, \mathbf{v})_{\Omega_i} - (p_h, \nabla \cdot \mathbf{v})_{\Omega_i} + \langle \lambda_H, \mathbf{v} \cdot v_i \rangle_{\Gamma_i} = 0 \quad \forall\, \mathbf{v} \in \mathbf{V}_{h,i}, \qquad (64)$$

$$(\nabla \cdot \mathbf{u}_h, w)_{\Omega_i} = (f, w)_{\Omega_i} \qquad\qquad \forall\, w \in W_{h,i}, \qquad (65)$$

$$\sum_i \langle \mathbf{u}_h \cdot v_i, \mu \rangle_{\Gamma_i} = 0 \qquad\qquad \forall\, \mu \in M_H. \qquad (66)$$

Now the last equation enforces only *weak* continuity of flux on Γ.

The usual way to solve this system is to reduce it to an interface problem [40]. Since it is an affine problem in λ, we define the bilinear and linear forms on M_H by

$$d_H(\lambda, \mu) := \sum_i d_{H,i}(\lambda, \mu) = -\sum_i \langle \hat{\mathbf{u}}_h(\lambda) \cdot v_i, \mu \rangle_{\Gamma_i},$$

$$g_H(\mu) := \sum_i g_{H,i}(\mu) = \sum_i \langle \tilde{\mathbf{u}}_h \cdot v_i, \mu \rangle_{\Gamma_i},$$

where $(\hat{\mathbf{u}}_h(\lambda), \hat{p}_h(\lambda)) \in \mathbf{V}_h \times W_h$ solves the linear part of the problem (i.e., with λ given and $f = 0$)

$$(a^{-1}\hat{\mathbf{u}}_h(\lambda), \mathbf{v})_{\Omega_i} - (\hat{p}_h(\lambda), \nabla \cdot \mathbf{v})_{\Omega_i} = -\langle \lambda, \mathbf{v} \cdot v_i \rangle_{\Gamma_i} \quad \forall\, \mathbf{v} \in \mathbf{V}_{h,i}, \qquad (67)$$

$$(\nabla \cdot \hat{\mathbf{u}}_h(\lambda), w)_{\Omega_i} = 0 \qquad\qquad \forall\, w \in W_{h,i}, \qquad (68)$$

and $(\tilde{\mathbf{u}}_h, \tilde{p}_h) \in \mathbf{V}_h \times W_h$ solves the constant part (i.e., with $\lambda = 0$ and f given)

$$(a^{-1}\tilde{\mathbf{u}}_h, \mathbf{v})_{\Omega_i} - (\tilde{p}_h, \nabla \cdot \mathbf{v})_{\Omega_i} = 0 \quad \forall\, \mathbf{v} \in \mathbf{V}_{h,i}, \qquad (69)$$

$$(\nabla \cdot \tilde{\mathbf{u}}_h, w)_{\Omega_i} = (f, w)_{\Omega_i} \qquad\qquad \forall\, w \in W_{h,i}. \qquad (70)$$

Theorem 12. *The interface bilinear form $d_H(\cdot,\cdot)$ is symmetric and positive definite on M_H. In fact,*

$$d_H(\lambda,\mu) = \left(a^{-1}\hat{\mathbf{u}}_h(\lambda), \hat{\mathbf{u}}_h(\mu)\right). \tag{71}$$

Moreover,

$$d_H(\lambda_H,\mu) = g_H(\mu) \quad \forall \mu \in M_H$$

if, and only if, the solution to (64)–(66) satisfies

$$p_h = \hat{p}_h(\lambda_H) + \tilde{p}_h \quad \text{and} \quad \mathbf{u}_h = \hat{\mathbf{u}}_h(\lambda_H) + \tilde{\mathbf{u}}_h.$$

Thus, our problem reduces to a symmetric and positive definite linear system, and it can be solved, for example, by conjugate gradient iteration. In that case, the computations involve once solving for $(\tilde{\mathbf{u}}_h, \tilde{p}_h)$ to get $g_H(\mu)$, and at each iteration k, solving for $\left(\hat{\mathbf{u}}_h(\lambda_H^k), \hat{p}_h(\lambda_H^k)\right)$ to get $d_H(\lambda_H^k, \mu)$.

Multiscale finite element formulation. Implicit in the method are multiscale finite elements. To see them, let $\{\mu_\ell\}$ be a basis for M_H and define

$$w_\ell := \hat{p}_h(\mu_\ell) \quad \text{and} \quad \mathbf{v}_\ell := \hat{\mathbf{u}}_h(\mu_\ell). \tag{72}$$

Then

$$\lambda_H = \sum_\ell \lambda_\ell \mu_\ell, \quad p_h = \sum_\ell \lambda_\ell w_\ell + \tilde{p}_h, \quad \text{and} \quad \mathbf{u}_h = \sum_\ell \lambda_\ell \mathbf{v}_\ell + \tilde{\mathbf{u}}_h,$$

and the method seeks $\{\lambda_\ell\}$ such that

$$\sum_\ell \lambda_\ell \, d_H(\mu_\ell, \mu_k) = g_H(\mu_k) \quad \forall k,$$

which, using (71) and (70) and taking $\lambda = \mu_k$ and $\mathbf{v} = \tilde{\mathbf{u}}_h$ in (67), is equivalent to

$$\sum_\ell \lambda_\ell (a^{-1}\mathbf{v}_\ell, \mathbf{v}_k) = (f, w_k) - (a^{-1}\tilde{\mathbf{u}}_h, \mathbf{v}_k) \quad \forall k.$$

The multiscale finite elements are now evident [11, 37]. Let

$$N_{h,H} := \mathrm{span}\left\{ \begin{pmatrix} w_\ell \\ \mathbf{v}_\ell \end{pmatrix} \right\}$$

$$= \mathrm{span}\left\{ \begin{pmatrix} \hat{p}_h(\mu_\ell) \\ \hat{\mathbf{u}}_h(\mu_\ell) \end{pmatrix} \right\} \subset \begin{pmatrix} W_h \\ \mathbf{V}_h \end{pmatrix} \tag{73}$$

Then the method is

Find $(p_h, \mathbf{u}_h) \in N_{h,H} + (\tilde{p}_h, \tilde{\mathbf{u}}_h)$ such that

$$(a^{-1}\mathbf{u}_h, \mathbf{v}) = (f, w) \quad \forall (w, \mathbf{v}) \in N_{h,H}.$$

This is an unusual multiscale finite element space. Not only do we couple pressures and velocities, but we allow nonzero normal flow on the boundary of the "coarse elements" $\partial \Omega_i$ (it is merely weakly zero). However, our multiscale finite elements are indeed locally defined over the subdomains.

Variational multiscale method formulation. We can also relate our method to the variational multiscale method in several ways, and thereby extract different sets of multiscale finite elements. If we decompose the discrete mortar space $L^2(\Gamma) = M_H \oplus M'_H$ and drop M'_H, we obtain the original formulation with multiscale finite elements (73).

Another approach is to decompose the velocity space. Define the *weakly continuous normal velocities*

$$\mathbf{V}_w := \left\{ \mathbf{v} \in \mathbf{V}_h : \sum_i \langle \mathbf{v} \cdot \nu_i, \mu \rangle_{\Gamma_i} = 0 \ \forall \mu \in M_H \right\}.$$

The method reduces to: Find $p_h \in W_h$ and $\mathbf{u}_h \in \mathbf{V}_w$, such that

$$(a^{-1}\mathbf{u}_h, \mathbf{v}) - \sum_i (p_h, \nabla \cdot \mathbf{v})_{\Omega_i} = 0 \quad \forall \mathbf{v} \in \mathbf{V}_w, \tag{74}$$

$$\sum_i (\nabla \cdot \mathbf{u}_h, w)_{\Omega_i} = (f, w) \qquad \forall w \in W_h. \tag{75}$$

Our Hilbert space decomposition of $\mathbf{V}_w$ involves the *weakly zero normal velocities*

$$\mathbf{V}'_w := \{ \mathbf{v} \in \mathbf{V}_w : \langle \mathbf{v} \cdot \nu, \mu \rangle_{\Gamma_{ij}} = 0 \ \forall \mu \in M_{H,ij} \text{ and } \forall i, j \}.$$

Then $\bar{\mathbf{V}}_w \simeq \mathbf{V}_w / \mathbf{V}'_w$ is defined by its degrees of freedom on the interfaces as

$$\bar{\mathbf{V}}_w \simeq \{ \langle \mathbf{v}, \mu_\ell \rangle : \mathbf{v} \in \mathbf{V}_w \ \forall \ell \}. \tag{76}$$

With $W'_w := \nabla \cdot \mathbf{V}'_w$ and $\bar{W}_w := (W'_w)^\perp$, we have the decomposition

$$W_h = \bar{W}_w \oplus W'_w \quad \text{and} \quad \mathbf{V}_w = \bar{\mathbf{V}}_w \oplus \mathbf{V}'_w.$$

Proceeding as before, we obtain coarse and fine scale equations analogous to (30)–(33), an upscaling operator analogous to (34)–(37), and the upscaled equation analogous to (38)–(39).

The point is that we obtain formally the same variational multiscale method as before, but now we use *nonconforming* elements with greater flexibility near $\partial \Omega_i$.

The greater flexibility results from the fact that we control the normal velocities only weakly. According to (76) a basis can be found by finding for each ℓ *any* function $\mathbf{v}_{w,\ell} \in \mathbf{V}_w$ such that $\langle \mathbf{v}_{w,\ell}, \mu_k \rangle = \delta_{\ell k}$. That is, we specify $\mathbf{v}_{w,\ell} \cdot \nu$ on the boundaries of the "course elements" $\partial \Omega_i$, but only in the weak sense. Moreover, the boundary condition on the other coarse edges $\Gamma_i \cup \Gamma_j \setminus \Gamma_{ij}$ is only weakly zero, so some flow between subdomains is allowed on the fine scale.

We have the following a-priori error estimates [11].

Theorem 13. *If p_h, $\mathbf{u}_h$, and λ_H are locally approximated by polynomials of degree $\ell - 1$, $k - 1$, and $m - 1$, respectively, then there exists C, independent of h and H, such that*

$$\|\nabla \cdot (\mathbf{u} - \mathbf{u}_h)\|_0 \leq C \|f\|_\ell h^\ell,$$

$$\|\mathbf{u} - \mathbf{u}_h\|_0 \leq C \{\|\mathbf{u}\|_k h^k + \|p\|_{m+1/2} H^{m-1/2} + \|\mathbf{u}\|_{k+1/2} h^k H^{1/2}\},$$

$$\|p - p_h\|_0 \leq C \{\|p\|_\ell h^\ell + \|p\|_{m+1/2} H^{m+1/2}$$
$$+ (\|f\|_\ell h^\ell + \|\mathbf{u}\|_k h^k) H + \|\mathbf{u}\|_{k+1/2} h^k H^{3/2}\},$$

where the last estimate requires that Ω support elliptic regularity.

The velocity estimate is formally $\mathcal{O}(h^k + H^{m-1/2})$, but of course it suffers from the problem of scale when $h < \epsilon < H$. Since λ is defined on Γ, we lose $1/2$-derivative going to Ω, or $H^{-1/2}$ in the estimates. Thus when h and H are of the same order, we need polynomials of degree 1 (actually 1/2) more for the mortar approximation [9].

The error estimate bounds given in Theorem 13 depend on the solution, and so may be very large for our heterogeneous problem. To deal with this problem of scale, one can use a-posteriori analysis to adapt the meshes to the scales of the system [9,58]. One can also define a mortar space based on homogenization, which results in an error estimate for the two-scale separation case that is optimal and has no numerical resonance term [12]. From Theorem 2,

$$\lambda \approx p_\epsilon^1 = \left(1 + \epsilon \sum_{j=1}^{d} \omega_j(x, x/\epsilon) \frac{\partial}{\partial x_j}\right) p_0(x), \tag{77}$$

so we define M_H by replacing p_0 above by piecewise polynomials and restrict back to Γ.

8   Some Numerical Results

We present two numerical test examples that model incompressible single phase flow in a porous medium. The permeability fields are geostatistically generated, with the first being statistically homogeneous. Since homogenization theory extends to this case, the multiscale convergence results relating the coarse grid spacing h to ϵ apply, as do the techniques mentioned that are defined using homogenization theory. The other test case has a permeability that is highly nonhomogeneous in the statistical sense, and so the convergence results do not strictly apply. That is, we are far from the case of scale separation. Nevertheless, the example is presented to demonstrate that the methods presented herein can work well even in these cases.

In each example, the domain is a rectangle in $\mathbb{R}^2$. The permeability is defined as a piecewise constant on a fine uniform rectangular grid. Moreover, the source function f is positive in the lower left corner element and of equal strength but negative in the upper right corner element.

We solve each problem nine times, divided into three sets of experiments. For the first set of experiments, we solve the problem on the fine grid using standard RT0 and BDM1 mixed elements. This gives us the "true" or reference solution that the multiscale techniques should approximate. We take the RT0 results as the reference solution.

For the second set of experiments, we solve the problem using the variational multiscale method on a coarsened grid that leaves a 10×10 subgrid; that is, we use a factor of 100 upscaling. We use the multiscale finite elements defined in Sect. 5, in particular, the standard ME0 and ME1 multiscale finite elements, as well as the multiscale dual support elements MD and the ones defined through homogenization theory HE.

For the final set of experiments, we solve each problem using the domain decomposition mortar method, using subdomains with the same 10×10 subgrid. We use a single mortar element along the subdomain (or coarse grid) edges, with the mortar space defined as the piecewise discontinuous linear or quadratic functions P1M and P2M, as well as the mortar space defined using homogenization HM (77).

The upscaling operator or subgrid problems are solved on the fine 10×10 grid using RT0.

8.1 Example 1: A Statistically Homogeneous Permeability

In the first example, the permeability field is a scalar field with a variation of about 5 orders of magnitude. It is depicted in Fig. 10 on a base-10 log scale. The fine grid is 50×50, and the coarse grid is only 5×5. We also show the fine-scale velocity using BDM1. Note that the color is the speed $|\mathbf{u}|$, on a log scale, while the arrows, barely visible, show the velocity itself. Thus nearly all the flow concentrates in channels, depicted in warmer colors, from the lower left injection well to the upper right extraction well.

In Fig. 11 we show results for the seven multiscale methods, and the fine-scale RT0 result in the center plot (which is nearly identical to the BDM1 result). The variational multiscale method using various multiscale elements have been plotted beside and above the RT0 result for easy comparison. The mortar results appear in a row below RT0. All the methods appear to do well to the eye.

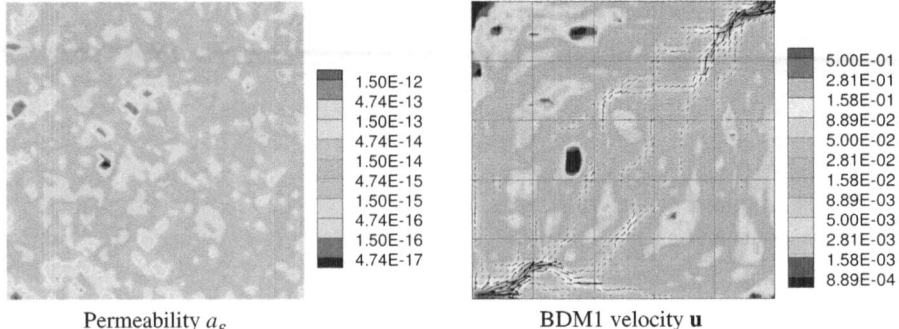

Permeability a_ε BDM1 velocity **u**

Fig. 10 Ex. 1. *Left*: The permeability on a log scale. *Right*: The velocity **u** as computed on the fine scale using BDM1. The arrows show the velocity and the color shows its magnitude (speed) on a log scale

Considering the fine-scale RT0 results as the reference solution, we give the relative errors of the other eight methods in Table 1. The methods differ in the number of degrees of freedom each has on the coarse interfaces, so this is given in one column of the table. Next we show errors in both pressure and velocity, though velocity is normally the more important quantity. The errors are measured in the L^2- and L^∞-norms. Note that RT0 and BDM1 disagree in velocity by 3.6% in L^2, indicating that the problem is difficult to resolve on our 50×50 mesh, and giving an idea of the error that we should tolerate in these applications.

The basic methods ME0, ME1, P1M, and P2M are actually relatively poor, giving 25% to 39% relative L^2 velocity errors. If one looks carefully at Fig. 11, one can see the differences between these methods and RT0. ME0 and ME1 are too numerically diffuse, and P1M and P2M mainly have difficulties in a few isolated points in the domain. The remaining three methods are quite reasonable, both in terms of the numerical error and in the velocity plots. The MD method has 17% L^2 velocity error, followed by the HE method with 9% error, and finally the HM mortar method does the best job at 4% relative velocity error.

8.2   *Example 2: A Statistically Nonhomogeneous Permeability*

The second numerical test example is based on the Tenth Society of Petroleum Engineers Comparative Solution Project [30]. The project includes a difficult three-dimensional permeability field. We take one 60×220 two-dimensional slice, the twentieth, which represents a near shore environment with definite local channeling. It is badly nonisotropic, as depicted in Fig. 12. The permeability is a diagonal tensor, so we show the two components of permeability, which vary on a log scale by about 5 orders of magnitude. As one can see from the BDM1 fine-scale solution (we show only the speed $|\mathbf{u}|$, again on a log scale), the flow field is quite complex. Generally

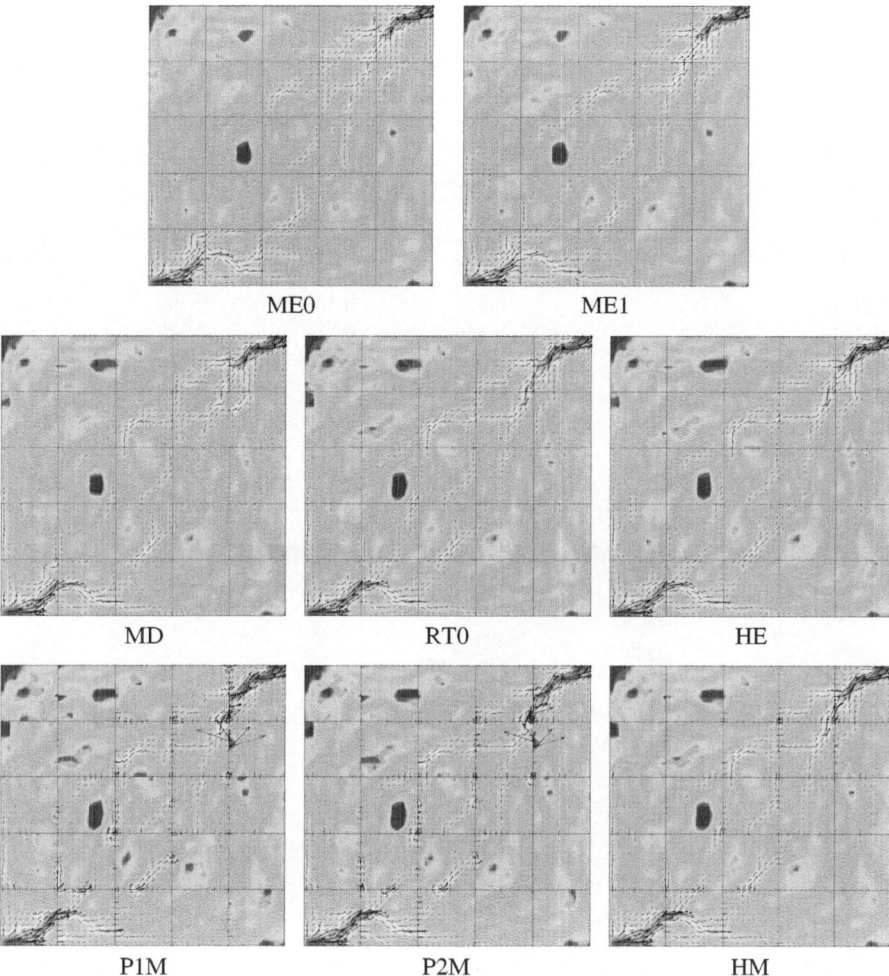

Fig. 11 Ex. 1. The *arrows* show the velocity **u**, and the color depicts its magnitude (speed) on a log scale. The center plot shows the fine-scale RT0 velocity, with the multiscale finite elements beside and above it. The mortar methods are on the bottom row

speaking, fluid that enters the domain at the lower left corner injection well cannot travel upwards in the y-direction until it travels in x a good third of the domain. It then experiences a greater y permeability, and so flows along the permeability channels to the upper right corner extraction well.

The other eight techniques' speeds are shown in Fig. 13, and the relative errors are given in Table 2. As we surmised from the permeability field, we have now quantified that this is a very difficult problem: the BDM1 solution has 8.6% relative velocity error in L^2 compared to RT0, even though each is solved on the fine grid.

Table 1 Ex. 1. Relative errors with respect to the fine-scale RT0 solution for pressure p and velocity $\mathbf{u}$, measured in both L^2 and L^∞ norms for three sets of methods. First, BDM1 is obtained on the fine grid, and differs by a small amount from RT0. Next are the multiscale finite element methods, and the number of coarse degrees of freedom (DOF) used by each method is noted. Finally, the mortar methods are shown with their number of DOF per mortar interface (i.e., coarse grid edge)

Method	DOF per coarse edge	Pressure error		Velocity error	
		L^2	L^∞	L^2	L^∞
BDM1	–	0.043	0.041	0.036	0.028
ME0	1	0.329	0.329	0.300	0.343
ME1	2	0.150	0.144	0.249	0.345
MD	1	0.068	0.105	0.170	0.136
HE	2	0.054	0.084	0.088	0.066
P1M	2	0.157	0.139	0.388	0.948
P2M	3	0.099	0.097	0.323	0.819
HM	3	0.006	0.012	0.041	0.044

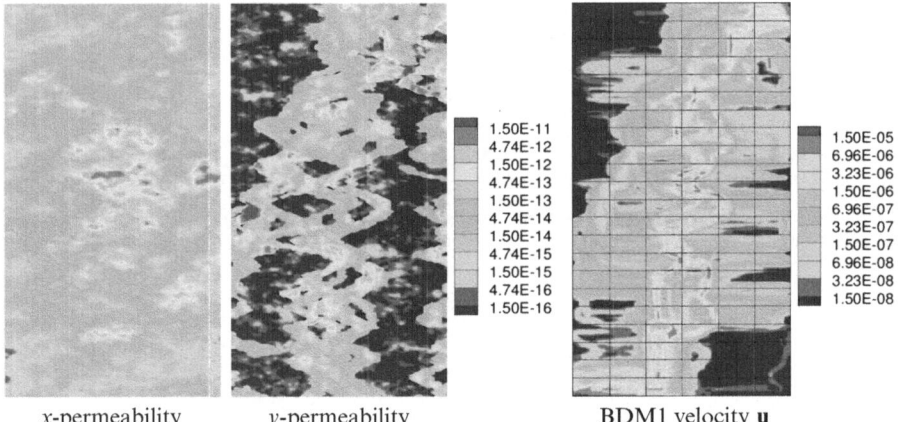

x-permeability y-permeability   BDM1 velocity $\mathbf{u}$

Fig. 12 Ex. 2. *Left*: The anisotropic x- and y-permeabilities on a log scale. *Right*: The speed $|\mathbf{u}|$ as computed on the fine scale using BDM1 on a log scale

The multiscale finite elements ME0 and ME1 have very large 72% and 63% errors, respectively. The figure shows that they give speeds that are much too diffuse compared to RT0.

The multiscale elements MD and HE do much better (45% and 30% error, respectively). The mortar methods P1M and P2M are quite similar (42% and 31% error, respectively). Each of these four methods have speed plots that match RT0 quite well to the eye, and in particular, match the high-speed channel flow quite well. These solutions would be sufficient for preliminary engineering analyses and some stochastic simulation studies. However, only the mortar method based on

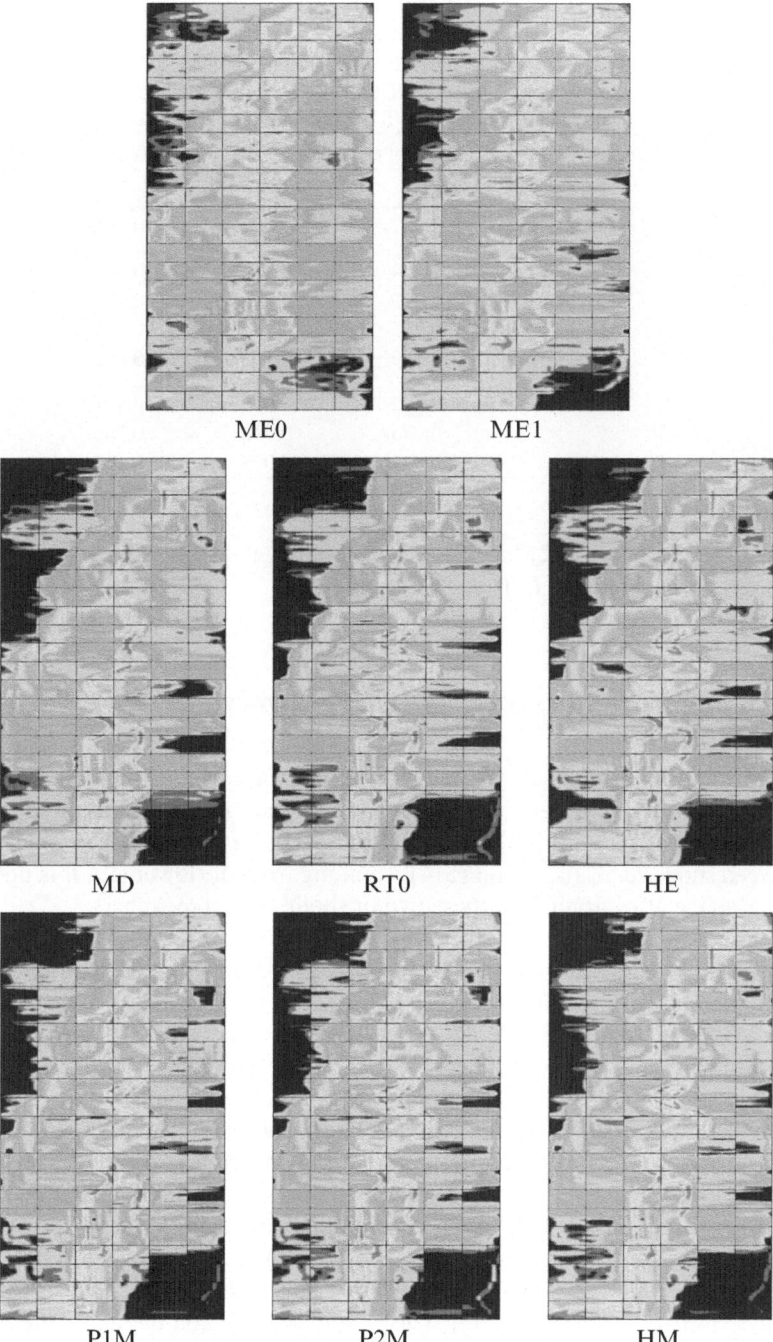

Fig. 13 Ex. 2. The speed $|\mathbf{u}|$ on a log scale. The center plot shows the fine-scale RT0 velocity, with the multiscale finite elements beside and above it. The mortar methods are on the bottom row

Table 2 Ex. 2. Relative errors with respect to the fine-scale RT0 solution for pressure p and velocity **u**, measured in both L^2 and L^∞ norms for three sets of methods. First, BDM1 is obtained on the fine grid, and differs somewhat from RT0. Next are the multiscale finite element methods, and the number of coarse degrees of freedom (DOF) used by each method is noted. Finally, the mortar methods are shown with their number of DOF per mortar interface (i.e., coarse grid edge)

Method	DOF per coarse edge	Pressure error		Velocity error	
		L^2	L^∞	L^2	L^∞
BDM1	–	0.047	0.052	0.086	0.164
ME0	1	2.454	0.414	0.717	0.598
ME1	2	0.799	0.194	0.625	0.443
MD	1	0.223	0.332	0.453	0.575
HE	2	0.149	0.197	0.303	0.261
P1M	2	0.132	0.056	0.418	0.748
P2M	3	0.075	0.038	0.306	0.441
HM	3	0.027	0.023	0.142	0.198

homogenization theory [12] has a reasonable error at 14%. Perhaps to the eye it is also the closest match to RT0.

8.3   *Some Techniques for Controlling Errors*

Because multiscale methods are reduced degree of freedom methods, they are subject to error in somewhat unforeseen ways. A clear example was seen in Fig. 11, where both P1M and P2M have trouble at a spot on the coarse interface parallel to y between subdomains $(4, 4)$ and $(5, 4)$, counting from the lower left. It is not clear why this spot causes trouble for these two methods but not the others.

There has been a great deal of recent work on trying to mitigate this problem. Several authors advocate the use of limited global information to improve the definition of multiscale finite elements [2, 3, 27, 34]. The method seems most useful in the case of nonlinear problems, in which case one solves a global fine-scale *linear* problem, and time dependent and stochastic problems, in which case one solves only one or a few global fine-scale problems. The global information can be used to better define the multiscale finite element on the boundaries of the coarse elements, and then one uses energy minimizing extension into the interior. Some works deal with adaptive methods and a-posteriori error estimation and control of errors [11, 52, 57, 58]. Basically, one includes more scales where the errors are estimated to be large. Other works deal with using the ideas of multiscale finite elements and domain decomposition to define better iterative solvers for the fine-scale system [41, 53–55, 59, 60, 65]. The idea is to iterate on the fine-scale system to convergence using multiscale ideas basically as a preconditioner or in defining prolongation and restriction operators in multigrid.

Acknowledgements This work was supported by the U.S. National Science Foundation and the Center for Frontiers of Subsurface Energy Security, an Energy Frontier Research Center funded by the U.S. Department of Energy, Office of Science, Office of Basic Energy Sciences under Award Number DE-SC0001114.

References

1. J. E. Aarnes and B.-O. Heimsund. Multiscale discontinuous Galerkin methods for elliptic problems with multiple scales. In Timothy J. Barth et al., editors, *Multiscale Methods in Science and Engineering*, volume 44 of *Lecture Notes in Computational Science and Engineering*, pages 1–20. Springer Berlin Heidelberg, 2005.
2. J. E. Aarnes, Y. Efendiev, and L. Jiang. Mixed multiscale finite element methods using limited global information. *Multiscale Model. Simul.*, 7(2):655–676, 2008.
3. J. E. Aarnes. On the use of a mixed multiscale finite element method for greater flexibility and increased speed or improved accuracy in reservoir simulation. *Multiscale Model. Simul.*, 2(3):421–439, 2004.
4. J. E. Aarnes, S. Krogstad, and K.-A. Lie. A hierarchical multiscale method for two-phase flow based upon mixed finite elements and nonuniform coarse grids. *Multiscale Model. Simul.*, 5:337–363, 2006.
5. T. Arbogast. Numerical subgrid upscaling of two-phase flow in porous media. In Z. Chen, R. E. Ewing, and Z.-C. Shi, editors, *Numerical treatment of multiphase flows in porous media*, volume 552 of *Lecture Notes in Physics*, pages 35–49. Springer, Berlin, 2000.
6. T. Arbogast. Analysis of a two-scale, locally conservative subgrid upscaling for elliptic problems. *SIAM J. Numer. Anal.*, 42:576–598, 2004.
7. T. Arbogast. Homogenization-based mixed multiscale finite elements for problems with anisotropy. *Multiscale Model. Simul.*, 9(2):624–653, 2011.
8. T. Arbogast and K. J. Boyd. Subgrid upscaling and mixed multiscale finite elements. *SIAM J. Numer. Anal.*, 44(3):1150–1171, 2006.
9. T. Arbogast, L. C. Cowsar, M. F. Wheeler, and I. Yotov. Mixed finite element methods on non-matching multiblock grids. *SIAM J. Numer. Anal.*, 37:1295–1315, 2000.
10. T. Arbogast, S. E. Minkoff, and P. T. Keenan. An operator-based approach to upscaling the pressure equation. In V. N. Burganos et al., editors, *Computational Methods in Water Resources XII, Vol. 1: Computational Methods in Contamination and Remediation of Water Resources*, pages 405–412, Southampton, U.K., 1998. Computational Mechanics Publications.
11. T. Arbogast, G. Pencheva, M. F. Wheeler, and I. Yotov. A multiscale mortar mixed finite element method. *Multiscale Model. Simul.*, 6(1):319–346, 2007.
12. T. Arbogast and H. Xiao. A multiscale mortar mixed space based on homogenization for heterogeneous elliptic problems. *Submitted*, 2011.
13. I. Babuška. The finite element method with Lagrangian multipliers. *Numer. Math.*, 20: 179–192, 1973.
14. I. Babuška, G. Caloz, and J. E. Osborn. Special finite element methods for a class of second order elliptic problems with rough coefficients. *SIAM J. Numer. Anal.*, 31:945–981, 1994.
15. I. Babuška and R. Lipton. Optimal local approximation spaces for generalized finite element methods with application to multiscale problems. Technical Report 10–12, Institute for Computational Engineering and Sciences, Univ. of Texas, Austin, Texas, USA, Mar. 2010.
16. I. Babuška and J. E. Osborn. Generalized finite element methods: their performance and their relation to mixed methods. *SIAM J. Numer. Anal.*, 20:510–536, 1983.
17. J. Bear. *Dynamics of Fluids in Porous Media*. Dover, New York, 1972.
18. J. Bear and A. H.-D. Cheng. *Modeling Groundwater Flow and Contaminant Transport*. Springer, New York, 2010.

19. A. Bensoussan, J. L. Lions, and G. Papanicolaou. *Asymptotic Analysis for Periodic Structure*. North-Holland, Amsterdam, 1978.
20. C. Bernardi, Y. Maday, and A. T. Patera. A new nonconforming approach to domain decomposition: The mortar element method. In H. Brezis and J. L. Lions, editors, *Nonlinear partial differential equations and their applications*. Longman Scientific & Technical, UK, 1994.
21. S. C. Brenner and L. R. Scott. *The Mathematical Theory of Finite Element Methods*. Springer-Verlag, New York, 1994.
22. F. Brezzi. On the existence, uniqueness and approximation of saddle-point problems arising from Lagrangian multipliers. *RAIRO*, 8:129–151, 1974.
23. F. Brezzi. Interacting with the subgrid world. In *Numerical Analysis, 1999*, pages 69–82. Chapman and Hall, 2000.
24. F. Brezzi, J. Douglas, Jr., R. Duràn, and M. Fortin. Mixed finite elements for second order elliptic problems in three variables. *Numer. Math.*, 51:237–250, 1987.
25. F. Brezzi, J. Douglas, Jr., and L. D. Marini. Two families of mixed elements for second order elliptic problems. *Numer. Math.*, 47:217–235, 1985.
26. F. Brezzi and M. Fortin. *Mixed and hybrid finite element methods*. Springer-Verlag, New York, 1991.
27. Y. Chen and L. J. Durlofsky. Adaptive local-global upscaling for general flow scenarios in heterogeneous formations. *Transp. Por. Med.*, 62:157–185, 2006.
28. Z. Chen and T. Y. Hou. A mixed multiscale finite element method for elliptic problems with oscillating coefficients. *Math. Comp.*, 72:541–576, 2003.
29. Z. Chen, G. Huan, and Y. Ma. *Computational Methods for Multiphase Flows in Porous Media*, volume 2 of *Computational Science and Engineering Series*. SIAM, Philadelphia, 2006.
30. M. A. Christie. Tenth spe comparative solution project: A comparison of upscaling techniques, SPE 72469. *SPE Res. Eval. & Engng.*, 2001. 2001 SPE Reservoir Simulation Symposium.
31. Ph. G. Ciarlet. *The Finite Element Method for Elliptic Problems*. North-Holland, Amsterdam, 1978.
32. Weinan E and B. Engquist. The heterogeneous multiscale methods. *Commun. Math. Sci.*, 1:87–132, 2003.
33. Y. Efendiev, J. Galvis, and X.-H. Wu. Multiscale finite element methods for high-contrast problems using local spectral basis functions. *J. Comput. Phys.*, 230(4):937–955, 2011.
34. Y. Efendiev, V. Ginting, T. Y. Hou, and R. E. Ewing. Accurate multiscale finite element methods for two-phase flow simulations. *J. Comput. Phys.*, 220(1):155–174, 2006.
35. Y. R. Efendiev, T. Y. Hou, and X.-H. Wu. Convergence of a nonconforming multiscale finite element method. *SIAM J. Numer. Anal.*, 37:888–910, 2000.
36. G. B. Folland. *Introduction to Partial Differential Equations*. Princeton, 1976.
37. B. Ganis and I. Yotov. Implementation of a mortar mixed finite element method using a multiscale flux basis. *Comput. Methods Appl. Mech. Engrg.*, 198:3989–3998, 2009.
38. D. Gilbarg and N. S. Trudinger. *Elliptic Partial Differential Equations of Second Order*. Springer-Verlag, Berlin, 1983.
39. V. Ginting. Analysis of two-scale finite volume element method for elliptic problem. *J. Numer. Math.*, 12(2):119–141, 2004.
40. R. Glowinski and M. F. Wheeler. Domain decomposition and mixed finite element methods for elliptic problems. In R. Glowinski et al., editors, *First International Symposium on Domain Decomposition Methods for Partial Differential Equations*, pages 144–172. SIAM, Philadelphia, 1988.
41. I. G. Graham and R. Scheichl. Robust domain decomposition algorithms for multiscale PDEs. *Numer. Meth. Partial Diff. Eqns.*, 23(4):859–878, 2007.
42. P. Grisvard. *Elliptic Problems in Nonsmooth Domains*. Pitman, Boston, 1985.
43. M. A. Hesse, B. T. Mallison, and H. A. Tchelepi. Compact multiscale finite volume method for heterogeneous anisotropic elliptic equations. *Multiscale Model. Simul.*, 7(2):934–962, 2008.
44. U. L. Hetmaniuk and R. B. Lehoucq. A special finite element methods based on component mode synthesis techniques. *ESAIM: Math. Modelling and Numer. Anal.*, 2010.

45. U. Hornung, editor. *Homogenization and Porous Media*. Interdisciplinary Applied Mathematics Series. Springer-Verlag, New York, 1997.
46. T. Y. Hou and X. H. Wu. A multiscale finite element method for elliptic problems in composite materials and porous media. *J. Comput. Phys.*, 134:169–189, 1997.
47. T. Y. Hou, X.-H. Wu, and Z. Cai. Convergence of a multiscale finite element method for elliptic problems with rapidly oscillating coefficients. *Math. Comp.*, 68:913–943, 1999.
48. T. J. R. Hughes. Multiscale phenomena: Green's functions, the Dirichlet-to-Neumann formulation, subgrid scale models, bubbles and the origins of stabilized methods. *Comput. Methods Appl. Mech. Engrg.*, 127:387–401, 1995.
49. T. J. R. Hughes, G. R. Feijóo, L. Mazzei, and J.-B. Quincy. The variational multiscale method – a paradigm for computational mechanics. *Comput. Methods Appl. Mech. Engrg.*, 166:3–24, 1998.
50. P. Jenny, S. H. Lee, and H. A. Tchelepi. Multi-scale finite-volume method for elliptic problems in subsurface flow simulation. *J. Comp. Phys.*, 187:47–67, 2003.
51. V. V. Jikov, S. M. Kozlov, and O. A. Oleinik. *Homogenization of Differential Operators and Integral Functions*. Springer-Verlag, New York, 1994.
52. M. G. Larson and A. Målqvis. Adaptive variational multiscale methods based on a posteriori error estimation: energy norm estimates for elliptic problems. *Comput. Methods Appl. Mech. Engrg.*, 196(21–24):2313–2324, 2007.
53. J. Van Lent, R. Scheichl, and I. G. Graham. Energy minimizing coarse spaces for two-level Schwarz methods for multiscale PDEs. *Numer. Lin. Alg. with Applic.*, 16(10):775–799, 2009.
54. S. P. MacLachlan and J. D. Moulton. Multilevel upscaling through variational coarsening. *Water Resour. Res.*, 42, 2006.
55. J. D. Moulton, Jr. J. E. Dendy, and J. M. Hyman. The black box multigrid numerical homogenization algorithm. *J. Comput. Phys.*, 142(1):80–108, 1998.
56. J. Nolen, G. Papanicolaou, and O. Pironneau. A framework for adaptive multiscale methods for elliptic problems. *Multiscale Model. Simul.*, 7(1):171–196, 2008.
57. J. M. Nordbotten. Adaptive variational multiscale methods for multiphase flow in porous media. *Multiscale Model. Simul.*, 7(3):1455–1473, 2009.
58. G. Pencheva, M. Vohralik, M. F. Wheeler, and T. Wildey. Robust a posteriori error control and adaptivity for multiscale, multinumerics, and mortar coupling. *Submitted*, 2010.
59. J. M. Rath. Darcy flow, multigrid, and upscaling. In et al. W. W. Hager, editor, *Multiscale Optimization Methods and Applications*, volume 82 of *Nonconvex Optimization and its Applications*, pages 337–366. Springer, New York, 2006.
60. J. M. Rath. *Multiscale Basis Optimization for Darcy Flow*. PhD thesis, Univ. of Texas, Austin, Texas, May 2007.
61. R. A. Raviart and J. M. Thomas. A mixed finite element method for 2nd order elliptic problems. In I. Galligani and E. Magenes, editors, *Mathematical Aspects of Finite Element Methods*, number 606 in Lecture Notes in Math., pages 292–315. Springer-Verlag, New York, 1977.
62. E. Sanchez-Palencia. *Non-homogeneous Media and Vibration Theory*. Number 127 in Lecture Notes in Physics. Springer-Verlag, New York, 1980.
63. H. A. Schwarz. Gesammelte mathematische adhandlungen. *Vierteljahrsschrift der Naturforschenden Gesellschaft in Zürich*, 15:272–286, 1870.
64. T. Strouboulis, K. Copps, and I. Babuška. The generalized finite element method. *Comput. Methods Appl. Mech. Engrg.*, 190:4081–4193, 2001.
65. Jinchao Xu and L. Zikatanov. On an energy minimizing basis for algebraic multigrid methods. *Comput. Vis. Sci.*, 7(3–4):121–127, 2004.

On Stability of Discretizations of the Helmholtz Equation

S. Esterhazy and J.M. Melenk

Abstract We review the stability properties of several discretizations of the Helmholtz equation at large wavenumbers. For a model problem in a polygon, a complete k-explicit stability (including k-explicit stability of the continuous problem) and convergence theory for high order finite element methods is developed. In particular, quasi-optimality is shown for a fixed number of degrees of freedom per wavelength if the mesh size h and the approximation order p are selected such that kh/p is sufficiently small and $p = O(\log k)$, and, additionally, appropriate mesh refinement is used near the vertices. We also review the stability properties of two classes of numerical schemes that use piecewise solutions of the homogeneous Helmholtz equation, namely, Least Squares methods and Discontinuous Galerkin (DG) methods. The latter includes the Ultra Weak Variational Formulation.

1 Introduction

A fundamental equation describing acoustic or electromagnetic phenomena is the time-dependent wave equation

$$\frac{\partial^2 w}{\partial t^2} - c^2 \Delta w = g,$$

given here for homogeneous, isotropic media whose propagation speed of waves is c. It arises in many applications, for example, radar/sonar detection, noise filtering, optical fiber design, medical imaging and seismic analysis. A commonly encountered setting is the time-harmonic case, in which the solution w (and

S. Esterhazy · J.M. Melenk (✉)
Vienna University of Technology, Institute for Analysis and Scientific Computing. Wiedner Hauptstrasse 8-10, A-1040 Vienna
e-mail: s.esterhazy@tuwien.ac.at; melenk@tuwien.ac.at

I.G. Graham et al. (eds.), *Numerical Analysis of Multiscale Problems*, Lecture Notes in Computational Science and Engineering 83, DOI 10.1007/978-3-642-22061-6_9,
© Springer-Verlag Berlin Heidelberg 2012

correspondingly the right-hand side g) is assumed to be of the form $\mathrm{Re}\left(e^{-i\omega t}u(x)\right)$ for a frequency ω. Upon introducing the *wavenumber* $k = \omega/c$ and the *wave length* $\lambda := 2\pi/k$, the resulting equation for the function u, which depends solely on the spatial variable x, is then the *Helmholtz equation*

$$-\Delta u - k^2 u = f. \tag{1}$$

In this article, we concentrate on numerical schemes for the Helmholtz equation at large wavenumbers k. Standard discretizations face several challenges, notably:

(I) For large wavenumber k, the solution u is highly oscillatory. Its resolution, therefore, requires fine meshes, namely, at least $N = k^d$ degrees of freedom, where d is the spatial dimension.

(II) The standard H^1-conforming variational formulation is indefinite, and stability on the discrete level is therefore an issue. A manifestation of this problem is the so-called "pollution", which expresses the observation that much more stringent conditions on the discretization have to be met than the minimal $N = O(k^d)$ to achieve a given accuracy.

The second point, which will be the focus of the article, is best seen in the following, one-dimension example:

Example 1.1. For the boundary value problem

$$-u'' - k^2 u = 1 \quad \text{in } (0,1), \quad u(0) = 0, \quad u'(1) - \mathbf{i}ku(1) = 0, \tag{2}$$

we consider the h-version finite element method (FEM) on uniform meshes with mesh size h for different approximation orders $p \in \{1, 2, 3, 4\}$ and wavenumbers $k \in \{1, 10, 100\}$. Figure 1 shows the relative error in the $H^1(\Omega)$-semi norm (i.e., $|u - u_N|_{H^1(\Omega)}/|u|_{H^1(\Omega)}$, where u_N is the FEM approximation) versus the number of degrees of freedom per wavelength $N_\lambda := N/\lambda = 2\pi N/k$ with N being the dimension of the finite element space employed. We observe several effects in Fig. 1: Firstly, since the solution u of (2) is smooth, higher order methods lead to higher accuracy for a given number of degrees of freedom per wavelength than lower order methods. Secondly, *asymptotically*, the FEM is quasioptimal with the finite element error $|u - u_N|_{H^1(\Omega)}$ satisfying

$$|u - u_N|_{H^1(\Omega)} \approx C_p N_\lambda^{-p} |u|_{H^1(\Omega)} \tag{3}$$

for a constant C_p independent of k. Thirdly, the performance of the FEM as measured in "error vs. number of degrees of freedom per wavelength" does depend on k: As k increases, the preasymptotic range with reduced FEM performance becomes larger. Fourthly, higher order methods are less sensitive to k than lower order ones, i.e., for given k, high order methods enter the asymptotic regime in which (3) holds for smaller values of N_λ than lower order methods.   ∎

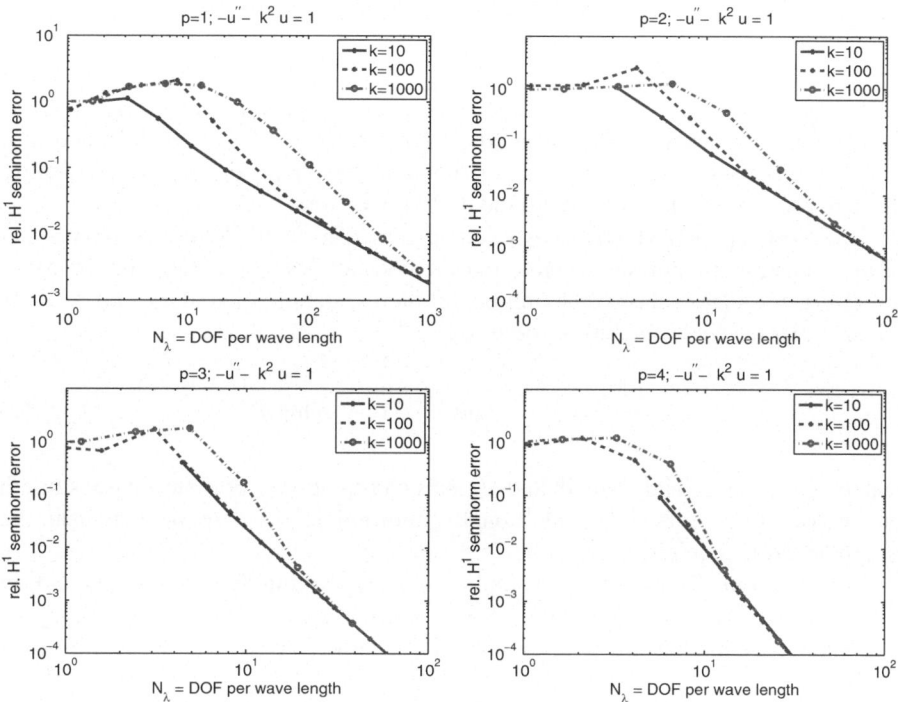

Fig. 1 Performance of h-FEM for (2). *Top*: $p = 1$, $p = 2$. *Bottom*: $p = 3$, $p = 4$ (see Example 1.1)

The behavior of the FEM in Example 1.1 has been analyzed in [36, 37], where error bounds of the form (see [36, Thm. 4.27])

$$|u - u_N|_{H^1(\Omega)} \le C_p \left(1 + k^{p+1} h^p\right) h^p |u|_{H^{p+1}(\Omega)} \tag{4}$$

are established for a constant C_p depending only on the approximation order p. In this particular example, it is also easy to see that $|u|_{H^{p+1}(\Omega)}/|u|_{H^1(\Omega)} \sim k^p$, so that (4) can be recast in the form

$$|u - u_N|_{H^1(\Omega)} \le C_p \left(1 + k^{p+1} h^p\right) (kh)^p |u|_{H^1(\Omega)} \sim \left(1 + kN_\lambda^{-p}\right) N_\lambda^{-p} |u|_{H^1(\Omega)}. \tag{5}$$

This estimate goes a long way to explain the above observations. The presence of the factor $1 + kN_\lambda^{-p}$ explains the "pollution effect", i.e., the observation that for fixed N_λ, the (relative) error of the FEM as compared with the best approximation (which is essentially proportional to N_λ^{-p} in this example) increases with k. The estimate (5) also indicates that the asymptotic convergence behavior (3) is reached for $N_\lambda = O(k^{1/p})$. This confirms the observation made above that higher order methods are less prone to pollution than lower order methods. Although Example 1.1 is restricted to 1D, similar observations have been made in the literature also for multi-d situations as early as [11]. We emphasize that for uniform meshes (as in Example 1.1) or, more generally, translation invariant meshes, complete and detailed dispersion analyses are available in an h-version setting, [1, 20, 36, 37], and in a p/hp-setting, [1–3], that give strong mathematical evidence for the superior ability of high order methods to cope with the pollution effect.

The present paper, which discusses and generalizes the work [49,50], proves that also on unstructured meshes, high order methods are less prone to pollution. More precisely, for a large class of Helmholtz problems, stability and quasi-optimality is given under the scale resolution condition

$$\frac{kh}{p} \leq c_1 \qquad \text{and} \qquad p \geq c_2 \log k, \tag{6}$$

where c_1 is sufficiently small and c_2 sufficiently large. For piecewise smooth geometries (e.g., polygons), additionally appropriate mesh refinement near the singularities is required.

We close our discussion of Example 1.1 by remarking that its analysis and, in fact, the analysis of significant parts of this article rests on H^1-like norms. Largely, this choice is motivated by the numerical scheme, namely, an H^1-conforming FEM.

1.1 Non-standard FEM

The limitations of the classical FEM mentioned above in (I) and (II) have sparked a significant amount of research in the past decades to overcome or at least mitigate them. This research focuses on two techniques that are often considered in tandem: firstly, the underlying approximation by classical piecewise polynomials is replaced with special, problem-adapted functions such as systems of plane waves; secondly, the numerical scheme is based on a variational formulation different from the classical H^1-conforming Galerkin approach. Before discussing these ideas in more detail, we point the reader to the interesting work [8], which shows for a model situation on regular, infinite grids in 2D that no 9-point stencil (i.e., a numerical method based on connecting the value at a node with those of the eight nearest neighbors) generates a completely pollution-free method; the 1D situation is special and discussed briefly in [21, Sec. 7].

Work that is based on a new or modified variational formulation but rests on the approximation properties of piecewise polynomials includes the Galerkin Least Squares Method [29, 30], the methods of [7], and Discontinuous Galerkin Methods ([24–26] and references there). Several methods have been proposed that are based on the approximation properties of special, problem-adapted systems of functions such as systems of plane waves. In an H^1-conforming Galerkin setting, this idea has been pursued in the Partition of Unity Method/Generalized FEM by several authors, e.g., [4, 34, 38, 39, 44, 48, 56, 57, 69]. A variety of other methods that are based on problem-adapted ansatz functions leave the H^1-conforming Galerkin setting and enforce the jump across element boundaries in a weak sense. This can be done by least squares techniques ([10, 41, 53, 58, 68] and references there), by Lagrange multiplier techniques as in the Discontinuous Enrichment Method [22, 23, 70] or by Discontinuous Galerkin (DG) type methods, [14–16, 27, 32, 33, 35, 43, 51, 52]; in these last references, we have included the work on the Ultra Weak Variational Formulation (UWVF) since it can be understood as a special DG method as discussed in [14, 27].

1.2 Scope of the Article

The present article focuses on the stability properties of numerical methods for Helmholtz problems and exemplarily discusses three different approaches in more detail for their differences in techniques. The first approach, studied in Sect. 4, is that of the classical H^1-based Galerkin method for Helmholtz problems. The setting is that of a Gårding inequality so that stability of a numerical method can be inferred from the stability of the continuous problem by perturbation arguments. This motivates us to study for problem (9), which will serve as our model Helmholtz problem in this article, the stability properties of the continuous problem in Sect. 2. In order to make the perturbation argument explicit in the wavenumber k, a detailed, k-explicit regularity analysis for Helmholtz problems is necessary. This is worked out in Sect. 3 for our model problem (9) posed on polygonal domains. These results generalize a similar regularity theory for convex polygons or domains with analytic boundary of [49, 50]. Structurally similar results have been obtained in connection with boundary integral formulations in [42, 47].

We discuss in Sects. 6.2 and 6.3 somewhat briefly a second and a third approach to stability of numerical schemes. In contrast to the setting discussed above, where stability is only ensured asymptotically for sufficiently fine discretizations, these methods are stable by construction and can even feature quasioptimality in appropriate residual norms. We point out, however, that relating this residual norm to a more standard norm such as the L^2-norm for the error is a non-trivial task. Our presentation for these methods will follow the works [14, 27, 33, 53].

Many aspects of discretizations for Helmholtz problems are not addressed in this article. Recent developments in boundary element techniques for this problem class are surveyed in [17]. We also refer to the extended version of the present article [21].

1.3 Some Notation

We employ standard notation for Sobolev spaces, [13, 55, 65]. For domains ω and $k \neq 0$ we denote

$$\|u\|_{1,k,\omega}^2 := k^2 \|u\|_{L^2(\omega)}^2 + \|\nabla u\|_{L^2(\omega)}^2. \tag{7}$$

This norm is equivalent to the standard H^1-norm. The presence of the weight k in the L^2-part leads to a balance between the H^1-seminorm and the L^2-norm for functions with the expected oscillatory behavior such as plane waves $e^{ik\mathbf{d}\cdot x}$ (with $\mathbf{d}$ being a unit vector). Additionally, the bilinear form B considered below is bounded uniformly in k with respect to this (k-dependent) norm.

Throughout this work, a standing assumption will be

$$|k| \geq k_0 > 0; \tag{8}$$

our frequently used phrase "independent of k" will still implicitly assume (8). We denote by C a generic constant. If not stated otherwise, C will be independent of the wavenumber k but may depend on k_0. For smooth functions u defined on a d-dimensional manifold, we employ the notation $|\nabla^n u(x)|^2 :=$ $\sum_{\alpha \in \mathbb{N}_0^d : |\alpha| = n} \frac{|\alpha|!}{\alpha!} |D^\alpha u(x)|^2.$

1.4 A Model Problem

In order to fix ideas, we will use the following, specific model problem: For a bounded Lipschitz domain $\Omega \subset \mathbb{R}^d$, $d \in \{2, 3\}$, we study for $k \in \mathbb{R}$, $|k| \geq k_0$,

$$-\Delta u - k^2 u = f \text{ in } \Omega, \tag{9a}$$

$$\partial_n u + iku = g \text{ on } \partial\Omega. \tag{9b}$$

Henceforth, to simplify the notation, we assume $k \geq k_0 > 0$ but point out that the choice of the sign of k is not essential. The weak formulation for (9) is:

$$\text{Find } u \in H^1(\Omega) \text{ s.t.} \quad B(u, v) = l(v) \quad \forall v \in H^1(\Omega), \tag{10}$$

where, for $f \in L^2(\Omega)$ and $g \in L^2(\partial\Omega)$, B and l are given by

$$B(u, v) := \int_\Omega (\nabla u \cdot \nabla \bar{v} - k^2 u\bar{v}) + ik \int_{\partial\Omega} u\bar{v}, \quad l(v) := (f, v)_{L^2(\Omega)} + (g, v)_{L^2(\partial\Omega)}. \tag{11}$$

As usual, if $f \in (H^1(\Omega))'$ and $g \in H^{-1/2}(\partial\Omega)$, then the L^2-inner products $(\cdot, \cdot)_{L^2(\Omega)}$ and $(\cdot, \cdot)_{L^2(\partial\Omega)}$ are understood as duality pairings. The multiplicative trace inequality proves continuity of B; in fact, there exists $C_B > 0$ independent of k such that (see, e.g., [50, Cor. 3.2] for details)

$$|B(u, v)| \leq C_B \|u\|_{1,k,\Omega} \|v\|_{1,k,\Omega} \quad \forall u, v \in H^1(\Omega). \tag{12}$$

2   Stability of the Continuous Problem

Helmholtz problems can often be cast in the form "coercive + compact perturbation" where the compact perturbation is k-dependent. In other words, a Gårding inequality is satisfied. For example, the sesquilinear form B of (11) is of this form since

$$\text{Re } B(u, u) + 2k^2(u, u)_{L^2(\Omega)} = \|u\|_{1,k,\Omega}^2 \tag{13}$$

and the embedding $H^1(\Omega) \subset L^2(\Omega)$ is compact by Rellich's theorem. Classical Fredholm theory (the "Fredholm alternative") then yields unique solvability of (10) for all $f \in (H^1(\Omega))'$ and $g \in H^{-1/2}(\partial\Omega)$, if one can show uniqueness. Uniqueness in turn is often obtained by exploiting analyticity of the solutions of homogeneous Helmholtz equation, or, more generally, the unique continuation principle for elliptic problems, (see, e.g., [40, Chap. 4.3]):

Example 2.1 (Uniqueness for (9)). Let $f = 0$ and $g = 0$ in (9). Then, any solution $u \in H^1(\Omega)$ of (9) satisfies $u|_{\partial\Omega} = 0$ since $0 = \text{Im } B(u, u) = k\|u\|_{L^2(\partial\Omega)}^2$ (see Lemma 2.2). Hence, the trivial extension $\widetilde{u}$ to $\mathbb{R}^2$ satisfies $\widetilde{u} \in H^1(\mathbb{R}^2)$. The observations $B(u, v) = 0$ for all $v \in H^1(\Omega)$ and $u|_{\partial\Omega} = 0$ show

$$\int_{\mathbb{R}^2} \nabla\widetilde{u} \cdot \nabla\overline{v} - k^2\widetilde{u}\overline{v} = 0 \qquad \forall v \in C_0^\infty(\mathbb{R}^2).$$

Hence, $\widetilde{u}$ is a solution of the homogeneous Helmholtz equation and $\widetilde{u}$ vanishes on $\mathbb{R}^2 \setminus \overline{\Omega}$. Analyticity of $\widetilde{u}$ (or, more generally, the unique continuation principle presented in [40, Chap. 4.3]) then implies that $\widetilde{u} \equiv 0$, which in turn yields $u \equiv 0$. ∎

The arguments based on the Fredholm alternative do not give any indication of how the solution operator depends on the wavenumber k. Yet, it is clearly of interest to know how k enters bounds for the solution operator. It turns out that both the geometry and the type of boundary conditions strongly affect these bounds. For example, for an exterior Dirichlet problem, [12] exhibits a geometry and a sequence of wavenumber $(k_n)_{n\in\mathbb{N}}$ tending to infinity such that the norm of the solution operator for these wavenumbers is bounded from below by an exponentially growing term Ce^{bk_n} for some $C, b > 0$. These geometries feature so-called "trapping" or near-trapping and are not convex. For convex or at least star-shaped geometries, the k-dependence is much better behaved. An important ingredient of the analysis on such geometries are special test functions in the variational formulation. For example, assuming in the model problem (10) that Ω is star-shaped with respect to the origin (and has a smooth boundary), one may take as the test function $v(x) = x \cdot \nabla u(x)$, where u is the exact solution. An integration by parts (more generally, the so-called "Rellich identities" [55, p. 261] or an identity due to Pohožaev, [59]) then leads to the following estimate for the model problem (10):

$$\|u\|_{1,k,\Omega} \le C \left[\|f\|_{L^2(\Omega)} + \|g\|_{L^2(\partial\Omega)}\right]; \tag{14}$$

this was shown in [44, Prop. 8.1.4] (for $d = 2$) and subsequently by [19] for $d = 3$. Uniform in k bounds were established in [31] for star-shaped domains and certain boundary conditions of mixed type by related techniques. The same test function was also crucial for a boundary integral setting in [18]. A refined version of this test function that goes back to Morawetz and Ludwig, [54] was used recently in a boundary integral equations context (still for star-shaped domains), [66].

While (14) does not make minimal assumptions on the regularity of f and g, the estimate (14) can be used to show that (for star-shaped domains) the sesquilinear form B of (10) satisfies an inf-sup condition with inf-sup constant $\gamma = O(k^{-1})$ – this can be shown using the arguments presented in the proof Theorem 2.5.

An important ingredient of the regularity and stability theory will be the concept of *polynomial well-posedness* by which we mean polynomial-in-k-bounds for the norm of the solution operator. The model problem (9) on star-shaped domains with the a priori bound (14) is an example. The following Sect. 2.1 shows polynomial well-posedness for the model problem (9) on general Lipschitz domains (Thm. 2.4). It is thus not the geometry but the type of boundary conditions in our model problem (9), namely, Robin boundary conditions that makes it polynomially well-posed. In contrast, the Dirichlet boundary conditions in conjunction with the lack of star-shapedness in the examples given in [12] make these problem not polynomially well-posed.

2.1 Polynomial Well-Posedness for the Model Problem (9)

Lemma 2.2. *Let $\Omega \subset \mathbb{R}^d$ be a bounded Lipschitz domain. Let $u \in H^1(\Omega)$ be a weak solution of (9) with $f = 0$ and $g \in L^2(\partial\Omega)$. Then $\|u\|_{L^2(\partial\Omega)} \leq k^{-1}\|g\|_{L^2(\partial\Omega)}$.*

Proof. Selecting $v = u$ in the weak formulation (10) and considering the imaginary part yields

$$k\|u\|_{L^2(\partial\Omega)}^2 = \operatorname{Im} \int_{\partial\Omega} g\overline{u} \leq \|g\|_{L^2(\partial\Omega)}\|u\|_{L^2(\partial\Omega)}.$$

This concludes the argument.   □

Next we use results on layer potentials for the Helmholtz equation from [47] to prove the following lemma:

Lemma 2.3. *Let $\Omega \subset \mathbb{R}^d$ be a bounded Lipschitz domain, $u \in H^1(\Omega)$ solve (9) with $f = 0$. Assume $u|_{\partial\Omega} \in L^2(\partial\Omega)$ and $\partial_n u \in L^2(\partial\Omega)$. Then there exists $C > 0$ independent of k and u such that*

$$\|u\|_{L^2(\Omega)} \leq Ck \left(\|u\|_{L^2(\partial\Omega)} + \|\partial_n u\|_{H^{-1}(\partial\Omega)}\right),$$

$$\|u\|_{1,k,\Omega} \leq C \left[k^2\|u\|_{L^2(\partial\Omega)} + k^2\|\partial_n u\|_{H^{-1}(\partial\Omega)} + k^{-2}\|\partial_n u\|_{L^2(\partial\Omega)}\right].$$

Proof. With the single layer and double layer potentials $\widetilde{V}_k$ and $\widetilde{K}_k$ we have the representation formula $u = \widetilde{V}_k \partial_n u - \widetilde{K}_k u$. From [47, Lemmata 2.1, 2.2, Theorems 4.1, 4.2] we obtain

$$\|\widetilde{V}_k \partial_n u\|_{L^2(\Omega)} \le C k \|\partial_n u\|_{H^{-1}(\partial\Omega)}, \qquad \|\widetilde{K}_k u\|_{L^2(\Omega)} \le C k \|u\|_{L^2(\partial\Omega)}.$$

Thus, $\|u\|_{L^2(\Omega)} \le C k \left(\|u\|_{L^2(\partial\Omega)} + \|\partial_n u\|_{H^{-1}(\partial\Omega)} \right)$. Next, using $v = u$ in the weak formulation (10) yields

$$\|\nabla u\|_{L^2(\Omega)}^2 \le C \left[k^2 \|u\|_{L^2(\Omega)}^2 + \|\partial_n u\|_{L^2(\partial\Omega)} \|u\|_{L^2(\partial\Omega)} \right]$$

and therefore

$$\|\nabla u\|_{L^2(\Omega)}^2 + k^2 \|u\|_{L^2(\Omega)}^2 \le C \left[k^4 \|u\|_{L^2(\partial\Omega)}^2 + k^4 \|\partial_n u\|_{H^{-1}(\partial\Omega)}^2 + k^{-4} \|\partial_n u\|_{L^2(\partial\Omega)}^2 \right],$$

which concludes the proof. $\qquad\square$

Theorem 2.4. *Let $\Omega \subset \mathbb{R}^d$, $d \in \{2, 3\}$ be a bounded Lipschitz domain. Then there exists $C > 0$ (independent of k) such that for $f \in L^2(\Omega)$ and $g \in L^2(\partial\Omega)$ the solution $u \in H^1(\Omega)$ of (9) satisfies*

$$\|u\|_{1,k,\Omega} \le C \left[k^2 \|g\|_{L^2(\partial\Omega)} + k^{5/2} \|f\|_{L^2(\Omega)} \right].$$

Proof. We first transform the problem to one with homogeneous right-hand side f in the standard way. A particular solution of (9a) is given by the Newton potential $u_0 := G_k \star f$; here, G_k is a Green's function for the Helmholtz equation and we tacitly extend f by zero outside Ω. Then $u_0 \in H^2_{loc}(\mathbb{R}^d)$ and by the analysis of the Newton potential given in [50, Lemma 3.5] we have

$$k^{-1} \|u_0\|_{H^2(\Omega)} + \|u_0\|_{H^1(\Omega)} + k \|u_0\|_{L^2(\Omega)} \le C \|f\|_{L^2(\Omega)}. \tag{15}$$

The difference $\widetilde{u} := u - u_0$ then satisfies

$$-\Delta \widetilde{u} - k^2 \widetilde{u} = 0 \quad \text{in } \Omega, \qquad \partial_n \widetilde{u} + ik\widetilde{u} = g - (\partial_n u_0 + iku_0) =: \widetilde{g}. \tag{16}$$

We have with the multiplicative trace inequality

$$\|\widetilde{g}\|_{L^2(\partial\Omega)} \le C \left[\|g\|_{L^2(\partial\Omega)} + \|u_0\|_{H^2(\Omega)}^{1/2} \|u_0\|_{H^1(\Omega)}^{1/2} + k \|u_0\|_{H^1(\Omega)}^{1/2} \|u_0\|_{L^2(\Omega)}^{1/2} \right]$$
$$\le C \left[\|g\|_{L^2(\partial\Omega)} + k^{1/2} \|f\|_{L^2(\Omega)} \right]. \tag{17}$$

To get bounds on $\widetilde{u}$, we employ Lemma 2.2 and (17) to conclude

$$\|\widetilde{u}\|_{L^2(\partial\Omega)} \le C k^{-1} \|\widetilde{g}\|_{L^2(\partial\Omega)} \le C \left[k^{-1} \|g\|_{L^2(\partial\Omega)} + k^{-1/2} \|f\|_{L^2(\Omega)} \right], \tag{18}$$

$$\|\partial_n \widetilde{u}\|_{L^2(\partial\Omega)} \le C \left[\|\widetilde{g}\|_{L^2(\partial\Omega)} + k \|\widetilde{u}\|_{L^2(\partial\Omega)} \right] \le C \left[\|g\|_{L^2(\partial\Omega)} + k^{1/2} \|f\|_{L^2(\Omega)} \right].$$
$$\tag{19}$$

Lemma 2.3 and the generous estimate $\|\partial_n \widetilde{u}\|_{H^{-1}(\partial\Omega)} \leq C \|\partial_n \widetilde{u}\|_{L^2(\partial\Omega)}$ produce

$$\|\widetilde{u}\|_{H^1(\Omega)} + k\|\widetilde{u}\|_{L^2(\Omega)} \leq C\left[k^2\|g\|_{L^2(\partial\Omega)} + k^{5/2}\|f\|_{L^2(\Omega)}\right]. \qquad (20)$$

Combining (15), (20) finishes the argument.   □

The *a priori* estimate of Theorem 2.4 does not make minimal assumptions on the regularity of f and g. However, it can be used to obtain estimates on the inf-sup and hence *a priori* bounds for $f \in (H^1(\Omega))'$ and $g \in H^{-1/2}(\partial\Omega)$ as we now show:

Theorem 2.5. *Let $\Omega \subset \mathbb{R}^d$, $d \in \{2,3\}$ be a bounded Lipschitz domain. Then there exists $C > 0$ (independent of k) such that the sesquilinear form B of (11) satisfies*

$$\inf_{0 \neq u \in H^1(\Omega)} \sup_{0 \neq v \in H^1(\Omega)} \frac{\operatorname{Re} B(u,v)}{\|u\|_{1,k,\Omega}\|v\|_{1,k,\Omega}} \geq Ck^{-7/2}. \qquad (21)$$

Furthermore, for every $f \in (H^1(\Omega))'$ and $g \in H^{-1/2}(\partial\Omega)$ the problem (10) is uniquely solvable, and its solution $u \in H^1(\Omega)$ satisfies the a priori bound

$$\|u\|_{1,k,\Omega} \leq Ck^{7/2}\left[\|f\|_{(H^1(\Omega))'} + \|g\|_{H^{-1/2}(\partial\Omega)}\right]. \qquad (22)$$

If Ω is convex or if Ω is star-shaped and has a smooth boundary, then the following, sharper estimate holds:

$$\inf_{0 \neq u \in H^1(\Omega)} \sup_{0 \neq v \in H^1(\Omega)} \frac{\operatorname{Re} B(u,v)}{\|u\|_{1,k,\Omega}\|v\|_{1,k,\Omega}} \geq Ck^{-1}. \qquad (23)$$

Proof. The proof relies on standard arguments for sesquilinear forms satisfying a Gårding inequality. For simplicity of notation, we write $\|\cdot\|_{1,k}$ for $\|\cdot\|_{1,k,\Omega}$.
   Given $u \in H^1(\Omega)$ we define $z \in H^1(\Omega)$ as the solution of

$$2k^2(\cdot, u)_{L^2(\Omega)} = B(\cdot, z).$$

Theorem 2.4 implies $\|z\|_{1,k} \leq Ck^{9/2}\|u\|_{L^2(\Omega)}$, and $v = u + z$ satisfies

$$\operatorname{Re} B(u,v) = \operatorname{Re} B(u,u) + \operatorname{Re} B(u,z) = \|u\|_{1,k}^2 - 2k^2\|u\|_{L^2(\Omega)}^2 + \operatorname{Re} B(u,z) = \|u\|_{1,k}^2.$$

Thus,

$$\operatorname{Re} B(u,v) = \|u\|_{1,k}^2,$$

$$\|v\|_{1,k} = \|u + z\|_{1,k} \leq \|u\|_{1,k} + \|z\|_{1,k} \leq \|u\|_{1,k} + Ck^{9/2}\|u\|_{L^2(\Omega)}$$

$$\leq Ck^{7/2}\|u\|_{1,k}.$$

Therefore,

$$\text{Re } B(u, v) = \|u\|_{1,k}^2 \geq \|u\|_{1,k} C k^{-7/2} \|v\|_{1,k},$$

which concludes the proof of (21). Example 2.1 provides unique solvability for (9) so that (21) gives the *a priori* estimate (22). Finally, (23) is shown by the same arguments using (14).   □

3 k-Explicit Regularity Theory

3.1 Regularity by Decomposition

Since the Sobolev regularity of elliptic problems is determined by the leading order terms of the differential equation and the boundary conditions, the Sobolev regularity properties of our model problem (9) are the same as those for the Laplacian. However, regularity results that are explicit in the wavenumber k are clearly of interest; for example, we will use them in Sect. 4.2 below to quantify how fine the discretization has to be (relative to k) so that the FEM is stable and quasi-optimal.

The k-explicit regularity theory developed in [49, 50] (and, similarly, for integral equations in [42, 47]) takes the form of an additive splitting of the solution into a part with finite regularity but k-independent bounds and a part that is analytic and for which k-explicit bounds for all derivatives are available. Below, we will present a similar regularity theory for the model problem (9) for polygonal $\Omega \subset \mathbb{R}^2$, thereby extending the results of [49], which restricted its analysis of polygons to the convex case. In order to motivate the ensuing developments, we quote from [50] a result that shows in a simple setting the key features of our k-explicit "regularity by decomposition":

Lemma 3.1 ([50, Lemma 3.5]). *Let $B_R(0) \subset \mathbb{R}^d$, $d \in \{1, 2, 3\}$ be the ball of radius R centered at the origin. Then, there exist $C, \gamma > 0$ such that for all k (with $k \geq k_0$) the following is true: For all $f \in L^2(\mathbb{R}^d)$ with supp $f \subset B_R(0)$ the solution u of*

$$-\Delta u - k^2 u = f \quad \text{in } \mathbb{R}^d,$$

subject to the Sommerfeld radiation condition

$$\lim_{|x| \to \infty} |x|^{\frac{d-1}{2}} \left(\frac{\partial u}{\partial |x|} - iku \right) = 0 \quad \text{for} \quad |x| \to \infty,$$

has the following regularity properties:

(i) $u|_{B_{2R}(0)} \in H^2(B_{2R}(0))$ and $\|u\|_{H^2(B_{2R}(0))} \leq Ck\|f\|_{L^2(B_R(0))}$.
(ii) $u|_{B_{2R}(0)}$ can be decomposed as $u = u_{H^2} + u_{\mathscr{A}}$ for a $u_{H^2} \in H^2(B_{2R})$ and an analytic $u_{\mathscr{A}}$ together with the bounds

$$k\|u_{H^2}\|_{1,k,B_{2R}(0)} + \|u_{H^2}\|_{H^2(B_{2R}(0))} \le C\|f\|_{L^2(B_R(0))},$$

$$\|\nabla^n u_{\mathscr{A}}\|_{L^2(B_{2R}(0))} \le C\gamma^n \max\{n,k\}^{n-1}\|f\|_{L^2(B_R(0))} \quad \forall n \in \mathbb{N}_0.$$

A few comments concerning Lemma 3.1 are in order. For general $f \in L^2(B_R(0))$, one cannot expect better regularity than H^2-regularity for the solution u and, indeed, both regularity results (i) and (ii) assert this. The estimate (3.1) is sharp in its dependence on k as the following simple example shows: For the fundamental solution G_k (with singularity at the origin) and a cut-off function $\chi \in C_0^\infty(\mathbb{R}^d)$ with supp $\chi \subset B_{2R}(0)$ and $\chi \equiv 1$ on $B_R(0)$, the functions $u := (1-\chi)G_k$ and $f := -\Delta u - k^2 u$ satisfy $\|u\|_{H^2(B_{2R}(0))} = O(k^2)$ and $\|f\|_{L^2(B_R(0))} = O(k)$. Compared to (3.1), the regularity assertion (3.1) is finer in that its H^2-part u_{H^2} has a better k-dependence. The k-dependence of the analytic part $u_{\mathscr{A}}$ is not improved (indeed, $\|u_{\mathscr{A}}\|_{H^2(B_{2R}(0))} \le Ck\|f\|_{L^2(B_R(0))}$), but the analyticity of $u_{\mathscr{A}}$ is a feature that higher order methods can exploit.

The decomposition in (ii) of Lemma 3.1 is obtained by a decomposition of the datum f using low pass and high pass filters, i.e., $f = L_{\eta k} f + H_{\eta k} f$, where the low pass filter $L_{\eta k}$ cuts off frequencies beyond ηk (here, $\eta > 1$) and $H_{\eta k}$ eliminates the frequencies small than ηk. Similar frequency filters will be important tools in our analysis below as well (see Sec. 3.3.1). The regularity properties stated in (ii) then follow from this decomposition and the explicit solution formula $u = G_k \star f$ (see [50, Lemma 3.5] for details).

Lemma 3.1 serves as a prototype for "regularity theory by decomposition". Similar decompositions have been developed recently for several Helmholtz problems in [49] and [42, 47]. Although they vary in their details, these decomposition are structurally similar in that they have the form of an additive splitting into a part with finite regularity with k-independent bounds and an analytic part with k-dependent bounds. The basic ingredients of these decomposition results are (a) (piecewise) analyticity of the geometry (or, more generally, the data of the problem) and (b) a priori bounds for solution operator. The latter appear only in the estimate for the analytic part of the decomposition, and the most interesting case is that of polynomially well-posed problems. We illustrate the construction of the decomposition for the model problem (9) in polygonal domains $\Omega \subset \mathbb{R}^2$. This result is an extension to general polygons of the results [49], which restricted its attention to the case of convex polygons. We emphasize that the choice of the boundary conditions (9b) is not essential for the form of the decomposition and other boundary conditions could be treated using similar techniques.

3.2 Setting and Main Result

Let $\Omega \subset \mathbb{R}^2$ be a bounded, polygonal Lipschitz domain with vertices A_j, $j = 1, \ldots, J$, and interior angles ω_j, $j = 1, \ldots, J$. We will require the countably normed spaces introduced in [6, 45]. These space are designed to capture important features of solutions of elliptic partial differential equations posed on polygons, namely, analyticity of the solution and the singular behavior at the vertices. Their characterization in terms of these countably normed spaces also permits proving exponential convergence of piecewise polynomial approximation on appropriately graded meshes.

These countably normed spaces are defined with the aid of weight functions $\Phi_{p,\vec{\beta},k}$ that we now define. For $\beta \in [0,1)$, $n \in \mathbb{N}_0$, $k > 0$, and $\vec{\beta} \in [0,1)^J$, we set

$$\Phi_{n,\beta,k}(x) = \min\left\{1, \frac{|x|}{\min\left\{1, \frac{|n|+1}{k+1}\right\}}\right\}^{n+\beta}$$

$$\Phi_{n,\vec{\beta},k}(x) = \prod_{j=1}^{J} \Phi_{n,\beta_j,k}(x - A_j). \tag{24}$$

Finally, we denote by $H_{pw}^{1/2}(\partial\Omega)$ the space of functions whose restrictions of the edges of $\partial\Omega$ are in $H^{1/2}$.

We furthermore introduce the constant $C_{sol}(k)$ as a suitable norm of the solution operator for (9). That is, $C_{sol}(k)$ is such that for all $f \in L^2(\Omega)$, $g \in L^2(\partial\Omega)$ and corresponding solution u of (9) there holds

$$\|u\|_{1,k,\Omega} \leq C_{sol}(k)\left[\|f\|_{L^2(\Omega)} + \|g\|_{L^2(\partial\Omega)}\right]. \tag{25}$$

We recall that Theorem 2.4 gives $C_{sol}(k) = O(k^{5/2})$ for general polygons and $C_{sol}(k) = O(1)$ by [44, Prop. 8.1.4] for convex polygons. Our motivation for using the notation $C_{sol}(k)$ is emphasize in the following theorem how *a priori* estimates for Helmholtz problems affect the decomposition result:

Theorem 3.2. *Let $\Omega \subset \mathbb{R}^2$ be a polygon with vertices A_j, $j = 1, \ldots, J$. Then there exist constants C, $\gamma > 0$, $\vec{\beta} \in [0,1)^J$ independent of $k \geq k_0$ such that for every $f \in L^2(\Omega)$ and $g \in H_{pw}^{1/2}(\partial\Omega)$ the solution u of (9) can be written as $u = u_{H^2} + u_{\mathscr{A}}$ with*

$$k\|u_{H^2}\|_{1,k,\Omega} + \|u_{H^2}\|_{H^2(\Omega)} \leq CC_{f,g}$$

$$\|u_{\mathscr{A}}\|_{H^1(\Omega)} \leq (C_{sol}(k) + 1)\, C_{f,g}$$

$$k\|u_{\mathscr{A}}\|_{L^2(\Omega)} \leq (C_{sol}(k) + k)\, C_{f,g}$$

$$\left\|\Phi_{n,\vec{\beta},k}\nabla^{n+2}u_{\mathscr{A}}\right\|_{L^2(\Omega)} \leq C(C_{sol}(k) + 1)k^{-1}\max\{n,k\}^{n+2}\gamma^n C_{f,g} \quad \forall n \in \mathbb{N}_0$$

with $C_{f,g} := \|f\|_{L^2(\Omega)} + \|g\|_{H_{pw}^{1/2}(\partial\Omega)}$ and $C_{sol}(k)$ introduced in (25).

Proof. The proof is relegated to Sect. 3.4. We mention that the k-dependence of our bounds on $\|u_{\mathscr{A}}\|_{L^2(\Omega)}$ is likely to be suboptimal due to our method of proof.   □

Theorem 3.2 may be viewed as the analog of Lemma 3.1, (ii); we conclude this section with the analog of Lemma 3.1, (i):

Corollary 3.3. *Assume the hypotheses of Theorem 3.2. Then there exist constants* $C > 0$, $\vec{\beta} \in [0, 1)^J$ *independent of* k *such that for all* $f \in L^2(\Omega)$, $g \in H_{pw}^{1/2}(\partial\Omega)$ *the solution* u *of* (9) *satisfies* $\|u\|_{1,k,\Omega} \leq CC_{sol}(k)\left[\|f\|_{L^2(\Omega)} + \|g\|_{L^2(\partial\Omega)}\right]$ *as well as*

$$\|\Phi_{0,\vec{\beta},k} \nabla^2 u\|_{L^2(\Omega)} \leq Ck(C_{sol}(k) + 1)\left[\|f\|_{L^2(\Omega)} + \|g\|_{H_{pw}^{1/2}(\partial\Omega)}\right].$$

Proof. The estimate for $\|u\|_{1,k,\Omega}$ expresses (25). The estimate for the second derivatives of u follows from Theorem 3.2 since $u = u_{H^2} + u_{\mathscr{A}}$. $\square$

3.3   Auxiliary Results

Just as in the proof of Lemma 3.1, an important ingredient of the proof of Theorem 3.2 are high and low pass filters. The underlying reason is that the Helmholtz operator $-\Delta - k^2$ acts very differently on low frequency and high frequency functions. Here, the dividing line between high and low frequencies is at $O(k)$. For this reason, appropriate high and low pass filters are defined and analyzed in Sect. 3.3.1. Furthermore, when applied to high frequency functions the Helmholtz operator behaves similarly to the Laplacian $-\Delta$ or the modified Helmholtz operator $-\Delta + k^2$. This latter operator, being positive definite, is easier to analyze and yet provides insight into the behavior of the Helmholtz operator restricted to high frequency functions. The modified Helmholtz operator will therefore be a tool for the proof of Theorem 3.2 and is thus analyzed in Sect. 3.3.3.

3.3.1   High and Low Pass Filters, Auxiliary Results

For the polygonal domain $\Omega \subset \mathbb{R}^2$ we introduce for $\eta > 1$ the following two low and high pass filters in terms of the Fourier transform $\mathscr{F}$:

1. The low and high pass filters $L_{\Omega,\eta} f : L^2(\Omega) \to L^2(\Omega)$ and $H_{\Omega,\eta} : L^2(\Omega) \to L^2(\Omega)$ are defined by

$$L_{\Omega,\eta} f = (\mathscr{F}^{-1}\chi_{B_{\eta k}(0)}\mathscr{F}(E_\Omega f))|_\Omega, \quad H_{\Omega,\eta} f = (\mathscr{F}^{-1}\chi_{\mathbb{R}^2\setminus B_{\eta k}(0)}\mathscr{F}(E_\Omega f))|_\Omega;$$

   here, $B_{\eta k}(0)$ is the ball of radius ηk with center 0, the characteristic function of a set A is χ_A, and E_Ω denotes the Stein extension operator of [67, Chap. VI].
2. Analogously, we define $L_{\partial\Omega,\eta} f : L^2(\partial\Omega) \to L^2(\partial\Omega)$ and $H_{\partial\Omega,\eta} : L^2(\partial\Omega) \to L^2(\partial\Omega)$ in an *edgewise* fashion. Specifically, identifying an edge e of Ω with an

interval and letting E_e be the Stein extension operator for the interval $e \subset \mathbb{R}$ to the real line $\mathbb{R}$, we can define with the univariate Fourier transformation $\mathscr{F}$ the operators $L_{e,\eta}$ and $H_{e,\eta}$ by

$$L_{e,\eta}g = (\mathscr{F}^{-1}\chi_{B_{\eta k}(0)}\mathscr{F}(E_e g))|_e, \qquad H_{e,\eta}g = (\mathscr{F}^{-1}\chi_{\mathbb{R}\setminus B_{\eta k}(0)}\mathscr{F}(E_e f))|_e;$$

the operators $L_{\partial\Omega,\eta}$ and $H_{\partial\Omega,\eta}$ are then defined edgewise by $(L_{\partial\Omega,\eta}g)|_e = L_{e,\eta}g$ and $(H_{\partial\Omega,\eta}g)|_e = H_{e,\eta}g$ for all edges $e \subset \partial\Omega$.

These operators provide stable decompositions of $L^2(\Omega)$ and $L^2(\partial\Omega)$. For example, one has $L_{\Omega,\eta} + H_{\Omega,\eta} = \mathrm{Id}$ on $L^2(\Omega)$ and the bounds

$$\|L_{\Omega,\eta}f\|_{L^2(\Omega)} + \|H_{\Omega,\eta}f\|_{L^2(\Omega)} \le C\|f\|_{L^2(\Omega)} \qquad \forall f \in L^2(\Omega),$$

where $C > 0$ depends solely on Ω (via the Stein extension operator E_Ω). The operators $H_{\Omega,\eta}$ and $H_{\partial\Omega,\eta}$ have furthermore approximation properties if the function they are applied to has some Sobolev regularity. We illustrate this for $H_{\partial\Omega,\eta}$:

Lemma 3.4. *Let $\Omega \subset \mathbb{R}^2$ be a polygon. Then there exists $C > 0$ independent of k and $\eta > 1$ such that for all $g \in H_{pw}^{1/2}(\partial\Omega)$*

$$k^{1/2}(1 + \eta^{1/2})\|H_{\partial\Omega,\eta}g\|_{L^2(\partial\Omega)} + \|H_{\partial\Omega,\eta}g\|_{H_{pw}^{1/2}(\partial\Omega)} \le C\|g\|_{H_{pw}^{1/2}(\partial\Omega)}.$$

Proof. We only show the estimate for $\|H_{\partial\Omega,\eta}g\|_{L^2(\partial\Omega)}$. We consider first the case of an interval $I \subset \mathbb{R}$. We define $H_{I,\eta}g$ by $H_{I,\eta}g = \mathscr{F}^{-1}\chi_{\mathbb{R}\setminus B_{\eta k}(0)}\mathscr{F}E_I g$, where $\chi_{\mathbb{R}\setminus B_{\eta k}(0)}$ is the characteristic function for $\mathbb{R} \setminus (-\eta k, \eta k)$ and E_I is the Stein extension operator for the interval I. Since, by Parseval, $\mathscr{F}$ is an isometry on $L^2(\mathbb{R})$ we have

$$\|H_{I,\eta}g\|_{L^2(I)}^2 \le \|H_{I,\eta}g\|_{L^2(\mathbb{R})}^2 = \int_{\mathbb{R}\setminus B_{\eta k}(0)} |\mathscr{F}E_I g|^2 \, d\xi$$

$$= \int_{\mathbb{R}\setminus B_{\eta k}(0)} \frac{(1 + |\xi|^2)^{1/2}}{(1 + |\xi|^2)^{1/2}}|\mathscr{F}E_I g|^2 \, d\xi \le \frac{1}{(1+(\eta k)^2)^{1/2}} \int_{\mathbb{R}}(1+|\xi|^2)^{1/2}|\mathscr{F}E_I g|^2 \, d\xi.$$

The last integral can be bounded by $C\|E_I g\|_{H^{1/2}(\mathbb{R})}^2$. The stability properties of the extension operator E_I then imply furthermore $\|E_I g\|_{H^{1/2}(\mathbb{R})} \le C\|g\|_{H^{1/2}(I)}$. In total,

$$\|H_{I,\eta}g\|_{L^2(I)} \le C\frac{1}{(1 + (\eta k)^2)^{1/4}}\|g\|_{H^{1/2}(I)} \le Ck^{-1/2}(1 + \eta)^{-1/2}\|g\|_{H^{1/2}(I)},$$

where, in the last estimate, the constant C depends additionally on k_0. From this estimate, we obtain the desired bound for $\|H_{\partial\Omega,\eta}g\|_{L^2(\partial\Omega)}$ by identifying each edge of Ω with an interval. $\qquad\square$

3.3.2   Corner Singularities

We recall the following result harking back to the work by Kondratiev and Grisvard:

Lemma 3.5. *Let* $\Omega \subset \mathbb{R}^d$ *be a polygon with vertices* A_j, $j = 1, \ldots, J$, *and interior angles* ω_j, $j = 1, \ldots, J$. *Define for each vertex* A_j *the singularity function* S_j *by*

$$S_j(r_j, \varphi_j) = r_j^{\pi/\omega_j} \cos\left(\frac{\pi}{\omega_j}\varphi_j\right), \tag{26}$$

where (r_j, φ_j) *are polar coordinates centered at the vertex* A_j *such that the edges of* Ω *meeting at* A_j *correspond to* $\varphi_j = 0$ *and* $\varphi_j = \omega_j$. *Then every solution* u *of*

$$-\Delta u = f \quad in \ \Omega, \qquad \partial_n u = g \quad on \ \partial\Omega,$$

can be written as $u = u_0 + \sum_{j=1}^{J} a_j^{\Delta}(f, g) S_j$ *with the a priori bounds*

$$\|u_0\|_{H^2(\Omega)} + \sum_{j=1}^{J} |a_j^{\Delta}(f, g)| \le C\left[\|f\|_{L^2(\Omega)} + \|g\|_{H_{pw}^{1/2}(\partial\Omega)} + \|u\|_{H^1(\Omega)}\right]. \tag{27}$$

The a_j^{Δ} *are linear functionals, and* $a_j^{\Delta} = 0$ *for convex corners* A_j *(i.e., if* $\omega_j < \pi$*).*

Proof. This classical result is comprehensively treated in [28].   □

3.3.3   The Modified Helmholtz Equation

We consider the modified Helmholtz equation in both a bounded domain with Robin boundary conditions and in the full space $\mathbb{R}^2$. The corresponding solution operators will be denoted S_Ω^+ and $S_{\mathbb{R}^2}^+$:

1. The operator $S_\Omega^+ : L^2(\Omega) \times H_{pw}^{1/2}(\partial\Omega) \to H^1(\Omega)$ is the solution operator for

$$-\Delta u + k^2 u = f \quad in \ \Omega, \qquad \partial_n u + iku = g \quad on \ \partial\Omega. \tag{28}$$

2. The operator $S_{\mathbb{R}^2}^+ : L^2(\mathbb{R}^2) \to H^1(\mathbb{R}^2)$ is the solution operator for

$$-\Delta u + k^2 u = f \quad in \ \mathbb{R}^2. \tag{29}$$

Lemma 3.6 (properties of S_Ω^+). *Let* $\Omega \subset \mathbb{R}^2$ *be a polygon and* $f \in L^2(\Omega)$, $g \in H_{pw}^{1/2}(\partial\Omega)$. *Then the solution* $u := S_\Omega^+(f, g)$ *satisfies*

$$\|u\|_{1,k,\Omega} \le k^{-1/2}\|g\|_{L^2(\partial\Omega)} + k^{-1}\|f\|_{L^2(\Omega)}. \tag{30}$$

Furthermore, there exists $C > 0$ independent of k and the data f, g, and there exists a decomposition $u = u_{H^2} + \sum_{i=1}^{J} a_i^+(f, g) S_i$ for some linear functionals a_i^+ with

$$\|u_{H^2}\|_{H^2(\Omega)} + \sum_{i=1}^{J} |a_i^+(f, g)| \leq C \left[\|f\|_{L^2(\Omega)} + \|g\|_{H_{pw}^{1/2}(\partial\Omega)} + k^{1/2} \|g\|_{L^2(\partial\Omega)} \right].$$

(31)

Proof. The estimate (30) for $\|u\|_{1,k,\Omega}$ follows by Lax-Milgram – see [49, Lemma 4.6] for details. Since u satisfies

$$-\Delta u = f - k^2 u \quad \text{in } \Omega, \qquad \partial_n u = g - iku \quad \text{on } \partial\Omega,$$

the standard regularity theory for the Laplacian (see Lemma 3.5) permits us to decompose $u = u_{H^2} + \sum_{i=1}^{J} a_i^\Delta(f - k^2 u, g - iku) S_i$. The continuity of the linear functionals a_i^Δ reads

$$\sum_{i=1}^{J} |a_i^\Delta(f - k^2 u, g - iku)| \leq C \left[\|f - k^2 u\|_{L^2(\Omega)} + \|g - iku\|_{H_{pw}^{1/2}(\partial\Omega)} \right].$$

Since $(f, g) \mapsto S_\Omega^+(f, g)$ is linear, the map $(f, g) \mapsto a_i^+(f, g) := a_i^\Delta(f - k^2 u, g - iku)$ is linear, and (30), (27) give the desired estimates for u_{H^2} and $a_i^+(f, g)$. □

Lemma 3.7 (properties of $S_{\mathbb{R}^2}^+$). *There exists $C > 0$ such that for every $\eta > 1$ and every $f \in L^2(\mathbb{R}^2)$ whose Fourier transform $\mathscr{F} f$ satisfies $\operatorname{supp} \mathscr{F} f \subset \mathbb{R}^2 \setminus B_{\eta k}(0)$, the solution $u = S_{\mathbb{R}^2}^+ f$ of (29) satisfies*

$$\|u\|_{1,k,\mathbb{R}^2} \leq k^{-1} \frac{1}{\sqrt{1 + \eta^2}} \|f\|_{L^2(\mathbb{R}^2)}, \qquad \|u\|_{H^2(\mathbb{R}^2)} \leq C \|f\|_{L^2(\mathbb{R}^2)}.$$

Proof. The result follows from Parseval's theorem and the weak formulation for u as follows (we abbreviate the Fourier transforms by $\widehat{f} = \mathscr{F} f$ and $\widehat{u} = \mathscr{F} u$):

$$\|u\|_{1,k,\mathbb{R}^2}^2 = (f, u)_{L^2(\mathbb{R}^2)} = (\widehat{f}, \widehat{u})_{L^2(\mathbb{R}^2)}$$

$$\leq \sqrt{\int_{\mathbb{R}^2} (|\xi|^2 + k^2)^{-1} |\widehat{f}|^2 \, d\xi} \sqrt{\int_{\mathbb{R}^2} (|\xi|^2 + k^2) |\widehat{u}|^2 \, d\xi}$$

$$= \sqrt{\int_{\mathbb{R}^2 \setminus B_{\eta k}(0)} (|\xi|^2 + k^2)^{-1} |\widehat{f}|^2 \, d\xi} \|u\|_{1,k,\mathbb{R}^2} \leq \frac{1}{k\sqrt{1 + \eta^2}} \|\widehat{f}\|_{L^2(\mathbb{R}^2)} \|u\|_{1,k,\mathbb{R}^2},$$

where, in the penultimate step, we used the support properties of $\widehat{f}$. Appealing again to Parseval, we get the desired claim for $\|u\|_{1,k,\mathbb{R}^2}$. The estimate for $\|u\|_{H^2(\mathbb{R}^2)}$ now follows from $f \in L^2(\mathbb{R}^2)$ and the standard interior regularity for the Laplacian. □

3.4   Proof of Theorem 3.2

We denote by $S : (f, g) \mapsto S(f, g)$ the solution operator for (9). Concerning some of its properties, we have the following result taken essentially from [49, Lemma 4.13]:

Lemma 3.8 (analytic regularity of $S(f, g)$). *Let Ω be a polygon. Let f be analytic on Ω and $g \in L^2(\partial\Omega)$ be piecewise analytic and satisfy for some constants $\widetilde{C}_f, \widetilde{C}_g, \gamma_f, \gamma_g > 0$*

$$\|\nabla^n f\|_{L^2(\Omega)} \leq \widetilde{C}_f \gamma_f^n \max\{n, k\}^n \qquad \forall n \in \mathbb{N}_0 \tag{32a}$$

$$\|\nabla_T^n g\|_{L^2(e)} \leq \widetilde{C}_g \gamma_g^n \max\{n, k\}^n \qquad \forall n \in \mathbb{N}_0 \quad \forall e \in \mathscr{E}, \tag{32b}$$

where $\mathscr{E}$ denotes the set of edges of Ω and ∇_T tangential differentiation. Then there exist $\vec{\beta} \in [0, 1)^J$ (depending only on Ω) and constants $C, \gamma > 0$ (depending only on $\Omega, \gamma_f, \gamma_g, k_0$) such that the following is true with the constant $C_{sol}(k)$ of (25):

$$\|u\|_{1,k,\Omega} \leq C_{sol}(k)(\widetilde{C}_f + \widetilde{C}_g) \tag{33}$$

$$\|\phi_{n,\vec{\beta},k} \nabla^{n+2} u\|_{L^2(\Omega)} \leq C C_{sol}(k) k^{-1} (\widetilde{C}_f + \widetilde{C}_g)\gamma^n \max\{n, k\}^{n+2} \quad \forall n \in \mathbb{N}_0. \tag{34}$$

Proof. The estimate (33) is simply a restatement of the definition of $C_{sol}(k)$. The estimate (34) will follow from [45, Prop. 5.4.5]. To simplify the presentation, we assume by linearity that g vanishes on all edges of Ω with the exception of one edge Γ. Furthermore, we restrict our attention to the vicinity of one vertex, which we take to be the origin; we assume $\Gamma \subset (0, \infty) \times \{0\}$, and that near the origin, Ω is above $(0, \infty) \times \{0\}$, i.e., $\{(r\cos\varphi, r\sin\varphi) : 0 < r < \rho, 0 < \varphi < \omega\} \subset \Omega$ for some $\rho, \omega > 0$.

Upon setting $\varepsilon := 1/k$, we note that u solves

$$-\varepsilon^2 \Delta u - u = \varepsilon^2 f \quad \text{on } \Omega, \qquad \varepsilon^2 \partial_n u = \varepsilon(\varepsilon g - iu) \quad \text{on } \partial\Omega.$$

On the edge Γ, the function g is the restriction of $G_{1,0}(x, y) := g(x)e^{-y/\varepsilon}$ to Γ. The assumptions on f and g then imply that [45, Prop. 5.4.5] is applicable with the following choice of constants appearing in [45, Prop. 5.4.5]:

$$C_f = \varepsilon^2 \widetilde{C}_f, \ C_{G_1} = \varepsilon \varepsilon^{1/2} \widetilde{C}_g, \ C_{G_2} = \varepsilon, \qquad C_b = 0, \ C_c = 1,$$
$$\gamma_f = O(1), \ \gamma_{G_1} = O(1), \qquad \gamma_{G_2} = O(1), \ \gamma_b = 0, \ \gamma_c = 0,$$

resulting in the existence of constants $C, K > 0$ and $\vec{\beta} \in [0, 1)^J$ with

$$\|\Phi_{n,\vec{\beta},k}\nabla^{n+2}u\|_{L^2(\Omega)}$$
$$\leq CK^{n+2}\max\{n+2,k\}^{n+2}\left(k^{-2}\widetilde{C}_f + k^{-1}\|u\|_{1,k,\Omega} + k^{-3/2}\widetilde{C}_g\right)$$

for all $n \in \mathbb{N}_0$. We conclude the argument by inserting (33) and estimating generously $k^{-1}\widetilde{C}_f + k^{-1/2}\widetilde{C}_g \leq C\left(\widetilde{C}_f + \widetilde{C}_g\right)$.

We remark that this last generous estimate comes from the precise form of our stability assumption (25). Its form (25) is motivated by the estimates *available* for the star-shaped case, but could clearly be replaced with other assumptions. □

Corollary 3.9 (analytic regularity of $S(L_{\Omega,\eta}f, L_{\partial\Omega,\eta}g)$). *Let Ω be a polygon and $\eta > 1$. Then there exist $\vec{\beta} \in [0,1)^J$ (depending only on Ω) and $C, \gamma > 0$ (depending only on Ω, k_0, and $\eta > 1$) such that for every $f \in L^2(\Omega)$ and $g \in L^2(\partial\Omega)$, the function $u = S(L_{\Omega,\eta}f, L_{\partial\Omega,\eta}g)$ satisfies with $C_{f,g} := \|f\|_{L^2(\Omega)} + \|g\|_{L^2(\partial\Omega)}$*

$$\|u\|_{1,k,\Omega} \leq CC_{sol}(k)C_{f,g} \tag{35}$$
$$\|\Phi_{n,\vec{\beta},k}\nabla^{n+2}u\|_{L^2(\Omega)} \leq CC_{sol}(k)k^{-1}\gamma^n\max\{n,k\}^{n+2}C_{f,g} \qquad \forall n \in \mathbb{N}_0. \tag{36}$$

Proof. The definitions of $L_{\Omega,\eta}f$ and $L_{\partial\Omega,\eta}$ imply with Parseval

$$\|\nabla^n L_{\Omega,\eta}f\|_{L^2(\Omega)} \leq C(\eta k)^n \|f\|_{L^2(\Omega)} \qquad \forall n \in \mathbb{N}_0,$$
$$\|\nabla_T^n L_{\partial\Omega,\eta}g\|_{L^2(\partial\Omega)} \leq C(\eta k)^n \|g\|_{L^2(\partial\Omega)} \qquad \forall n \in \mathbb{N}_0,$$

where again ∇_T is the (edgewise) tangential gradient. The desired estimates now follow from Lemma 3.8. □

Key to the proof of Theorem 3.2 is the following contraction result:

Lemma 3.10 (contraction lemma). *Let $\Omega \subset \mathbb{R}^2$ be a polygon. Fix $q \in (0,1)$. Then one can find $\vec{\beta} \in [0,1)^J$ (depending solely on Ω) and constants $C, \gamma > 0$ independent of k such that for every $f \in L^2(\Omega)$ and every $g \in H_{pw}^{1/2}(\partial\Omega)$, the solution u of (9) can be decomposed as $u = u_{H^2} + \sum_{i=1}^J a_i(f,g)S_i + u_{\mathscr{A}} + r$, where $u_{H^2} \in H^2(\Omega)$, the a_i are linear functionals, and $u_{\mathscr{A}} \in C^\infty(\Omega)$. These functions satisfy*

$$k\|u_{H^2}\|_{1,k,\Omega} + \|u_{H^2}\|_{H^2(\Omega)} + \sum_{i=1}^J |a_i(f,g)| \leq C\left[\|f\|_{L^2(\Omega)} + \|g\|_{H_{pw}^{1/2}(\partial\Omega)}\right],$$

$$\|u_{\mathscr{A}}\|_{1,k,\Omega} \leq CC_{sol}(k)\left[\|f\|_{L^2(\Omega)} + \|g\|_{L^2(\partial\Omega)}\right],$$

$$\|\Phi_{n,\vec{\beta},k}\nabla^{n+2}u_{\mathscr{A}}\|_{L^2(\Omega)} \leq CC_{sol}(k)k^{-1}\gamma^n\max\{n,k\}^{n+2}\left[\|f\|_{L^2(\Omega)} + \|g\|_{L^2(\partial\Omega)}\right]$$

for all $n \in \mathbb{N}_0$. Finally, the remainder r satisfies

$$-\Delta r - k^2 r = \widetilde{f} \quad \text{on } \Omega, \qquad \partial_n r + ikr = \widetilde{g}$$

for some $\widetilde{f} \in L^2(\Omega)$ *and* $\widetilde{g} \in H_{pw}^{1/2}(\partial\Omega)$ *with*

$$\|\widetilde{f}\|_{L^2(\Omega)} + \|\widetilde{g}\|_{H_{pw}^{1/2}(\partial\Omega)} \leq q \left(\|f\|_{L^2(\Omega)} + \|g\|_{H_{pw}^{1/2}(\partial\Omega)} \right).$$

Proof. We start by decomposing $(f, g) = (L_{\Omega,\eta}f, L_{\partial\Omega,\eta}g) + (H_{\Omega,\eta}f, H_{\partial\Omega,\eta}g)$ with a parameter $\eta > 1$ that will be selected below. We set

$$u_{\mathscr{A}} := S(L_{\Omega,\eta}f, L_{\partial\Omega,\eta}g), \qquad u_1 := S_{\mathbb{R}^2}^+(H_{\Omega,\eta}f),$$

where we tacitly extended $H_{\Omega,\eta}f$ (which is only defined on Ω) by zero outside Ω. Then $u_{\mathscr{A}}$ satisfies the desired estimates by Corollary 3.9. For u_1 we have by Lemma 3.7 and the stability $\|H_{\Omega,\eta}f\|_{L^2(\Omega)} \leq C\|f\|_{L^2(\Omega)}$ (we note that $C > 0$ is independent of k and η) the *a priori* estimates

$$\|u_1\|_{1,k,\mathbb{R}^2} \leq Ck^{-1}(1+\eta^2)^{-1/2}\|H_{\Omega,\eta}f\|_{L^2(\Omega)} \leq Ck^{-1}(1+\eta)^{-1}\|f\|_{L^2(\Omega)},$$

$$\|u_1\|_{H^2(\mathbb{R}^2)} \leq C\|H_{\Omega,\eta}f\|_{L^2(\Omega)} \leq C\|f\|_{L^2(\Omega)}.$$

The trace and the multiplicative trace inequalities imply for $g_1 := \partial_n u_1 + iku_1$:

$$k^{1/2}(1+\eta)^{1/2}\|g_1\|_{L^2(\partial\Omega)} + \|g_1\|_{H_{pw}^{1/2}(\partial\Omega)} \leq C\|f\|_{L^2(\Omega)}.$$

For $g_2 := H_{\partial\Omega,\eta}g - g_1$ we then get from Lemma 3.4 and the triangle inequality

$$k^{1/2}(1+\eta)^{1/2}\|g_2\|_{L^2(\partial\Omega)} + \|g_2\|_{H_{pw}^{1/2}(\partial\Omega)} \leq C \left[\|g\|_{H_{pw}^{1/2}(\partial\Omega)} + \|f\|_{L^2(\Omega)} \right].$$

Lemma 3.6 yields for $u_2 := S_\Omega^+(0, g_2)$,

$$\|u_2\|_{1,k,\Omega} \leq Ck^{-1/2}\|g_2\|_{L^2(\partial\Omega)} \leq Ck^{-1}(1+\eta)^{-1/2}\left[\|f\|_{L^2(\Omega)} + \|g\|_{H_{pw}^{1/2}(\partial\Omega)} \right],$$

and furthermore we can write $u_2 = u_{H^2} + \sum_{i=1}^{J} a_i^+(0, g_2)S_i$, with

$$\|u_{H^2}\|_{H^2(\Omega)} + \sum_{i=1}^{J} |a_i^+(0, g_2)| \leq C \left[\|f\|_{L^2(\Omega)} + \|g\|_{H_{pw}^{1/2}(\partial\Omega)} \right].$$

We then define $a_i(f, g) := a_i^+(0, g_2)$ and note that $(f, g) \mapsto a_i(f, g)$ is linear by linearity of the maps a_i^+ and $(f, g) \mapsto g_2$. The above shows that u_{H^2} and the a_i satisfy the required estimates. Finally, the function $\widetilde{u} := u - (u_{\mathscr{A}} + u_1 + u_2)$ satisfies

$$-\Delta\widetilde{u} - k^2\widetilde{u} = 2k^2(u_1 + u_2) =: \widetilde{f}, \qquad \partial_n\widetilde{u} + \mathbf{i}k\widetilde{u} = 0 =: \widetilde{g},$$

together with

$$\|\widetilde{f}\|_{L^2(\Omega)} \le 2k^2\left(\|u_1\|_{L^2(\Omega)} + \|u_2\|_{L^2(\Omega)}\right) \le C(1+\eta)^{-1/2}\left[\|f\|_{L^2(\Omega)} + \|g\|_{H_{pw}^{1/2}(\partial\Omega)}\right].$$

Hence, selecting $\eta > 1$ sufficiently large so that for the chosen $q \in (0,1)$ we have $C(1+\eta)^{-1/2} \le q$ allows us to conclude the proof. $\qquad\square$

Proof of Theorem 3.2. The contraction property of Lemma 3.10 can be restated as $S(f,g) = u_{H^2} + \sum_{i=1}^{J} a_i(f,g)S_i + u_{\mathscr{A}} + S(\widetilde{f},\widetilde{g})$, where, for a chosen $q \in (0,1)$, we have $\|\widetilde{f}\|_{L^2(\Omega)} + \|\widetilde{g}\|_{H_{pw}^{1/2}(\partial\Omega)} \le q\left[\|f\|_{L^2(\Omega)} + \|g\|_{H_{pw}^{1/2}(\partial\Omega)}\right]$ together with appropriate estimates for u_{H^2}, $a_i(f,g)$, and $u_{\mathscr{A}}$. This consideration can be repeated for $S(\widetilde{f},\widetilde{g})$. We conclude that a geometric series argument can be employed to write $u = S(f,g) = u_{H^2} + \sum_{i=1}^{J} \widetilde{a}_i(f,g)S_i + \widetilde{u}_{\mathscr{A}}$, where $u_{H^2} \in H^2(\Omega), \widetilde{u}_{\mathscr{A}} \in C^\infty(\Omega)$, and the coefficients $\widetilde{a}_i$ are in fact linear functionals of the data (f,g). Furthermore, we have with the abbreviation $C_{f,g} := \|f\|_{L^2(\Omega)} + \|g\|_{H_{pw}^{1/2}(\partial\Omega)}$

$$\|\widetilde{u}_{\mathscr{A}}\|_{1,k,\Omega} \le CC_{f,g}$$

$$\|\Phi_{n,\vec{\beta},k}\nabla^{n+2}\widetilde{u}_{\mathscr{A}}\|_{L^2(\Omega)} \le CC_{sol}(k)k^{-1}C_{f,g}\gamma^n\max\{n,k\}^{n+2} \qquad \forall n \in \mathbb{N}_0,$$

$$k\|u_{H^2}\|_{1,k,\Omega} + \|u_{H^2}\|_{H^2(\Omega)} + \sum_{i=1}^{J}|\widetilde{a}_i(f,g)| \le CC_{f,g}.$$

Finally, Lemma 3.11 below allows us to absorb the contribution $\sum_{i=1}^{J}\widetilde{a}_i(f,g)S_i$ in the analytic part by setting $u_{\mathscr{A}} := \widetilde{u}_{\mathscr{A}} + \sum_{i=1}^{J}\widetilde{a}_i(f,g)S_i$. In view of $\beta_i < 1$, we have $2 - \beta_i \ge 1$ and arrive at

$$\|u_{\mathscr{A}}\|_{H^1(\Omega)} \le C(C_{sol}(k)+1)C_{f,g}, \quad k\|u_{\mathscr{A}}\|_{L^2(\Omega)} \le CC_{f,g}(C_{sol}(k)+k),$$

$$\|\Phi_{n,\vec{\beta},k}\nabla^{n+2}u_{\mathscr{A}}\|_{L^2(\Omega)} \le CC_{f,g}\left[C_{sol}(k)k^{-1} + k^{-1}\right]\max\{n,k\}^{n+2} \quad \forall n \in \mathbb{N}_0,$$

which concludes the argument. $\qquad\square$

Lemma 3.11. *Let $\beta_i \in [0,1)$ satisfy $\beta_i > 1 - \frac{\pi}{\omega_i}$. Then, for some $C, \gamma > 0$ independent of k, the singularity functions S_i of (26) satisfy $\|S_i\|_{H^1(\Omega)} \le C$ and*

$$\|\Phi_{n,\vec{\beta},k}\nabla^{n+2}S_i\|_{L^2(\Omega)} \le Ck^{-(2-\beta_i)}\gamma^n\max\{n,k\}^{n+2} \quad \forall \in \mathbb{N}_0$$

Proof. Follows by a direct calculation. See [21] for details. $\qquad\square$

4 Stability of Galerkin Discretizations

4.1 Abstract Results

We consider the model problem (9) and a sequence $(V_N)_{N \in \mathbb{N}} \subset H^1(\Omega)$ of finite-dimensional spaces. Furthermore, we assume that $(V_N)_{N \in \mathbb{N}}$ is such that for every $v \in H^1(\Omega)$ we have $\lim_{N \to \infty} \inf_{v_N \in V_N} \|v - v_N\|_{H^1(\Omega)} = 0$. The conforming approximations u_N to the solution u of (9) are then defined by:

$$\text{Find } u_N \in V_N \text{ s.t. } \quad B(u_N, v) = l(v) \qquad \forall v \in V_N. \tag{37}$$

Since the sesquilinear form B satisfies a Gårding inequality, general functional analytic argument show that *asymptotically*, the discrete problem (37) has a unique solution u_N and are quasi-optimal (see, e.g., [61, Thm. 4.2.9], [62]). More precisely, there exist $N_0 > 0$ and $C > 0$ such that for all $N \geq N_0$

$$\|u - u_N\|_{1,k,\Omega} \leq C \inf_{v \in V_N} \|u - v\|_{1,k,\Omega}. \tag{38}$$

This general functional analytic argument does not give any indication of how C and N_0 depend on discretization parameters and the wavenumber k. Inspection of the arguments reveals that it is the approximation properties of the spaces V_N for the approximation of the solution of certain adjoint problems that leads to the quasi-optimality result (38). For the reader's convenience, we repeat the argument, which has been used previously in, e.g., [5, 9, 44, 49, 50, 60, 62] and is often attributed to Schatz, [62]:

Lemma 4.1 ([49, Thm. 3.2]). *Let $\Omega \subset \mathbb{R}^d$ be a bounded Lipschitz domain and B be defined in (11). Denote by $S^\star : L^2(\Omega) \to H^1(\Omega)$ the solution operator for the problem*

$$\text{Find } u^\star \in H^1(\Omega) \text{ s.t. } \quad B(v, u^\star) = (v, f)_{L^2(\Omega)} \qquad \forall v \in H^1(\Omega). \tag{39}$$

Define the adjoint approximation property $\eta(V_N)$ by

$$\eta(V_N) := \sup_{f \in L^2(\Omega)} \inf_{v \in V_N} \frac{\|S^\star(f) - v\|_{1,k,\Omega}}{\|f\|_{L^2(\Omega)}}.$$

If, for the continuity constant C_B of (12), the space V_N satisfies

$$2 C_B k \eta(V_N) \leq 1, \tag{40}$$

then the solution u_N of (37) exists and satisfies

$$\|u - u_N\|_{1,k,\Omega} \le 2C_B \inf_{v \in V_N} \|u - v\|_{1,k,\Omega}. \tag{41}$$

Proof. We will not show existence of u_N but restrict our attention on the quasi-optimality result (41); we refer to [42, Thm. 3.9] for the demonstration that (41) in fact implies existence and uniqueness of u_N. Letting $e = u - u_N$ be the error, we start with an estimate for $\|e\|_{L^2(\Omega)}$: Using the definition of the operator $S^\star$ and the Galerkin orthogonality satisfied by e, we have for arbitrary $v \in V_N$

$$\|e\|^2_{L^2(\Omega)} = (e,e)_{L^2(\Omega)} = B(e, S^\star e) = B(e, S^\star e - v) \le C_B \|e\|_{1,k,\Omega} \|S^\star e - v\|_{1,k,\Omega}.$$

Infimizing over all $v \in V_N$ yields with the adjoint approximation property $\eta(V_N)$

$$\|e\|_{L^2(\Omega)} \le C_B \eta(V_N) \|e\|_{1,k,\Omega}.$$

The Gårding inequality and the Galerkin orthogonality yield for arbitrary $v \in V_N$:

$$\|e\|^2_{1,k,\Omega} = \mathrm{Re}\, B(e,e) + 2k^2 \|e\|^2_{L^2(\Omega)} = \mathrm{Re}\, B(e, u - v) + 2k^2 \|e\|^2_{L^2(\Omega)}$$

$$\le C_B \|e\|_{1,k,\Omega} \|u - v\|_{1,k,\Omega} + (C_B k \eta(V_N))^2 \|e\|^2_{1,k,\Omega}.$$

The assumption $C_B k \eta(V_N) \le 1/2$ allows us to rearrange this bound to get $\|e\|_{1,k,\Omega} \le 2C_B \|u - v\|_{1,k,\Omega}$. Since $v \in V_N$ is arbitrary, we arrive at (41). $\qquad\square$

Lemma 4.1 informs us that the convergence analysis for the Galerkin discretization of (9) can be reduced to the study of the adjoint approximation property $\eta(V_N)$, which is purely a question of approximation theory. In the context of piecewise polynomial approximation spaces V_N this requires a good regularity theory for the operator $S^\star$. The strong form of the equation satisfied by $u^\star := S^\star f$ is

$$-\Delta u^\star - k^2 u^\star = f \quad \text{in } \Omega, \qquad \partial_n u^\star - iku^\star = 0 \quad \text{on } \partial\Omega, \tag{42}$$

which is again a Helmholtz problem of the type considered in Sect. 3. More formally, with the solution operator S of Sect. 3, we have $S^\star f = \overline{S(\overline{f}, 0)}$, where an overbar denotes complex conjugation. Thus, the regularity theory of Sect. 3 is applicable.

4.2 Stability of *hp*-FEM

The estimates of Theorem 3.2 suggest that the effect of the corner singularities is essentially restricted to an $O(1/k)$-neighborhood of the vertices. This motivates us to consider meshes that are refined in a small neighborhood of the vertices. To fix ideas, we restrict our attention to meshes $\mathscr{T}^{geo}_{h,L}$ that are obtained in the following way: First, a quasi-uniform triangulation $\mathscr{T}_h$ with mesh size h is selected. Then, the elements abutting the vertices A_j, $j = 1, \ldots, J$, are refined further with a mesh that is geometrically graded towards these vertices. These geometric meshes have L layers and use a grading factor $\sigma \in (0,1)$ (see [65, Sec. 4.4.1] for a precise formal definition). Furthermore, for any regular, shape-regular mesh $\mathcal{T}$, we define

$$S^p(\mathcal{T}) := \{u \in H^1(\Omega): u|_K \in \mathscr{P}_p \quad \forall K \in \mathcal{T}\}, \tag{43}$$

where $\mathscr{P}_p$ denotes the space of polynomials of degree p. We now show that on the geometric meshes $\mathcal{T}_{h,L}^{geo}$, stability of the FEM is ensured if the mesh size h and the polynomial degree p satisfy the scale resolution condition (6) and, additionally, $L = O(p)$ layers of geometric refinement are used near the vertices:

Theorem 4.2 (quasi-optimality of hp-FEM). *Let $\mathcal{T}_{h,L}^{geo}$ denote the geometric meshes on the polygon $\Omega \subset \mathbb{R}^2$ as described above. Fix $c_3 > 0$. Then there are constants c_1, $c_2 >, 0$ depending solely on Ω and the shape-regularity of the mesh $\mathcal{T}_{h,L}^{geo}$ such that the following is true: If h, p, and L satisfy the conditions*

$$\frac{kh}{p} \leq c_1 \quad and \quad p \geq c_2 \log k \quad and \quad L \geq c_3 p \tag{44}$$

then the hp-FEM based on the space $S^p(\mathcal{T}_{h,L}^{geo})$ has a unique solution $u_N \in S^p(\mathcal{T}_{h,L}^{geo})$ and

$$\|u - u_N\|_{1,k,\Omega} \leq 2C_B \inf_{v \in S^p(\mathcal{T}_{h,L}^{geo})} \|u - v\|_{1,k,\Omega}. \tag{45}$$

Proof. By Lemma 4.1, we have to estimate $k\eta(V_N)$ with $V_N = S^p(\mathcal{T}_{h,L}^{geo})$. Recalling the definition of $\eta(V_N)$ we let $f \in L^2(\Omega)$ and observe that we can decompose $S^* f = u_{H^2} + u_{\mathscr{A}}$, where u_{H^2} and $u_{\mathscr{A}}$ satisfy the bounds

$$\|u_{H^2}\|_{H^2(\Omega)} \leq C \|f\|_{L^2(\Omega)},$$

$$\|\Phi_{n,\vec{\beta},k} \nabla^{n+2} u_{\mathscr{A}}\|_{L^2(\Omega)} \leq C(C_{sol}(k) + 1)k^{-1}\gamma^n \max\{k,n\}^{n+2} \|f\|_{L^2(\Omega)} \quad \forall n \in \mathbb{N}_0.$$

Piecewise polynomial approximation on $\mathcal{T}_{h,L}^{geo}$ as discussed in [49, Prop. 5.6] gives under the assumptions $kh/p \leq C$ and $L \geq c_3 p$: (inspection of the proof of [49, Prop. 5.6] shows that only bounds on the derivatives of order ≥ 2 are needed):

$$\inf_{v \in V_N} \|u_{H^2} - v\|_{1,k,\Omega} \leq C \frac{h}{p} \|f\|_{L^2(\Omega)},$$

$$\inf_{v \in V_N} \|u_{\mathscr{A}} - v\|_{1,k,\Omega} \leq C \left[(kh)^{1-\beta_{max}} e^{ckh-bp} + \left(\frac{kh}{\sigma_0 p}\right)^p \right] (C_{sol}(k) + 1) \|f\|_{L^2(\Omega)},$$

where $\beta_{max} = \max_{j=1,\ldots,J} \beta_j < 1$, and $C, c, b > 0$ are constants independent of h, p, and k. From this, we can easily infer

$$k\eta(V_N) \leq C \left\{ \frac{kh}{p} + k(C_{sol}(k) + 1) \left[(kh)^{1-\beta_{max}} e^{ckh-bp} + \left(\frac{kh}{\sigma_0 p} \right)^p \right] \right\}.$$

Noting that Theorem 2.4 gives $C_{sol}(k) = O(k^{5/2})$, and selecting c_1 sufficiently small as well as c_2 sufficient large allows us to make $k\eta(V_N)$ so small that the condition (40) in Lemma 4.1 is satisfied. □

Corollary 4.3 (exponential convergence on geometric meshes). *Let f be analytic on $\overline{\Omega}$ and g be piecewise analytic, i.e., f, g satisfy (32). Given $c_3 > 0$, there exist c_1, $c_2 > 0$ such that under the scale resolution conditions (44) of Theorem 4.2, the finite element approximation $u_N \in S^p(\mathscr{T}_{h,L}^{geo})$ exists, and there are constants C, $b > 0$ independent of k such that the error $u - u_N$ satisfies*

$$\|u - u_N\|_{1,k,\Omega} \leq Ce^{-bp}.$$

Proof. In view of Theorem 4.2, estimating $\|u - u_N\|_{1,k,\Omega}$ is purely a question of approximability for c_1 sufficiently small and c_2 sufficiently large. Lemma 3.8 gives that the solution $u = S(f,g)$ satisfies the bounds given there and, as in the proof of Theorem 4.2, we conclude from [49, Prop. 5.6] (more precisely, this follows from its proof)

$$\inf_{v \in V_N} \|u_{\mathscr{A}} - v\|_{1,k,\Omega} \leq C \left[(kh)^{1-\beta_{max}} e^{ckh-bp} + \left(\frac{kh}{\sigma_0 p} \right)^p \right] (C_{sol}(k) + 1)(\widetilde{C}_f + \widetilde{C}_g).$$

Theorem 2.4 asserts $C_{sol}(k) = O(k^{5/2})$, which implies the result by suitably adjusting c_1 and c_2 if necessary. □

Remark 4.4. 1. The problem size $N = \dim S^p(\mathscr{T}_{h,L}^{geo})$ is $N = O((L + h^{-2})p^2)$. The particular choice of $L = c_3 p$ layers of geometric refinement, approximation order $p = c_2 \log k$, and mesh size $h = c_1 p/k$ in Theorem 4.2 ensures quasi-optimality of the hp-FEM with problem size $N = O(k^2)$, i.e., quasi-optimality can be achieved with a fixed number of degrees of freedom per wavelength.

2. The sparsity pattern of the system matrix is that of the classical hp-FEM, i.e., each row/column has $O(p^2)$ non-zero entries. Noting that the scale resolution conditions (6), (44) require $p = O(\log k)$, we see that the number of non-zero entries per row/column is not independent of k. It is worth relating this observation to [8]. It is shown there for a model problem in 2D that *no* 9 point stencil can be found that leads to a pollution-free method.

3. Any space V_N that contains $S^p(\mathscr{T}_{h,L}^{geo})$, where h, p, and L satisfy the scale resolution condition (44) also features quasi-optimality.

4. The factor 2 on the right-hand side of (45) is arbitrary and can be replaced by any number greater than 1.

5. The stability analysis of Theorem 4.2 requires quite a significant mesh refinement near the vertices, namely, $L \sim p$. It is not clear whether this is an artifact of the proof. For a more careful numerical analysis of this issue, more detailed

information about the stability properties of the solution operator S is needed, e.g., estimates for $\|S(f,g)\|_{1,k,B_{1/k}(A_j)}$.

4.3 Numerical Examples: hp-FEM

All calculations reported in this section are performed with the hp-FEM code NETGEN/NGSOLVE by J. Schöberl, [63,64].

Example 4.5. In this 2D analog of Example 1.1, we consider the model problem (9) with exact solution being a plane wave $e^{i(k_1 x + k_2 y)}$, where $k_1 = -k_2 = \frac{1}{\sqrt{2}}k$ and $k \in \{4, 40, 100, 400\}$. For fixed $p \in \{1, 2, 3\}$, we show in Fig. 2 the performance of the h-version FEM for $p \in \{1, 2, 3\}$ on quasi-uniform meshes by displaying the relative error in the H^1-seminorm versus the number of degrees of freedom per wavelength. We observe that higher order methods are less prone to pollution. We

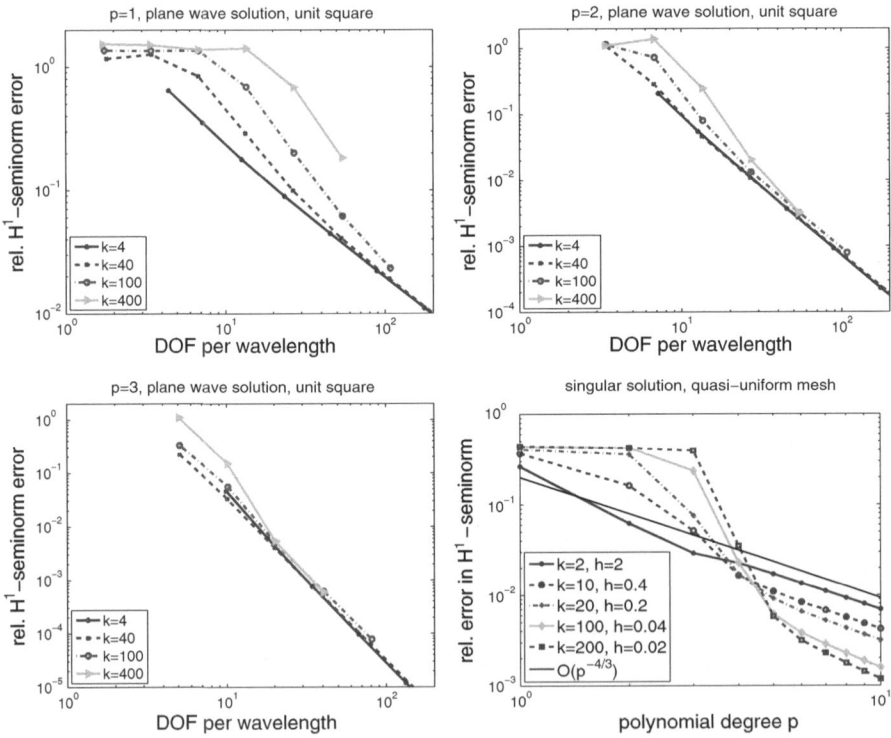

Fig. 2 *Top*: h-FEM with $p = 1$ (*left*) and $p = 2$ (*right*) as described in Example 4.5. *Bottom left*: h-FEM with $p = 3$ as described in Example 4.5. *Bottom right*: p-FEM for singular solution on quasi-uniform mesh as described in Example 4.7

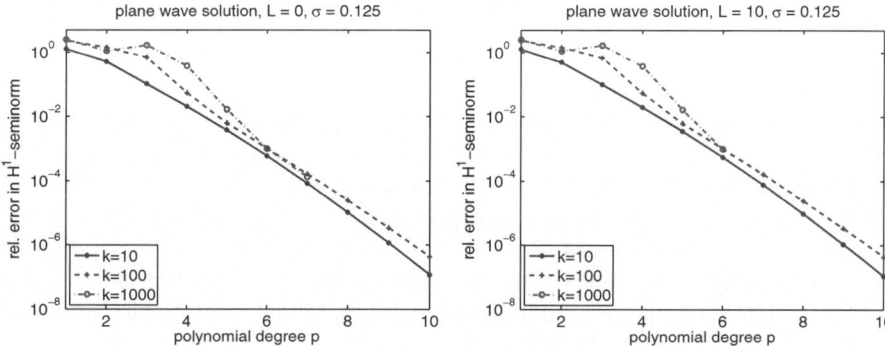

Fig. 3 p-FEM for plane wave solution as described in Example 4.6. *Left*: quasiuniform mesh $\mathcal{T}_h$ with $kh \approx 4$. *Right*: Mesh $\mathcal{T}^{geo}$ obtained from $\mathcal{T}_h$ by strong geometric refinement near origin

note that the meshes are quasi-uniform, i.e., no geometric mesh refinement near the vertices is performed in contrast to the requirements of Theorem 4.2. ∎

Example 4.6. On the L-shaped domain $\Omega = (-1, 1)^2 \setminus (0, 1) \times (-1, 0)$ with Γ being the union of the two edges meeting at $(0, 0)$, we consider

$$-\Delta u - k^2 u = 0 \quad \text{in } \Omega, \quad \partial_n u = 0 \quad \text{on } \Gamma, \quad \partial_n u - iku = g \quad \text{on } \partial\Omega \setminus \Gamma, \quad (46)$$

where the Robin data g are such that the exact solution is $u(x, y) = e^{i(k_1 x + k_2 y)}$ with $k_1 = -k_2 = \frac{1}{\sqrt{2}}k$, and $k \in \{10, 100, 1,000\}$. We consider two kinds of meshes, namely, quasi-uniform meshes $\mathcal{T}_h$ with mesh size h such that $kh \approx 4$ and meshes $\mathcal{T}^{geo}$ that are geometrically refined near the origin. The meshes $\mathcal{T}^{geo}$ are derived from the quasi-uniform mesh $\mathcal{T}_h$ by introducing a geometric grading on the elements abutting the origin; the grading factor is $\sigma = 0.125$ and the number of refinement levels is $L = 10$. Figure 3 shows the relative errors in the H^1-seminorm for the p-version of the FEM where for fixed mesh the approximation order p ranges from 1 to 10. It is particularly noteworthy that the refinement near the origin has hardly any effect on the convergence behavior of the FEM; this is quite in contrast to the stability result Theorem 4.2, which requires geometric refinement near all vertices of Ω. ∎

Example 4.7. The geometry and the boundary conditions are as described in Example 4.6. The data g are selected such that the exact solution is $u = J_{2/3}(kr) \cos \frac{2}{3}\varphi$, where (r, φ) denote polar coordinates and J_α is a first kind Bessel function. $k \in \{1, 10, 20, 100, 200\}$. Our calculations are p-FEMs with $p \in \{1, \ldots, 10\}$ on the quasiuniform mesh $\mathcal{T}_h$ described in Example 4.7. The results are displayed in the bottom right part of Fig. 2. The numerics illustrate that the singularity at the origin is rather weak: we observe that the asymptotic algebraic convergence behavior is $|u - u_N|_{H^1(\Omega)} \approx C_k p^{-4/3} |u|_{H^1(\Omega)}$, where the constant C_k depends favorably on k. ∎

4.4   Stability of Partition of Unity Method/Generalized FEM

The abstract stability result of Lemma 4.1 only assumes certain approximation properties of the spaces V_N. Particularly in an "h-version" setting, even non-polynomial, operator-adapted spaces may have sufficient approximation properties to ensure the important condition (40) for stability. We illustrate this effect for the PUM/gFEM, [44, 48] with local approximation spaces consisting of systems of plane waves or generalized harmonic polynomials (see Sect. 5 below) and the classical FEM shape functions as the partition of unity. The key observation is that for h sufficiently small, the resulting space has approximation properties similar to the classical (low order) FEM space:

Lemma 4.8. *Let $\mathcal{T}$ be a shape-regular triangulation of the polygon $\Omega \subset \mathbb{R}^2$. Let h be its mesh size; let $(x_i)_{i=1}^M$ be the nodes of the triangulation and $(\varphi_i)_{i=1}^M$ be the piecewise linear hat functions associated with the nodes $(x_i)_{i=1}^M$. Assume $kh \leq C_1$ for some $C_1 > 0$. Let V^{master} be either the space V_{GHP}^p with $p \geq 0$ (see (47) below) or the space W_{PW}^p with $p \geq 2$ (see (48) below). Define, for each $i = 1, \ldots, M$, the local approximation V_i by $V_i := \mathrm{span}\{b(x - x_i) : b \in V^{master}\}$. Then the space $V_N := \sum_{i=1}^M \varphi_i V_i$ has the following approximation property: There exists $C > 0$ depending only on the shape regularity of $\mathcal{T}$, the constant C_1, and V^{master} such that*

$$\inf_{v \in V_N} \|u - v\|_{L^2(\Omega)} + h\|u - v\|_{H^1(\Omega)} \leq C \left[h^2 \|u\|_{H^2(\Omega)} + (kh)^2 \|u\|_{L^2(\Omega)} \right] \quad \forall u \in H^2(\Omega).$$

Proof. The proof exploits the smoothness of the functions in V^{master}. Specifically, one can find an element $\psi \in V^{master}$ with $\psi = 1 + O((kh)^2)$. Then, the approximation properties of the space $\mathrm{span}\{\varphi_i : i = 1, \ldots, M\}$ can be exploited. We refer to [21] for details.   □

Lemma 4.8 shows that the space V_N, which is derived from solutions of the homogeneous Helmholtz equation, nevertheless has some approximation power for arbitrary functions with some Sobolev regularity. Hence, the condition (40) can be met for sufficiently small mesh sizes:

Corollary 4.9 ([44, Prop. 8.2.7]). *Assume the hypotheses of Lemma 4.8; in particular, let the space V_N be constructed from systems of plane waves or generalized harmonic polynomials. Assume additionally that Ω is a convex polygon. Then there exists $C > 0$ independent of k such that for $k^2 h \leq C$ the Galerkin method for (9) with $f = 0$ is quasi-optimal, i.e., the solution $u_N \in V_N$ of (37) exists and satisfies*

$$\|u - u_N\|_{1,k,\Omega} \leq 2C_B \inf_{v \in V_N} \|u - v\|_{1,k,\Omega}.$$

Proof. In view of Lemma 4.1, we have to estimate $\eta(V_N)$. To that end, we consider (9) with $f \in L^2(\Omega)$ and $g = 0$. In view of the convexity of Ω, we have $C_{sol}(k) = O(1)$ and elliptic regularity then yields for the solution u of (9)

$$\|u\|_{1,k,\Omega} + k^{-1}\|u\|_{H^2(\Omega)} \leq C\|f\|_{L^2(\Omega)}.$$

This allows us to conclude with Lemma 4.8 that

$$\inf_{v \in V_N} \|u - v\|_{1,k,\Omega} \leq C\left[(kh^2 + h)\|u\|_{H^2(\Omega)} + (k(kh)^2 + k^2h)\|u\|_{L^2(\Omega)}\right]$$

$$\leq C((kh)^2 + kh))\|f\|_{L^2(\Omega)} \leq Ckh(1 + kh)\|f\|_{L^2(\Omega)}.$$

Hence, $k\eta(V_N)$ can be made sufficiently small if k^2h is sufficiently small. We point out that convexity of Ω is assumed for convenience – under more stringent conditions on the mesh size h, quasioptimality holds for general polygons. □

5 Approximation with Plane, Cylindrical, and Spherical Waves

Systems of functions that are solutions of a (homogeneous) differential equation are often called "Trefftz systems". Prominent examples in the context of the Helmholtz equation are, in the two-dimensional setting, "generalized harmonic polynomials" and systems of plane waves given by

$$V_{GHP}^p := \mathrm{span}\{J_n(kr)e^{in\varphi} : -p \leq n \leq n\}, \tag{47}$$

$$W_{PW}^p := \mathrm{span}\{e^{ik\omega_n \cdot (x,y)} : n = 0, \ldots, p - 1\}, \qquad \omega_n = (\cos\frac{2\pi n}{p}, \sin\frac{2\pi n}{p}); \tag{48}$$

here, J_n is a first kind Bessel function, the functions in V_{GHP}^p are described in polar coordinates and the functions of W_{PW}^p in Cartesian coordinates. We point out that analogous systems can be developed in 3D. These functions are solutions of the homogeneous Helmholtz equation. For the approximation of a function u that satisfies the homogeneous Helmholtz equation on a domain $\Omega \subset \mathbb{R}^2$, one may study the "$p$-version", i.e., study how well u can be approximated from the spaces V_{GHP}^p or W_{PW}^p as $p \to \infty$; alternatively, one may study the "h-version", in which, for fixed p, the approximation properties of the spaces V_{GHP}^p or W_{PW}^p are expressed in terms of the diameter $h = \mathrm{diam}\,\Omega$ of a domain under consideration. In the way of illustration, we present two types of results:

Lemma 5.1 ([44]). *Let $\Omega \subset \mathbb{R}^2$ be a simply connected domain and $\Omega' \subset\subset \Omega$ be a compact subset. Let u solve $-\Delta u - k^2 u = 0$ on Ω. Then there exist constants C, $b > 0$ (possibly depending on k) such that for all $p \geq 2$:*

$$\inf_{v \in V_{GHP}^p} \|u - v\|_{H^1(\Omega')} \leq Ce^{-bp}, \qquad \inf_{v \in W_{PW}^p} \|u - v\|_{H^1(\Omega')} \leq Ce^{-bp/\ln p}.$$

Proof. See, e.g., [44] or [46, Thm. 5.3].   □

Remark 5.2. Analogs of Lemma 5.1 hold if u has only some finite Sobolev regularity. Then, the convergence rates are algebraic, [44], [46, Thm. 5.4], [32].   ∎

The approximation properties of the spaces V_{GHP}^p and W_{PW}^p can be also be studied in an h-version setting:

Proposition 5.3 ([32, Thm. 3.2.2]). *Let $\Omega \subset \mathbb{R}^2$ be a domain with diameter h and inscribed circle of radius ρh. Let $p = 2\mu + 1$. Assume $kh \leq C_1$. Then there exist $C_p > 0$ (depending only on C_1, $\rho > 0$, m, and p) and $v \in W_{PW}^{2\mu+1}$ such that*

$$\|u - v\|_{j,k,\Omega,\Sigma} \leq C_p h^{\mu-j+1} \|u\|_{\mu+1,k,\Omega,\Sigma}, \qquad 0 \leq j \leq \mu + 1,$$

where $\|v\|_{j,k,\Omega,\Sigma}^2 = \sum_{m=0}^{j} k^{2(j-m)} |v|_{H^m(\Omega)}^2$.

Several comments concerning Proposition 5.3 are in order:

1. The constant C_p in Proposition 5.3 depends favorably on p, and its dependence on p can be found in [32, Thm. 3.2.3].
2. Proposition 5.3 is formulated for the space W_{PW}^p of plane waves – analogous results are valid for generalized harmonic polynomials, see [32, Thm. 2.2.1] for both the h and hp-version.
3. Proposition 5.3 is formulated for the two-dimensional case. Similar results are available in 3D, [32].
4. The approximation properties of plane waves in terms of the element size have previously been studied in slightly different norms in [15].

6   Stability of Least Squares and DG Methods

Discrete stability in Sect. 4 is obtained from stability of the continuous problem by a perturbation argument. This approach does not seem to work very well if one aims at using approximation spaces that have special features linked to the differential equation under consideration. The reason can be seen from the proof of Lemma 4.1: The adjoint approximation property $\eta(V_N)$ (which needs to be small) measures how well certain solutions to the *in*homogeneous equation can be approximated from the test space. If the ansatz space is based on solutions of the homogeneous equation, then its capabilities to approximate solutions of the inhomogeneous equation are clearly limited. In an h-version, the situation is not as severe as we have just seen in Sect. 4.4 for the PUM/gFEM. In a pure p-version setting, however, the techniques of Sect. 4.4 do not seem applicable.

An option is to leave the setting of Galerkin methods and to work with formulations with built-in stability properties. Such approaches can often be understood as minimizing some residual norm, which then provides automatically stability and quasi-optimality (in this residual norm). We will illustrate this procedure here by two examples, namely, Least Squares methods and DG-methods. Our presentation will highlight an issue stemming from this approach, namely, the fact that error estimates in this residual norm do not easily lead to error estimates in more classical norms such as the $L^2(\Omega)$-norm.

6.1   Some Notation for Spaces of Piecewise Smooth Functions

Let $\mathcal{T}$ be a regular, shape-regular triangulation of the polygon $\Omega \subset \mathbb{R}^2$. We decompose the set of edges $\mathcal{E}$ as $\mathcal{E} = \mathcal{E}_I \dot\cup \mathcal{E}_B$, where $\mathcal{E}_I$ is the set of edges in Ω and $\mathcal{E}_B$ consists of the edges on $\partial\Omega$. For functions $u : \Omega \to \mathbb{R}$ and $\sigma : \Omega \to \mathbb{R}^2$ that are smooth on the elements $K \in \mathcal{T}$, we define the jumps and averages as it is customary in DG-settings:

- For $e \in \mathcal{E}_I$, let K_e^+ and K_e^- be the two elements sharing e and denote by $\mathbf{n}^+$ and $\mathbf{n}^-$ the normal vectors on e pointing out of K_e^+ and K_e^-. Correspondingly, we let u^+, u^- and σ^+ and σ^- be traces on e of u and σ from K_e^+ and K_e^-. We set:

$$\{\!\!\{u\}\!\!\}|_e := \frac{1}{2}\left(u^+ + u^-\right), \qquad \{\!\!\{\sigma\}\!\!\}|_e := \frac{1}{2}\left(\sigma^+ + \sigma^-\right),$$

$$[\![u]\!]|_e := u^+\mathbf{n}^+ + u^-\mathbf{n}^-, \qquad [\![\sigma]\!]|_e := \sigma^+ \cdot \mathbf{n}^+ + \sigma^- \cdot \mathbf{n}^-.$$

- For boundary edges $e \in \mathcal{E}_B$ we define

$$\{\!\!\{\sigma\}\!\!\}|_e := \sigma|_e \qquad [\![u]\!]|_e := u|_e \mathbf{n}$$

With this notation, one can conveniently rearrange certain sums over edges:

Lemma 6.1 ("DG magic formula"). *Let* $v : \Omega \to \mathbb{R}$ *and* $\sigma : \Omega \to \mathbb{R}^2$ *be piecewise smooth on the triangulation* $\mathcal{T}$. *Then:*

$$\sum_{K \in \mathcal{T}} \int_{\partial K} v\sigma \cdot \mathbf{n} = \int_{\mathcal{E}_I} [\![v]\!] \cdot \{\!\!\{\sigma\}\!\!\} + \int_{\mathcal{E}_I} \{\!\!\{v\}\!\!\} \cdot [\![\sigma]\!] + \int_{\mathcal{E}_B} [\![v]\!] \cdot \{\!\!\{\sigma\}\!\!\},$$

where $\int_{\mathcal{E}_I}$ *and* $\int_{\mathcal{E}_B}$ *are shorthand notations for the sums of integrals over all edges in* $\mathcal{E}_I$ *and* $\mathcal{E}_B$.

Finally, for piecewise smooth functions, ∇_h denotes the piecewise defined gradient.

6.2   Stability of Least Squares Methods

Although Least Squares methods could be based on any space of approximation spaces, we will concentrate here on the approximation by piecewise solutions of the homogeneous Helmholtz equation. With varying focus, this is the setting of [10,41,53,58,68] and references therein. We illustrate the procedure for the model problem (9) with $f = 0$. The approximation space has the form

$$V_N = \{u \in L^2(\Omega) : u|_K \in V_{N,K} \quad \forall K \in \mathcal{T}\}, \tag{49}$$

where the spaces $V_{N,K}$ are spaces of solutions of the homogeneous Helmholtz equation, e.g., systems of plane waves. For each edge $e \in \mathcal{E}$, we select weights $w_{1,e}, w_{2,e} > 0$ and define the functional $J : V_N \to \mathbb{R}$ by

$$J(v) := \sum_{e \in \mathcal{E}_I} w_{1,e}^2 \|[v]\|_{L^2(e)}^2 + w_{2,e}^2 \|[\partial_n u]\|_{L^2(e)}^2 + \sum_{e \in \mathcal{E}_B} w_{2,e}^2 \|g - (\partial_n v + \mathbf{i}kv)\|_{L^2(e)}^2;$$

here $[v]|_e := [v]|_e$ and $[\partial_n v]|_e := [\nabla_h v]|_e$ represent the jumps of v and $\partial_n v$ across the edge e. If the exact solution u of (9) is sufficiently regular, then it is a minimizer of J with $J(u) = 0$. In a Least Squares method, J is minimizer over a finite dimensional space V_N of the form (49). Its variational form reads:

$$\text{find } u_N \in V_N \text{ s.t.} \langle u_N, v \rangle_{J,N} = \sum_{e \in \mathcal{E}_B} (g, \partial_n v + \mathbf{i}kv)_{L^2(e)} \qquad \forall v \in V_N, \tag{50}$$

where

$$\langle u, v \rangle_{J,N} := \sum_{e \in \mathcal{E}_I} w_{1,e}^2 ([u], [v])_{L^2(e)} + w_{2,e}^2 ([\partial_n u], [\partial_n v])_{L^2(e)}$$

$$+ \sum_{e \in \mathcal{E}_B} w_{2,e}^2 (\partial_n u + \mathbf{i}ku, \partial_n v + \mathbf{i}kv)_{L^2(e)}.$$

The positive semidefinite sesquilinear form $\langle \cdot, \cdot \rangle_{J,N}$ induces in fact a norm on V_N: To see the definiteness of $\langle \cdot, \cdot \rangle_{J,N}$, we note that $v \in V_N$ and $J(v) = 0$ implies that v is in $C^1(\Omega)$ and elementwise a solution of the homogeneous Helmholtz equation. Thus, it is a classical solution of the Helmholtz equation on Ω and satisfies $\partial_n v + \mathbf{i}kv = 0$ on $\partial\Omega$. The uniqueness assertion for (9) with $f = 0$ and $g = 0$ worked out in Example 2.1 then implies $v = 0$. Therefore, the minimization problem (50) is well-defined. If the solution u of (9) satisfies $u \in H^{3/2+\varepsilon}(\Omega)$ for some $\varepsilon > 0$, then $J(u) = 0$, and we get quasi-optimality of the Least Squares method in the norm $\|\cdot\|_{J,N} = J(\cdot)^{1/2}$:

$$\|u - u_N\|_{J,N}^2 = J(u - u_N) = J(u_N) = \min_{v \in V_N} J(v) = \|u - v\|_{J,N}^2. \tag{51}$$

We mention here that estimates for this minimum can be obtained from (local) estimates in classical Sobolev norms as given in Sect. 5 using appropriate trace estimates. Turning estimates for $\|u - u_N\|_{J,N} = J(u_N)^{1/2}$ into estimates in terms of more familiar norms such as $\|u - u_N\|_{L^2(\Omega)}$ is not straight forward. It may be expected that the norm of the solution operator of the continuous problem appears again; the next result, which is closely related to [14, 32, 33, 51], illustrates the kind of result one can obtain, in particular in a p-version setting:

Lemma 6.2 ([53, Thm. 3.1]). *Let $\Omega \subset \mathbb{R}^2$ be a polygon. Let $w_{1,e} = k$ and $w_{2,e} = 1$ for all edges and $g \in L^2(\partial\Omega)$. Let $u_N \in V_N$ be the minimizer of J, where V_N, given by (49), consists of elementwise solutions of the homogeneous Helmholtz equation.*

(i) If Ω is convex, then $\|u - u_N\|^2_{L^2(\Omega)} \leq Ck^{-1}\left((kh)^{-1} + (kh)^1\right) J(u_N).$
(ii) If Ω is not convex, then

$$\|u - u_N\|^2_{L^2(\Omega)} \leq$$

$$Ck^{-1}\left[(kh)^{-1} + (kh)^1 \left\{1 + \min\{1, kh\}^{-2\beta_{max}}\right\}\right](C_{sol}(k) + 1)^2 J(u_N),$$

where $C_{sol}(k)$ is defined in (25) and satisfies $C_{sol}(k) = O(k^{5/2})$ by Theorem 2.4. The parameter $\beta_{max} \geq 0$ can be selected arbitrarily to satisfy the condition $\beta_{max} > 1 - \min_i \frac{\pi}{\omega_i}$, where the ω_i are the interior angles of the polygon.

Proof. The result (i) is essentially the statement of [53, Thm. 3.1] in a refined form as given in [33, Lemma 3.7]. While (ii) is a novel result, it is only a slight modification of (i). We refer to [21] for the proof. $\qquad\square$

Remark 6.3. Lemma 6.2 assumes quasi-uniform meshes and the weights $w_{1,e}$, $w_{2,e}$ do not take the edge length into account. This limits somewhat it applicability in an h-version context. However, the result is very suitable for a p-version setting. We point out that in a case where the p-version features only algebraic rates of convergence, one would have to give the parameters $w_{1,e}$, $w_{2,e}$ a p-dependent relative weight as opposed to the situation studied in Lemma 6.2. $\qquad\blacksquare$

6.3 Stability of Plane Wave DG and UWVF

The framework of Discontinuous Galerkin (DG) methods permits another way of deriving numerical schemes that are inherently stable. In a classical, piecewise polynomial setting, this is pursued in [24–26]; related work is in [52]. Here, we concentrate on a setting where the ansatz functions satisfy the homogeneous Helmholtz equation. In particular, we study the plane wave DG method, [27, 33, 51], and the closely related Ultra Weak Variational Formulation (UWVF), [14–16, 35, 43]. We point out that the UWVF can be derived in different way. Here, we follow [14, 27] in viewing it as a special DG method.

Our model problem (9) can be reformulated as a first order system by introducing the flux $\sigma := (1/ik)\nabla u$:

$$ik\sigma = \nabla u \text{ in } \Omega, \qquad iku - \nabla \cdot \sigma = 0 \text{ in } \Omega, \qquad ik\sigma \cdot \mathbf{n} + iku = g \text{ on } \partial\Omega. \quad (52)$$

The weak elementwise formulation of the first two equations is for each $K \in \mathcal{T}$:

$$\int_K \mathbf{ik\sigma} \cdot \overline{\boldsymbol{\tau}} + \int_K u \nabla \cdot \overline{\boldsymbol{\tau}} - \int_{\partial K} u \overline{\boldsymbol{\tau}} \cdot \mathbf{n} = 0 \qquad \forall \boldsymbol{\tau} \in H(\mathrm{div}, K),$$

$$\int_K \mathbf{ik} u \overline{v} + \int_K \boldsymbol{\sigma} \cdot \nabla \overline{v} - \int_{\partial K} \boldsymbol{\sigma} \cdot \mathbf{n} \overline{v} = 0 \qquad \forall v \in H^1(K),$$

where $H(\mathrm{div}, K) = \{u \in L^2(K) : \mathrm{div}\, u \in L^2(K)\}$ and $\mathbf{n}$ is the outward pointing normal vector. Replacing the spaces $H^1(K)$ and $H(\mathrm{div}, K)$ by finite-dimensional subsets $V_{N,K} \subset H^1(K)$ and $\boldsymbol{\Sigma}_{N,K} \subset H(\mathrm{div}, K)$ and, additionally, imposing a coupling between neighboring elements by replacing the multivalued traces u and σ on the element edges by single-valued numerical fluxes $\widehat{u}_N$, $\widehat{\sigma}_N$ to be specified below, leads to the problem: Find $(u_N, \sigma_N) \in V_{N,K} \times \boldsymbol{\Sigma}_{N,K}$ such that

$$\int_K \mathbf{ik\sigma}_N \cdot \overline{\boldsymbol{\tau}} + \int_K u_N \nabla \cdot \overline{\boldsymbol{\tau}} - \int_{\partial K} \widehat{u}_N \overline{\boldsymbol{\tau}} \cdot \mathbf{n} = 0 \quad \forall \boldsymbol{\tau} \in \boldsymbol{\Sigma}_{N,K},$$

$$\int_K \mathbf{ik} u_N \overline{v} + \int_K \boldsymbol{\sigma}_N \cdot \nabla \overline{v} - \int_{\partial K} \widehat{\sigma}_N \cdot \mathbf{n} \overline{v} = 0 \quad \forall v \in V_{N,K}.$$

The variable σ_N can be eliminated by making the assumption that $\nabla V_{N,K} \subset \boldsymbol{\Sigma}_{N,K}$ for all $K \in \mathcal{T}$ and then selecting the test function $\boldsymbol{\tau} = \nabla v$ on each element. This yields after an integration by parts:

$$\int_K \nabla u_N \nabla \overline{v} - k^2 u_N \overline{v} - \int_{\partial K} (u_N - \widehat{u}_N) \partial_n \overline{v} - \mathbf{ik}\widehat{\sigma}_N \cdot \mathbf{n} \overline{v} = 0 \qquad \forall K \in \mathcal{T}. \quad (53)$$

Since $V_N = \{u \in L^2(\Omega) : u|_K \in V_{N,K} \forall K \in \mathcal{T}\}$ consists of discontinuous functions without any interelement continuity imposed across the element edges, (53) is equivalent to the sum over the elements: Find $u_N \in V_N$ such that for all $v \in V_N$

$$\sum_{K \in \mathcal{T}} \int_K \nabla u_N \cdot \nabla \overline{v} - k^2 u_N \overline{v} + \int_{\partial K} (\widehat{u}_N - u_N) \nabla \overline{v} \cdot \mathbf{n} - \int_{\partial K} \mathbf{ik}\widehat{\sigma}_N \cdot \mathbf{n} \overline{v} = 0. \quad (54)$$

This formulation is now completed by specifying the fluxes $\widehat{u}_N$ and $\widehat{\sigma}_N$, which at the same time takes care of the boundary condition in (52):

For $e \in \mathcal{E}_I$:
$$\begin{cases} \widehat{\sigma}_N = \frac{1}{\mathbf{ik}} \{\!\{\nabla_h u_N\}\!\} - \alpha [\![u_N]\!], \\ \widehat{u}_N = \{\!\{u_N\}\!\} - \beta \frac{1}{\mathbf{ik}} [\![\nabla_h u_N]\!] \end{cases}$$

For $e \in \mathcal{E}_B$:
$$\begin{cases} \widehat{\sigma}_N = \frac{1}{\mathbf{ik}} \nabla_h u_N - \frac{1-\delta}{\mathbf{ik}} (\nabla_h u_N + \mathbf{ik} u_N \mathbf{n} - g\mathbf{n}). \\ \widehat{u}_N = u_N - \frac{\delta}{\mathbf{ik}} (\nabla_h u \cdot \mathbf{n} + \mathbf{ik} u_N - g). \end{cases}$$

Different choices of the parameters α, β, δ lead to different methods analyzed in the literature. For example:

1. $\alpha = \beta = \delta = 1/2$: this is the UWVF as analyzed in [14–16,35,43] if the spaces $V_{N,K}$ consist of a space W_{PW}^p of plane waves.
2. $\alpha = O(p/(kh \log p))$, $\quad \beta = O((kh \log p)/p)$, $\quad \delta = O((kh \log p)/p)$: this choice is introduced and advocated in [33,51] in conjunction with $V_{N,K} = W_{PW}^p$.

With these choices of fluxes, the formulation (54) takes the form

$$\text{Find } u_N \in V_N \text{ s.t.} \quad A_N(u_N, v) = l(v) \qquad \forall v \in V_N, \tag{56}$$

where the sesquilinear form A_N and the linear form l are given by

$$A_N(u, v) = \int_\Omega \nabla_h u \cdot \nabla_h \bar{v} - k^2 u \bar{v} - \int_{\mathcal{E}_I} [u] \{\nabla_h \bar{v}\} - \int_{\mathcal{E}_I} \{\nabla_h u\} [\bar{v}] - \int_{\mathcal{E}_B} \delta u \partial_n \bar{v} - \int_{\mathcal{E}_B} \delta \partial_n u \bar{v}$$
$$- \frac{1}{ik} \int_{\mathcal{E}_I} \beta [\nabla_h u][\nabla_h \bar{v}] - \frac{1}{ik} \int_{\mathcal{E}_B} \delta \partial_n u \partial_n \bar{v} + ik \int_{\mathcal{E}_I} \alpha [u][\bar{v}] + ik \int_{\mathcal{E}_B} (1 - \delta) u \bar{v} \tag{57}$$

$$l(v) = -\frac{1}{ik} \int_{\mathcal{E}_B} \delta g \partial_n \bar{v} + \int_{\mathcal{E}_B} (1 - \delta) g \bar{v}.$$

So far, the choice of the spaces $V_{N,K}$ is arbitrary. If the approximation spaces $V_{N,K}$ (more precisely: the test spaces) consist of piecewise solutions of the homogeneous Helmholtz equation, then a further integration by parts is possible to eliminate all volume contributions in A_N. Indeed, Lemma 6.1 produces

$$\sum_{K \in \mathcal{T}} \int_K \nabla u \cdot \nabla \bar{v} - k^2 u \bar{v} = \sum_{K \in \mathcal{T}} \int_{\partial K} u \nabla \bar{v} \mathbf{n} = \int_{\mathcal{E}_I} [u] \{\nabla \bar{v}\} + \{u\} [\nabla \bar{v}] + \int_{\mathcal{E}_B} [u] \{\nabla \bar{v}\}$$

so that A_N simplifies to

$$A_N(u, v) = \int_{\mathcal{E}_I} \{u\} [\nabla_h \bar{v}] + i\frac{1}{k} \int_{\mathcal{E}_I} \beta [\nabla_h u][\nabla_h \bar{v}] - \int_{\mathcal{E}_I} \{\nabla_h u\} [\bar{v}] + ik \int_{\mathcal{E}_I} \alpha [u][\bar{v}]$$
$$+ \int_{\mathcal{E}_B} (1 - \delta) u \partial_n \bar{v} + i\frac{1}{k} \int_{\mathcal{E}_B} \delta \partial_n u \partial_n \bar{v} - \int_{\mathcal{E}_B} \delta \partial_n u \bar{v} + ik \int_{\mathcal{E}_B} (1 - \delta) u \bar{v}.$$

Next, we make the important observation that $\text{Im } A_N$ induces a norm on the space V_N if $\alpha, \beta > 0$ and $\delta \in (0, 1)$. Indeed:

1. $\alpha, \beta > 0$ and $\delta \in (0, 1)$ implies $\text{Im } A_N(v, v) \geq 0$ $\quad \forall v \in V_N$ by inspection of (57).
2. $\text{Im } A_N(v, v) = 0$ and the fact that V_N consists of elementwise solutions of the homogeneous Helmholtz equation implies as in the case of $\langle \cdot, \cdot \rangle_{J,N}$ in Sect. 6.2 that $v \in C^1(\Omega)$ solves the homogeneous Helmholtz equation and $\partial_n v = v = 0$ on $\partial \Omega$; the uniqueness assertion of Example 2.1 then proves $v \equiv 0$.

This is at the basis of the convergence analysis. Introducing

$$\|u\|_{DG}^2 := \sqrt{\operatorname{Im} A_N(u,u)} = \frac{1}{k}\|\beta^{1/2}[\![\nabla_h u]\!]\|_{L^2(\mathscr{E}_I)}^2 + \|\alpha^{1/2}[\![u]\!]\|_{L^2(\mathscr{E}_I)}^2$$

$$+ \frac{1}{k}\|\delta^{1/2}\partial_n u\|_{L^2(\mathscr{E}_B)}^2 + k\|(1-\delta)^{1/2}u\|_{L^2(\mathscr{E}_B)}^2,$$

$$\|u\|_{DG,+}^2 := \|u\|_{DG}^2 + k\|\beta^{-1/2}\{\!\!\{u\}\!\!\}\|_{L^2(\mathscr{E}_I)}^2 + k^{-1}\|\alpha^{-1/2}\{\!\!\{u\}\!\!\}\|_{L^2(\mathscr{E}_I)}^2 + k\|\delta^{-1/2}u\|_{L^2(\mathscr{E}_B)}^2,$$

we can formulate coercivity and continuity results:

Proposition 6.4 ([14, 33]). *Let V_N consist of piecewise solutions of the homogeneous Helmholtz equation. Then $\|\cdot\|_{DG}$ is a norm on V_N and for some $C > 0$ depending solely on the choice of α, $\beta > 0$, and $\delta \in (0,1)$:*

$$\operatorname{Im} A_N(u,u) = \|u\|_{DG}^2 \qquad \forall u \in V_N,$$

$$|A_N(u,v)| \leq C\|u\|_{DG,+}\|v\|_{DG} \qquad \forall u,v \in V_N$$

Let the solution of u of (9) (with $f = 0$) satisfy $u \in H^{3/2+\varepsilon}(\Omega)$ for some $\varepsilon > 0$. Then, by consistency of A_N, the solution $u_N \in V_N$ of (56) satisfies the following quasioptimality estimate for some $C > 0$ independent of k:

$$\|u - u_N\|_{DG} \leq C \inf_{v \in V_N} \|u - v\|_{DG,+}. \tag{58}$$

Several comments are in order:

1. The UWVF of [15] featured quasi-optimality in a residual type norm. We recall that the UWVF is a DG method for the particular choice $\alpha = \beta = \delta = 1/2$.
2. When V_N consists (elementwise) of systems of plane waves or generalized harmonic polynomials, then the infimum in (58) can be estimated using approximation results on the elements by taking appropriate traces. This is worked out in detail in [32, 33, 51] and earlier in an h-version setting in [15] (see also [14]).
3. The $\|\cdot\|_{DG}$-norm controls the error on the skeleton $\mathscr{E}$ only. The proof of Lemma 6.2 shows how error estimates in such norms can be used to obtain estimates for $\|u - u_N\|_{L^2(\Omega)}$; we refer again to [14] where this worked out for the UWVF and to [32, 33, 51] where the case of the plane wave DG is studied. As pointed out in Remark 6.3, quasi-uniformity of the underlying mesh $\mathscr{T}$ is an important ingredient for the arguments of Lemma 6.2.

It is noteworthy that Proposition 6.4 does not make any assumptions on the mesh size h and the space V_N except that it consist of piecewise solutions of the homogeneous Helmholtz equation. Optimal error estimates are possible in an h-version setting, where the number of plane waves per element is kept fixed:

Proposition 6.5 ([27]). *Let Ω be convex. Assume that $V_{N,K} = W_{Pw}^{2\mu+1}$ ($\mu \geq 1$ fixed) for all $K \in \mathscr{T}$. Assume that α is of the form $\alpha = \mathsf{a}/(kh)$ and that $\beta > 0$,*

$\delta \in (0, 1/2)$. *Then there exist* a_0, c_0, $C > 0$ *(all independent of h and k) such that if* $a \geq a_0$ *and* $k^2 h \leq c_0$, *then following error bound is true:*

$$\|u - u_N\|_{1,DG} \leq C \inf_{v \in V_N} \|u - v\|_{1,DG,+};$$

here, $\| \cdot \|_{1,DG}$ *and* $\| \cdot \|_{1,DG,+}$ *are given by* $\|v\|_{1,DG}^2 := \sum_{K \in \mathcal{T}} |v|_{H^1(K)}^2 + k^2 \|v\|_{L^2(K)}^2 + \|v\|_{DG}^2$ *and* $\|v\|_{1,DG,+}^2 := \sum_{K \in \mathcal{T}} |v|_{H^1(K)}^2 + k^2 \|v\|_{L^2(K)}^2 + \|v\|_{DG,+}^2$.

Proof. The proof follows by inspection of the procedure in [27, Sec. 5] and is stated in [51, Props. 4.2, 4.3]. The essential ingredients of the proof are: (a) inverse estimates for systems of plane waves that have been made in available in [27] so that techniques of standard DG methods can be used to treat A_N; (b) use of duality arguments as in Lemma 4.1 to treat the L^2-norm of the error; (c) the fact that in an h-version setting, plane waves have some approximation power for arbitrary functions in H^2 (this is analogous to Lemma 4.8). □

Acknowledgements Financial support by the *Vienna Science and Technology Fund* (WWTF) is gratefully acknowledged.

References

1. M. Ainsworth, P. Monk, and W. Muniz. Dispersive and dissipative properties of discontinuous Galerkin finite element methods for the second-order wave equation. *J. Sci. Comput.*, 27(1-3):5–40, 2006.
2. M. Ainsworth. Discrete dispersion relation for hp-version finite element approximation at high wave number. *SIAM J. Numer. Anal.*, 42(2):553–575, 2004.
3. M. Ainsworth and H.A. Wajid. Dispersive and dissipative behavior of the spectral element method. *SIAM J. Numer. Anal.*, 47(5):3910–3937, 2009.
4. R. J. Astley and P. Gamallo. Special short wave elements for flow acoustics. *Comput. Methods Appl. Mech. Engrg.*, 194(2-5):341–353, 2005.
5. A. K. Aziz, R. B. Kellogg, and A. B. Stephens. A two point boundary value problem with a rapidly oscillating solution. *Numer. Math.*, 53(1-2):107–121, 1988.
6. I. Babuška and B.Q. Guo. The $h - p$ version of the finite element method. Part 1: The basic approximation results. *Computational Mechanics*, 1:21–41, 1986.
7. I. Babuška, F. Ihlenburg, E. Paik, and S. Sauter. A generalized finite element method for solving the Helmholtz equation in two dimensions with minimal pollution. *Comput. Meth. Appl. Mech. Engrg.*, 128:325–360, 1995.
8. I. Babuška and S. Sauter. Is the pollution effect of the FEM avoidable for the Helmholtz equation? *SIAM Review*, 42:451–484, 2000.
9. L. Banjai and S. Sauter. A refined Galerkin error and stability analysis for highly indefinite variational problems. *SIAM J. Numer. Anal.*, 45(1):37–53, 2007.
10. A. H. Barnett and T. Betcke. An exponentially convergent nonpolynomial finite element method for time-harmonic scattering from polygons. *SIAM J. Sci. Stat. Comp.*, 32:1417–1441, 2010.
11. A. Bayliss, C.I. Goldstein, and E. Turkel. On accuracy conditions for the numerical computation of waves. *J. Comput. Physics*, 59:396–404, 1985.

12. T. Betcke, S. Chandler-Wilde, I. Graham, S. Langdon, and M. Lindner. Condition number estimates for combined potential integral operators in acoustics and their boundary element discretization. *Numer. Meths. PDEs*, 27:31–69, 2011.

13. S.C. Brenner and L.R. Scott. *The mathematical theory of finite element methods*. Springer Verlag, 1994.

14. A. Buffa and P. Monk. Error estimates for the ultra weak variational formulation of the Helmholtz equation. *M2AN (Math. Modelling and Numer. Anal.)*, 42(6):925–940, 2008.

15. O. Cessenat and B. Després. application of the ultra-weak variational formulation to the 2d Helmholtz problem. *SIAM J. Numer. Anal.*, 35:255–299, 1998.

16. O. Cessenat and B. Després. Using plane waves as base functions for solving time harmonic equations with the ultra weak variational formulation. *J. Comput. Acoust.*, 11:227–238, 2003.

17. S. Chandler-Wilde and I. Graham. Boundary integral methods in high frequency scattering. In B. Engquist, A. Fokas, E. Hairer, and A. Iserles, editors, *highly oscillatory problems*. Cambridge University Press, 2009.

18. S.N. Chandler-Wilde and P. Monk. Wave-number-explicit bounds in time-harmonic scattering. *SIAM J. Math. Anal*, 39:1428–1455, 2008.

19. P. Cummings and X. Feng. Sharp regularity coefficient estimates for complex-valued acoustic and elastic Helmholtz equations. *Math. Models Methods Appl. Sci.*, 16(1):139–160, 2006.

20. A. Deraemaeker, I. Babuška, and P. Bouillard. Dispersion and pollution of the FEM solution for the Helmholtz equation in one, two and three dimensions. *Int. J. Numer. Meth. Eng.*, 46(4), 1999.

21. S. Esterhazy and J.M. Melenk. On stability of discretizations of the Helmholtz equation (extended version). Technical Report 01, Inst. for Analysis and Sci. Computing, Vienna Univ. of Technology, 2011. Available at http://www.asc.tuwien.ac.at. arXiv:1105.2112

22. C. Farhat, I. Harari, and U. Hetmaniuk. A discontinuous Galerkin method with Lagrange multipliers for the solution of Helmholtz problems in the mid-frequency regime. *Comp. Meth. Appl. Mech. Eng.*, 192:1389–1419, 2003.

23. C. Farhat, R. Tezaur, and P. Weidemann-Goiran. Higher-order extensions of discontinuous Galerkin method for mid-frequency Helmholtz problems. *Int. J. Numer. Meth. Eng.*, 61, 2004.

24. X. Feng and H. Wu. Discontinuous Galerkin methods for the Helmholtz equation with large wave number. *SIAM J. Numer. Anal.*, 47(4):2872–2896, 2009.

25. X. Feng and H. Wu. *hp*-discontinuous Galerkin methods for the Helmholtz equation with large wave number. *Math. Comp.*, 80:1997–2025, 2011.

26. X. Feng and Y. Xing. Absolutely stable local discontinuous Galerkin methods for the Helmholtz equation with large wave number. 2010. arXiv:1010.4563v1 [math.NA].

27. C. Gittelson, R. Hiptmair, and I. Perugia. Plane wave discontinuous Galerkin methods. *M2AN (Mathematical Modelling and Numerical Analysis)*, 43:297–331, 2009.

28. P. Grisvard. *Elliptic Problems in Nonsmooth Domains*. Pitman, 1985.

29. I. Harari. A survey of finite element methods for time-harmonic acoustics. *Comput. Methods Appl. Mech. Engrg.*, 195(13-16):1594–1607, 2006.

30. I. Harari and T. J. R. Hughes. Galerkin/least-squares finite element methods for the reduced wave equation with nonreflecting boundary conditions in unbounded domains. *Comput. Methods Appl. Mech. Engrg.*, 98(3):411–454, 1992.

31. U. Hetmaniuk. Stability estimates for a class of Helmholtz problems. *Commun. Math. Sci.*, 5(3):665–678, 2007.

32. R. Hiptmair, A. Moiola, and I. Perugia. Approximation by plane waves. Technical Report 2009-27, Seminar für Angewandte Mathematik, ETH Zürich, 2009.

33. R. Hiptmair, A. Moiola, and I. Perugia. Plane wave discontinuous Galerkin methods for the 2d Helmholtz equation: analysis of the *p*-version. *SIAM J. Numer. Anal.*, 49:264–284, 2011.

34. T. Huttunen, P. Gamallo, and R. J. Astley. Comparison of two wave element methods for the Helmholtz problem. *Comm. Numer. Methods Engrg.*, 25(1):35–52, 2009.

35. T. Huttunen and P. Monk. The use of plane waves to approximate wave propagation in anisotropic media. *J. Computational Mathematics*, 25:350–367, 2007.

36. F. Ihlenburg. *Finite Element Analysis of Acoustic Scattering*, volume 132 of *Applied Mathematical Sciences*. Springer Verlag, 1998.
37. F. Ihlenburg and I. Babuška. Finite element solution to the Helmholtz equation with high wave number. Part II: The hp-version of the FEM. *SIAM J. Numer. Anal.*, 34:315–358, 1997.
38. O. Laghrouche and P. Bettess. Solving short wave problems using special finite elements; towards an adaptive approach. In J. Whiteman, editor, *Mathematics of Finite Elements and Applications X*, pages 181–195. Elsevier, 2000.
39. O. Laghrouche, P. Bettess, and J. Astley. Modelling of short wave diffraction problems using approximation systems of plane waves. *Internat. J. Numer. Meths. Engrg.*, 54:1501–1533, 2002.
40. R. Leis. *Initial Boundary Value Problems in Mathematical Physics*. Teubner, Wiley, 1986.
41. Z. C. Li. The Trefftz method for the Helmholtz equation with degeneracy. *Appl. Numer. Math.*, 58(2):131–159, 2008.
42. M. Löhndorf and J.M. Melenk. Wavenumber-explicit hp-BEM for high frequency scattering. Technical Report 02/2010, Institute for Analysis and Scientific Computing, TU Wien, 2010.
43. T. Luostari, T. Huttunen, and P. Monk. Plane wave methods for approximating the time harmonic wave equation. In *Highly oscillatory problems*, volume 366 of *London Math. Soc. Lecture Note Ser.*, pages 127–153. Cambridge Univ. Press, Cambridge, 2009.
44. J. M. Melenk. *On Generalized Finite Element Methods*. PhD thesis, Univ. of Maryland, 1995.
45. J.M. Melenk. *hp finite element methods for singular perturbations*, volume 1796 of *Lecture Notes in Mathematics*. Springer Verlag, 2002.
46. J.M. Melenk. On approximation in meshless methods. In J. Blowey and A. Craig, editors, *Frontier in Numerical Analysis, Durham 2004*. Springer Verlag, 2005.
47. J.M. Melenk. Mapping properties of combined field Helmholtz boundary integral operators. Technical Report 01/2010, Institute for Analysis and Scientific Computing, TU Wien, 2010.
48. J.M. Melenk and I. Babuška. The partition of unity finite element method: Basic theory and applications. *Comput. Meth. Appl. Mech. Engrg.*, 139:289–314, 1996.
49. J.M. Melenk and S. Sauter. Wavenumber explicit convergence analysis for finite element discretizations of the Helmholtz equation. *SIAM J. Numer. Anal.*, 49:1210–1243, 2011.
50. J.M. Melenk and S. Sauter. Convergence analysis for finite element discretizations of the Helmholtz equation with Dirichlet-to-Neumann boundary conditions. *Math. Comp.*, 79:1871–1914, 2010.
51. A. Moiola. Approximation properties of plane wave spaces and application to the analysis of the plane wave discontinuous Galerkin method. Technical Report 2009-06, Seminar für Angewandte Mathematik, ETH Zürich, 2009.
52. P. Monk, J. Schöberl, and A. Sinwel. Hybridizing Raviart-Thomas elements for the Helmholtz equation. *Electromagnetics*, 30:149–176, 2010.
53. P. Monk and D.Q. Wang. A least squares methods for the Helmholtz equation. *Comput. Meth. Appl. Mech. Engrg.*, 175:121–136, 1999.
54. C. S. Morawetz and D. Ludwig. An inequality for the reduced wave operator and the justification of geometrical optics. *Comm. Pure Appl. Math.*, 21:187–203, 1968.
55. J. Nečas. *Les méthodes directes en théorie des équations elliptiques*. Masson, 1967.
56. P. Ortiz. Finite elements using a plane-wave basis for scattering of surface water waves. *Philos. Trans. R. Soc. Lond. Ser. A Math. Phys. Eng. Sci.*, 362(1816):525–540, 2004.
57. E. Perrey-Debain, O. Laghrouche, P. Bettess, and J. Trevelyan. Plane wave basis finite elements and boundary elements for three-dimensional wave scattering. *Phil. Trans. R. Soc. Lond. A*, 362:561–577, 2004.
58. B. Pluymers, B. van Hal, D. Vandepitte, and W. Desmet. Trefftz-based methods for time-harmonic acoustics. *Arch. Comput. Methods Eng.*, 14(4):343–381, 2007.
59. S. I. Pohožaev. On the eigenfunctions of the equation $\Delta u + \lambda f(u) = 0$. *Dokl. Akad. Nauk SSSR*, 165:36–39, 1965.
60. S.A. Sauter. A refined finite element convergence theory for highly indefinite Helmholtz problems. *Computing*, 78(2):101–115, 2006.
61. S.A. Sauter and C. Schwab. *Boundary element methods*. Springer Verlag, 2010.

62. Alfred H. Schatz. An observation concerning Ritz-Galerkin methods with indefinite bilinear forms. *Math. Comp.*, 28:959–962, 1974.
63. J. Schöberl. *FE Software Netgen/NGSolve vers. 4.13*. http://sourceforge.net/projects/ngsolve/
64. J. Schöberl. NETGEN - an advancing front 2d/3d-mesh generator based on abstract rules. *Computing and Visualization in Science*, 1(1):41–52.
65. C. Schwab. *p- and hp-Finite Element Methods*. Oxford University Press, 1998.
66. E.A. Spence, S.N. Chandler-Wilde, I.G. Graham, and V.P. Smyshlyaev. A new frequency-uniform coercive boundary integral equation for acoustic scattering. *Comm. Pure Appl. Math.*, 60:1384–1415, 2011.
67. E.M. Stein. *Singular integrals and differentiability properties of functions*. Princeton University Press, 1970.
68. M. Stojek. Least squares Trefftz-type elements for the Helmholtz equation. *Internat. J. Numer. Meths. Engrg.*, 41:831–849, 1998.
69. T. Strouboulis, I. Babuška, and R. Hidajat. The generalized finite element method for Helmholtz equation: theory, computation, and open problems. *Comput. Methods Appl. Mech. Engrg.*, 195(37-40):4711–4731, 2006.
70. R. Tezaur and C. Farhat. Three-dimensional discontinuous Galerkin elements with plane waves and Lagrange multipliers for the solution of mid-frequency Helmholtz problems. *Internat. J. Numer. Meths. Engrg.*, 66:796–815, 2006.

Why it is Difficult to Solve Helmholtz Problems with Classical Iterative Methods

O.G. Ernst and M.J. Gander

Abstract In contrast to the positive definite Helmholtz equation, the deceivingly similar looking indefinite Helmholtz equation is difficult to solve using classical iterative methods. Simply using a Krylov method is much less effective, especially when the wave number in the Helmholtz operator becomes large, and also algebraic preconditioners such as incomplete LU factorizations do not remedy the situation. Even more powerful preconditioners such as classical domain decomposition and multigrid methods fail to lead to a convergent method, and often behave differently from their usual behavior for positive definite problems. For example increasing the overlap in a classical Schwarz method degrades its performance, as does increasing the number of smoothing steps in multigrid. The purpose of this review paper is to explain why classical iterative methods fail to be effective for Helmholtz problems, and to show different avenues that have been taken to address this difficulty.

1 Introduction

We consider in this paper the iterative solution of linear systems of equations arising from the discretization of the *indefinite Helmholtz equation*,

$$\mathscr{L}u := -(\Delta + k^2)u = f, \tag{1}$$

M.J. Gander (✉)
University of Geneva, Section of Mathematics, 2-4 Rue du Lievre, CP 64, 1211 Geneva 4, Switzerland
e-mail: martin.gander@unige.ch

O.G. Ernst
TU Bergakademie Freiberg, Institut für Numerische Mathematik und Optimierung, 09596 Freiberg, Germany
e-mail: ernst@math.tu-freiberg.de

I.G. Graham et al. (eds.), *Numerical Analysis of Multiscale Problems*, Lecture Notes in Computational Science and Engineering 83, DOI 10.1007/978-3-642-22061-6_10, © Springer-Verlag Berlin Heidelberg 2012

with suitable boundary conditions yielding a well-posed problem. For $k > 0$ solutions of the Helmholtz equation, also known as the *reduced wave equation*, describe the variation in space of linear propagating waves with wave number k. The performance of standard iterative methods is much worse for such problems than for the deceivingly similar looking equation

$$- (\Delta - \eta)u = f, \qquad \eta > 0, \tag{2}$$

which describes stationary reaction-diffusion phenomena but is often also called Helmholtz equation in certain communities. For example in meteorology, the early seminal papers [51, 59] led an entire community to call equations of the form (2) Helmholtz equations, see for example [14]. Even standard texts in applied mathematics now sometimes use the term Helmholtz equation for both (1) and (2), see for example [69]. The subject of this paper is exclusively the indefinite Helmholtz equation (1), which is much harder to solve with classical iterative methods than equation (2), see also the recent review [20].

Discretizations of the indefinite Helmholtz equation (1) using, e.g., finite differences or a finite element or spectral element method and appropriate boundary conditions result in a linear system of equations

$$A\mathbf{u} = \mathbf{f}, \tag{3}$$

which, for k sufficiently large, possesses an indefinite coefficient matrix A.

Often an approximation of the Sommerfeld radiation condition, which in d space dimensions reads $\partial_r u - iku = o(r^{(1-d)/2})$ as the radial variable r tends to infinity, is imposed on part of the boundary, specifying that wave motion should be outgoing along physically open boundaries. The Sommerfeld condition prescribes the asymptotic behavior of the solution, and its representation on finite boundaries leads to nonlocal operators. For this reason localized approximations of the Sommerfeld condition are used, the simplest of which is the Robin condition $\partial_n u - iku = 0$. As a result, the linear system (3) has a complex-symmetric, but non-Hermitian coefficient matrix as well as a complex-valued solution. The iterative solution of the discrete Helmholtz problem (3) is difficult, even for constant wave number k, and we will illustrate this in the first part of this paper, for Krylov methods, preconditioned Krylov methods, domain decomposition methods and multigrid. We then try to explain where these difficulties come from, and show what types of remedies have been developed over the last two decades in the literature. We will conclude the paper with some more recent ideas.

2 Problems of Classical Iterative Methods

2.1 Krylov Subspace Methods

Krylov subspace methods seek an approximate solution of the linear system (3) in the *Krylov space*

$$\mathscr{K}_m(A, \mathbf{f}) = \text{span}\{\mathbf{f}, A\mathbf{f}, A^2\mathbf{f}, \ldots, A^{m-1}\mathbf{f}\} = \text{span}\{\mathbf{q}_0, \mathbf{q}_1, \mathbf{q}_2, \ldots, \mathbf{q}_{m-1}\}, \qquad (4)$$

where $\mathbf{q}_j$ denotes the jth Arnoldi vector of $\mathscr{K}_m$, i.e. the vectors obtained by orthonormalization of the vectors $\mathbf{f}, A\mathbf{f}, A^2\mathbf{f}, \ldots$ defining the Krylov space. We have made the common choice of a zero initial guess for the solution, as is recommended in the absence of any additional information, see for example [53]. We show in Fig. 1 how the wave number k fundamentally influences the solution of the Helmholtz equation. We have set homogeneous Dirichlet conditions on all boundaries, except on the left, where the Robin condition $\partial_n u - iku = 0$ was imposed, and we used a point source in the upper right corner, i.e., a Dirac delta distribution concentrated at this point, as the source term. In the case of Laplace's equation ($k = 0$) the solution is large only near the point source in the corner, whereas for $k = 25$, the solution is large throughout the domain. The Krylov space constructed in (4), however, is very similar for both problems: due to the local connectivity (we used a five-point finite difference discretization for the Laplacian), the vector $\mathbf{f}$ is zero everywhere, except for the grid point associated with the upper right corner point, and thus the Arnoldi vector $\mathbf{q}_0$ is just a canonical basis vector $(1, 0, \ldots, 0)^T$. The next vector in the Krylov space, $A\mathbf{f}$, is then non-zero only for the points connected with the first point, and the corresponding Arnoldi vector $\mathbf{q}_1$ will have only two non-zero entries, and so on. In the case of Laplace's equation we see that the first Arnoldi vectors are precisely non-zero where the solution is large, and thus it can be well

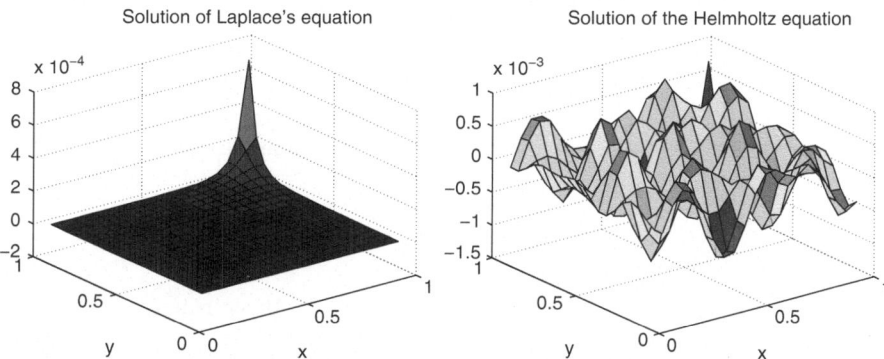

Fig. 1 Solution of Laplace's equation on the left, with a point source on the boundary, and on the right the solution of the Helmholtz equation, with the same boundary conditions

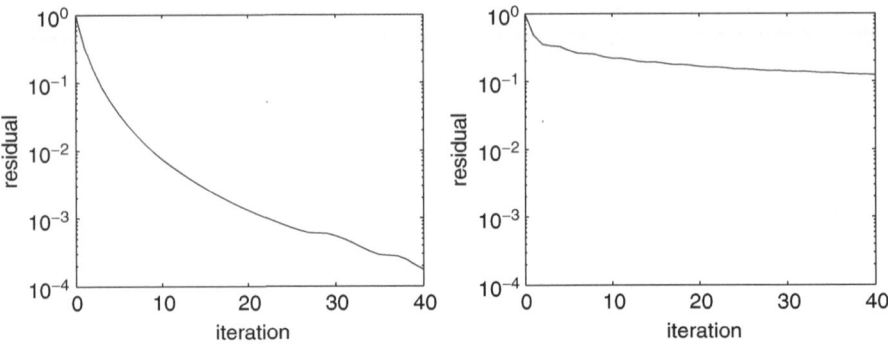

Fig. 2 Evolution of the residual for GMRES, on the left for the case of Laplace's equation, $k = 0$, and on the right for the Helmholtz equation, $k = 25$

approximated in the Krylov space. By contrast, in the indefinite Helmholtz case, where the solution is of the same size throughout the domain, these vectors do not have an appropriate support to approximate the solution. We show in Fig. 2 how this influences the convergence of GMRES. While the residual decreases well in the Laplace case over the first $2 \times n$ iterations, where n is the number of grid points in one direction, convergence stagnates in the Helmholtz case. For a more precise quantitative analysis of this phenomenon, see [36]. Similar effects are also observed in the advection dominated case of advection diffusion equations, see [24, 53]. It is therefore important to have a preconditioner, a Krylov method alone is not an effective iterative solver.

2.2 Algebraic Preconditioners Based on Factorization

The idea of preconditioning is as follows: instead of solving the original discretized system $A\mathbf{u} = \mathbf{f}$, we solve the preconditioned system

$$M^{-1}A\mathbf{u} = M^{-1}\mathbf{f}, \tag{5}$$

where M is the so-called preconditioner. Preconditioners often arise from a stationary iterative method

$$M\mathbf{u}^{k+1} = N\mathbf{u}^k + \mathbf{f} \tag{6}$$

derived from a *matrix splitting* $A = M - N$ with M nonsingular. It is well known that this method converges asymptotically rapidly, if the spectral radius of the iteration matrix $M^{-1}N$ is small. This implies that the preconditioned matrix in (5),

$$M^{-1}A = M^{-1}(M - N) = I - M^{-1}N$$

Table 1 Iteration counts for QMR with and without preconditoner, applied to an indefinite Helmholtz equation with increasing wave number

k	QMR		ILU('0')		ILU(1e-2)	
	It	Mflops	It	Mflops	It	Mflops
5	197	120.1	60	60.4	22	28.3
10	737	1858.2	370	1489.3	80	421.4
15	1,775	10185.2	> 2,000	> 18133.2	220	2615.1
20	> 2,000	> 20335.1	–	–	> 2,000	> 42320.1

has a spectrum clustered around 1 in the complex plane, which leads to fast asymptotic convergence also for a Krylov method applied to the preconditioned system (5). Clearly, the best preconditioner would be A^{-1}, since this makes the spectral radius of $M^{-1}N$ vanish since then $M^{-1}N=A^{-1}0 = 0$, and all the eigenvalues of the preconditioned system $M^{-1}A = I$ equal 1. But then one could directly solve the system without iteration.

The idea of factorization preconditioners is to use an approximation of A^{-1} by computing an approximate LU factorization of the matrix A, $A \approx LU$, and then in each iteration step of (5), a forward and a backward substitution need to be performed. Two popular algebraic variants are the ILU(0) and ILU(tol) preconditioners, see [60]. For ILU(0), one computes an approximate LU factorization, retaining entries in the LU factors only if the corresponding entry in the underlying matrix A is non-zero. In the ILU(tol) variant, elements are kept, provided they are bigger than the tolerance tol. We compare in Table 1 the performance of this type of preconditioner when applied to the Helmholtz equation for growing wave number k. We solve an open cavity problem as in the previous example in Sect. 2.1, but now with a point source in the center. For this experiment, we keep the number of grid points per wavelength constant, i.e. $kh = 10$, which means that the grid is refined proportionally with increasing wave number. We observe that the ILU preconditioners are quite effective for small wave numbers, but their performance deteriorates when k becomes larger: the situation with ILU('0') is worse than without preconditioning, and even ILU(tol) with a small drop tolerance does not permit the solution of the problem. Here the Krylov subspace solver used was QMR [30], but similar results are observed when using GMRES and other Krylov methods, see [38].

2.3 Domain Decomposition Methods

The oldest and simplest domain decomposition method is due to Schwarz [62]. He invented his alternating method in order to prove the Dirichlet principle, on which Riemann had based his theory of analytic functions of a complex variable (See [39] for a historical overview, and also [32] for an overview over the different continuous and discrete variants of the Schwarz method). The idea of the alternating Schwarz

Fig. 3 Drawing for the
original Schwarz method
using the notation in the text

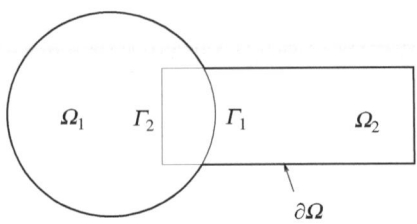

Table 2 Performance of a classical Schwarz domain decomposition method for a discretized
Helmholtz equation

k	10π	20π	40π	80π	160π	10π	20π	40π	80π	160π
	Overlap $= h$					Overlap fixed				
Iterative	**Div**	**Div**	**Div**	**Div**	**Div**	**Div**	**Div**	**Div**	**Div**	**Div**
Preconditioner	20	33	45	69	110	16	23	43	86	155

method is illustrated in Fig. 3. One simply solves the original partial differential
equation alternatingly in overlapping subdomains, and uses as interface condition
the trace of the previously computed solution in the neighboring subdomain. For the
case of the Helmholtz equation and the two-subdomain decomposition in Fig. 3, the
algorithm is

$$-(\Delta + k^2)u_1^{n+1} = 0 \quad \text{in } \Omega_1, \quad -(\Delta + k^2)u_2^{n+1} = 0 \quad \text{in } \Omega_2,$$
$$u_1^{n+1} = u_2^n \text{ on } \Gamma_1, \qquad\qquad u_2^{n+1} = u_1^{n+1} \text{ on } \Gamma_2. \tag{7}$$

We show in Table 2 numerical experiments for growing wave number k for the case
of a unit square cavity, open both on the left and on the right, using two subdomains
obtained by partitioning the cavity in the middle. We used the alternating Schwarz
method both as an iterative solver, as in (6), and as a preconditioner, as in (5), for
GMRES. We see that the alternating Schwarz method is not convergent for the
indefinite Helmholtz equation. Used as a preconditioner, we obtain a convergent
method, but iteration numbers grow with increasing wave number k. For diffusive
problems the alternating Schwarz method converges better when the overlap is
increased, which is also intuitively understandable. This is, however, not the case
for the Helmholtz equation, as we see comparing the case with overlap h, the mesh
size, and with fixed overlap, equal to $2h$ on the coarsest grid, and then $4h$, $8h$ etc
when the mesh is refined: at the beginning, for small wave numbers, overlap seems
to help, but later, bigger overlap is detrimental to the performance of the Schwarz
preconditioner when applied to the Helmholtz equation.

2.4   Fictitious Domain Methods

While domain decomposition methods arrive at more manageable subproblems by
dividing a given problem region into smaller subregions, *fictitious domain methods*

are based on imbedding the former in a larger domain for which a more efficient solver may be available. The first such techniques [11, 46, 58, 61], also known as *domain imbedding* or *capacitance matrix methods*, were developed to extend the efficiency of *fast Poisson solvers* based on the Fast Fourier Transform or cyclic reduction also to problems for which these methods are not directly applicable, as they require some form of separation of variables. In [22] (see also [23]) this idea was applied to exterior boundary value problems for the Helmholtz equation in two dimensions, and it was shown how the Sommerfeld radiation condition can be incorporated into a fast Poisson solver. Large-scale scattering calculations using this approach can be found in [45].

Computationally, fictitious-domain methods represent the original discrete problem as a low-rank modification of a larger problem amenable to fast methods. The fast solver plays the role of a discrete Green's function much in the same way as its continuous counterpart is used in the integral equation method for solving the Helmholtz equation using layer potentials [13]. In fact, fictitious domain methods require the solution of an auxiliary system of equations which is a discretization of an integral operator on the boundary of the problem (scattering) domain. If a suitable formulation is chosen these operators are often compact perturbations of the identity, which can be exploited to obtain mesh-independent convergence for iterative solution methods. The dependence on the wave number, however, is typically linear. Convergence independent of the wave number and mesh size would require more efficient preconditioning schemes for the discrete integral operator, which are currently not available. Recent developments on the spectral analysis of such operators necessary for the design of effective preconditioners can be found in [6].

2.5 Multigrid Methods

Two fundamental observations led to the invention of multigrid methods:

- When applied to the Poisson equation, classical stationary iterative methods such as Gauss-Seidel or damped Jacobi iteration effectively remove high-frequency components of the error, but are very ineffective for low-frequency components. Stiefel points this out very vividly in his 1952 paper [64] on precursors of the conjugate gradient method, remarking that, after a few iterations of one such basic iterative method, in which the residual is reduced significantly, subsequent iteration steps decrease the residual only by very little, as if the approximation were confined to a "cage". [1]

---

[1]"Das Auftreten von Käfigen ist eine allgemeine Erscheinung bei Relaxationsverfahren und sehr unerwünscht. Es bewirkt, dass eine Relaxation am Anfang flott vorwärts geht, aber dann immer weniger ausgiebig wird . . .".

• The remaining low-frequency components in the error can be well represented on a coarser grid,[2] as Federenko points out in his 1961 paper presenting the first complete multigrid method [29]:

> We shall speak of the eigenfunctions as "good" and "bad"; the good ones include those that are smooth on the net and have few changes of sign in the domain; the bad ones often change sign and oscillate rapidly [...] After a fairly small number of iterations, the error will consist of "good" eigenfunctions [...] We shall use the following method to annihilate the "good" components of the error. We introduce into the domain an auxiliary net, the step of which is q times greater than the step of the original net.

The simplest multigrid scheme to which these developments led is the classical "V-cycle", which, applied to the system $A\mathbf{u} = \mathbf{f}$, reads:

$$
\begin{aligned}
&\text{function } u = \text{Multigrid}(A, f, u_0); \\
&\text{if isSmall}(A) \text{ then } u = A \backslash f \text{ else} \\
&\quad u = \text{DampedJacobi}(\nu, A, f, u_0); \\
&\quad r = \text{Restrict}(f - Au); \\
&\quad e = \text{Multigrid}(A_c, r, 0); \\
&\quad u = u + \text{Interpolate}(e); \\
&\quad u = \text{DampedJacobi}(\nu, A, f, u); \\
&\text{end};
\end{aligned}
$$

Here u_0 denotes the initial approximation and A_c the coarse-grid representation of A.

We show in Table 3 the performance of this multigrid algorithm when applied to a discretized Helmholtz equation, in our example a closed cavity without resonance for the discretized problem.[3] We observe that the multigrid method fails to converge as a stand-alone iterative solver except for a very small wave number. When multigrid is used as a preconditioner, we obtain a convergent method, as in the

Table 3 Performance of a classical geometric multigrid method with optimally damped Jacobi smoother applied to a discretized Helmholtz equation. ν denotes the number of smoothing steps

k	2.5π	5π	10π	20π	2.5π	5π	10π	20π	2.5π	5π	10π	20π
	$\nu = 2$				$\nu = 5$				$\nu = 10$			
Iterative	7	Div	Div	Div	7	Stag	Div	Div	8	Div	Div	Div
Preconditioner	6	12	41	127	5	13	41	223	5	10	14	87

---

[2] The idea of beginning the iteration on a coarse grid with a subsequent "advance to a finer net", not unlike the modern *full multigrid* approach, was in use already in the early days of "relaxation methods", as evidenced, e.g., in the book of Southwell [63, Sect. 52] from 1946.

[3] In a closed cavity, i.e., with homogeneous Dirichlet conditions imposed on all sides, it is important to ensure that k^2 is not an eigenvalue of the *discrete* Laplacian, since otherwise one obtains a singular matrix. In the case of a multigrid solver then, one must be careful that k^2 is not an eigenvalue of the discrete Laplacian on each of the grids used in the multigrid hierarchy, which we did for this experiment (see also Sect. 3.4).

case of the Schwarz domain decomposition method, but again the iteration numbers grow substantially when the wave number increases. We used again about 10 points per wavelength in these experiments. Often one increases the number of smoothing steps in the multigrid method to improve performance, and we see in Table 3 that, for small wave numbers, this seems to help the preconditioned version, but for large wave numbers, adding more smoothing steps can both improve and diminish performance. Again, we observe that standard multigrid methods are not suitable for solving the Helmholtz equation.

3   Iterative Methods for Helmholtz Problems

We now describe several iterative methods and preconditioners which have been developed especially for solving discrete Helmholtz problems. In each case we first give an explanation of why the classical iterative method or preconditioner fails, and then show possible remedies.

3.1   Analytic Incomplete LU

The incomplete LU (ILU) preconditioners are based on the fact that the linear system (3) could be solved by a direct factorization, the so-called LU factorization

$$A = LU, \qquad L \text{ lower triangular,} \quad U \text{ upper triangular.} \tag{8}$$

The solution of the linear system $A\mathbf{u} = LU\mathbf{u} = \mathbf{f}$ is then obtained by solving

$$L\mathbf{v} = \mathbf{f} \quad \text{by forward substitution,}$$

$$U\mathbf{u} = \mathbf{v} \quad \text{by backward substitution.}$$

If the matrix A is a discretization of the Helmholtz operator $-(\Delta + k^2)$ in two dimensions, and we use the lexicographic ordering[4] of the unknowns indicated in Fig. 4, we can interpret the forward and backward substitutions geometrically: the forward substitution process $L\mathbf{v} = \mathbf{f}$ determines first the variables in the leftmost column of the domain, see Fig. 4, then in the second leftmost, and so on, until the last column on the right. The process is sequential, and could be interpreted as a time-stepping in the positive x-direction, solving an evolutionary problem. The backward substitution process $U\mathbf{u} = \mathbf{v}$, on the other hand, starts with the variables in the rightmost column in Fig. 4, and then computes the second rightmost column, and so

---

[4]This presupposes a tensor-product grid structure.

Fig. 4 Ordering of the
unknowns in the
discretization of the
Helmholtz operator

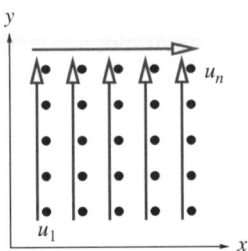

on, until the first column on the left is determined. Again the process is sequential, and could be interpreted as a time-stepping, but this time in the negative x-direction.

From the explanation of the convergence of Krylov methods without preconditioning given in Sect. 2, we see that efficient transport of information in the preconditioner is important for Helmholtz problems. We have, however, also seen that the classical ILU preconditioners do not seem to bring about this transport effectively enough: even the rather accurate approximate ILU(1e-2) factorization does not suffice.

In order to find what the evolution problems described by the LU factorization could correspond to for the underlying Helmholtz equation, it was sought in [38] to determine a factorization of the Helmholtz operator in the x direction,

$$-(\Delta + k^2) = -(\partial_x + \Lambda_1)(\partial_x - \Lambda_2), \tag{9}$$

where Λ_1 and Λ_2 are (non-local) operators to be determined such that (9) holds. Given such a factorization at the continuous level one can solve $-(\Delta + k^2)u = -(\partial_x + \Lambda_1)(\partial_x - \Lambda_2)u = f$ by solving two evolution problems:

$$-(\partial_x + \Lambda_1)v = f \quad \text{evolution problem in the forward } x \text{ direction,}$$

$$(\partial_x - \Lambda_2)u = v \quad \text{evolution problem in the backward } x \text{ direction.}$$

Taking a Fourier transform of the Helmholtz operator (ignoring boundary conditions) in the y-direction with Fourier variable ξ, we obtain

$$\mathscr{F}_y(-(\Delta + k^2)) = -\partial_{xx} + \xi^2 - k^2 = -(\partial_x + \sqrt{\xi^2 - k^2})(\partial_x - \sqrt{\xi^2 - k^2}), \tag{10}$$

and thus we have the continuous analytic factorization of the Helmholtz operator

$$-(\Delta + k^2) = -(\partial_x + \Lambda_1)(\partial_x - \Lambda_2), \tag{11}$$

where $\Lambda_1 = \Lambda_2 = \mathscr{F}_y^{-1}(\sqrt{\xi^2 - k^2})$. Note that Λ_j, $j = 1, 2$, are non local operators in y due to the square root in their symbol $\sqrt{\xi^2 - k^2}$.

The discrete analogue of this factorization at the continuous level is the block LDL^T factorization of the discrete Helmholtz matrix A. In the case of a five point finite difference discretization, this matrix has the block structure

$$A = \frac{1}{h^2} \begin{bmatrix} A_1 & -I & \\ -I & A_2 & \ddots \\ & \ddots & \ddots \end{bmatrix}, \quad A_j = \begin{bmatrix} 4 - kh^2 & -1 & \\ -1 & 4 - kh^2 & \ddots \\ & \ddots & \ddots \end{bmatrix}.$$

A direct calculation shows that the block LDL^T factorization of A is given by

$$L = \frac{1}{h} \begin{bmatrix} I & & \\ -T_1^{-1} & I & \\ & -T_2^{-1} & \ddots \\ & & \ddots & \ddots \end{bmatrix}, \quad D = \begin{bmatrix} T_1 & & \\ & T_2 & \\ & & \ddots \end{bmatrix},$$

where the matrices T_j satisfy the recurrence relation

$$T_{j+1} = A_{j+1} - T_j^{-1}, \quad T_1 = A_1, \tag{12}$$

as is easily verified by multiplying the matrices. We observe that in this exact factorization the matrices T_j are no longer sparse, since the recurrence relation (12) which determines them involves an inverse. This fill-in at the discrete level corresponds to the non-local nature of the operators Λ_j. Using a local approximation of the matrices T_j with tridiagonal structure gives only an approximate LDL^T factorization of A, which we call AILU('0') (Analytic Incomplete LU). In order to obtain a good approximation, the relation to the continuous factorization was used in [38], and the spectral radius of the corresponding iteration matrix was minimized. The performance of this preconditioner, which is now tuned for the Helmholtz nature of the problem, is shown in Table 4, for the same open cavity problem as before. We clearly see that this approximate factorization contains much more of the physics of the underlying Helmholtz equation, and leads to a better preconditioner. Nonetheless, the iteration counts are still seen to increase with growing wave number k.

Table 4 Performance comparison of the specialized AILU('0') preconditioner, compared to the other ILU variants

k	QMR		ILU('0')		ILU(1e-2)		AILU('0')	
	It	Mflops	It	Mflops	It	Mflops	It	Mflops
5	197	120.1	60	60.4	22	28.3	23	28.3
10	737	1858.2	370	1489.3	80	421.4	36	176.2
15	1,775	10185.2	2,000	18133.2	220	2615.1	43	475.9
20	2,000	20335.1	–	–	2,000	42320.1	64	1260.2
30	–	–	–	–	–	–	90	3984.1
40	–	–	–	–	–	–	135	10625.0
50	–	–	–	–	–	–	285	24000.4

The AILU preconditioner goes back to the analytic factorization idea, see [56] and references therein. It is very much related to the Frequency Filtering Decomposition, as proposed by Wittum in [67, 68] and analyzed for symmetric positive problems in [65], and for non-symmetric problems in [66]. There was substantial research activity for these kinds of preconditioners around the turn of the century, see [42], [12], [37], [1], and for Helmholtz problems this is one of the best incomplete factorization preconditioners available. For more recent work, see [2], [57], and for Helmholtz problems in particular [17] and [18], where this type of preconditioner is called a "sweeping preconditioner", and an optimal approximation is proposed in the sense that iteration numbers no longer depend on the wave number k, see also [19].

3.2 Domain Decomposition Methods for Helmholtz Problems

In the late 1980s researchers realized that classical domain decomposition methods were not effective for Helmholtz problems, and the search for specialized methods began. In his PhD thesis [15], Bruno Després summarizes this situation precisely:

> L'objectif de ce travail est, après construction d'une méthode de décomposition de domaine adaptée au problème de Helmholtz, d'en démontrer la convergence.[5]

The fundamental new ingredient for such an algorithm turned out to be the transmission condition between subdomains, as in the non-overlapping variant of the Schwarz algorithm proposed by Lions [54]. The algorithm proposed by Bruno Després reads

$$
\begin{aligned}
-(\Delta + k^2)u_j^{n+1} &= f, & &\text{in } \Omega_j \\
(\partial_{n_j} - ik)u_j^{n+1} &= (\partial_{n_j} - ik)u_l^n, & &\text{on interface } \Gamma_{jl},
\end{aligned}
\tag{13}
$$

and, on comparing with the classical alternating Schwarz algorithm in (7), we see that now a Robin transmission condition is used at the interfaces. The algorithm was considered by Després for many subdomains, but only without overlap, so that its convergence can be proved using energy estimates.

To obtain further insight into why the transmission conditions are important, we show in Fig. 5 on the vertical axis the convergence factor of the algorithm for the simple model problem of a square decomposed into two rectangles, plotted against the index ξ of the Fourier modes. In this case, we can use Fourier series in the direction of the interface to explicitly compute how each Fourier mode converges, see for example [35]. We see on the left for the classical alternating Schwarz method that low frequency modes are not converging at all, their convergence factor equals

---

[5]The goal of this work is to design a special domain decomposition method for Helmholtz problems, and to prove that it converges.

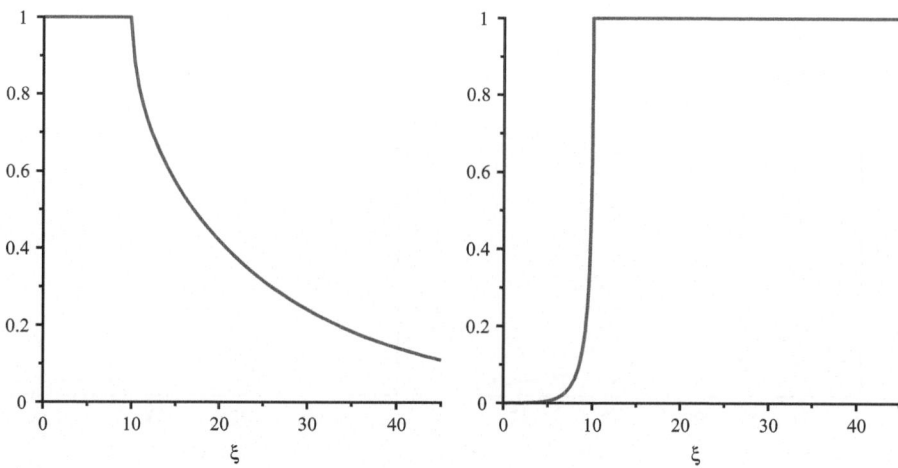

Fig. 5 Comparison of how each Fourier mode ξ in the error converges, on the *left* for the classical alternating Schwarz method with overlap, and on the *right* for the variant designed for the Helmholtz equation, without overlap. The vertical axis denotes the convergence factor of a Fourier mode

one. These modes correspond to the oscillatory, or propagating modes in the solution of the Helmholtz equation, as are clearly visible, e.g., in the example in Fig. 1 on the right. High-frequency components, however, converge well in the classical alternating Schwarz method. These modes correspond to evanescent modes, usually only well visible for diffusive problems, as in Fig. 1 on the left. The situation for the non-overlapping method of Després on the right is reversed: the new transmission conditions lead to a rapidly converging method for the propagating modes in the low-frequency part of the spectrum, but now high frequency components are not converging.

Després wanted to prove convergence of the algorithm, and the technique of energy estimates generally works only for the non-overlapping variants of the algorithm. But Fig. 5 suggests that one could use the overlap for the high-frequency modes, and the transmission condition for the low-frequency modes, in order to obtain a method effective for all modes in a Helmholtz problem. In addition, it might be possible to choose an even better transmission condition, as indicated toward the end in Lions' work [54], and also by Hagström et al. in [44]. All these observations and further developments led at the turn of the century to the invention of the new class of optimized Schwarz methods [34], with specialized variants for Helmholtz problems [33, 35]. For an overview for symmetric coercive problems, see [31].

Using optimized transmission conditions of zeroth order, which means choosing the best complex constant in place of ik in the Robin condition, we obtain for the same model problem as in Fig. 5 the contraction factors shown in Fig. 6 on the left.

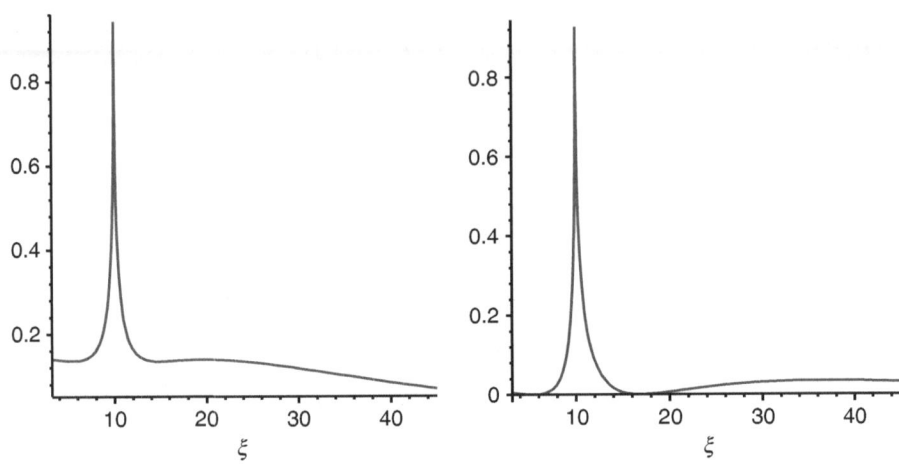

Fig. 6 Comparison of how each Fourier mode ξ in the error converges, on the left for an optimized Schwarz method of order zero (OO0), and on the right for a second order optimized Schwarz method (OO2), both with overlap

Table 5 Asymptotic convergence factors obtained for a model problem

	k fixed	$k^\gamma h$ const
Overlap 0	$1 - O(h^{\frac{1}{4}})$	$1 - O(k^{\frac{1-2\gamma}{8}})$
Overlap $C_L h$	$1 - O(h^{\frac{1}{5}})$	$\begin{cases} 1 - O(k^{-\frac{1}{8}}) & 1 \le \gamma \le \frac{9}{8} \\ 1 - O(k^{\frac{1-2\gamma}{10}}) & \gamma > \frac{9}{8} \end{cases}$
Overlap const	$1 - const$	$1 - O(k^{-\frac{1}{8}})$

We can see that all modes, except for the resonance mode,[6] now converge well. On the right in the same figure, we show a second-order optimized Schwarz method, in which one also uses the Laplace-Beltrami operator at the interface to obtain an even more effective transmission condition. Using this operator in no way increases the sparsity pattern of the subdomain solver, since second-order derivatives are already present in the underlying discretization of the Laplacian.

A general convergence analysis of optimized Schwarz methods for Helmholtz problems currently only exists for the non-overlapping case, using energy estimates. This approach, however, does not allow us to obtain convergence factor estimates. In addition, to prove convergence for the general overlapping case is an open problem. For the model situation of two subdomains however, one can quantify precisely the dependence of the convergence factor on the wave number k and the mesh parameter h. We show in Table 5 the resulting convergence factors from [33]. We see that for a fixed wave number k and constant overlap, independent of the mesh size h,

---

[6]We denote by resonance mode that value of the Fourier parameter for which the transformed Helmholtz operator becomes singular.

Table 6 Numerical experiment for a two-subdomain decomposition

| h | Iterative | Krylov | | k | Krylov | |
	Optimized	Deprés	Optimized		Deprés	Optimized
1/50	322	26	14	10π	24	13
1/100	70	34	17	20π	33	18
1/200	75	44	20	40π	43	20
1/400	91	57	23	80π	53	21
1/800	112	72	27	160π	83	32

the algorithm converges with a contraction factor independent of h. In the important case where k scales with h as $k^\gamma h$ in order to avoid the pollution effect, see [47,48], we see that the contraction factor only depends very weakly on the growing wave number: for example if the overlap is held constant, all Fourier modes of the error contract at least with a factor $1 - O(k^{-\frac{1}{8}})$.

In Table 6 we show a numerical experiment for a square cavity open on two sides and the non-overlapping optimized Schwarz method in order to illustrate the asymptotic results from Table 5. We used a fixed wave number k on the left, and a growing wave number k on the right, while again keeping ten points per wavelength. We show in the leftmost column the stand-alone iterative variant of the algorithm in order to illustrate the sensitivity of the algorithm with respect to the peak of the convergence factor at the resonance frequency. Since the discretization modifies the continuous spectrum, a discretization with insufficient resolution may have eigenvalues close to the resonance frequency, which are not taken into account by the continuous optimization based on Fourier analysis, which in turn can result in an arbitrarily large iteration count, as we see for example for $h = \frac{1}{50}$. Such problems, however, disappear once the mesh is sufficiently refined, or when Krylov acceleration is added, as one can observe in Table 6. This issue is therefore not of practical concern. We also see that it clearly pays to use optimized parameters, as the iteration count is substantially lower than with the first choice of ik in the transmission conditions.

We finally show two numerical experiments, in order to illustrate that optimized Schwarz methods for Helmholtz equations also work well in more practical situations. We first show the acoustic pressure in two spatial dimensions for the approach of an Airbus A340 over the silhouette of a city, computed with a decomposition into 16 subdomains, as shown in Fig. 7 on the left. In this case, using a Robin transmission condition with ik as parameter required 172 iterations, whereas the optimized Schwarz method needed only 58 iterations to converge to the same tolerance. For more details, see [35]. The second example is the interior of a Twingo car from Renault, shown in Fig. 8. Here, the Robin transmission condition with ik as parameter required 105 iterations, and the optimized Schwarz method 34. For further details, see [33].

There is a second type of domain decomposition methods for Helmholtz problems from the FETI family of methods (Finite Element Tearing and Interconnect, see [28]). These methods are based on a dual Schur complement formulation,

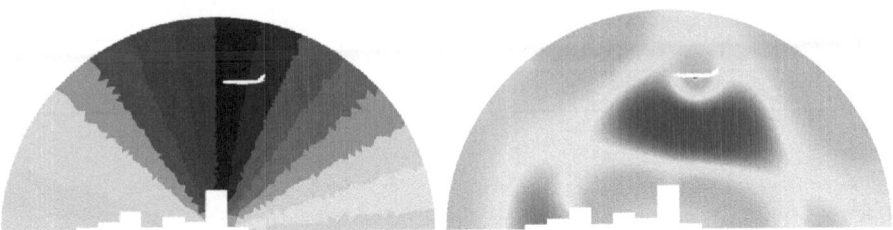

Fig. 7 Airbus A340 in approach over a city, domain decomposition on the *left*, and result of one simulation on the *right*

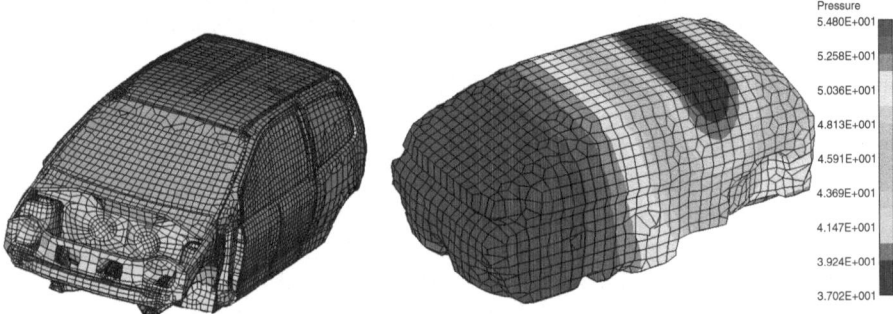

Fig. 8 Simulation of the noise in the passenger cabin of a Twingo car from Renault: the pressure range on the right goes from 37.02 to 54.8

which means that interior variables in the subdomains are eliminated, assuming that Neumann traces are continuous across interfaces, and then a substructured system is obtained by requiring that Dirichlet traces along interfaces match. A primal Schur formulation would do the opposite: eliminate interior unknowns of subomains, assuming that Dirichlet traces are continuous across interfaces, and then impose continuity of the Neumann traces along interfaces in order to obtain a substructured formulation. These methods usually require an additional preconditioner in order to obtain convergence rates independent of (or only weakly dependent on) the mesh parameter h. An optimal choice is to use the primal Schur complement method for the dual Schur complement formulation, and vice versa. In order to scale with the number of subdomains, also a coarse grid is needed. For the case of Laplace's equation, the classical coarse grid is to use a constant per subdomain, since if FETI is used to solve Laplace's equation, interior subdomain problems containing Neumann conditions all around have precisely the constant as a kernel. This idea transformed an inconvenience of the original FETI idea, namely that interior subdomains are singular, into a benefit: a natural coarse grid.

In order to adapt this class of methods to Helmholtz problems, the first variant was the FETI-H method (for FETI-Helmholtz), see [27]. Instead of using Neumann transmission conditions in the dual Schur complement formulation, Robin conditions $\partial_n - ik$ are used, but then still Dirichlet traces are matched in order to obtain a substructured formulation. This approach is thus very much related to an optimized Schwarz method without overlap; however, only one type of Robin conditions can be imposed, since the other one is Dirichlet. This means that always on one side of the interface, a Robin condition with the good sign is used, whereas on the other side, a Robin condition with the bad sign is imposed. For checkerboard type partitions, one can ensure that subdomains have only Robin conditions with constant sign all around. Otherwise, an algorithm was proposed to generate a choice of sign which guarantees that subdomain problems are not singular. The original formulation has no additional preconditioner, but a coarse grid in form of plane waves.

The second algorithm in the FETI class specialized for Helmholtz problems is FETI-DPH, see [26]. This is a FETI-DP formulation, which means that some interface unknowns are kept as primal variables, where continuity is enforced, and which serve at the same time as coarse space components. These are usually cross points, and in FETI-DPH additional primal constraints are enforced at the interfaces, using planar waves. Furthermore, a Dirichlet preconditioner is used on top, like in the classical FETI formulation. A convergence analysis exists for this algorithm, see [25], but it needs the assumption that subdomains are small enough. A systematic comparison of all currently existing domain decomposition algorithms for Helmholtz problems is in preparation, see [40].

3.3   Multigrid for Helmholtz Problems

We will see in this section that neither of the two fundamental observations made by Stiefel and Federenko (cf. Sect. 2.5) hold for the case of the Helmholtz equation. In an early theoretical paper about multigrid methods [4], Bakhvalov first advertises the method also for indefinite problems:

> For instance it is used in the case of the equation $\Delta u + \lambda u = f$ with large positive $\lambda(x_1, x_2)$. Previously no methods of solving this equation with good asymptotics for the number of operations were known

but then later in the paper he discovers potential problems:

> In the case of the equation $\Delta u + \lambda u = f$ with large positive λ we do not exclude the possibility that the evaluation of (3.21) may be attained in order. Then the increase in the number m in comparison with that calculated can lead to a deterioration in the discrepancy of the approximation.

Three decades later Brandt and Livshits [8] take on the difficult Helmholtz case again, and they try to explain the origin of the difficulties of the multigrid algorithm:

> On the fine grids, where [the characteristic components] are accurately approximated by the discrete equations, they are invisible to any local relaxation, since their errors can have very small residuals. On the other hand, on coarser grids such components cannot be approximated, because the grid does not resolve their oscillations.

Similarly, Lee, Manteuffel, McCormick and Ruge [50] explain the problem as follows:

Helmholtz problems tax multigrid methods by admitting certain highly oscillatory error components that yield relatively small residuals. Because these components are oscillatory, standard coarse grids cannot represent them well, so coarsening cannot eliminate them effectively. Because they yield small residuals, standard relaxation methods cannot effectively reduce them.

In order to more precisely illustrate the problems of the multigrid algorithm when applied to the Helmholtz equation, we consider now the Helmholtz equation in two dimensions on the unit square,

$$- (\Delta + k^2)u = f, \quad \text{in } \Omega := (0, 1) \times (0, 1). \tag{14}$$

We show two numerical experiments following the common strategy (cf. [7], [10, Chap. 4]) that, in order to investigate the behavior of multigrid methods, one should replace one of the two components (smoother or coarse grid correction) by a component which one knows to be effective (even if it is not feasible to use this component in practice), to test the other one. In a first experiment, we use a Fourier smoother in order to explicitly remove the high-frequency components of the error, and try to compute the solution shown in Fig. 9, on top, which corresponds to the choice of parameters $f = -\frac{1}{20}$, $k = \sqrt{19.7}$ and fine-grid parameter $h = \frac{1}{32}$. We use a random initial guess u_0, and a two-grid cycle. The result is shown in Fig. 9.

We clearly observe the following in this experiment: while the error on the coarse grid is well resolved, the correction calculated on the coarse grid is 100% incorrect, it has the wrong sign! Hence the problem does not seem to be that certain (high) frequency components in the error are left to the coarse grid and cannot be approximated accurately there: the mesh here is largely fine enough to represent the component left. However, the correction calculated is incorrect: it is the operator itself which is not well approximated. This had already been discovered in an earlier paper by Brandt and Ta'asan [9] on slightly indefinite problems:

Usual multigrid for indefinite problems is sometimes found to be very inefficient. A strong limitation exists on the coarsest grid to be used in the process. The limitation is not so much a result of the indefiniteness itself, but of the nearness to singularity, that is, the existence of nearly zero eigenvalues. These eigenvalues are badly approximated (e.g. they may even have a different sign) on coarse grids, hence the corresponding eigenfunctions, which are usually smooth ones, cannot efficiently converge.

For our second numerical experiment, we now use a damped Jacobi smoother and compute the exact coarse-grid correction by computing it on the fine grid, then restricting it to the coarse grid and prolongating it again back to the fine grid to ensure that the coarse-grid correction is working properly (this would obviously not make sense in a real multigrid code, but allows us to illustrate the reason why the smoother fails). We try to compute the solution shown in Fig. 10, in the top left graph, which corresponds to the parameters $f = -1{,}000$, $k = 20$ and fine mesh size $h = \frac{1}{32}$, and we use again a random initial guess u_0 and a two-grid cycle. Its behavior is shown in the remaining graphs of Fig. 10. We clearly see that, even though the coarse-grid correction is very effective, the smoother is responsible for

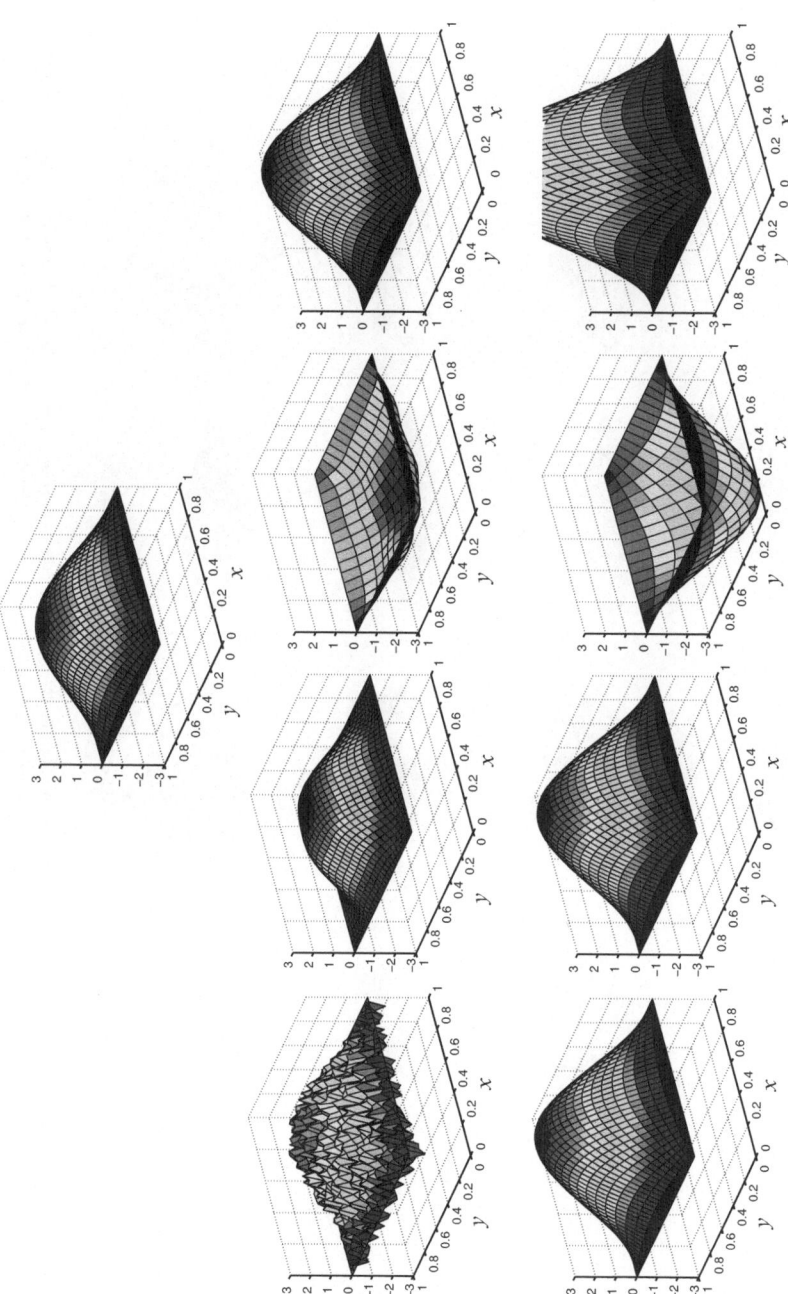

Fig. 9 Problem of the coarse-grid correction when multigrid is used for the Helmholtz equation. *Top row:* solution we want to compute. *Following two rows:* error before presmoothing, error after presmoothing, coarse-grid correction that needs to be *subtracted,* error after coarse-grid correction, for two consecutive iterations

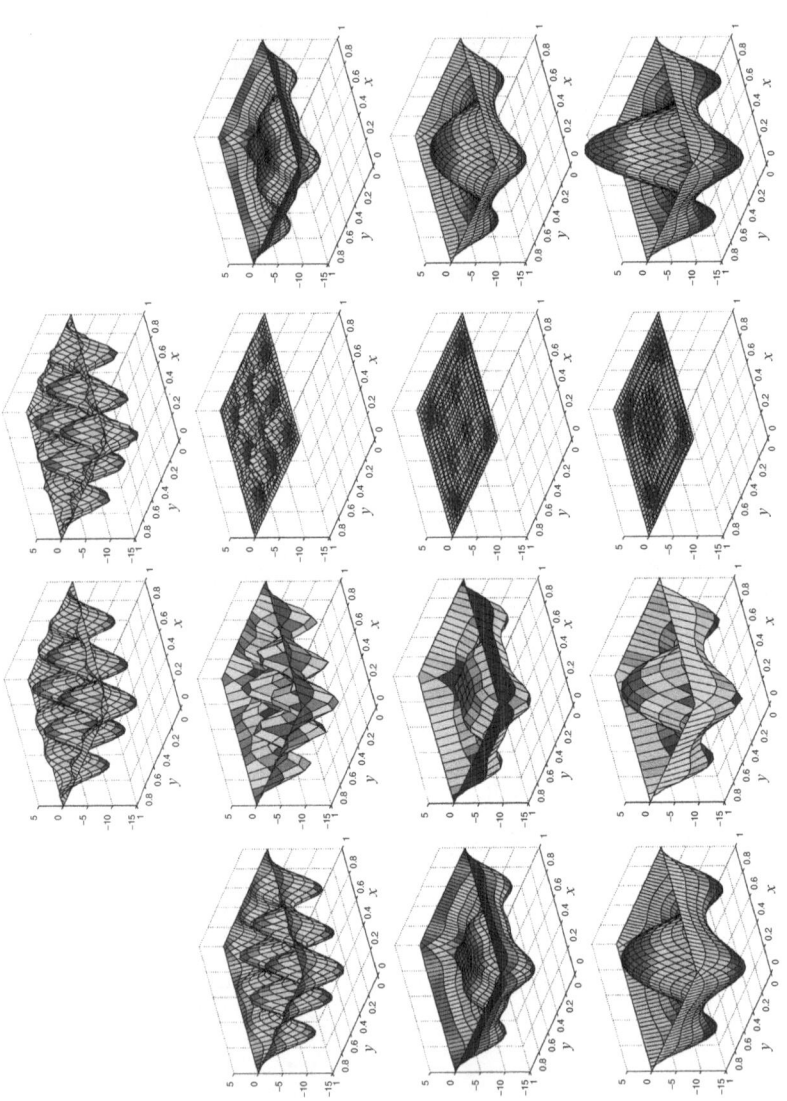

Fig. 10 Problem of the smoother when multigrid is used for the Helmholtz equation. *Top row*: solution we want to calculate, and the initial error. *Following three rows*: error after presmoothing, coarse-grid correction that needs to be *subtracted*, error after coarse-grid correction, error after postsmoothing, for three consecutive iterations. One can clearly see how the smoother amplifies the error

a growing low-frequency mode, and the two-grid method does not converge. We explain these two observations in the next section with a detailed two-grid analysis.

3.4   Two-Grid Analysis for the 1D Model Problem

To explain the difficulties of multigrid applied to the Helmholtz equation, we consider the simplest possible case of the one-dimensional problem

$$- u'' - k^2 u = f \quad \text{in } \Omega = (0,1), \qquad u(0) = u(1) = 0, \tag{15}$$

with constant wave number k and perform a spectral analysis, much along the lines of [43, Chap. 2] and [10, Chap. 5].

We assume that k^2 is not an eigenvalue of the Dirichlet-Laplacian for this domain and therefore the continuous problem possesses a unique solution, as do sufficiently accurate discrete approximations. When multigrid is applied to cavity problems like (15) one must always be careful that all coarse-grid problems are nonsingular. This is, however, no longer an issue when damping is present, either in the form of an absorbing medium or radiation boundary conditions.

Using the standard three-point centered finite-difference approximation of the second derivative on a uniform mesh with N interior grid points and mesh width $h = 1/(N+1)$, (15) is approximated by the linear system of equations $A\mathbf{u} = \mathbf{f}$ for the function values $u(x_j) \approx u_j$, $j = 1, \ldots, N$, at the grid points $x_j = jh$, where

$$A = \frac{1}{h^2} \text{tridiag} \left(-1, 2 - k^2 h^2, -1\right) \in \mathbb{R}^{N \times N}. \tag{16}$$

The matrix A is symmetric and has the complete set of orthogonal eigenvectors

$$\mathbf{v}_j = [\sin j \ell \pi h]_{\ell=1}^N, \qquad j = 1, \ldots, N. \tag{17}$$

When it is necessary to rescale these eigenvectors to have unit Euclidean norm this is achieved by the factor $\sqrt{2h}$ (for all j). The associated eigenvalues are given by

$$\lambda_j = \frac{2(1 - \cos j\pi h)}{h^2} - k^2 = \frac{4}{h^2} \sin^2 \frac{j\pi h}{2} - k^2, \qquad j = 1, \ldots, N.$$

The form of the eigenvectors (17) reveals that these become more oscillatory with increasing index j.

When N is odd, which we shall always assume for the pure Dirichlet problem, we set $n := (N-1)/2$ and refer to the eigenpairs associated with the indices $1 \le j \le n$ as the *smooth* part I_{sm} of the spectrum and the remainder as the *oscillatory* part I_{osc}. Note that the eigenvalue with index $j = (N+1)/2 = n+1$ lies exactly in the middle, with an associated eigenvector of wavelength $4h$.

3.4.1 Smoothing

The (damped) Jacobi smoother is based on the splitting $A = \frac{1}{\omega}D - (\frac{1}{\omega}D - A)$ of the matrix A in (16), where $D = \text{diag}(A)$ and ω is the damping parameter, resulting in the iteration

$$\mathbf{u}_{m+1} = \mathbf{u}_m + \omega D^{-1}(\mathbf{f} - A\mathbf{u}_m) \qquad (18)$$

with associated error propagation operator

$$S_\omega = I - \omega D^{-1}A. \qquad (19)$$

Noting that $D = (2/h^2 - k^2)I$, we conclude that A and D are simultaneously diagonalizable, which gives for S_ω the eigenvalues

$$\sigma_j = \sigma_j(\omega) = 1 - \omega\left(1 - \frac{2\cos j\pi h}{2 - k^2 h^2}\right) =: 1 - \omega\frac{\lambda_j}{\delta}, \qquad j = 1, \ldots, N, \quad (20)$$

where we have introduced $\delta = \delta(k, h) := (2 - k^2 h^2)/h^2$ to denote the diagonal entry of D, which is constant for this model problem.

In multigrid methods the smoothing parameter ω is chosen to maximize damping on the oscillatory part I_{osc} of the spectrum. For the Laplace operator ($k = 0$) the eigenvalues of $D^{-1}A$ are given by $\lambda_j/\delta = 1 - \cos(j\pi h)$, $j = 1, \ldots, N$, resulting in, up to order h^2, the spectral interval $[0, 2]$, with $I_{\text{osc}} = [1, 2]$. Maximal damping on I_{osc} thus translates to the requirement of equioscillation, i.e.,

$$1 - \omega = -(1 - 2\omega), \qquad \text{obtained for} \qquad \omega = \omega_0 := \tfrac{2}{3}. \qquad (21)$$

For this optimal value of ω each eigenmode belonging to the oscillatory modes $\text{span}\{\mathbf{v}^h_{n+1}, \ldots, \mathbf{v}^h_N\}$ is reduced by at least a factor of $\sigma_{n+1}(\omega_0) = 1 - \omega_0 = \frac{1}{3}$ in each smoothing step, independently of the mesh size h. Figure 11 displays the spectrum $\Lambda(S_\omega)$ of S_ω for the discrete 1D-Laplacian on the unit interval with mesh width $h = 1/50$ for the values $\omega = 1$ (undamped) and the optimal value $\omega = \omega_0 = 2/3$, plotted against the eigenvalues of $D^{-1}A$, where we have scaled A in order to fix the spectral interval independently of h. The smooth and oscillatory parts of the spectrum I_{sm} and I_{osc} are highlighted, and it can be seen that the eigenvalues of S_ω lie on a straight line and that the spectral radius of S_ω is minimized on I_{osc} for $\omega = \omega_0$.

For the 1D Helmholtz operator ($k > 0$) the eigenvalues of $D^{-1}A$ are

$$\frac{\lambda_j}{\delta} = 1 - \frac{2\cos j\pi h}{2 - k^2 h^2}, \qquad j = 1, \ldots, N,$$

and therefore, up to $O(h^2)$, these range between the extremal values

$$\frac{\lambda_1}{\delta} = \frac{-k^2 h^2}{2 - k^2 h^2}, \qquad \text{and} \qquad \frac{\lambda_N}{\delta} = \frac{4 - k^2 h^2}{2 - k^2 h^2}.$$

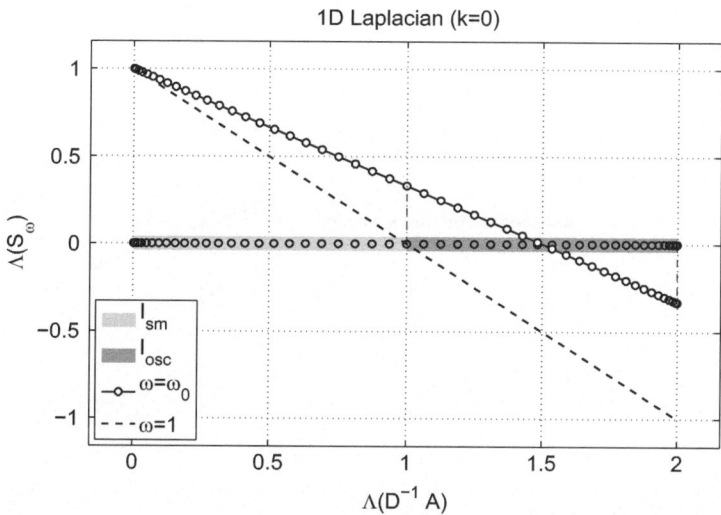

Fig. 11 Eigenvalues of the undamped ($\omega = 1$) and optimally damped ($\omega = \omega_0$) Jacobi smoother plotted against those of the associated diagonally scaled 1D-Laplacian $-\Delta_h$, $h = 1/50$, divided into smooth and oscillatory parts I_{sm} and I_{osc}. The *Vertical dashed lines* indicate the spectral radius of S_ω restricted to the space of oscillatory eigenfunctions

Assuming the midpoint $(\lambda_1 + \lambda_N)/2$ is still positive, maximal smoothing of the oscillatory modes is again obtained by equioscillation, which fixes ω by requiring

$$1 - \omega \frac{\lambda_N}{\delta} = -\left(1 - \omega \frac{\lambda_1 + \lambda_N}{2\delta}\right), \quad \text{i.e.,} \quad \omega = \omega_k := \frac{2 - k^2 h^2}{3 - k^2 h^2}. \quad (22)$$

Figure 12 shows the analogous quantities of Fig. 11 for the Helmholtz equation with wave number $k = 10\pi$. In contrast with the Laplacian case, the spectrum of A now extends into the negative real axis. By consequence, any choice of the relaxation parameter ω will result in amplification of some modes, as we have seen in our example in Fig. 10. In the case shown, these are precisely the eigenmodes of A associated with negative eigenvalues. If this constitutes only a small portion of $\Lambda(A)$, then the coarse grid correction, the second component of multigrid methods which eliminates smooth error components, can be expected to compensate for this amplification. It is clear, however, that the amplification will both grow unacceptably strong and extend over too large a portion of the spectrum for diminishing wave resolution, i.e., for kh large.

Therefore, fundamentally different smoothing iterations are needed for Helmholtz problems. For this reason Brandt and Ta'asan [9] proposed using the Kazmarcz relaxation, which is essentially Gauss-Seidel iteration applied to the normal equations. This smoother has the advantage of not amplifying any modes, but at the cost of very weak smoothing. Elman, Ernst and O'Leary [16] proposed using Krylov subspace methods as smoothers. The difficulty here is that different

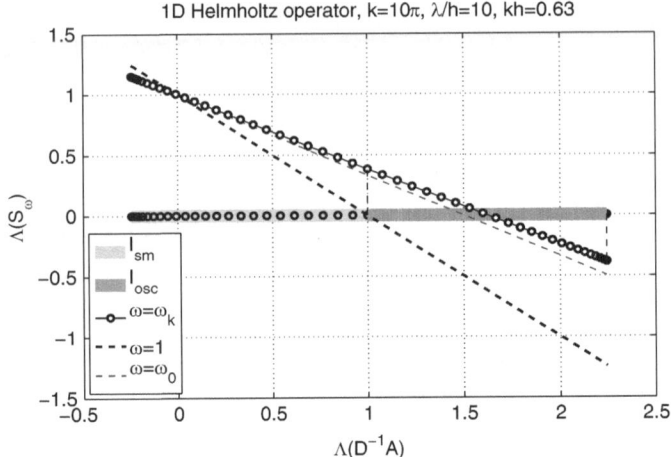

Fig. 12 Same as Fig. 11, here for the Helmholtz operator. Eigenvalues of Jacobi smoother for $\omega = \omega_k$, $\omega = 1$ and $\omega = \omega_0$ against those of the associated diagonally scaled 1D Helmholtz operator $-\Delta_h - k^2$, $h = 1/50$ with wavelength-to-mesh ratio $\lambda/h = 10$

numbers of smoothing steps are necessary at different grid levels, and their optimal determination is challenging.

3.4.2 Coarse-Grid Correction

Besides the finite difference discretization on the mesh

$$\Omega^h := \{x_j = jh : j = 0, \dots, N+1\}$$

we consider the 1D model problem (15) discretized on a coarser grid with only n interior mesh points

$$\Omega^H := \{x_j = jH : j = 0, \dots, n+1\}$$

with twice the mesh width $H = 2h$, where $N = 2n + 1$ denotes the number of fine-grid interior points. We transfer grid functions $\mathbf{u}^H = [u_1^H, \dots, u_n^H]$ (we omit the zero boundary values) from Ω^H to the fine grid Ω^h using linear interpolation, which defines the linear mapping

$$I_h^H : \Omega^H \to \Omega^h, \qquad \mathbf{u}^H \mapsto I_h^H \mathbf{u}^H$$

defined by

$$[I_H^h \mathbf{u}^H]_j = \begin{cases} [\mathbf{u}^H]_{j/2} & \text{if } j \text{ is even,} \\ \frac{1}{2}\left([\mathbf{u}^H]_{(j-1)/2} + [\mathbf{u}^H]_{(j+1)/2}\right) & \text{if } j \text{ is odd,} \end{cases} \quad j = 0, \dots, N+1,$$

$$(23)$$

with matrix representation

$$
I_H^h = \frac{1}{2}
\begin{bmatrix}
1 & & & & \\
2 & & & & \\
1 & 1 & & & \\
& 2 & & & \\
& 1 & & & \\
& & \ddots & 1 & \\
& & & 2 & \\
& & & 1 &
\end{bmatrix}
\in \mathbb{R}^{N \times n}
$$

with respect to the standard unit coordinate bases in $\mathbb{R}^n$ and $\mathbb{R}^N$, respectively.

Following [43], we analyze the mapping properties of the linear interpolation operator I_H^h on the coarse-grid eigenvectors

$$
\mathbf{v}_j^H = [\sin(j \ell \pi H)]_{\ell=1}^n, \qquad j = 1, \dots, n,
$$

of the discrete 1D Dirichlet-Laplacian by way of elementary trigonometric manipulations.

Proposition 1. *The coarse-grid eigenvectors are mapped by the interpolation operator I_H^h according to*

$$
I_H^h \mathbf{v}_j^H = c_j^2 \mathbf{v}_j^h - s_j^2 \mathbf{v}_{N+1-j}^h, \qquad j = 1, \dots, n,
$$

where we define

$$
c_j := \cos \frac{j \pi h}{2}, \quad s_j := \sin \frac{j \pi h}{2}, \qquad j = 1, \dots, n. \tag{24}
$$

In particular, $\mathbf{v}_{n+1}^h$ is not in the range of interpolation.

The coarse-grid modes $\mathbf{v}_j^H$ are thus mapped to a linear combination of their fine-grid counterparts $\mathbf{v}_j^h$ and a *complementary mode* $\mathbf{v}_{j'}^h$, with index $j' := N + 1 - j$. Note the relations

$$
c_{j'} = s_j \quad s_{j'} = c_j, \qquad j = 1, \dots, n,
$$

between complementary s_j and c_j. Interpolating coarse-grid functions therefore always activates high-frequency modes on the fine grid, with a factor that is less than one but grows with j (cf. Fig. 13).

To transfer fine-grid functions to the coarse grid we employ the *full weighting* restriction operator

Fig. 13 Coefficients c_j^2 and s_j^2 of the eigenvectors of the discrete 1D Dirichlet-Laplacian under the linear interpolation operator for $N = 31$, i.e., $n = 15$

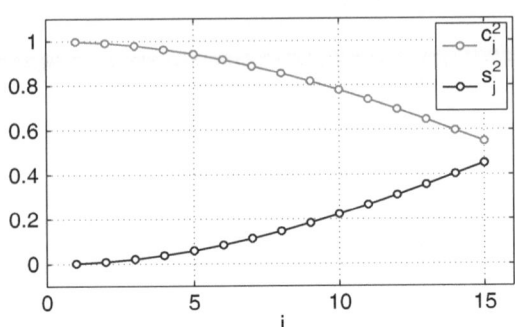

$$I_h^H : \Omega^h \to \Omega^H, \qquad \mathbf{u}^h \mapsto I_h^H \mathbf{u}^h$$

defined by

$$\left[I_h^H \mathbf{u}^h \right]_j = \frac{1}{4} \Big([\mathbf{u}^h]_{2j-1} + 2[\mathbf{u}^h]_{2j} + [\mathbf{u}^h]_{2j+1} \Big), \qquad j = 1, \dots, n. \tag{25}$$

The associated matrix representation is given by $I_h^H = \frac{1}{2}[I_H^h]^{\mathsf{T}}$. The restriction operator is thus seen to be the adjoint of the interpolation operator if one introduces on $\mathbb{R}^n$ and $\mathbb{R}^N$ the Euclidean inner product weighted by the mesh size H and h, respectively. Denoting by $\mathcal{N}(\cdot)$ and $\mathcal{R}(\cdot)$ the null-space and range of a matrix, the basic relation

$$\mathbb{R}^N = \mathcal{R}(I_H^h) \oplus \mathcal{N}([I_H^h]^{\mathsf{T}}) = \mathcal{R}(I_H^h) \oplus \mathcal{N}(I_h^H) \tag{26}$$

reveals that the range of interpolation and the null-space of the restriction are complementary linear subspaces of $\mathbb{R}^N$, which are also orthogonal with respect to the Euclidean inner product. Since the columns of I_H^h are seen to be linearly independent, the interpolation operator has full rank, which together with (26) implies

$$\dim \mathcal{R}(I_H^h) = n, \qquad \dim \mathcal{N}(I_h^H) = N - n = n + 1.$$

Elementary trigonometric relations also yield the following characterization of I_h^H.

Proposition 2. *The fine-grid eigenvectors are mapped by the restriction operator I_h^H according to*

$$I_h^H \mathbf{v}_j^h = c_j^2 \mathbf{v}_j^H, \qquad\qquad j = 1, \dots, n, \tag{27a}$$

$$I_h^H \mathbf{v}_{N+1-j}^h = -s_j^2 \mathbf{v}_j^H, \qquad\qquad j = 1, \dots, n, \tag{27b}$$

$$I_h^H \mathbf{v}_{n+1}^h = \mathbf{0}. \tag{27c}$$

The coarse-grid correction of an approximation $\mathbf{u}^h$ to the solution of (15) on the fine grid Ω^h is obtained by solving the error equation $A_h \mathbf{e}^h = \mathbf{b} - A_h \mathbf{u}^h = \mathbf{r}^h$ on the coarse grid. To this end, the residual is first restricted to the coarse grid and a coarse-grid representation A_H of the differential operator is used to obtain the approximation $A_H^{-1} I_h^H \mathbf{r}^h$ of the error $A_h^{-1} \mathbf{r}^h$ on Ω^H. The update is then obtained after interpolating this error approximation to Ω^h as

$$\mathbf{u}^h \leftarrow \mathbf{u}^h + I_H^h A_H^{-1} I_h^H (\mathbf{b} - A_h \mathbf{u}^h) \tag{28}$$

with associated error propagation operator

$$C := I - I_H^h A_H^{-1} I_h^H A_h. \tag{29}$$

In view of Propositions 1 and 2 the coarse-grid correction operator C is seen to possess the invariant subspaces

$$\mathrm{span}\{\mathbf{v}_{n+1}^h\} \quad \text{and} \quad \mathrm{span}\{\mathbf{v}_j^h, \mathbf{v}_{j'}^h\}, \quad j' = N+1-j, \quad j = 1, \ldots, n. \tag{30}$$

Denoting the eigenvalues of the discrete 1D Helmholtz operators on Ω^h and Ω^H by

$$\lambda_j^h = \frac{4}{h^2} \sin^2 \frac{j\pi h}{2} - k^2, \qquad j = 1, \ldots, N$$

and

$$\lambda_j^H = \frac{4}{H^2} \sin^2 \frac{j\pi H}{2} - k^2, \qquad j = 1, \ldots, n,$$

respectively, the action of the coarse-grid correction operator on these invariant subspaces is given by

$$C \, [\mathbf{v}_j^h \; \mathbf{v}_{j'}^h] = [\mathbf{v}_j^h \; \mathbf{v}_{j'}^h] \, C_j, \qquad j = 1, \ldots, n,$$

where

$$C_j = \begin{bmatrix} 1 & 0 \\ 0 & 1 \end{bmatrix} - \begin{bmatrix} c_j^2 \\ -s_j^2 \end{bmatrix} \frac{1}{\lambda_j^H} \begin{bmatrix} c_j^2 & -s_j^2 \end{bmatrix} \begin{bmatrix} \lambda_j^h & 0 \\ 0 & \lambda_{j'}^h \end{bmatrix} = \begin{bmatrix} 1 - c_j^4 \frac{\lambda_j^h}{\lambda_j^H} & c_j^2 s_j^2 \frac{\lambda_{j'}^h}{\lambda_j^H} \\ c_j^2 s_j^2 \frac{\lambda_j^h}{\lambda_j^H} & 1 - s_j^4 \frac{\lambda_{j'}^h}{\lambda_j^H} \end{bmatrix} \tag{31}$$

in addition to $C \mathbf{v}_{n+1}^h = \mathbf{v}_{n+1}^h$.

For $k = 0$ we observe as in [43]

$$\frac{\lambda_j^h}{\lambda_j^H} = \frac{4s_j^2}{(2s_j c_j)^2} = \frac{1}{c_j^2} \quad \text{as well as} \quad \frac{\lambda_{j'}^h}{\lambda_j^H} = \frac{4c_j^2}{(2s_j c_j)^2} = \frac{1}{s_j^2}, \qquad j = 1, \ldots, n, \tag{32}$$

and therefore

$$C_j = \begin{bmatrix} 1 - c_j^2 & c_j^2 \\ s_j^2 & 1 - s_j^2 \end{bmatrix} = \begin{bmatrix} s_j^2 & c_j^2 \\ s_j^2 & c_j^2 \end{bmatrix}, \qquad j = 1, \ldots, n.$$

A matrix of the form $X = \begin{bmatrix} \xi & \eta \\ \xi & \eta \end{bmatrix}$ has the eigenvalues and spectral norm

$$\Lambda(X) = \{0, \xi + \eta\}, \tag{33a}$$

$$\|X\| = \|XX^\mathsf{T}\|^{1/2} = \sqrt{\xi^2 + \eta^2} \left\| \begin{bmatrix} 1 & 1 \\ 1 & 1 \end{bmatrix} \right\|^{1/2} = \sqrt{\xi^2 + \eta^2} \cdot \sqrt{2}. \tag{33b}$$

For C_j we thus obtain in the case of the Laplacian

$$\rho(C_j) = s_j^2 + c_j^2 = 1, \quad \|C_j\| = \sqrt{2(s_j^4 + c_j^4)}, \qquad j = 1, \ldots, n.$$

From $s_j^2 \in [0, \frac{1}{2}]$ for $j = 1, \ldots, n$ we infer the bound

$$\|C_j\| \leq \max_{0 \leq t \leq \frac{1}{2}} \sqrt{2[t^2 + (1-t)^2]} = \sqrt{2}, \qquad j = 1, \ldots, n.$$

In the Helmholtz case $k > 0$ the spectral analysis of the coarse grid correction operator C_j becomes more tedious and no simple closed-form expression exists for the spectral radius and norm of the 2×2 blocks C_j. We therefore resort to computation and consider the case of a fine mesh with $N = 31$ interior points. The left of Fig. 14 shows a stem plot of the eigenvalues of the 2×2 blocks of C for the Laplacian, which consist of ones and zeros, as C is an orthogonal projection in this case, see (33a). On the right of Fig. 14 we see the analogous plot for $k = 6.3\pi$. Note that the unit eigenvalues remain, but that the second eigenvalue of each pair is no longer zero. In particular, mode number 13 is amplified by a factor of nearly -4. This mode is well outside the oscillatory part of the spectrum, so that smoothing cannot be expected to offset such an error amplification. In the example we have shown in Fig. 9, we had chosen the parameters precisely such that the corresponding mode was multiplied by the factor -1, which led to the correct shape of the coarse grid correction, but with the wrong sign.

A simple device for obtaining a more effective coarse-grid correction for Helmholtz operators results from taking into account the dispersion properties of the discretization scheme. For our uniform centered finite-difference discretization of the 1D Helmholtz operator with constant k

$$\mathcal{L}u \approx \frac{1}{h^2} \left(-u_{j-1} + 2u_j - u_{j+1} \right) - k^2 u_j,$$

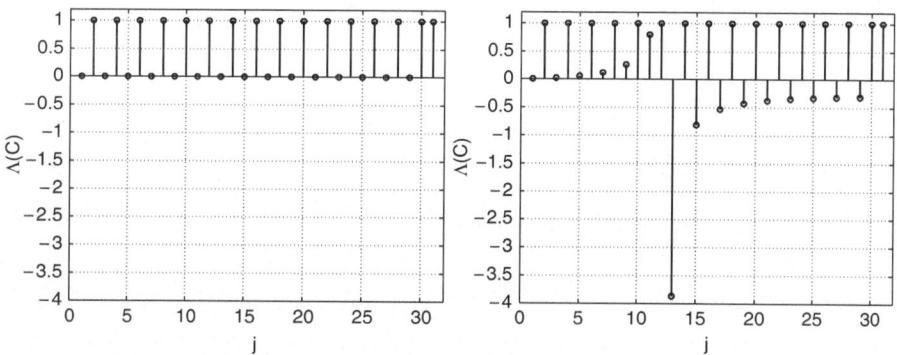

Fig. 14 Eigenvalues of the coarse-grid correction operator with respect to a fine mesh with $h = 1/32$ for $k = 0$ (*left*) and $k = 6.3\pi$ (*right*)

plane-wave solutions $e^{ik^h x_j}$ of the *discrete* homogeneous Helmholtz equation possess a discrete wave number k^h characterized by

$$\frac{k^h}{k} = \frac{1}{kh} \arccos\left(1 - \frac{k^2 h^2}{2}\right) > 1.$$

As a result, the discrete solution exhibits a *phase lead* with respect to the true solution, which grows with h. In the same way, coarse grid approximations in a multigrid hierarchy will be out of phase with fine-grid approximations. This suggests "slowing down" the waves on coarse grids in order that the coarse-grid correction again be in phase with the fine-grid approximation. For our example, this is achieved by using a modified wave number $\tilde{k}$ in the coarse-grid Helmholtz operator defined by the requirement

$$\tilde{k}^H = k, \quad \text{which is achieved by} \quad \tilde{k} = \sqrt{\frac{2(1 - \cos(kh))}{h^2}}.$$

An even better adjustment of the coarse-grid correction results from matching the coarse-grid discrete wave number k^H to the fine-grid discrete wave number k^h by choosing the modified wave number $\tilde{k}$ on the coarse grid to satisfy

$$\tilde{k}^H = k^h \quad \text{which is achieved by} \quad \tilde{k} = k\sqrt{1 - k^2 h^2/4}. \tag{34}$$

Choosing a modified wave number according to (34) is also equivalent to avoiding a possible "singularity" in the term $\lambda_j^h / \lambda_j^H$ in (31) by forcing the vanishing of λ_j^H as a continuous function of j to occur in the same location as for λ_j^h.

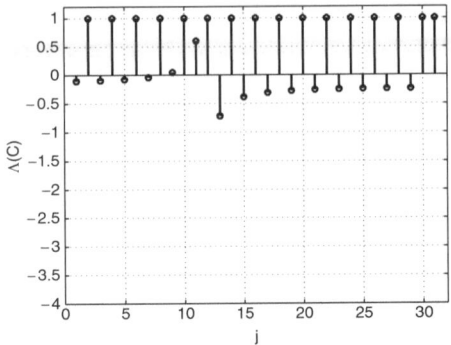

Fig. 15 Eigenvalues of the coarse-grid operator with respect to a fine mesh with $h = 1/32$ for $k = 6.3\pi$ using the modified wave number $\tilde{k}$ given in (34) in the coarse-grid Helmholtz operator

Figure 15 shows the eigenvalues of the coarse-grid correction operator depicted on the right of Fig. 14 with the modified wave number (34) used on the coarse grid. The strong amplification of mode number 13 is seen to be much less severe, all non-unit eigenvalues now being less than one in modulus.

Such a dispersion analysis can be carried out for all standard discretization schemes, and it is found that higher order schemes have much lower phase error (cf., e.g., [3]), making them a favorable choice also from the point of view of multigrid solvers. In higher dimensions higher order methods also possess nearly isotropic dispersion relations, a necessary requirement for (scalar) dispersion correction.

3.4.3 Two-Grid Operator

Two-grid iteration combines one or more smoothing steps with coarse-grid correction. If ν_1 and ν_2 denote the number of *presmoothing* and *postsmoothing* steps carried out before and after coarse-grid correction, the error propagation operator of the resulting *two-grid operator* is obtained as $T = S^{\nu_2}CS^{\nu_1}$. Choosing damped Jacobi iteration with relaxation factor ω as the smoothing operator, the results on the spectral analysis of the damped Jacobi smoother and coarse-grid correction allow us to decompose the analysis of the two-grid operator into the subspaces

$$\mathrm{span}\{\mathbf{v}_1, \mathbf{v}_N\}, \mathrm{span}\{\mathbf{v}_2, \mathbf{v}_{N-1}\}, \ldots, \mathrm{span}\{\mathbf{v}_n, \mathbf{v}_{n+2}\}, \mathrm{span}\{\mathbf{v}_{n+1}\}$$

of n pairs of complementary modes and the remaining "middle mode" $\mathbf{v}_{n+1}$. The action of T on these one- and two-dimensional subspaces is represented by the block diagonal matrix

$$T = \mathrm{diag}(T_1, \ldots, T_n, T_{n+1})$$

with

$$T_j = \begin{bmatrix} \sigma_j & 0 \\ 0 & \sigma_{j'} \end{bmatrix}^{\nu_2} \begin{bmatrix} 1 - c_j^4 \frac{\lambda_j^h}{\lambda_j^H} & c_j^2 s_j^2 \frac{\lambda_{j'}^h}{\lambda_j^H} \\ c_j^2 s_j^2 \frac{\lambda_j^h}{\lambda_j^H} & 1 - s_j^4 \frac{\lambda_{j'}^h}{\lambda_j^H} \end{bmatrix} \begin{bmatrix} \sigma_j & 0 \\ 0 & \sigma_{j'} \end{bmatrix}^{\nu_1} \qquad j = 1, \ldots, n, \qquad (35)$$

and

$$T_{n+1} = (1 - \omega)^{\nu_1 + \nu_2},$$

the latter resulting from $\sigma_{n+1} = 1 - \omega$ (cf. (20)).

We begin again with the case $k = 0$, in which, due to (32), the 2×2 blocks in (35) simplify to (see also [43])

$$T_j = \begin{bmatrix} \sigma_j & 0 \\ 0 & \sigma_{j'} \end{bmatrix}^{\nu_2} \begin{bmatrix} s_j^2 & c_j^2 \\ s_j^2 & c_j^2 \end{bmatrix} \begin{bmatrix} \sigma_j & 0 \\ 0 & \sigma_{j'} \end{bmatrix}^{\nu_1} \qquad \text{with} \quad \sigma_j = 1 - 2\omega s_j^2, \ \sigma_{j'} = 1 - 2\omega c_j^2.$$

Fixing $\nu_1 = \nu$ and $\nu_2 = 0$ (pre-smoothing only) and $\omega = \omega_0$ (cf. (21)), this becomes

$$T_j = \begin{bmatrix} s_j^2 \sigma_j^\nu & c_j^2 \sigma_{j'}^\nu \\ s_j^2 \sigma_j^\nu & c_j^2 \sigma_{j'}^\nu \end{bmatrix}, \quad j = 1, \ldots, n, \qquad T_{n+1} = \left(\frac{1}{3} \right)^\nu,$$

where

$$\sigma_j = \frac{1}{3} \left(3 - 4s_j^2 \right), \quad \sigma_{j'} = \frac{1}{3} \left(4s_j^2 - 1 \right), \qquad j = 1, \ldots, n.$$

Using (33a) we obtain for the spectral radius

$$\rho(T_j) = s_j^2 \sigma_j^\nu + c_j^2 \sigma_{j'}^\nu, \quad j = 1, \ldots, n, \qquad \rho(T_{n+1}) = 3^{-\nu}.$$

Noting that $c_j^2 = 1 - s_j^2$ and $s_j^2 \in [0, \frac{1}{2}]$ for all j, we obtain the upper bound

$$\rho(T_j) \leq R_\nu := \max_{0 \leq t \leq \frac{1}{2}} R_\nu(t), \quad R_\nu(t) := t \left(\frac{3 - 4t}{3} \right)^\nu + (1 - t) \left(\frac{4t - 1}{3} \right)^\nu$$

for $j = 1, \ldots, n$. Since $R_\nu(\frac{1}{2}) = \left(\frac{1}{3} \right)^\nu$ this bound holds also for T_{n+1}. A common upper bound for the spectral norms $\|T_j\|$ is obtained in an analogous way using (33b) as

$$\|T_j\| \leq N_\nu := \max_{0 \leq t \leq \frac{1}{2}} N_\nu(t), \quad N_\nu(t) := \sqrt{2 \left[t^2 \left(\frac{3 - 4t}{3} \right)^{2\nu} + (1 - t)^2 \left(\frac{4t - 1}{3} \right)^{2\nu} \right]},$$

Table 7 Spectral radius of the two-grid operator for the Helmholtz equation with $h = 1/32$ for varying wave number k and (pre) smoothing step number ν. *Left*: standard coarse-grid operator, *right*: with modified wave number on coarse grid

$\nu \backslash \rho(T)$	$k = 0$	$k = 1.3\pi$	$k = 4.3\pi$	$k = 6.3\pi$	$\nu \backslash \rho(T)$	$k = 1.3\pi$	$k = 4.3\pi$	$k = 6.3\pi$
1	0.3333	0.3364	0.4093	0.8857	1	0.3365	0.5050	0.6669
2	0.1111	0.1170	0.2391	1.8530	2	0.1173	0.1648	0.1999
3	0.0787	0.0779	0.2623	1.6455	3	0.0779	0.1012	0.1542
4	0.0617	0.0613	0.2481	1.6349	4	0.0614	0.0568	0.1761
5	0.0501	0.0493	0.2561	1.5832	5	0.0493	0.0591	0.2012
10	0.0263	0.0256	0.2668	1.3797	10	0.0257	0.0790	0.3916

which holds for all $j = 1 \ldots, n + 1$ due to $N_\nu(\frac{1}{2}) = \left(\frac{1}{3}\right)^\nu$.

Table 7 (left) gives the spectral radius of the two-grid operator for the Helmholtz equation with ν steps of presmoothing using damped Jacobi for a range of wave numbers k. We observe that the iteration is divergent for $k = 6.3\pi$, which corresponds to a resolution of roughly 10 points per wavelength. Moreover, while additional smoothing steps resulted in a faster convergence rate for k close to zero, this is no longer the case for higher wave numbers. Table 7 (right) gives the spectral radius of the same two-grid operator using the modified wave number according to (34) on the coarse grid. We observe that, even for the unstable damped Jacobi smoother, this results in a convergent two-grid method in this example. A more complete analysis of the potential and limitations of this approach is the subject of a forthcoming paper.

3.5 The Shifted Laplacian Preconditioner

An idea proposed in [21], going back to [5], which has received a lot of attention over the last few years, see for example the references in [41], is to precondition the Helmholtz equation (1) using a Helmholtz operator with a rescaled complex wave number,

$$\mathcal{L}_s := -(\Delta + (\alpha + i\beta)k^2), \tag{36}$$

i.e., where damping has been added. The main idea here is that if the imaginary shift β is large enough, standard multigrid methods are known to work again, and, if the shift is not too large and $\alpha \approx 1$, the shifted operator should still be a good preconditioner for the original Helmholtz operator, where $\alpha = 1$ and $\beta = 0$. We show here quantitatively these two contradicting requirements for the one-dimensional case on the unit interval with homogeneous Dirichlet boundary conditions and a finite difference discretization. In that case, both the Helmholtz operator and the shifted Helmholtz preconditioner can be diagonalized using a Fourier sine series, as we have seen in Sect. 3.4, and we obtain for the corresponding symbols (or eigenvalues)

$$\hat{\mathscr{L}}^h = \frac{2}{h^2}(1 - \cos j\pi h) - k^2, \quad \hat{\mathscr{L}}^h_s = \frac{2}{h^2}(1 - \cos j\pi h) - (\alpha + i\beta)k^2,$$

$$j = 1, \ldots, N.$$

Hence the preconditioned operator $(\mathscr{L}^h_s)^{-1}\mathscr{L}^h$ has the symbol

$$\frac{\hat{\mathscr{L}}^h}{\hat{\mathscr{L}}^h_s} = \frac{-2 + 2\cos j\pi h + h^2 k^2}{-2 + 2\cos j\pi h + h^2 k^2(\alpha + i\beta)}.$$

The spectrum of the preconditioned operator therefore lies on a circle in the complex plane, which passes through $(0,0)$, and if $\alpha = 1$, the center is at $(\frac{1}{2}, 0)$ and the radius equals $\frac{1}{2}$, as one can see using a direct calculation. Examples are shown in Fig. 16. Since the circle passes through $(0,0)$ when the numerator of the symbol of the preconditioned operator vanishes, i.e., when

$$2\cos j\pi h + h^2 k^2 = 2, \tag{37}$$

the spectrum of the preconditioned operator is potentially unfavorable for a Krylov method, as one can see in Fig. 16 on the right. For $\alpha = 1$ and β small, we have

$$\frac{\hat{\mathscr{L}}^h}{\hat{\mathscr{L}}^h_s} = 1 - i\frac{k^2 h^2}{-2 + 2\cos j\pi h + k^2 h^2}\beta + O(\beta^2),$$

which shows that the spectrum is clustered on an arc of the circle around $(1,0)$, as illustrated in Fig. 16 on the left, provided $\beta \ll \min_{j=1,\ldots,n} |-2 + 2\cos j\pi h + h^2 k^2|$. How small must we therefore choose β? A direct calculation shows that we must choose $\beta < \frac{1}{k}$ in order to obtain a spectrum clustered about $(1,0)$. We show in

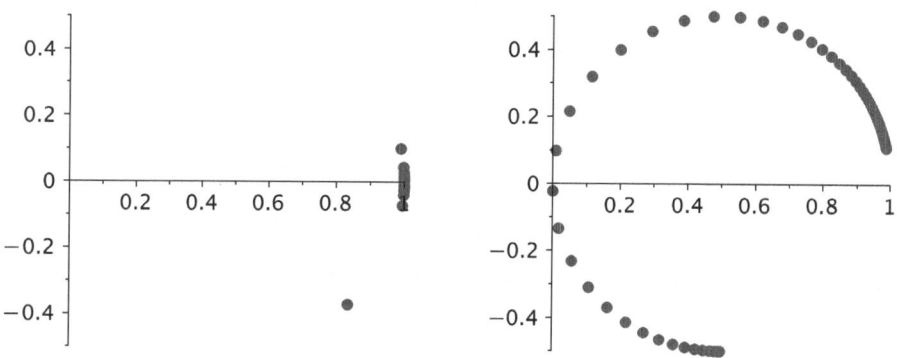

Fig. 16 Spectrum of the Helmholtz operator preconditioned with the shifted Laplacian preconditioner with $\alpha = 0$ and $\beta = 0.01$ on the *left*, and $\beta = 1$ on the *right*. The spectrum clustered around the point $(1,0)$ on the *left* is favorable for a Krylov method, while the spectrum on the *right* is not, due to the eigenvalues close to zero

Fig. 17 an illustration of this fact: from (37), we can compute a critical j where the
spectrum vanishes,

$$j_c = \frac{1}{\pi h}(\pi - \arccos(-1 + \frac{1}{2}k^2 h^2)).$$

The spectrum being restricted to integer j, we can plot

$$d := -2 + 2\cos j_c \pi h + k^2 h^2,$$

in order to get an impression of the size of this quantity. We see in Fig. 17 that the
minimum distance d (oscillatory curve) behaves like $1/k$ (smooth curve), and thus
β needs to be chosen smaller than $1/k$ for a given problem if one wants to obtain a
spectrum of the preconditioned operator close to $(1, 0)$.

Now, is it possible to solve the shifted Helmholtz equation effectively with
multigrid for this choice of β? In order to investigate this, we use the two-grid
analysis from Sect. 3.4, now applied to the shifted Laplace problem. We show in
Fig. 18 the spectral radius of the two grid iteration operator for each frequency pair
in (30), for $k = 10, 100, 1000$ using 10 points per wavelength on the fine grid,
choosing in each case $\beta = 1/k$. This numerical experiment shows clearly that,

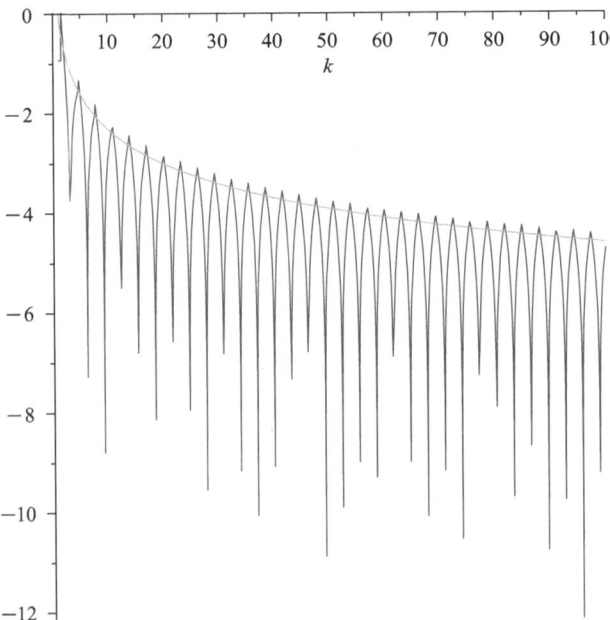

Fig. 17 Illustration of how small β has to be chosen in the shifted Helmholtz preconditioner in
order to remain an effective preconditioner for the Helmholtz equation. Note the log scale on the
y-axis

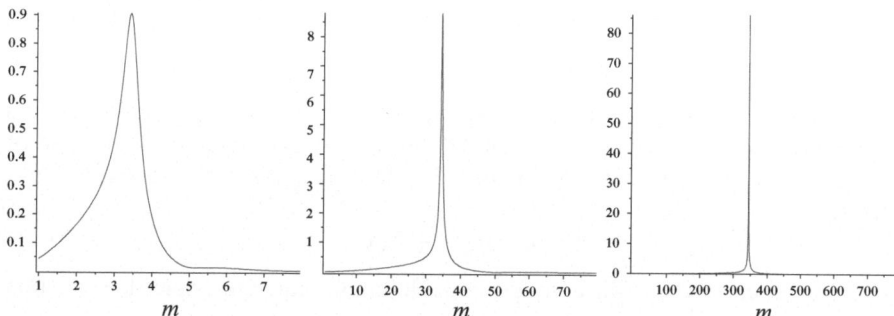

Fig. 18 Spectral radius of the two grid iteration operator for all frequency pairs. On the *left* for $k = 10$, in the *middle* for $k = 100$ and on the *right* for $k = 1,000$, with the shift $\beta = 1/k$ in order to guarantee a spectrum away from $(0,0)$ of the Helmholtz operator preconditioned by the shifted Laplace preconditioner

unfortunately, for the multigrid method to converge when applied to the shifted Laplace operator, β can not be chosen to satisfy $\beta < 1/k$, since already for $\beta = 1/k$ the contraction factor ρ of multigrid grows like $\rho \sim k$ (note the different scaling on the axes in Fig. 18) and thus the method is not convergent. One can furthermore show that β must be a constant, independent of k, in order to obtain a contraction factor $\rho < 1$ and a convergent multigrid algorithm. These results suggest a linear dependence on the wave number k of such a method, which is also observed numerically, see for example [20].

3.6 Wave-Ray Multigrid

In [8] Brandt and Livshits proposed a variant of multigrid especially tailored to the Helmholtz equation by exploiting the structure of the error components which standard multigrid methods fail to eliminate. These are the so-called *characteristic components*, which are discrete representations of functions of the form

$$u(x, y) = v(x, y)e^{ik_1 x + ik_2 y}, \qquad k_1^2 + k_2^2 = k^2. \tag{38}$$

Such factorizations are common in geometrical optics (see, e.g. [49, 52]), and from there the terminology *ray function* for the amplitude term $v(x, y)$ and *phase* for the exponent $k_1 x + k_2 y$ is adopted. Characteristic components of the error are nearly invisible to standard smoothing techniques since they have very small residuals on grids which resolve these oscillations. On coarser grids they are contaminated by phase errors and ultimately by approximation errors.

The ray functions, however, are smooth, and satisfy a convection-diffusion-type PDE, called the *ray equation*, which is obtained by inserting (38) into the Helmholtz equation. In their *wave-ray multigrid method*, Brandt and Livshits add so-called *ray cycles* to the standard multigrid scheme, in which the ray functions of principal components are approximated by performing smoothing with respect to the ray equation on auxiliary grids which they call *ray grids*.

We describe the basic idea for the simple 1D model problem (15) with constant wave number k as first described in Livshits' Ph.D. thesis [55]. Multidimensional generalizations such as described in [8] introduce a considerable number of technical and algorithmic complications. In 1D principal error components have the form

$$v(x) = a(x)e^{ikx} + b(x)e^{-ikx},$$

which, when inserted into the homogeneous Helmholtz equation, yields the equation

$$\left(a''(x) + 2ika'(x)\right)e^{ikx} + \left(b''(x) - 2ikb'(x)\right)e^{-ikx} = 0$$

which we separate into

$$L_+a = a'' + 2ika' = 0, \qquad L_-b = b'' - 2ikb' = 0.$$

The wave-ray method employs a standard multigrid scheme, say, a V-cycle, to first eliminate the non-characteristic components from the error $\mathbf{e}^h$, such that the associated residual $\mathbf{r}^h = A^h\mathbf{e}^h$ is approximately of the form

$$\mathbf{r}_j^h = (r_a^h)_j e^{ikx_j} + (r_b^h)_j e^{-ikx_j},$$

with smooth ray grid functions r_a^h and r_b^h. By a process called *separation* the two components of the residual are first isolated, resulting in the right hand sides of the two ray equations

$$L_+^h a^h = f_+^h, \quad L_-^h b^h = f_b^h,$$

which are each solved on separate grids and then used to construct a correction of the current approximation.

Details of the separation technique, the treatment of multidirectional rays necessary for higher space dimensions, suitable cycling schedules as well as the incorporation of radiation boundary conditions can be found in [8, 55].

4 Conclusions

Solving the indefinite Helmholtz equation by iterative methods is a difficult task. In all classical methods, the special oscillatory and non-local structure of the associated Green's function leads to severe convergence problems. Specialized methods exist for all well known classes of iterative methods, like preconditioned Krylov methods by incomplete factorizations, domain decomposition and multigrid, but they need additional components tailored for the indefinite Helmholtz problem, which can become very sophisticated and difficult to implement, especially if one wants to achieve a performance independent of the wave number k.

Acknowledgements The authors would like to acknowledge the support of the Swiss National Science Foundation Grant number 200020-121828.

References

1. Y. Achdou and F. Nataf. Dimension-wise iterated frequency filtering decomposition. *Num. Lin. Alg. and Appl.*, 41(5):1643–1681, 2003.
2. Y. Achdou and F. Nataf. Low frequency tangential filtering decomposition. *Num. Lin. Alg. and Appl.*, 14:129–147, 2007.
3. M. Ainsworth. Discrete dispersion relation for hp-version finite element approximation at high wave number. *SIAM J. Numer. Anal.*, 42(2):553–575, 2004.
4. N. Bakhvalov. Convergence of one relaxation method under natural constraints on the elliptic operator. *Zh. Vychisl. Mat. Mat. Fiz.*, 6:861–883, 1966.
5. A. Bayliss, C. Goldstein, and E. Turkel. An iterative method for the Helmholtz equation. *J. Comput. Phys.*, 49:443–457, 1983.
6. T. Betcke, S.N., Chandler-Wilde, I. Graham, S. Langdon, and M. Lindner. Condition number estimates for combined potential operators in acoustics and their boundary element discretisation. *Numerical Methods for PDEs*, 27(1):31–69, 2010.
7. A. Brandt. Multigrid techniques: 1984 guide with applications to fluid dynamics. GMD Studien 55, Gesellschaft für Mathematik und Datenverarbeitung, St. Augustin, Bonn, 1984.
8. A. Brandt and I. Livshits. Wave-ray multigrid method for standing wave equations. *Electron. Trans. Numer. Anal.*, 6:162–181, 1997.
9. A. Brandt and S. Ta'asan. Multigrid methods for nearly singular and slightly indefinite problems. In *Multigrid Methods II*, pages 99–121. Springer, 1986.
10. W. L. Briggs, V. E. Henson, and S. F. McCormick. *A Multigrid Tutorial*. SIAM, 2000.
11. B. Buzbee, F. W. Dorr, J. A. George, and G. Golub. The direct solution of the discrete Poisson equation on irregular regions. *SIAM J. Numer. Anal.*, 8:722–736, 1974.
12. A. Buzdin. Tangential decomposition. *Computing*, 61:257–276, 1998.
13. D. Colton and R. Kress. *Integral Equation Methods in Scattering Theory*. Wiley, New York, 1983.
14. L. Debreu and E. Blayo. On the Schwarz alternating method for ocean models on parallel computers. *J. Computational Physics*, 141:93–111, 1998.
15. B. Després. *Méthodes de décomposition de demains pour les problèms de propagation d'ondes en régime harmonique*. PhD thesis, Université Paris IX Dauphine, 1991.
16. H. C. Elman, O. G. Ernst, and D. P. O'Leary. A multigrid method enhanced by Krylov subspace iteration for discrete Helmholtz equations. *SIAM J. Sci. Comp.*, 23:1290–1314, 2001.
17. B. Engquist and L. Ying. Sweeping preconditioner for the Helmholtz equation: Hierarchical matrix representation. *Preprint*, 2010.
18. B. Engquist and L. Ying. Sweeping preconditioner for the Helmholtz equation: Moving perfectly matched layers. *Preprint*, 2010.
19. B. Engquist and L. Ying. Fast algorithms for high-frequency wave propagation. *this volume*, page 127, 2011.
20. Y. Erlangga. Advances in iterative methods and preconditioners for the Helmholtz equation. *Archives Comput. Methods in Engin.*, 15:37–66, 2008.
21. Y. Erlangga, C. Vuik, and C. Oosterlee. On a class of preconditioners for solving the Helmholtz equation. *Applied Numerical Mathematics*, 50:409–425, 2004.

22. O. G. Ernst. *Fast Numerical Solution of Exterior Helmholtz Problems with Radiation Boundary Condition by Imbedding*. PhD thesis, Stanford University, 1994.

23. O. G. Ernst. A finite element capacitance matrix method for exterior Helmholtz problems. *Numer. Math.*, 75:175–204, 1996.

24. O. G. Ernst. Residual-minimizing Krylov subspace methods for stabilized discretizations of convection-diffusion equations. *SIAM J. Matrix Anal. Appl.*, 22(4):1079–1101, 2000.

25. C. Farhat, P. Avery, R. Tesaur, and J. Li. FETI-DPH: a dual-primal domain decomposition method for acoustic scattering. *Journal of Computational Acoustics*, 13(3):499–524, 2005.

26. C. Farhat, P. Avery, R. Tezaur, and J. Li. FETI-DPH: a dual-primal domain decomposition method for accoustic scattering. *Journal of Computational Acoustics*, 13:499–524, 2005.

27. C. Farhat, A. Macedo, and R. Tezaur. FETI-H: a scalable domain decomposition method for high frequency exterior Helmholtz problem. In C.-H. Lai, P. Bjørstad, M. Cross, and O. Widlund, editors, *Eleventh International Conference on Domain Decomposition Method*, pages 231–241. DDM.ORG, 1999.

28. C. Farhat and F.-X. Roux. A method of Finite Element Tearing and Interconnecting and its parallel solution algorithm. *Int. J. Numer. Meth. Engrg.*, 32:1205–1227, 1991.

29. R. Federenko. A relaxation method for solving elliptic difference equations. *USSR Comput. Math. and Math. Phys.*, 1(5):1092–1096, 1961.

30. R. W. Freund and N. M. Nachtigal. QMR: a quasi-minimal residual method for non-Hermitian linear systems. *Numer. Math.*, 60:315–339, 1991.

31. M. J. Gander. Optimized Schwarz methods. *SIAM J. Numer. Anal.*, 44(2):699–731, 2006.

32. M. J. Gander. Schwarz methods over the course of time. *Electronic Transactions on Numerical Analysis (ETNA)*, 31:228–255, 2008.

33. M. J. Gander, L. Halpern, and F. Magoulès. An optimized Schwarz method with two-sided robin transmission conditions for the Helmholtz equation. *Int. J. for Num. Meth. in Fluids*, 55(2):163–175, 2007.

34. M. J. Gander, L. Halpern, and F. Nataf. Optimized Schwarz methods. In T. Chan, T. Kako, H. Kawarada, and O. Pironneau, editors, *Twelfth International Conference on Domain Decomposition Methods, Chiba, Japan*, pages 15–28, Bergen, 2001. Domain Decomposition Press.

35. M. J. Gander, F. Magoulès, and F. Nataf. Optimized Schwarz methods without overlap for the Helmholtz equation. *SIAM J. Sci. Comput.*, 24(1):38–60, 2002.

36. M. J. Gander, V. Martin, and J.-P. Chehab. GMRES convergence analysis for diffusion, convection diffusion and Helmholtz problems. *In preparation*, 2011.

37. M. J. Gander and F. Nataf. AILU: A preconditioner based on the analytic factorization of the elliptic operator. *Num. Lin. Alg. and Appl.*, 7(7-8):543–567, 2000.

38. M. J. Gander and F. Nataf. An incomplete LU preconditioner for problems in acoustics. *Journal of Computational Acoustics*, 13(3):1–22, 2005.

39. M. J. Gander and G. Wanner. From Euler, Ritz and Galerkin to modern computing. *SIAM Review*, 2011. to appear.

40. M. J. Gander and H. Zhang. Domain Decomposition Methods for the Helmholtz equation: A Numerical Study, submitted to Domain Decomposition Methods in Science and Engineering XX, Lecture Notes in Computational Science and Engineering, Springer-Verlag, 2012.

41. M. V. Gijzen, Y. Erlangga, and C. Vuik. Spectral analysis of the discrete Helmholtz operator preconditioned with a shifted Laplacian. *SIAM J. Sci. Comput.*, 29(5):1942–1958, 2007.

42. E. Giladi and J. B. Keller. Iterative solution of elliptic problems by approximate factorization. *Journal of Computational and Applied Mathematics*, 85:287–313, 1997.

43. W. Hackbusch. *Multi-Grid Methods and Applications*. Springer-Verlag, 1985.

44. T. Hagstrom, R. P. Tewarson, and A. Jazcilevich. Numerical experiments on a domain decomposition algorithm for nonlinear elliptic boundary value problems. *Appl. Math. Lett.*, 1(3), 1988.

45. E. Heikkola, J. Toivanen, and T. Rossi. A parallel fictitious domain method for the three-dimensional Helmholtz equation. *SIAM J. Sci. Comput*, 24(5):1567–1588, 2003.

46. M. A. Hyman. Non-iterative numerical solution of boundary-value problems. *Appl. Sci. Res. Sec. B2*, 2:325–351, 1952.
47. F. Ihlenburg and I. Babuška. Finite element solution to the Helmholtz equation with high wave number. Part I: The *h*-version of the FEM. *Computer Methods in Applied Mechanics and Engineering*, 39:9–37, 1995.
48. F. Ihlenburg and I. Babuška. Finite element solution to the Helmholtz equation with high wave number. Part II: The *h-p* version of the FEM. *SIAM Journal on Numerical Analysis*, 34:315–358, 1997.
49. J. B. Keller. Rays, waves and asymptotics. *Bull. Amer. Math. Soc.*, 84(5):727–750, 1978.
50. B. Lee, T. Manteuffel, S. McCormick, and J. Ruge. First-order system least squares (FOSLS) for the Helmholtz equation. *SIAM J. Sci. Comp.*, 21:1927–1949, 2000.
51. L. M. Leslie and B. J. McAveney. Comparative test of direct and iterative methods for solving Helmholtz-type equations. *Mon. Wea. Rev.*, 101:235–239, 1973.
52. R. M. Lewis. Asymptotic theory of wave-propagation. *Archive for Rational Mechanics and Analysis*, 20(3):191–250, 1965.
53. J. Liesen and Z. Strakoš. GMRES convergence analysis for a convection-diffusion model problem. *SIAM J. Sci. Comput*, 26(6):1989–2009, 2005.
54. P.-L. Lions. On the Schwarz alternating method. III: a variant for nonoverlapping subdomains. In T. F. Chan, R. Glowinski, J. Périaux, and O. Widlund, editors, *Third International Symposium on Domain Decomposition Methods for Partial Differential Equations , held in Houston, Texas, March 20-22, 1989*, Philadelphia, PA, 1990. SIAM.
55. I. Livshits. *Multigrid Solvers for Wave Equations*. PhD thesis, Bar-Ilan University, Ramat-Gan, Israel, 1996.
56. F. Nataf. Résolution de l'équation de convection-diffusion stationaire par une factorisation parabolique. *C. R. Acad. Sci.*, I 310(13):869–872, 1990.
57. Q. Niu, L. Grigori, P. Kumar, and F. Nataf. Modified tangential frequency filtering decomposition and its Fourier analysis. *Numerische Mathematik*, 116(1), 2010.
58. W. Proskurowski and O. B. Widlund. On the numerical solution of Helmholtz's equation by the capacitance matrix method. *Math. Comp.*, 30:433–468, 1976.
59. T. E. Rosmond and F. D. Faulkner. Direct solution of elliptic equations by block cyclic reduction and factorization. *Mon. Wea. Rev.*, 104:641–649, 1976.
60. Y. Saad. *Iterative Methods for Sparse Linear Systems*. PWS Publishing Company, 1996.
61. V. Saul'ev. On solution of some boundary value problems on high performance computers by fictitious domain method. *Siberian Math. J.*, 4:912–925, 1963 (in Russian).
62. H. A. Schwarz. Über einen Grenzübergang durch alternierendes Verfahren. *Vierteljahrsschrift der Naturforschenden Gesellschaft in Zürich*, 15:272–286, May 1870.
63. R. V. Southwell. *Relaxation Methods in Theoretical Physics*. Oxford University Press, 1946.
64. E. Stiefel. Über einige Methoden der Relaxationsrechnung. *Z. Angew. Math. Phys.*, 3:1–33, 1952.
65. C. Wagner. Tangential frequency filtering decompositions for symmetric matrices. *Numer. Math.*, 78(1):119–142, 1997.
66. C. Wagner. Tangential frequency filtering decompositions for unsymmetric matrices. *Numer. Math.*, 78(1):143–163, 1997.
67. G. Wittum. An ILU-based smoothing correction scheme. In *Parallel algorithms for partial differential equations*, volume 31, pages 228–240. Notes Numer. Fluid Mech., 1991. 6th GAMM-Semin., Kiel/Ger.
68. G. Wittum. *Filternde Zerlegungen. Schnelle Löser für grosse Gleichungssysteme*. Teubner Skripten zur Numerik, Stuttgart, 1992.
69. E. Zauderer. *Partial Differential Equations of Applied Mathematics*. John Wiley & Sons, second edition, 1989.

Editorial Policy

1. Volumes in the following three categories will be published in LNCSE:

i)   Research monographs
ii) Tutorials
iii) Conference proceedings

Those considering a book which might be suitable for the series are strongly advised to contact the publisher or the series editors at an early stage.

2. Categories i) and ii). Tutorials are lecture notes typically arising via summer schools or similar events, which are used to teach graduate students. These categories will be emphasized by Lecture Notes in Computational Science and Engineering. **Submissions by interdisciplinary teams of authors are encouraged.** The goal is to report new developments – quickly, informally, and in a way that will make them accessible to non-specialists. In the evaluation of submissions timeliness of the work is an important criterion. Texts should be well-rounded, well-written and reasonably self-contained. In most cases the work will contain results of others as well as those of the author(s). In each case the author(s) should provide sufficient motivation, examples, and applications. In this respect, Ph.D. theses will usually be deemed unsuitable for the Lecture Notes series. Proposals for volumes in these categories should be submitted either to one of the series editors or to Springer-Verlag, Heidelberg, and will be refereed. A provisional judgement on the acceptability of a project can be based on partial information about the work: a detailed outline describing the contents of each chapter, the estimated length, a bibliography, and one or two sample chapters – or a first draft. A final decision whether to accept will rest on an evaluation of the completed work which should include

– at least 100 pages of text;
– a table of contents;
– an informative introduction perhaps with some historical remarks which should be accessible to readers unfamiliar with the topic treated;
– a subject index.

3. Category iii). Conference proceedings will be considered for publication provided that they are both of exceptional interest and devoted to a single topic. One (or more) expert participants will act as the scientific editor(s) of the volume. They select the papers which are suitable for inclusion and have them individually refereed as for a journal. Papers not closely related to the central topic are to be excluded. Organizers should contact the Editor for CSE at Springer at the planning stage, see *Addresses* below.

In exceptional cases some other multi-author-volumes may be considered in this category.

4. Only works in English will be considered. For evaluation purposes, manuscripts may be submitted in print or electronic form, in the latter case, preferably as pdf- or zipped ps-files. Authors are requested to use the LaTeX style files available from Springer at http:// www. springer.com/authors/book+authors?SGWID=0-154102-12-417900-0.

For categories ii) and iii) we strongly recommend that all contributions in a volume be written in the same LaTeX version, preferably LaTeX2e. Electronic material can be included if appropriate. Please contact the publisher.

Careful preparation of the manuscripts will help keep production time short besides ensuring satisfactory appearance of the finished book in print and online.

5. The following terms and conditions hold. Categories i), ii) and iii):

Authors receive 50 free copies of their book. No royalty is paid.
Volume editors receive a total of 50 free copies of their volume to be shared with authors, but no royalties.

Authors and volume editors are entitled to a discount of 33.3 % on the price of Springer books purchased for their personal use, if ordering directly from Springer.

6. Commitment to publish is made by letter of intent rather than by signing a formal contract. Springer-Verlag secures the copyright for each volume.

Addresses:

Timothy J. Barth
NASA Ames Research Center
NAS Division
Moffett Field, CA 94035, USA
barth@nas.nasa.gov

Michael Griebel
Institut für Numerische Simulation
der Universität Bonn
Wegelerstr. 6
53115 Bonn, Germany
griebel@ins.uni-bonn.de

David E. Keyes
Mathematical and Computer Sciences
and Engineering
King Abdullah University of Science
and Technology
P.O. Box 55455
Jeddah 21534, Saudi Arabia
david.keyes@kaust.edu.sa

and

Department of Applied Physics
and Applied Mathematics
Columbia University
500 W. 120 th Street
New York, NY 10027, USA
kd2112@columbia.edu

Risto M. Nieminen
Department of Applied Physics
Aalto University School of Science
and Technology
00076 Aalto, Finland
risto.nieminen@tkk.fi

Dirk Roose
Department of Computer Science
Katholieke Universiteit Leuven
Celestijnenlaan 200A
3001 Leuven-Heverlee, Belgium
dirk.roose@cs.kuleuven.be

Tamar Schlick
Department of Chemistry
and Courant Institute
of Mathematical Sciences
New York University
251 Mercer Street
New York, NY 10012, USA
schlick@nyu.edu

Editor for Computational Science
and Engineering at Springer:
Martin Peters
Springer-Verlag
Mathematics Editorial IV
Tiergartenstrasse 17
69121 Heidelberg, Germany
martin.peters@springer.com

Lecture Notes
in Computational Science
and Engineering

23. L.F. Pavarino, A. Toselli (eds.), *Recent Developments in Domain Decomposition Methods.*

24. T. Schlick, H.H. Gan (eds.), *Computational Methods for Macromolecules: Challenges and Applications.*

25. T.J. Barth, H. Deconinck (eds.), *Error Estimation and Adaptive Discretization Methods in Computational Fluid Dynamics.*

26. M. Griebel, M.A. Schweitzer (eds.), *Meshfree Methods for Partial Differential Equations.*

27. S. Müller, *Adaptive Multiscale Schemes for Conservation Laws.*

28. C. Carstensen, S. Funken, W. Hackbusch, R.H.W. Hoppe, P. Monk (eds.), *Computational Electromagnetics.*

29. M.A. Schweitzer, *A Parallel Multilevel Partition of Unity Method for Elliptic Partial Differential Equations.*

30. T. Biegler, O. Ghattas, M. Heinkenschloss, B. van Bloemen Waanders (eds.), *Large-Scale PDE-Constrained Optimization.*

31. M. Ainsworth, P. Davies, D. Duncan, P. Martin, B. Rynne (eds.), *Topics in Computational Wave Propagation.* Direct and Inverse Problems.

32. H. Emmerich, B. Nestler, M. Schreckenberg (eds.), *Interface and Transport Dynamics.* Computational Modelling.

33. H.P. Langtangen, A. Tveito (eds.), *Advanced Topics in Computational Partial Differential Equations.* Numerical Methods and Diffpack Programming.

34. V. John, *Large Eddy Simulation of Turbulent Incompressible Flows.* Analytical and Numerical Results for a Class of LES Models.

35. E. Bänsch (ed.), *Challenges in Scientific Computing - CISC 2002.*

36. B.N. Khoromskij, G. Wittum, *Numerical Solution of Elliptic Differential Equations by Reduction to the Interface.*

37. A. Iske, *Multiresolution Methods in Scattered Data Modelling.*

38. S.-I. Niculescu, K. Gu (eds.), *Advances in Time-Delay Systems.*

39. S. Attinger, P. Koumoutsakos (eds.), *Multiscale Modelling and Simulation.*

40. R. Kornhuber, R. Hoppe, J. Périaux, O. Pironneau, O. Wildlund, J. Xu (eds.), *Domain Decomposition Methods in Science and Engineering.*

41. T. Plewa, T. Linde, V.G. Weirs (eds.), *Adaptive Mesh Refinement – Theory and Applications.*

42. A. Schmidt, K.G. Siebert, *Design of Adaptive Finite Element Software.* The Finite Element Toolbox ALBERTA.

43. M. Griebel, M.A. Schweitzer (eds.), *Meshfree Methods for Partial Differential Equations II.*

44. B. Engquist, P. Lötstedt, O. Runborg (eds.), *Multiscale Methods in Science and Engineering.*

45. P. Benner, V. Mehrmann, D.C. Sorensen (eds.), *Dimension Reduction of Large-Scale Systems.*

46. D. Kressner, *Numerical Methods for General and Structured Eigenvalue Problems.*

47. A. Boriçi, A. Frommer, B. Joó, A. Kennedy, B. Pendleton (eds.), *QCD and Numerical Analysis III.*

48. F. Graziani (ed.), *Computational Methods in Transport.*

49. B. Leimkuhler, C. Chipot, R. Elber, A. Laaksonen, A. Mark, T. Schlick, C. Schütte, R. Skeel (eds.), *New Algorithms for Macromolecular Simulation.*

50. M. Bücker, G. Corliss, P. Hovland, U. Naumann, B. Norris (eds.), *Automatic Differentiation: Applications, Theory, and Implementations.*

51. A.M. Bruaset, A. Tveito (eds.), *Numerical Solution of Partial Differential Equations on Parallel Computers.*

52. K.H. Hoffmann, A. Meyer (eds.), *Parallel Algorithms and Cluster Computing.*

53. H.-J. Bungartz, M. Schäfer (eds.), *Fluid-Structure Interaction.*

54. J. Behrens, *Adaptive Atmospheric Modeling.*

55. O. Widlund, D. Keyes (eds.), *Domain Decomposition Methods in Science and Engineering XVI.*

56. S. Kassinos, C. Langer, G. Iaccarino, P. Moin (eds.), *Complex Effects in Large Eddy Simulations.*

57. M. Griebel, M.A Schweitzer (eds.), *Meshfree Methods for Partial Differential Equations III.*

58. A.N. Gorban, B. Kégl, D.C. Wunsch, A. Zinovyev (eds.), *Principal Manifolds for Data Visualization and Dimension Reduction.*

59. H. Ammari (ed.), *Modeling and Computations in Electromagnetics: A Volume Dedicated to Jean-Claude Nédélec.*

60. U. Langer, M. Discacciati, D. Keyes, O. Widlund, W. Zulehner (eds.), *Domain Decomposition Methods in Science and Engineering XVII.*

61. T. Mathew, *Domain Decomposition Methods for the Numerical Solution of Partial Differential Equations.*

62. F. Graziani (ed.), *Computational Methods in Transport: Verification and Validation.*

63. M. Bebendorf, *Hierarchical Matrices. A Means to Efficiently Solve Elliptic Boundary Value Problems.*

64. C.H. Bischof, H.M. Bücker, P. Hovland, U. Naumann, J. Utke (eds.), *Advances in Automatic Differentiation.*

65. M. Griebel, M.A. Schweitzer (eds.), *Meshfree Methods for Partial Differential Equations IV.*

66. B. Engquist, P. Lötstedt, O. Runborg (eds.), *Multiscale Modeling and Simulation in Science.*

67. I.H. Tuncer, Ü. Gülcat, D.R. Emerson, K. Matsuno (eds.), *Parallel Computational Fluid Dynamics 2007.*

68. S. Yip, T. Diaz de la Rubia (eds.), *Scientific Modeling and Simulations.*

69. A. Hegarty, N. Kopteva, E. O'Riordan, M. Stynes (eds.), *BAIL 2008 – Boundary and Interior Layers.*

70. M. Bercovier, M.J. Gander, R. Kornhuber, O. Widlund (eds.), *Domain Decomposition Methods in Science and Engineering XVIII.*

71. B. Koren, C. Vuik (eds.), *Advanced Computational Methods in Science and Engineering.*

72. M. Peters (ed.), *Computational Fluid Dynamics for Sport Simulation.*

73. H.-J. Bungartz, M. Mehl, M. Schäfer (eds.), *Fluid Structure Interaction II - Modelling, Simulation, Optimization.*

74. D. Tromeur-Dervout, G. Brenner, D.R. Emerson, J. Erhel (eds.), *Parallel Computational Fluid Dynamics 2008.*

75. A.N. Gorban, D. Roose (eds.), *Coping with Complexity: Model Reduction and Data Analysis.*

76. J.S. Hesthaven, E.M. Rønquist (eds.), *Spectral and High Order Methods for Partial Differential Equations.*

77. M. Holtz, *Sparse Grid Quadrature in High Dimensions with Applications in Finance and Insurance.*

78. Y. Huang, R. Kornhuber, O.Widlund, J. Xu (eds.), *Domain Decomposition Methods in Science and Engineering XIX.*

79. M. Griebel, M.A. Schweitzer (eds.), *Meshfree Methods for Partial Differential Equations V.*

80. P.H. Lauritzen, C. Jablonowski, M.A. Taylor, R.D. Nair (eds.), *Numerical Techniques for Global Atmospheric Models.*

81. C. Clavero, J.L. Gracia, F.J. Lisbona (eds.), *BAIL 2010 – Boundary and Interior Layers, Computational and Asymptotic Methods.*

82. B. Engquist, O. Runborg, Y.R. Tsai (eds.), *Numerical Analysis and Multiscale Computations.*

83. I.G. Graham, T.Y. Hou, O. Lakkis, R. Scheichl (eds.), *Numerical Analysis of Multiscale Problems.*

For further information on these books please have a look at our mathematics catalogue at the following URL: www.springer.com/series/3527

Monographs in Computational Science and Engineering

1. J. Sundnes, G.T. Lines, X. Cai, B.F. Nielsen, K.-A. Mardal, A. Tveito, *Computing the Electrical Activity in the Heart.*

For further information on this book, please have a look at our mathematics catalogue at the following URL: www.springer.com/series/7417

Texts in Computational Science and Engineering

1. H. P. Langtangen, *Computational Partial Differential Equations.* Numerical Methods and Diffpack Programming. 2nd Edition

2. A. Quarteroni, F. Saleri, P. Gervasio, *Scientific Computing with MATLAB and Octave.* 3rd Edition

3. H. P. Langtangen, *Python Scripting for Computational Science.* 3rd Edition

4. H. Gardner, G. Manduchi, *Design Patterns for e-Science.*

5. M. Griebel, S. Knapek, G. Zumbusch, *Numerical Simulation in Molecular Dynamics.*

6. H. P. Langtangen, *A Primer on Scientific Programming with Python.* 2nd Edition

7. A. Tveito, H. P. Langtangen, B. F. Nielsen, X. Cai, *Elements of Scientific Computing.*

8. B. Gustafsson, *Fundamentals of Scientific Computing.*

For further information on these books please have a look at our mathematics catalogue at the following URL: www.springer.com/series/5151